Distributed Systems
An Algorithmic Approach

Second Edition

CHAPMAN & HALL/CRC
COMPUTER and INFORMATION SCIENCE SERIES

Series Editor: Sartaj Sahni

PUBLISHED TITLES

ADVERSARIAL REASONING: COMPUTATIONAL APPROACHES TO READING THE OPPONENT'S MIND
Alexander Kott and William M. McEneaney

DELAUNAY MESH GENERATION
Siu-Wing Cheng, Tamal Krishna Dey, and Jonathan Richard Shewchuk

DISTRIBUTED SENSOR NETWORKS, SECOND EDITION
S. Sitharama Iyengar and Richard R. Brooks

DISTRIBUTED SYSTEMS: AN ALGORITHMIC APPROACH, SECOND EDITION
Sukumar Ghosh

ENERGY-AWARE MEMORY MANAGEMENT FOR EMBEDDED MULTIMEDIA SYSTEMS: A COMPUTER-AIDED DESIGN APPROACH
Florin Balasa and Dhiraj K. Pradhan

ENERGY EFFICIENT HARDWARE-SOFTWARE CO-SYNTHESIS USING RECONFIGURABLE HARDWARE
Jingzhao Ou and Viktor K. Prasanna

FUNDAMENTALS OF NATURAL COMPUTING: BASIC CONCEPTS, ALGORITHMS, AND APPLICATIONS
Leandro Nunes de Castro

HANDBOOK OF ALGORITHMS FOR WIRELESS NETWORKING AND MOBILE COMPUTING
Azzedine Boukerche

HANDBOOK OF APPROXIMATION ALGORITHMS AND METAHEURISTICS
Teofilo F. Gonzalez

HANDBOOK OF BIOINSPIRED ALGORITHMS AND APPLICATIONS
Stephan Olariu and Albert Y. Zomaya

HANDBOOK OF COMPUTATIONAL MOLECULAR BIOLOGY
Srinivas Aluru

HANDBOOK OF DATA STRUCTURES AND APPLICATIONS
Dinesh P. Mehta and Sartaj Sahni

HANDBOOK OF DYNAMIC SYSTEM MODELING
Paul A. Fishwick

HANDBOOK OF ENERGY-AWARE AND GREEN COMPUTING
Ishfaq Ahmad and Sanjay Ranka

HANDBOOK OF PARALLEL COMPUTING: MODELS, ALGORITHMS AND APPLICATIONS
Sanguthevar Rajasekaran and John Reif

HANDBOOK OF REAL-TIME AND EMBEDDED SYSTEMS
Insup Lee, Joseph Y-T. Leung, and Sang H. Son

PUBLISHED TITLES CONTINUED

Distributed Systems
An Algorithmic Approach

Second Edition

Sukumar Ghosh

University of Iowa
Iowa City, USA

CRC Press
Taylor & Francis Group
Boca Raton London New York

CRC Press is an imprint of the
Taylor & Francis Group, an **informa** business

A CHAPMAN & HALL BOOK

Cover designed by Vani Murarka.

CRC Press
Taylor & Francis Group
6000 Broken Sound Parkway NW, Suite 300
Boca Raton, FL 33487-2742

© 2015 by Taylor & Francis Group, LLC
CRC Press is an imprint of Taylor & Francis Group, an Informa business

No claim to original U.S. Government works

Printed on acid-free paper
Version Date: 20140611

International Standard Book Number-13: 978-1-4665-5297-5 (Hardback)

Library of Congress Cataloging-in-Publication Data

Ghosh, Sukumar.
  Distributed systems : an algorithmic approach / Sukumar Ghosh. -- Second edition.
    pages cm. -- (Chapman & Hall/CRC computer & information science series)
  "A CRC title."
  Includes bibliographical references and index.
  ISBN 978-1-4665-5297-5 (hardcover : alk. paper) 1. Electronic data processing--Distributed processing. 2. Computer algorithms. I. Title.

QA76.9.D5G487 2014
004'.36--dc23   2014016491

Visit the Taylor & Francis Web site at
http://www.taylorandfrancis.com

and the CRC Press Web site at
http://www.crcpress.com

This book is dedicated to
Swami Dayatmananda,
President, Vedanta Center, United Kingdom

Contents

Section II Foundational Topics

Section IV Faults and Fault-Tolerant Systems

SECTION V **Real-World Issues**

Preface

DISTRIBUTED SYSTEMS HAVE WITNESSED PHENOMENAL GROWTH in the past few years. The declining cost of hardware, the advancements in communication technology, the explosive growth of the Internet, and our ever-increasing dependence on networks for a wide range of applications ranging from social communication to financial transactions have contributed to this growth. The breakthroughs in embedded systems, nanotechnology, and wireless communication have opened up new frontiers of applications like sensor networks and wearable computers. The rapid growth of cloud computing and the growing importance of big data have changed the landscape of distributed computing.

Most applications in distributed computing center around a set of core subproblems. A proper understanding of these subproblems requires a background of the underlying theory and algorithmic issues. This book provides a balanced coverage of the foundational topics and their relationship to real-world applications. The language has been kept as unobfuscated as possible—clarity has been given priority over formalism. The second edition fixes many of the problems in the first edition, adds new topics, and significantly upgrades the contents. The 21 chapters have been divided into five sections: Section I (Chapters 1 and 2) deals with *background materials* that include various cloud computing platforms. Section II (Chapters 3 through 6) presents *foundational topics*, which address system models, correctness criteria, and proof techniques. Section III (Chapters 7 through 11) presents the core paradigms in distributed systems—these include logical clocks, distributed snapshots and debugging, deadlock and termination detection, election, and distributed graph algorithms. Section IV (Chapters 12 through 17) addresses failures and fault-tolerance techniques in various applications—it covers consensus, transactions, group communication, replicated data management, and self-stabilization. Group communication and consensus have been included in this section since they are two of the primary beneficiaries of fault-tolerant designs. Finally, Section V (Chapters 18 through 21) addresses a few real-world issues—these include distributed discrete-event simulation, security, sensor networks, and social and peer-to-peer networks. Each chapter has a list of exercises that will challenge the readers (those tagged with * are the more challenging ones). A small number of these are programming exercises. Some exercises will encourage the readers to learn about outside materials.

The book is intended for use in a one-semester course at the senior undergraduate or the first-year graduate level. About 75% of the material can be covered in one semester. Accordingly, the chapters can be picked and packaged in several different ways.

A theory-oriented offering is possible using Chapters 1 through 17. For a more practical flavor, use Chapters 1 and 2, selected topics from Chapters 3 through 16, and Chapters 18 through 21, supplemented by a semester-long project chosen from replicated data management, sensor networks, group communication, discrete-event simulation, and social or peer-to-peer networks. Additional material is available from the author's website: http://homepage.cs.uiowa.edu/~ghosh/thebook.html.

Here is a disclaimer: this book is *not* about programming distributed systems. Chapter 2 is only a high-level description that we expect everyone to know, but it is *not* an introduction to programming. If programming is the goal, then I encourage readers to look for other resources. There are several good books available on this topic.

Sukumar Ghosh
Iowa City, Iowa

Acknowledgments

It is a pleasure to acknowledge the help and support of my friends and colleagues from all over the world in completing this project. Ted Herman has been a constant source of wisdom. Discussions with Sriram Pemmaraju on several topics have been helpful. Amlan Bhatacharya, Sridhar Dighe, and Kajari Ghosh Dastidar designed several example programs. Various parts of this book have been used in several offerings of the courses of 22C:166 and 22C:196 in the Computer Science Department at the University of Iowa— special thanks to the students of these courses for their constructive feedback. Simin Nadjm-Tehrani sent numerous corrections that helped improve the presentation of several sections. Andrew Berns carefully reviewed several chapters of the book, including the exercises. The feedbacks from Kayhan Erciyes, Konstantin Solomatov, Carl Hauser, Anand Padmanabhan, and Shrisha Rao are much appreciated. Thanks to Randi Cohen for her patience—it has been a pleasure to work with her.

Author

Sukumar Ghosh is a professor in the Department of Computer Science, University of Iowa, Iowa City, Iowa, since 1995. He earned his PhD (computer science and engineering) from Calcutta University, Calcutta, India, in 1971 and completed his postdoctoral research at the University of Dortmund, Germany, as a fellow of the Alexander von Humboldt Foundation. His research interests are in distributed systems, with special emphasis on dynamic distributed systems; fault-tolerant, self-stabilizing, and autonomic distributed systems; and peer-to-peer networks. He has published more than 100 research papers and 5 book chapters on these topics and has supervised 16 PhD students.

I

Background Materials

CHAPTER 1

Introduction

1.1 WHAT IS A DISTRIBUTED SYSTEM?

Life in the twenty-first century has a growing dependence on networked services that have changed the fabric of the society. Starting from web searching, video conferencing, stock trading, and net banking to keeping in touch with friends and peers through various kinds of social networks, network-based services play a dominant role. While networks provide the basic connectivity, the various services built on top of these networks are examples of distributed systems.

Leslie Lamport once noted, "A distributed system is one in which the failure of a computer you didn't even know existed can render your own computer unusable." While this is certainly not a definition, it characterizes the challenges in coming up with an appropriate definition of a distributed system. *What* is distributed in a distributed system? If the processor of computer system is located 100 yards away from its main memory, then is it a distributed system? What if the input–output devices are located 3 miles away from the processor? If physical distribution is taken into account, then the definition of a distributed system becomes uncomfortably dependent on the degree of physical distribution of the hardware components, which is certainly not acceptable. To alleviate this problem, it is customary to characterize a distributed system using the logical or functional distribution of the processing capabilities.

A distributed system typically satisfies the following criteria:

Multiple processes: The system consists of *more than one* sequential process. These processes can be either system or user processes, but each process should have an independent thread of control—either explicit or implicit.

Interprocess communication: Processes communicate with one another using *messages* that take a finite time to travel from one process to another. The actual nature or order of the delay will depend on the physical characteristics of the message links. These message links are also called *channels*.

Disjoint address spaces: Processes have *disjoint address spaces*. We will thus not take into account a shared-memory multiprocessor as a true representation of a distributed computing system. Note that programmers often represent interprocess communication using shared-memory primitives, but the abstraction of shared memory can be implemented using messages. The relationship between shared memory and message passing will be discussed in Chapter 3.

Collective goal: Processes must *interact* with one another to meet a common goal. Consider two processes P and Q in a network of processes. If P computes $f(x) = x^2$ for a given set of values of x, and Q multiplies a set of numbers by π, then we hesitate to call (P, Q) a distributed system, since there is no interaction between P and Q. However, if P and Q cooperate with one another to compute the areas of a set of circles of radius x, then (P, Q) collectively represent a meaningful distributed system. Similarly, if a set of sellers advertise the cost of their products, and a set of buyers post the list of the goods that they are interested in buying as well as the prices they are willing to pay, then individually, neither the buyers nor the sellers are meaningful distributed systems, but when they are coupled into an auction system through the Internet, then it becomes a meaningful distributed system.

The aforementioned definition is a minimal one. It does not take into consideration the system-wide executive control for interprocess cooperation or security issues, which are certainly important concerns in the run-time management and support of user computations. The definition highlights the simplest possible characteristics for a computation to be logically distributed. Physical distribution is often a prerequisite for logical distribution.

1.2 WHY DISTRIBUTED SYSTEMS

Over the past years, distributed systems have gained substantial importance. The reasons of their growing importance are manifold:

Geographically distributed environment: First, in many situations, the computing environment itself is *geographically distributed*. As an example, consider a banking network. Each bank is supposed to maintain the accounts of its customers. In addition, banks communicate with one another to monitor interbank transactions or record fund transfers from geographically dispersed automated teller machines (ATMs). Another common example of a geographically distributed computing environment is the Internet, which has deeply influenced our way of life. The mobility of the users has added a new dimension to the geographic distribution.

Speed up: Second, there is the need for *speeding up* computation. The speed of computation in traditional uniprocessors is fast approaching the physical limit. While multicore, superscalar, and very large instruction word (VLIW) processors stretch the limit by introducing parallelism at the architectural level, the techniques do not scale well beyond a certain level. An alternative technique of deriving more computational power is to use multiple processors. Dividing a total problem into smaller subproblems and assigning these subproblems to separate physical processors that can operate concurrently are potentially an attractive method of enhancing the speed of computation. Moreover, this approach promotes better scalability, where the users or administrators can incrementally increase the computational power by purchasing additional processing elements or resources. This concept is extensively used by the social networking sites for the concurrent upload and download of the photos and videos of millions of customers.

Resource sharing: Third, there is a need for *resource sharing*. Here, the term resource represents both hardware and software resources. The user of computer A may want to use a fancy laser printer connected with computer B, or the user of computer B may need some extra disk space available with computer C for storing a large file. In a network of workstations, workstation A may want to use the idle computing powers of workstations B and C to enhance the speed of a particular computation. Through Google Docs, Google lets you share their software for word processing, spreadsheet application, and presentation creation without anything else on your machine. Cloud computing essentially outsources the computing infrastructure of a user or an organization to data centers—these centers allow thousands of their clients to share their computing resources through the Internet for efficient computing at an affordable cost.

Fault tolerance: Fourth, powerful uniprocessors or computing systems built around a single central node are prone to a complete collapse when the processor fails. Many users consider this to be risky. Distributed systems have the potential to remedy this by using appropriate fault-tolerance techniques—when a fraction of the processors fail, the remaining processes take over the tasks of the failed processors and keep the application running. For example, in a system having triple modular redundancy (TMR), three identical functional units are used to perform the same computation, and the correct result is determined by a majority vote. In many fault-tolerant distributed systems, processors routinely check one another at predefined intervals, allowing for automatic failure detection, diagnosis, and eventual recovery. Some users of non-critical applications are willing to compromise with a partial degradation in system performance when a failure cripples a fraction of the processing elements or the communication links of a distributed system. This is the essence of graceful degradation. A distributed system thus provides an excellent opportunity for incorporating *fault tolerance* and *graceful degradation*.

1.3 EXAMPLES OF DISTRIBUTED SYSTEMS

There are numerous examples of distributed systems that are used in everyday life in a variety of applications. A fraction of these services are data intensive, and the computational component is very small. Examples are database-oriented applications (think about Google searching and collecting information from computers all over the world). Others are computation intensive. Most systems are structured as *client–server* systems, where the server machine is the custodian of data or resources and provides service to a number of geographically distributed clients. A few applications, however, do not rely on a central server—these are *peer-to-peer* (P2P) systems. Here are a few examples of distributed systems:

World Wide Web: The World Wide Web (WWW) is a popular service running on the Internet. It allows documents in one computer to refer to textual or nontextual information stored in other computers. For example, a document in the United States may contain references to the photograph of a rainforest in Africa or a music recording in Australia. Such references are highlighted on the user's monitor, and when selected by the user, the system fetches the item from a remote server using appropriate protocols and displays the picture or plays the music on the client machine.

The Internet and the WWW have changed the way we perform our daily activities or carry out business. For example, a large fraction of airline and hotel reservations are now done through the Internet. Shopping through the Internet has dramatically increased during the past few years. Millions of people now routinely trade stocks through the Internet. Music lovers download and exchange CD-quality music, pushing old-fashioned CD and DVD purchases to near obsolescence. Digital libraries provide users instant access to archival information from the comfort of their homes.

Social networks: Internet-mediated social interaction has witnessed a dramatic growth in recent times. Millions of users now use their desktop or laptop computers or smartphones to post and exchange messages, photos, and video clips with their buddies using these networking sites. Members of social networks can socialize by reading the profile pages of other members and contacting them. Interactions among members often lead to the formation of virtual communities or clubs sharing common interests. The inputs from the members are handled by thousands of servers at one or more geographic locations, and these servers carry out specific tasks on behalf of the members—some deal with photos, some handle videos, some handle membership changes, etc. As membership grows, more servers are added to the pool, and powerful servers replace the slower ones.

Interbank networks: Amy needs $300 on a Sunday morning, so she walks to a nearby ATM to withdraw some cash. Amy has a checking account in Iowa City, but she has two savings accounts—one in Chicago and the other in Denver. Each bank has set an upper limit of $100 on the daily cash withdrawal, so Amy uses three different bank cards to withdraw the desired cash. These debits are immediately registered in her bank accounts in three different cities, and her new balances are recomputed. The ATMs are registered with their respective financial institutions, and the interbank network carries out the entire operation.

Peer-to-peer networks: P2P systems are quite popular for file sharing, content distribution, and Internet telephony. Historically, Napster was the pioneer in the use of P2P technology for sharing the personal music collections of its clients. Instead of storing the songs on a central computer, the songs live on users' machines. There are millions of them scattered all over the world. When you want to download a song using Napster, you are downloading it from another person's machine, and that person could be your next-door neighbor or someone halfway around the world. This led to the development of P2P data sharing. Napster was not a true P2P system, since it used a centralized directory. But many subsequent systems providing similar service (e.g., Gnutella, KaZaA) avoided the use of a central server or a central directory. P2P systems are now finding applications in areas beyond exchanging music files. For example, the OceanStore project at the University of California, Berkeley, built an online data archiving mechanism on top of an underlying P2P network Tapestry. Millions use Skype that is built around P2P technology for Internet telephony and video chats.

Real-time distributed systems: Real-time distributed systems deal with time-critical coordination of events in a geographically distributed setting. Industrial plants extensively use networks of controllers to oversee production and maintenance. Consider a chemical plant, in which a controller maintains the pressure of a certain chamber to 200 psi. As the vapor pressure increases, the temperature has a tendency to increase—so there is another controller 300 ft away that controls the flow of a coolant. This coolant ensures that the temperature of the chamber never exceeds 180°F. Furthermore, the safety of the plant requires that the product of the pressure and the temperature does not exceed 35,000. Here, the distributed computing system maintains an invariance relationship on system parameters monitored and controlled by independent controllers. In urban *traffic control networks*, the computers at the different control points control traffic signals to minimize traffic congestion and maximize traffic flow. *Vehicular networks* allow spontaneous intervehicle wireless communication, which not only enables vehicles to share information about road conditions with other vehicles but also opens up the possibility of improved mobility and collective safety.

Sensor networks: The declining cost of hardware and the growth of wireless technology have led to new opportunities in the design of application-specific or special-purpose distributed systems. One such application is a *sensor network* [ASSC02]. Each node is a miniature processor equipped with a few sensors that sense various environmental parameters, and is capable of wireless communication with other nodes. Such networks can be potentially used in a wide class of problems: These range from battlefield surveillance, biological and chemical attack detection, to healthcare, home automation, ecological, and habitat monitoring. This is a part of a larger vision of *ubiquitous computing*.

Grid and cloud computing: *Grid computing* is a form of distributed computing that supports parallel programming on a network of computers. At the low end, a computational grid can use a fraction of the computational resources of one or two organizations, whereas at the high end, it can combine millions of computers worldwide to work on extremely large computational projects. The goal is to solve difficult computational problems more quickly and less expensively than by conventional methods. We provide two examples here.

The first example is the *Large Hadron Collider* (LHC), a particle accelerator and a complex experimental testbed built at the European Organization of Nuclear Research (CERN) started operation from 2008, and it is being used to answer fundamental questions of science. The scientific experiments generate approximately 15 PB of data each year (1 PB = 10^{15} bytes, and this is the equivalent of more than 20 million CDs). These data are distributed around the globe, to 11 large computer centers located in Canada, France, Germany, Italy, the Netherlands, the Nordic countries, Spain, Taipei, the United Kingdom, and two sites in the United States forming a worldwide virtual organization. These Tier-1 centers make the data available to more than 150 Tier-2 centers for specific analysis tasks. Individual scientists at the various countries can locally access the LHC data using local computer clusters and PCs.

The second example is the SETI@home project. Do extraterrestrials exist? SETI (acronym for search for extraterrestrial intelligence) is a massive project aimed at discovering the existence of extraterrestrial life in this universe. The large volume of data that is constantly being collected from hundreds of radio telescopes needs to be analyzed to draw any conclusion about the possible existence of extraterrestrial life. This requires massive computing power. Rather than using supercomputers, the University of California, Berkeley, SETI team decided to harness the idle computing power of the millions of PCs and workstations belonging to you and me, computing power that is otherwise wasted by running useless screensaver programs. Currently, about 40 GB of data are pulled down daily by the telescope and sent to over three million computers all over the world to be analyzed. The results are sent back through the Internet, and the program then collects a new segment of radio signals for the PC to work on. The system executes 14 trillion floating-point operations per second and has garnered over 500,000 years of PC time in the past year and a half. It would normally cost millions of dollars to achieve that type of power on one or even two supercomputers.

Cloud computing enables clients to outsource their software usage, data storage, and even the computing infrastructure to remote data centers. Clients interact with the cloud storage and applications through the Internet using their web browsers. Users need not acquire or maintain expensive hardware and software but nevertheless get their jobs done by paying for the usage of the resources. With the extensive availability of high-speed networks and the declining cost of computers, cloud computing provides an economic alternative to conventional computing. It presents a flexible approach where users can increase their computing capacity on the fly as a pay-per-use service. The task of maintaining the hardware and software at the data centers rests with the cloud service provider, who is responsible for overseeing the seamless sharing of these resources and maintaining privacy of user data.

1.4 IMPORTANT ISSUES IN DISTRIBUTED SYSTEMS

This book will mostly deal with process models and distributed computations supported by these models. A model is an abstract view of a system. It ignores many physical details of the system. That does not mean that the implementation issues are unimportant, but these are outside the scope of this book. Thus, when discussing about a network of processes, we will never describe the type of processors running the processes, or the characteristics of the physical memory, or the rate at which the bits of a message are being pumped across a particular channel. Our emphasis is on computational activities represented by the concurrent or interleaved execution of actions on a network of sequential processes. Some of the important issues in the study of the computational models of distributed systems are as follows:

Knowledge of a process: The knowledge of a process is local. No process is ordinarily expected to have global knowledge about either the network topology or the global state. Each process thus has a myopic view of the system. It is fair to expect that a process knows (1) its own identity, (2) its own state, and (3) the identity of its immediate neighbors. In some special cases, a process may also have exact or approximate knowledge about the size (i.e., the number of nodes) of the network. Any other knowledge that a process might need has to be acquired from time to time through appropriate algorithmic actions.

Network topology: A network of processes may either be completely connected or sparsely connected. In a completely connected network, a *channel* (also called a *link*) exists between every pair of processes in the system. This condition does not hold for a sparsely connected topology. As a result, message routing is an important activity. A link between a pair of processes may be unidirectional or bidirectional. Examples of sparse topologies are trees, rings, arrays, and hypercubes (Figure 1.1).

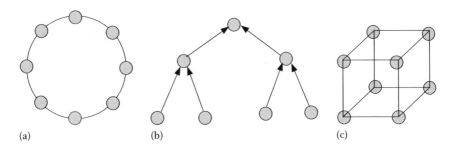

FIGURE 1.1 Examples of network topologies: (a) ring, (b) directed tree, and (c) 3D cube. Each node represents a process, and each edge connecting a pair of nodes represents a channel.

Degree of synchronization: Some of the deeper issues in distributed systems center around the notion of synchrony and asynchrony. According to the laws of astronomy, real time is defined in terms of the rotation of the Earth in the solar system. However, the international time standard now is the Coordinated Universal Time (UTC). UTC is the current term for what was commonly referred to as Greenwich Meridian Time (GMT). Zero hour in UTC is midnight in Greenwich, England, which lies on the zero longitudinal meridian. UTC is based on a 24 h clock; therefore, afternoon hours such as 6 p.m. UTC are expressed as 18:00 UTC. Each second in UTC is precisely the time for 9,192,631,770 orbital transitions of the *cesium 133* atom. The time keeping in UTC is based on atomic clocks. UTC signals are regularly broadcast from satellites as well as many radio stations. In the United States, this is done from the WWV radio station in Fort Collins, Colorado, where satellite signals are received through Global Positioning System (GPS). A useful aspect of atomic clocks is the fact that these can, unlike solar clocks, be made available anywhere in the universe.

Assume that each process in a distributed system has a local clock. If these clocks represent the UTC (static differences due to time zones can be easily taken care of and ignored from this equation), then every process has a common notion of time, and the system can exhibit synchronous behavior by the simultaneous scheduling of their actions. Unfortunately, in practical distributed systems, this is difficult to achieve, since the drift of the local physical clocks is a fact of life. One approach to handle this is to use a time server that keeps all the local clocks synchronized with one another. This is good enough for some applications, but not all.

The concept of a synchronous system has evolved over many years. There are many facets of synchrony. One is the existence of an upper bound on the propagation delay of messages. If the message sent by process A is not received by process B within the expected interval of real time, then process B suspects some kind of failure. Another feature of a synchronous system is the first-in-first-out (FIFO) behavior of the channels connecting the processes. With these various possibilities, it seems prudent to use the attribute *synchronous* to separately characterize the behaviors of clocks, or communication, or channels.

In a fully asynchronous system, not only there is clock drift, but also there is no upper bound on the message propagation delays. Processes can be arbitrarily slow, and out-of-order message delivery between any pair of processes is considered feasible. In other words, such systems may completely disregard the rule of time, and processes schedule events at an arbitrary pace. The properties of a distributed system depend on the type of synchrony. Results about one system often completely fall apart when assumptions about synchrony change from synchronous to asynchronous. In Chapter 6, we will find out how the lack of a common basis of time complicates the notion of global state and consequently the ordering of events in a distributed system.

Failures: The handling of failures is an important area of study in distributed systems. A failure occurs when a system as a whole or one or more of its components do not behave according to their specifications. Numerous failure models have been studied. The most common failure model is *crash*, where a process ceases to produce any output. In another case of failure, a process does not stop but simply fails to send one or more messages or execute one or more steps. This is called *omission failure*. This includes the case when a message is sent but lost in transit. Sometimes, the failure of a process or a link may alter the topology by partitioning the network into disjoint subnetworks. In the *byzantine failure model*, a process may behave in a completely arbitrary manner—for example, it can send inconsistent or conflicting message to its neighboring processes—or may execute a program that is different from the designated one.

Along with the type of failure, its duration is of significance. It is thus possible that a process exhibits byzantine failure for 5 s, then resumes normal behavior, and after 30 min, fails by stopping. We will discuss more about various fault models in Chapter 12.

Scalability: An implementation of a distributed system is considered *scalable* when its performance is not impaired regardless of the final scale of the system. The need for additional resources to cope with the increased scale should be manageable. Scalability is an important issue since many distributed systems have witnessed tremendous growth in size over the past decade—it is quite common for current social networks to have millions of registered users. Scalability suffers when the resource requirement grows alarmingly with the scale of the system. Some systems deliver the expected performance when the number of nodes is small, but fail to deliver when the number of nodes increases. From an algorithmic perspective, the scalability is excellent when the space or time complexity of a distributed algorithm is $O(\log n)$ or lower, where n is the number of processes in the system—however, when it is $O(n)$ or higher, the scalability is considered poor. Well-designed distributed systems usually exhibit good scalability.

1.5 COMMON SUBPROBLEMS

Most applications in distributed computing center around a set of common subproblems. If we can solve these common subproblems in a satisfactory way, then we have a good handle on system design. Here are a few examples of common subproblems:

Leader election: When a number of processes cooperate with one another for solving a problem, many implementations prefer to elect one of them as the leader and the remaining processes as followers. The leader assumes the role of a coordinator and runs a program that is different from that of the followers. If the leader crashes, then one of the followers is elected the leader, after which the system runs as usual.

Mutual exclusion: Access to certain hardware resources is restricted to one process at a time: an example is a printer. There are also software resources where concurrent accesses run the risk of producing inconsistent results: for example, multiple processes are not ordinarily allowed to update a shared data structure. Mutual exclusion guarantees that at most one process acquires the resource or performs a critical operation on a shared data at any time and concurrent access attempts to such resources are serialized.

Time synchronization: Local clocks invariably drift and need periodic resynchronization to support a common notion of time across the entire distributed system.

Global state: The global state of a distributed system consists of the local states of its component processes. Any computation that needs to compute the global state at a given time has to read the local states of every component process at that time. However, given the facts that local clocks are never perfectly synchronized and message propagation delays are finite, computation of the global state is a nontrivial problem.

Multicasting: Sending of a given data to multiple processes in a distributed system is a common subtask in many applications. As an example, in group communication, one may want to send some breaking news to millions of members as quickly as possible. The important issues here are efficiency, reliability, and scalability.

Replica management: To support fault tolerance and improve system availability, the use of process replicas is quite common. When the main server is down, one of the replica servers replaces the main server. Data replication (also known as caching) is widely used for conserving system bandwidth. However, replication requires that the replicas be appropriately updated. Since such updates can never be instantaneously done, it leaves open the possibility of inconsistent replicas. How to update the replicas and what kind of response can a client expect from these replicas? Are there different notions of consistency in replica management? How are these related to the cost of the update operation?

1.6 IMPLEMENTING A DISTRIBUTED SYSTEM

A model is an abstract view of a system. Any implementation of a distributed computing model must involve the implementation of processes, message links, routing schemes, and timing. The most natural implementation of a distributed system is a network of computers, each of which runs one or more processes. Using the terminology from computer architecture, such implementations belong to the class of loosely coupled multiple instruction multiple data (MIMD) machines, where each processor has a private address space. The best example of a large-scale implementation of a distributed system is the WWW. A cluster of workstations connected to one another via a local area network (LAN) serves as a medium-scale implementation. In a smaller scale, mobile ad hoc networks and a wireless sensor network are appropriate examples.

Distributed systems can also be implemented on a tightly coupled MIMD machine, where processes running on separate processors are connected to a globally shared memory. In this implementation, the shared memory *simulates* the interprocess communication channels. Finally, a multiprogrammed uniprocessor can be used to simulate a shared-memory multiprocessor and hence a distributed system. For example, the very old RC4000 was the first message-based operating system designed and implemented by Brinch Hansen [BH73] on a uniprocessor. Amoeba, Mach, and Windows NT are examples of microkernel-based operating systems where processes communicate via messages.

Distributed systems have received significant attention from computer architects because of their potential for better scalability. In a *scalable* architecture, resources can be continuously added to improve performance, and there is no appreciable bottleneck in this process. Bus-based multiprocessors do not scale beyond 8–16 processors because the bus bandwidth acts as a bottleneck. Shared-memory symmetric multiprocessors (also called SMPs) built around multistage interconnection networks suffer from some degree of contention when the number of processors reaches 1000 or more. Recent trends in scalable architecture show reliance on multicomputers, where a large number of autonomous machines (i.e., processors with private memories) are used as building blocks. For the ease of programming, various forms of *distributed shared memory* (DSM) are then implemented on it, since programmers do not commonly use message passing in developing application programs.

Social networks implement a version of a distributed system that can be viewed as a graph, with the nodes representing members and the edges representing friendship relation between members. The servers maintain the friendship relationships and oversee the access control.

Another implementation of a distributed system is a *neural network*, which is a system mimicking the operation of a human brain. A neural network contains a number of processors operating in parallel, each with its own small sphere of knowledge and access to data in its local memory. Such networks are initially *trained* by rules about data relationships (e.g., "A mother is older than her daughter"). A program can then tell the network how to behave in response to input from a computer user. The results of the interaction can be used to enhance the training. Some important applications of neural networks include stock market prediction, weather prediction, oil exploration, and the interpretation of nucleotide sequences.

These different architectures merely serve as platforms for implementation or simulation. A large number of system functions are necessary to complete the implementation of a particular system. For example, many models assume communication channels to be FIFO. Therefore, if the architecture does not naturally support FIFO communication between a pair of processes, then FIFO communication has to be implemented first. Similarly, many models assume there will be no loss or corruption of messages. If the architecture does not guarantee these features, then appropriate protocols have to be used to remedy this shortcoming. No system can be blamed for not performing properly, if the model specifications are not appropriately satisfied.

1.7 PARALLEL VERSUS DISTRIBUTED SYSTEMS

What is the relationship between a parallel system and a distributed system? Like distributed systems, parallel systems are yet to be clearly defined. The folklore is that any system in which the events can at best be partially ordered is a parallel system. This naturally includes every distributed system and all shared-memory systems with multiple threads of control. According to this view, distributed systems form a subclass of parallel systems, where the state spaces of processes do not overlap. Processes have greater autonomy. This view is not universally accepted. Some distinguish parallel systems from distributed systems on the basis of their objectives: *parallel systems* focus on *increasing performance*, whereas *distributed systems* focus on tolerating *partial failures*. As an alternative view, *parallel systems* consist of processes in a single instruction single data (SIMD) type of *synchronous* environment or a synchronous MIMD environment, *asynchronous* processes in a shared-memory environment are the building blocks of *concurrent systems*, and cooperating processes with *private address spaces* constitute a *distributed system*.

1.8 BIBLIOGRAPHIC NOTES

The book by Coulouris et al. [CDK11] contains a good overview of distributed systems and their applications. Tel [T00] covers numerous algorithmic aspects of distributed systems. Tannenbaum and van Steen's book [TS07] addresses practical aspects and implementation issues of distributed systems. Distributed operating systems have been presented by Singhal and Shivaratri [SS94]. Greg Andrew's book [A00] provides a decent coverage of concurrent and distributed programming methodologies. The SETI@home project and its current status are described in [SET02]. Akyildiz et al. [ASSC02] presents a survey of wireless sensor networks.

EXERCISES

To solve these problems, identify all sequential processes involved in your solution and give an informal description of the messages exchanged among them. No code is necessary—pseudocode is ok.

1.1 A distributed system is charged with the responsibility of deciding whether a given integer N is a prime number. The system has a fixed number of processes. Initially, only a designated process called the *initiator* knows N, and the final answer must be available to the initiator. Informally describe what each process will do and what interprocess messages will be exchanged.

1.2 In a network of processes, every process knows about itself and its immediate neighbors only. Illustrate with an example how these processes can exchange information to gain knowledge about the global topology of the network.

1.3 A robot B wants to cross a road, while another robot A is moving from left to right (Figure 1.2). Assuming that each robot can determine the (x, y) coordinates of both the robots, outline the program for each robot, so that they do not collide with each other.

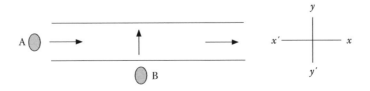

FIGURE 1.2   Robot B crossing a road and trying to avoid collision with robot A.

You can assume that (1) the clocks are synchronized and (2) the robots advance in discrete steps—with each tick of the clock, they move one foot at a time.

1.4 On a Friday afternoon, a passenger asks a travel agent to reserve the earliest flight next week from Cedar Rapids to Kathmandu via Chicago and London. United Airlines operates hourly flights in the sector Cedar Rapids to Chicago. In the sector Chicago–London, British Airways operates daily flights. The final sector is operated by the Royal Nepal Airlines on Tuesdays and Thursdays only. Assuming that each of these airlines has an independent agent to schedule its flights, outline the interactions between the travel agent and these three airline agents, so that the passenger eventually books a flight to Kathmandu.

1.5 Alice plans to call Bob from a pay phone in Bangkok using a credit card. The call is successful only if (1) Alice's credit card is still valid, (2) Alice does not have any past due in her account, and (3) Bob's telephone number is correctly dialed by Alice. Assuming that a process *card* checks the validity of the calling card, a second process *bill* takes care of billing, and a third process *switch* routes the call to Bob, outline the sequence of actions during the call establishment period.

1.6 In many distributed systems, resource sharing is a major goal. Provide examples of systems where the shared resource is (1) a disk, (2) a network bandwidth, and (3) a processor.

1.7 KaZaA is a system that allows users to download music files in a transparent way from another computer that may belong to a next-door neighbor or to someone halfway around the world. Investigate how this file sharing is implemented.

1.8 A customer wants to fly from airport A to airport B within a given period of time by paying the cheapest fare. She submits the query for flights and expects to receive the reply in a few seconds. Travelocity.com, expedia.com, and orbitz.com already have such services in place. Investigate how these services are implemented.

1.9 Sixteen motes (miniature low-cost processors with built-in sensors) are being used to monitor the average temperature of a furnace. Each mote has limited communication ability and can communicate with two other motes only. The wireless network of motes is not partitioned. Find out how each mote can determine the average temperature of the furnace.

1.10 How can a single processor system be used to implement a unidirectional ring of N processes?

1.11 Can we view a digital circuit as a distributed system? Justify your answer.

CHAPTER 2

Interprocess Communication

An Overview

2.1 INTRODUCTION

Interprocess communication is at the heart of distributed computing. User processes run on host machines that are connected to one another through a network, and the network carries signals that propagate from one process to another. These signals represent *data*.* We separate interprocess communication into two parts:

Networking: This deals with how processes communicate with one another via the various protocol layers. The important issues in networking are routing, error control, flow control, authentication, etc. This is the internal view.

Users' view: User processes have an abstract high-level view of the interprocess communication medium. This is the external view. An average user does not bother about *how* the communication takes place. The processes may communicate across a LAN, or through the Internet, or a combination of these, or via shared (virtual) address space, an abstraction that is created on a message-passing substrate to facilitate programming. Most working distributed systems use the *client–server* model. A few systems also adopt the *P2P* model of interprocess communication, where there is no difference between servers and clients. User interfaces rely on programming tools available to a client for communicating with a server or to a peer for communicating with another peer. Some tools are general, while others are proprietary. In this chapter, we will focus mostly on the users' view of communication.

2.1.1 Processes and Threads

A *process* is the execution of a program. The operating system supports multiple processes on a processor, so multiple logical processes (LPs) can execute on the same physical processor. Threads are lightweight processes. Like a process, each thread maintains a separate flow of control, but threads share a common address space. Multiple threads improve the

---

* The word *data* here represents anything that can be used by a string of bits.

transparency of implementation of clients and servers and the overall performance. In a multithreaded server, while one thread is blocked on an event, other threads carry out the pending unrelated operations. In multicore processors, each core can support one or more threads. Today's multicore processors have 4–16 cores (and this number will grow), so these processors can support a large number of threads.

2.1.2 Client–Server Model

The client–server model is a widely accepted model for designing distributed systems. Client processes request for service, and server processes provide the desired service. A simple example of client–server communication is the Domain Name Service (DNS)—clients request for network addresses of Internet domain names, and DNS returns the addresses to the clients. Another example is that of a search engine like Google. When a client submits a query about a document, the search engine looks up its servers and returns pointers to the web pages that can possibly contain information about that document. Note that the designation of clients and servers is not unique and a server can be a client of another server.

2.1.3 Middleware

Complex distributed systems have significant degrees of heterogeneity. Processes, processors, and objects may be scattered anywhere in a network. To simplify the task of software development, the users should not be bothered about the locations of these entities or the kind of machines that they are on. Developers should not have to worry about integrating enterprise software applications developed at different times, by different vendors, or even communicating via different protocols. The layer of software that simplifies the task of tying complex subsystems together or connecting software components is called *middleware*. It is an extension of the services offered by the operating system and is logically positioned between the application layer and the operating system layer of the individual machines (Figure 2.1).

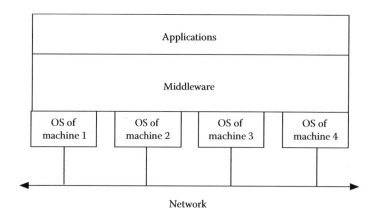

FIGURE 2.1 Understanding middleware.

With the rapid growth of distributed applications, middleware services are one of the fastest-growing services. There are many services under this category. Some important middleware services address the following issues:

1. How does a process locate another named process or object anywhere on the Internet?

2. How does a process send queries to multiple distributed databases anywhere on the Internet?

3. How to isolate the application programs from differences in programming languages and communication protocols?

4. How is the security of the communication guaranteed without any knowledge of the trustworthiness of the operating systems at the two endpoints?

5. How does a mobile personal device (like a patient's smartphone) switch to 3G/4G network during travel and then switch to Bluetooth when it comes close to a Bluetooth-enabled PC at home?

We begin with an overview of networking. The following section is an outline of networking and a summary of some commonly used communication protocols. Further details are available in any textbook on networking.

2.2 NETWORK PROTOCOLS

We begin with a brief description of Ethernet and IEEE 802.11 since they are the most widely used protocols for wired and wireless LANs.

2.2.1 Ethernet

Bob Metcalfe at Xerox PARC developed Ethernet in 1973. The network consists of a number of computers connected with one another through a common high-speed bus. Every machine constantly listens to the signals propagating through the bus. Because the bus is common, at most one sender is allowed to send data at any time. However, no machine is aware of when other machines want to send data—so several senders may try to send simultaneously and it leads to a collision. Senders detect the collision, back off for a random interval of time, and make another attempt to transmit data. This protocol is known as carrier sensing multiple access with collision detection (CSMA/CD). The protocol guarantees that, eventually, exactly one of the contending processes becomes the bus master and is able to use the bus for sending data.

The CSMA/CD protocol is quite similar to the informal protocol used by students to speak with the instructor in a classroom. Under normal conditions, at most one student should speak at any time. However, no student has an a priori knowledge of when another student wants to speak, so a number of students may simultaneously try to raise their hands to express their intention to speak. At the same time, every student raising a hand constantly watches if any other hand is raised—this is collision detection. When a collision is detected, they back off and try later. Eventually, only one hand is raised, and that student speaks—all others wait until that student is done.

To detect collision in the Ethernet, every sender needs to wait for a minimum period of time $2T$ after starting transmission, where T is the maximum time required by the signal to travel from one node to another. If the maximum distance between a pair of nodes is 1 KM, then $T \geq (10^3/3 \times 10^8)$ s* = 3.33 μs. So, after attempting a send operation, a node has to wait for at least 6.66 μs to ensure that no one else has started data transmission. After a node detects a collision, it waits for a period qT before attempting a retransmission, where q is a random number. It can be mathematically demonstrated that this strategy leads to collision avoidance with probability 1.

The data transfer rate for the original Xerox PARC Ethernet was only 3 Mb/s (three million bits per second). In the later version of widely used Ethernets, the transfer rate was 10 Mb/s. Technological improvements have led to the emergence of *fast Ethernets* (100 Mb/s) and *gigabit Ethernets* (1 GB/s). Gigabit Ethernets can be used as the backbone of a very high-speed network.

The latency of message propagation in an Ethernet depends on the degree of contention. The term *channel efficiency* is used to specify the ratio of the number of packets successfully transmitted to the theoretical maximum number of packets that could be transmitted without collision. Typical Ethernets can accommodate up to 1024 machines.

2.2.2 Wireless Networks

The dramatic increase in the number of portable or handheld devices like smartphones has increased the emphasis on mobile computing, also known as nomadic computing. The applications are manifold: from accessing the Internet through your smartphone or PDA to disaster management and communication in the battlefield. Mobile users do not always rely on wireless connection between the two endpoints. For example, a mobile user carrying a laptop computer may connect to a fixed network as he or she moves from one place to another. However, for many other applications, wireless connection becomes necessary. There are several issues that are unique to mobile communication: The portable unit is often powered off for conserving battery power. The activities are generally short and bursty in nature, like checking an email or making a query about whether the next flight is on time. The potential for radio interference is much higher, making error control difficult. The administrative backbone divides the areas of coverage into *cells*, and a user roaming from one cell to another has to successfully *hand off* the application to the new cell without the loss or corruption of data. Finally, the portable device may easily be lost or stolen, which poses new security problems.

In this section, we only discuss those systems that have no wired infrastructure for communication. This includes mobile ad hoc networks but excludes networks where the wireless communication replaces the last hop of the wired communication. The salient features of wireless transmission are as follows:

Limited range: The range measured by the Euclidean distance across which a communication can take place is limited. It is determined by the power of the transmitter and the power of the battery.

---

* The speed of signal propagation is 3×10^8 m/s.

Dynamic topology: If the nodes are mobile, then the neighborhood relationship is not fixed. The connectivity is also altered when the sender increases the power of transmission. Algorithms running on wireless networks must be able to adapt to changes in the network topology.

Collision: The communication from two different processes can collide and garble the message. Transmissions need to be coordinated so that no process is required to receive a message from two different senders concurrently.

IEEE 802.11 defines the standards of wireless communication. The physical transmission uses either direct sequence spread spectrum (DSSS) or frequency hopping spread spectrum (FHSS), or infrared (IR). It defines a family of specifications (802.11a, 802.11b, 802.11g) (commonly called Wi-Fi) each with different capabilities. The basic IEEE 802.11 makes provisions for data rates of either 1 or 2 Mb/s in the 2.4 GHz frequency band using DSSS or FHSS transmission. IEEE 802.11b extends the basic 802.11 standard by allowing data rates of 11 Mb/s in the 2.4 GHz band and uses only DSSS. 802.11g further extends the data rate to 54 Mb/s. 802.11a also extends the basic data rate of 802.11 to 54 Mb/s, but uses a different type of physical transmission called *orthogonal frequency division multiplexing*. It has a shorter transmission range and a lesser chance of radio-frequency interference. 802.11a is not interoperable with 802.11b or 802.11g. More recently, the 802.11n protocol uses multiple antennas to increase data rates from 54 Mb/s (offered by 802.11a and 802.11g) to 600 Mb/s.

To control access to the shared medium, 802.11 standard specifies a *carrier sense multiple access with collision avoidance* (CSMA/CA) protocol that allows at most one process in a neighborhood to transmit. As in CSMA/CD, when a node has a packet to transmit, it first listens to ensure that no other node is transmitting. If the channel is clear, that is, no one else within the sender's radio range is transmitting, then it transmits the packet. However, collision detection, as is employed in Ethernet, cannot be used for radio-frequency transmissions of 802.11. This is because when a node is transmitting, it cannot hear any other node in the system that may start transmitting, since its own signal will drown out others.

To resolve this, as an optional mechanism, whenever a packet is to be transmitted, the transmitting node first sends out a short *ready-to-send* (RTS) packet containing information on the length of the packet. If the receiving node hears the RTS, it responds with a short *clear-to-send* (CTS) packet. After this exchange, the transmitting node sends its packet. When the packet is received successfully as certified by a cyclic redundancy check (CRC), the receiving node transmits an acknowledgment (ack) packet. This back-and-forth exchange is necessary to avoid the *hidden node problem*. Consider three nodes A, B, C, such that node A can communicate with node B and node B can communicate with node C, but node A cannot communicate with node C. As a result, although node A may sense the channel to be clear, node C may in fact be transmitting to node B. The protocol described earlier alerts node A that node B is busy, and hence it must wait before transmitting its packet. CSMA/CA significantly improves bandwidth utilization.

A *wireless sensor network* consists of a set of primitive computing elements (called *sensor nodes* or *motes*) capable of communicating by radio waves. Each node has one or more sensors that can sense the physical parameters of the environment around it and report

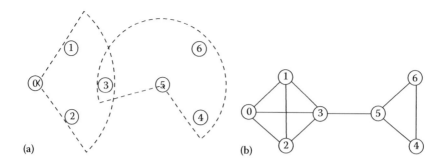

FIGURE 2.2   (a) Nodes 0, 1, 2 are within the range of node 3, but outside the ranges of nodes 4, 5, 6. Nodes 3, 4, 6 are within the range of node 5. Nodes 0, 1, 2 are within one another's range and so are nodes 4, 6. (b) The topology of the network.

it to the neighbors within its transmission range. The use of wireless sensor networks has substantially grown in the past 10 years. In some large-scale applications, the number of motes can scale to a few thousand. Protocols pay special attention to the power consumption and battery life. To conserve power, processes adopt a variety of techniques, including periodically switching to the *sleep mode* (that consumes very little battery power). Maintaining good clock synchronization and sender–receiver coordination turns out to be an important issue.

The topology of a wireless network depends on the power level used for transmission. Consider the nodes 0–6 in Figure 2.2a. The nodes 0, 1, 2 are within the radio range of node 3, but outside the ranges of nodes 4, 5, 6. Nodes 0, 1, 2 are within the radio range of one another. Nodes 3, 4, 6 are within the range of node 5. Finally, nodes 4 and 6 are within each other's range. The corresponding topology of the network is shown in Figure 2.2b. To broadcast across the entire network, at most three *broadcast steps* are required—this corresponds to the diameter of the network. The topology will change if a different level of power is used. A useful goal here is to minimize power consumption while maximizing connectivity.

2.2.3 OSI Model

Communication between users belonging to the same network or different networks requires the use of appropriate protocols. A protocol is a collection of data encoding standards and message exchange specifications that the sender and the receiver processes follow for completing a specific task. Protocols should be logically correct and well documented, and their implementations should be error-free. The International Standards Organization (ISO) recommends such protocols for a variety of networking applications. The *Open System Interconnection* (OSI) model adopted by ISO is a framework for such a protocol. It has seven *layers* (Figure 2.3) and serves as a reference for discussing other network protocols.

To understand the idea behind a protocol layer, consider the president of a company sending a proposal to the president of another company. The president explains the idea to her secretary. The secretary converts it to a proposal in the appropriate format and then hands it over to the person in charge of dispatching letters. Finally, that person

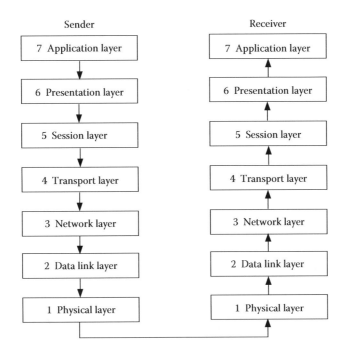

FIGURE 2.3 The seven-layer OSI model.

sends out the letter. In the receiving company, the proposal papers follow the reverse direction, that is, a person in the dispatch section first receives it and then gives the proposal to the secretary to the president, who delivers it to the president.

Communication between two processes follows a similar pattern. Each protocol layer can be compared to a secretary or a clerk. Breaking a large task into layers of abstraction is an accepted way of mastering software complexity. Furthermore, appropriate error control mechanisms at the lower layers ensure that the best possible information percolates to the top layer or the application layer. The roles of the different layers in OSI are summarized in the following:

Application layer: This layer caters to specific application needs of the user processes. Examples are *email, bulletin boards, chat rooms, web applications, and directory services.*

Presentation layer: Data representation formats may vary between the sender and the receiver machines. This layer resolves the compatibility problems by addressing the syntactic differences in data representation. Mime encoding/decoding, data compression/decompression, and encryption/decryption are addressed in this layer. The presentation layer also contains user interface components—these include ASP.NET web forms or Microsoft® Windows forms that contain codes to perform the functions of configuring the visual appearance of controls, acquiring and rendering data for business components, etc.

Session layer: The connection between the processes at the two endpoints is established and maintained at this level for all connection-oriented communications. Once a connection is established, all references to the remote machine use a session address. Such a session can be used for *ftp, telnet*, etc.

Transport layer: The goal of the transport layer is to provide *end-to-end communication* between the sender and the receiver processes. Depending on the error control mechanism used, such communications may be unreliable or reliable. Well-known examples of transport layer protocols are transmission control protocol (TCP) and user datagram protocol (UDP). Each message is targeted towards a destination process with a unique IP address on the Internet. The IP address may either be a permanent or a temporary one when a DHCP server is used. Note that the messages are not yet divided into packets.

Network layer: The network layer provides *machine-to-machine communication* and is responsible for message routing. Messages are broken down into packets of a prescribed size and format. Each packet is treated as an autonomous entity and is routed through the routers to the destination node. Two kinds of communications are possible: In *virtual circuits*, first, a connection is established from the sender to the receiver, and then packets arrive at the destination in the same order in which they are sent. *Datagrams* do not require a connection establishment phase. Out-of-order delivery of messages is possible, and the transport layer handles the task of packet resequencing. Some important issues in this layer are routing via shortest path or minimum hops, avoiding deadlocks during packet switching, etc. In a LAN, routing activity is nonexistent. IP is a network layer protocol in WAN.

Data link layer: This layer assembles the stream of bits into frames and appends error control bits (like cyclic redundancy codes) to safeguard against corruption of messages in transit. The receiver acknowledges the receipt of each frame, following which the next frame is sent out. Different data link protocols use different schemes for sending acks. Requests for retransmission of lost or corrupted frames are handled through an appropriate dialogue.

Physical layer: This layer deals with how a bit is sent across a channel. In electrical communication, the issue is what voltage levels (or what frequencies) are to be used to represent a 0 or a 1. In optical communication, the corresponding issue is what kind of light signals (i.e., amplitude and wavelength) is to be sent across fiber-optic links to represent a 0 or a 1.

The layers of protocols form a *protocol stack* or *a protocol suite*. The protocol stack defines the division of responsibilities. In addition, in each layer, some error control mechanism is incorporated to safeguard against possible malfunctions of that layer. This guarantees that the best possible information is forwarded to the upper layers. The OSI protocol stack provides a framework meant to encourage the development of nonproprietary software in open systems. Real protocol suites do not always follow the OSI guidelines. However, in most cases, the activities belonging to a particular layer of a real protocol can be mapped to the activities of one or more layers of the OSI protocol.

2.2.4 IP

Internetworking aims at providing a seamless communication system across a pair of computers on the Internet. The *Internet protocol* (IP) defines the method for sending data from one computer (also called a *host*) to another. Addressing is a critical part of Internet abstraction. Each computer has at least one IP address that uniquely distinguishes it from all other computers on the Internet. The message to be sent is divided into packets. Each packet contains the Internet address of both the sender and the receiver. IP is a *connectionless protocol*, which means it requires no continuing connection between the end hosts. Every packet is sent to a *router* (also called a *gateway*) that reads the destination address and forwards the packet to an adjacent router, which in turn reads the destination address and so forth, until one router recognizes that the destination of the packet is a computer in its immediate neighborhood or domain. That router then forwards the packet directly to the computer whose address is specified. Each packet can take a different route across the Internet, and packets can arrive in an order that is different from the order they were transmitted. This kind of service is a best-effort service. If the reception has to be error-free and the reception order has to be the same as the transmission order, then one needs another protocol like TCP.

The most widely used version of IP today is still IP Version 4 (IPv4) with a 32-bit address. However, the adoption of IP Version 6 (IPv6) is growing. IPv6 uses 128-bit addresses and can potentially accommodate many more Internet users. IPv6 includes the capabilities of IPv4 but does not provide downward compatibility—although the two implementations can coexist via dual-stack servers. In addition to the expanded address space, the deployment of IPv6 is targeted at the better handling of mobility, quality of service (QoS), multicasting, privacy extension, and so on. Although the global IP6 traffic is currently a little above 1%, it is scheduled to replace IP4 in the next few years.

On an Ethernet, each source and destination node has a 48-bit hardware address, called the medium access control address (MAC address) stored in its network interface card. The MAC address is used by the data link layer protocols. The IP addresses are mapped to the 48-bit hardware addresses using the *address resolution protocol* (ARP). The ARP client and server processes operate on all computers using IP over Ethernet.

2.2.5 Transport Layer Protocols

Two common transport layer protocols are UDP and TCP.

UDP/IP: UDP uses IP to send and receive data packets, but packet reception may not follow the transmission order. UDP is a *connectionless* protocol. The application program must ensure that the entire message arrived and in the right order. UDP provides two services not provided by the IP layer. It provides port numbers to help distinguish different user requests and a checksum capability to verify that the data arrived intact. UDP can detect errors but drops packets with errors. Network applications that want to save processing time because messages are short or occasional errors have no significant impact prefer UDP. Others use TCP.

TCP/IP: The TCP running over IP is responsible for overseeing the reliable and efficient delivery of data between a sender and a receiver. Data can be lost in transit. TCP adds support to the recovery of lost data by triggering retransmission, until the data are correctly and completely received in the proper sequence. By doing so, TCP implements a reliable stream service between a pair of ports in the sender and the receiver processes.

Unlike UDP, TCP is a *connection-oriented protocol*—so actual data communication is preceded by a connection establishment phase and terminated by a connection termination phase. The basic idea behind error control in TCP is to add sequence number to the packets prior to transmission and monitor the ack received for each packet from the

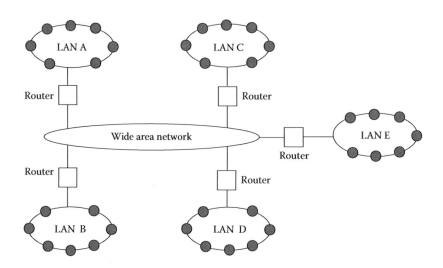

FIGURE 2.4   Five LANs connected to WANs that serve as the backbone.

destination process. If the ack is not received within a window of time that is a reasonable estimate of the turnaround delay, then the message is retransmitted. The receiving process sends an ack only when it receives a packet with the expected sequence number.

Transmission errors can occur for a variety of reasons. Figure 2.4 shows a slice of the Internet. Assume that each of the two users 1 and 2 are simultaneously sending messages to user 3 at the rate of 8 MB/s. However, LAN D, to which user 3 is connected, is unable to handle incoming data at a rate exceeding 10 MB/s. In this situation, there are two possible options:

Option 1: The router can drop the packets that LAN D is unable to handle.

Option 2: The router can save the extra packets in its local memory and transmit them to LAN D at a future time.

Option 2 does not rule out packet loss, since each router has a limited amount of memory. Among many tasks, TCP/IP helps recover lost packets, reorder them, and reject duplicate packets before they are delivered to the destination process. For the efficient use of the transmission medium, TCP allows multiple data packets to be transmitted before the ack to the first packet is received. TCP has a mechanism to estimate the round-trip time and limit data rates to clear out congestions.

TCP/IP provides a general framework over which many services can be built and message-passing distributed algorithms can be implemented. Despite occasional criticisms, it has successfully carried out its mission for many years, tolerating many changes in network technology.

2.2.6 Interprocess Communication Using Sockets

Sockets are abstract endpoints of communication between a pair of processes. Developed by Berkeley Software Distribution as a part of BSD UNIX, sockets are integrated into the I/O part of the operating system. Sockets are of two types: *stream sockets* and *datagram sockets.* Stream sockets use TCP/IP, and datagram sockets use UDP/IP. The steps involved in a client–server communication using sockets are as follows:

First, a *client* process requests the operating system to create a socket by making the *socket* system call, and a descriptor identifying the socket is returned to the client. Then it uses the *connect* system call to connect to a *port* of the server. Once the connection is established, the client communicates with the server.

Like the client, the *server* process also uses the *socket* system call to create a socket. Once created, the *bind* system call will assign a port number to the socket. If the protocol is a connectionless protocol like UDP, then the server is ready to receive packets. For connection-oriented protocols like TCP, the server first waits to receive a connection request by calling a *listen* procedure. The server eventually accepts the connection using the *accept* system call. Following this, the communication begins. After the communication is over, the connection is closed.

A port number is a 16-bit entity. Of the 2^{16} possible ports, some (0–1023) are reserved for standard services. For example, server port number 21 is for FTP, port 22 is for *Secure Shell (SSH) remote login*, and port number 23 is for *telnet*. An example of a socket program in Java is shown in the following:

```
/* Client.java sends a request to Server to do a computation */
/*Author: Amlan Bhattacahrya*/
import java.net.*;
import java.io.*;

class Client {
public static void main (String []args) {
try {
Socket sk = new Socket(args[0], Integer.parseInt(args[1]));
/* BufferedReader to read from the socket */
BufferedReader in = new BufferedReader (new InputStreamReader(sk.
  getInputStream ()));
/* PrintWriter to write to the socket */
PrintWriter out = new PrintWriter(sk.getOutputStream(), true);
/*Sending request for computation to the server from an imaginary
  getRequest() method */
String request = getRequest();
/* Sending the request to the server */
out.println(request);
/* Reading the result from the server */
String result = in.readLine();
System.out.println(" The result is " + result);
}
catch(IOException e) {
System.out.println(" Exception raised: " + e);
}
}
}
```

```
/* Server.java computes the value of a request from a client */
import java.net.*;
import java.io.*;

class Server {
public static void main (String []args) {
try {
ServerSocket ss = new ServerSocket(2000);
while (true) {
/* Waiting for a client to connect */
Socket sk = ss.accept();
/* BufferedReader to read from the socket */
BufferedReader in = new BufferedReader(new InputStreamReader(sk.
  getInputStream ()));
/* PrintWriter to write to the socket */
PrintWriter out = new PrintWriter(sk.getOutputStream(), true);
/*Reading the string which arrives from the client */
String request = in.readLine();
/*Performing the desired computation via an imaginary
  doComputation() method */
String result = doComputation(request);
/* Sending the result back to the client */
out.println(result);
}
}
catch(IOException e) {
System.out.println(" Exception raised: " + e);
}
}
}
```

2.3 NAMING

In this section, we look into the important middleware service of *naming* built on top of TCP/IP or UDP/IP. A *name* is a string of characters or bytes, and it identifies an *entity*. The entity can be just about anything—it can be a user, or a machine, or a process, or a file. An example of a name is the URL of a website on the WWW. Naming is a basic service using which entities can be identified and accessed only by name, regardless of where they are located. A user can access a file by providing the *filename* to the file manager. To log in to a remote machine, a user has to provide a *login name*. On the WWW, the Domain Name Service (DNS) maps domain names to IP addresses. Users cannot send emails to one another unless they can name one another by their email addresses. Note that an address can also be viewed as a name. The ARP uses the IP address to look up the MAC address of the receiving machine.

Each name must be associated with a unique entity, although a given entity may have different aliases. As an example, consider the name Alice. Can it be associated with a unique person in the universe? No. Perhaps Alice can be identified using a unique phone number, but Alice may have several telephone lines, so she may have several phone numbers. If Alice switches her job, then she can move to a new city, and her phone number will change. How do we know if we are communicating with the same Alice?

This highlights the importance of location independence in naming. Notice that mobile phone numbers are location independent. The email addresses in Gmail or Hotmail accounts are also not tied to any physical machine and are therefore location independent. A consistent naming scheme, location independence, and a scalable naming service are three cornerstones of a viable naming system for distributed applications.

A naming service keeps track of a name and the attributes of the object that it refers to. Given a name, the service looks up the corresponding attributes. This is called *name resolution*. For a single domain like a LAN, this is quite simple. However, on the Internet, there are many networks and multiple domains,* so the implementation of a name service is not trivial.

Names follow a tree structure, with the parent being a common suffix of the names of the subtree under it. Thus, *cs.uiowa.edu* is a name that belongs to the domain *uiowa.edu*. The prefix *cs* for the computer science department can be assigned only with permission from the administrators of the domain uiowa.edu. However, for the names *hp.cs.uiowa.edu* or *linux.cs.uiowa.edu*, it is the system administrator of the computer science department who assigns the prefixes *hp* or *linux*. Two different name spaces can be merged into a single name space by appending their name trees as subtrees to a common parent at a higher level. Figure 2.5 shows a naming tree for the generation of email ids. The leaves reflect the names of the users.

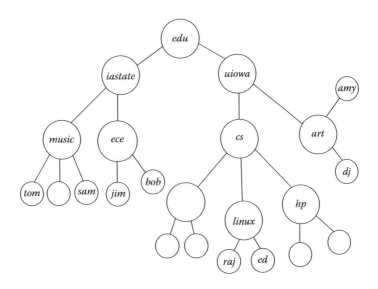

FIGURE 2.5 A naming hierarchy.

---

* On a single domain, a single administrative authority has the jurisdiction of assigning names.

2.3.1 Domain Name Service

Consider the task of translating a fully qualified name *cs.uiowa.edu*. DNS servers are logically arranged in a hierarchy that matches the naming hierarchy. Each server has authority over a part of the naming hierarchy. A root server has authority over the top-level domains like *.edu*, *.com*, *.org*, and *.gov*. The root server for *.edu* does not know the names of the computers at *uiowa.edu*, but it knows how to reach the server. An organization can place all its domain names on a single server or place them on several different servers.

When a DNS client needs to look up a name, it queries DNS servers to resolve the name. Visualize the DNS as a central table that contains more than a billion entries of the form (name, IP address). Clearly, no single copy of such a table can be accessed by millions of users at the same time. In fact, no one maintains a single centralized database of this size anywhere. It operates as a large distributed database. For the sake of availability, different parts of this table are massively replicated and distributed over various geographic locations. There is a good chance that a copy of a part of the DNS service is available with your Internet service provider or the system administrator of your organization. Each server contains links to other domain name servers. DNS is an excellent example of how replication enhances availability and scalability.

To resolve a name, each resolver places the specified name into a DNS request message and forwards it to the local server. If the local server has sole authority over the requested name, then it immediately returns the corresponding IP address. This is likely when you want to send a message to your colleague in the same office or the same institute. However, consider a user from *uiowa.edu* looking for a name like *boomerang.com* that is outside the authority of uiowa's local DNS server. In this case, the local server becomes a client and forwards the request to an upper-level DNS server that deals with .edu. However, since no server in .edu knows which other server has the authority over *boomerang.com*, a simple solution is to send the request to the *root* server. The root server may not have direct authority, but it knows the address of the next level server that may have the authority. In this manner, the request propagates through a hierarchy of servers and finally reaches the one that has the authority over the requested name. The returned IP address eventually reaches the resolver via the return path.

The name servers for the top-level names do not change frequently. In fact, the IP addresses of the root servers will rarely change. How does the root server know which server might have authority over the desired domain *boomerang.com*? At the time of domain registration, the registrant informs the DNS registry about the IP address of the name servers that will have authority over this domain. As new subdomains are added, the registry is updated.

The mechanism presented earlier will be unworkable without replication and caching at various levels. With millions of users simultaneously using the WWW, the traffic at or near the root servers will be so enormous that the system will break down. Replication of the root servers is the first step to reduce the overload. Depending on the promptness of the response, the DNS server responsible for the local node uses one of the root servers. More importantly, each server has a cache that stores a copy of the names that have been recently resolved. With every new lookup, the cache is updated unless the cache already contains that name. Before forwarding a name to a remote DNS server, the client cache is first looked up. If the name is found there, then the translation is the fastest.

The naming service exhibits good locality—a name that has just been resolved is likely to be accessed again in the near future. Naturally, caching expedites name resolution, and local DNS servers are able to provide effective service.

The number of levels in a domain name has an upper bound of 127, and each label in the name can contain up to 63 characters, as long as the whole domain name does not exceed a total length of 255 characters. But in practice, some domain registries require a shorter limit.

2.3.2 Naming Service for Mobile Clients

Mapping the names of mobile clients to addresses is tricky, since their addresses can change without notice. A simple method for translation uses broadcasting. To look up Alice, broadcast the query *Where is Alice?* The machine that currently hosts Alice or the access point to which Alice is connected will return its address in response to the query. On Ethernet-based LANs, this is the essence of the ARP, where the IP address is used to look up the MAC address of a machine having that IP address. While this is acceptable on Ethernets where broadcasts are easily implemented by the available hardware, on larger networks, this is inefficient. In such cases, a *location service* fills the gap.

Location service: The naming service will first convert each name to a *unique identifier*. For example, the naming service implemented through a telephone directory will always map your name to your local telephone number. A separate *location service* will accept this unique identifier as the input and return the current address of the client (in this example, the client's *current* telephone number).

While the ARP implements a simple location service, a general location service relies on *message redirection*. The implementation is comparable to *call forwarding*. Mobile IP uses a *home agent* for location service. While moving away from home or from one location to another, the client updates the home agent about her current location or address. Communications to the client are directed as a *care-of address* to the home agent. The home agent forwards the message to the client and also updates the sender with the current address of the client.

2.4 REMOTE PROCEDURE CALL

Consider a server providing service to a set of clients. In a trivial setting when the clients and the server are distinct processes residing on the same machine, the communication uses nothing more than a system call. As an example, assume that a server allocates memory to a number of concurrent processes. The client can use two procedures: *allocate* and *free*. Procedure *allocate(m)* allocates m blocks of memory to the calling client by returning a pointer to a free block in the memory. Procedure *free(addr, m)* releases m blocks of memory from the designated address *addr*. Being on the same machine, the implementation is straightforward. However, in a distributed system, there is no guarantee (and in fact it is unlikely) that clients and servers will run on the same machine. Procedure calls can cross machine boundaries and domain boundaries. This makes the implementation of the procedure call much more complex. A *remote procedure call* (RPC) is a procedure call that helps a client communicate with a server running on a different machine that may belong to a different network and a different administrative domain.* Gone is the support for shared memory. Any implementation of RPC has to take into account the passing of client's call parameters to the server machine and returning the response to the client machine.

2.4.1 Implementing RPC

The client calling a remote procedure blocks itself until it receives the result (or a message signaling completion of the call) from the server. Note that clients and servers have different address spaces. To make matters worse, crashes are not ruled out. To achieve transparency, the client's call is redirected to a *client stub* procedure in its local operating system. The stub function provides the local procedure call interface to the remote function. First, the client stub (1) packs the parameters of the call into a message and (2) sends the message to the server. Then the client blocks itself. The task of packing the parameters of the call into a message is called *parameter marshaling*.

On the server side, a *server stub* handles the message. Its role is complementary to the role of the *client stub*. The server stub first unpacks the parameters of the call (this is known as unmarshaling) and then calls the local procedure. The result is marshaled and sent back to the client's machine. The client stub unmarshals the parameters and returns the result to the client (Figure 2.6). The operation of the stubs in an RPC is summarized as follows:

Client Stub	Server Stub (Repeats the Following)
Pack parameters into a message	**Do** no message → *skip* **od**
Send message to remote machine	Unpack the call parameters
Do no result → *skip* **od***	**Call** the server procedure
Receive result and unpack it	Pack result into a message
Return to the client program	Send it to the client

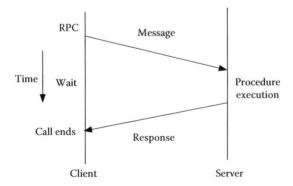

FIGURE 2.6 An RPC.

---

* Of course, RPC can be trivially used to communicate with a server on the same machine.

The lack of similarity between the client and server machines adds a level of complexity to the implementation of RPC. For example, the client machine may use the big-endian format, and the server machine may use the little-endian format. Passing pointers is another headache since the address spaces are different. To resolve such issues, the client and the server need to have a prior agreement regarding RPC protocols.

The synchronization between the client and the server can take various forms. While the previous outline blocks the client until the result becomes available, there are non-blocking versions too. In a nonblocking RPC, the client continues with unrelated activities after initiating the RPC. When the server sends the result, the client is notified, and it accepts the result.

Finally, network failures add a new twist to the semantics of RPC. Assume that the message containing a client's RPC is lost en route the server. The client has little alternative to waiting and reinitiating the RPC after a time-out period. However, if for some reason the first RPC is not lost but stuck somewhere in the network, then there is a chance that the RPC may be executed twice. In some cases, it makes little difference. For example, if the RPC reads the blocks of a file from a remote server, then the client may not care if the same blocks are retrieved more than once. However, if the RPC debits your account in a remote bank and the operation is carried out more than once, then you will be unhappy. Some applications need the *at-least-once* semantics, whereas some others need the *at-most-once* semantics or the *exactly once* semantics.

2.4.2 Sun ONC/RPC

To implement client–server communication in the network file service, Sun Microsystems* developed the open network computing RPC (ONC/RPC) mechanism, which is also packaged with many UNIX installations. Client–server communication is possible through either UDP/IP or TCP/IP. An interface definition language (called external data representation [XDR]) defines a set of procedures as a part of the service interface. Each interface has a unique *program number* and a *version number*, and these are available from a central authority. Clients and servers must verify that they are using the same version number. These numbers are passed on as a parameter in the request message from the client to the server. Sun ONC uses an RPC compiler (*rpcgen*) that automatically generates the client and server stubs.

Sun RPC allows a single argument to be passed from the client to the server and a single result from the server to the client. Accordingly, if there are multiple parameters, then they have to be passed on as a structured variable. At the server end, a service demon process called port mapper receives the RPC request. Clients can easily locate the port mapper since it is always on a designated port (port 111). ONC identifies each procedure by three parameters: *program number, version number,* and *procedure number.* Procedure 0 is the null procedure. The client queries the port mapper to find out the port number of the program and its version number. The port mapper binds the RPC to

---

* Now acquired by Oracle.

designated ports on the client and the server machines. Many different types of authentications can be incorporated in ONC/RPC—these include the traditional (uid, gid) of UNIX to the Kerberos type authentication service (see Chapter 19).

2.5 REMOTE METHOD INVOCATION

Remote method invocation (RMI) is a generalization of RPC in an object-oriented environment. The object resides on the server's machine that is different from the client's machine. This is known as *remote object*. An object for which the instance of the data associated with it is distributed across machines is known as a *distributed object*. An example of a distributed object is an object that is replicated over two or more machines. A remote object is a special case of a distributed object where the associated data are available on one remote machine.

To realize the scope of RMI (vis-à-vis RPC), recall the implementation of an RPC using sockets. In RPC, objects are passed *by value*; thus, the current state of the remote object is copied and passed from the server to the client, necessary updates are done, and the modified state of the object is sent back to the server. If multiple clients try to concurrently access/update the remote object by invoking methods in this manner, then the updates made by one client may not be reflected in the updates made by another client, unless such updates are serialized. In addition, the propagation of multiple copies of the remote object between the server and the various clients will consume significant bandwidth of the network.

RMI solves these problems in a transparent way. The various classes of the *java.rmi* package allow the clients to access objects residing on remote hosts, as if, *by reference*, instead of by value. Once a client obtains a reference to a remote object, it can invoke the methods on these remote objects as if they existed locally. All modifications made to the object through the remote object reference are reflected on the server and are available to other clients. The client is not required to know where the server containing the remote object is located—it only invokes a method through an interface called a *proxy*. The proxy is a client stub responsible for marshaling the invocation parameters and unmarshaling the results from the server (Figure 2.7). On the server side, a server stub called a *skeleton* unmarshals the client's invocations, invokes the desired method, and marshals the results back to the client. For each client, there is a separate proxy, which is a separate object in the client's address space.

When multiple clients concurrently access a remote object, the invocations of the methods are serialized, as in a *monitor*.* Some clients are blocked until their turns come, while others make progress. The implementation of the serialization mechanism is trivial for local objects, but for remote objects, it is tricky. For example, if the server handles the blocking and the current client accessing the remote object crashes, then all clients will be blocked forever. On the other hand, if clients handle blocking, then a client needs to block itself before its proxy sends out the method call. How will a client

---

* A synchronization mechanism originally invented by C.A.R. Hoare and Per Brinch Hansen.

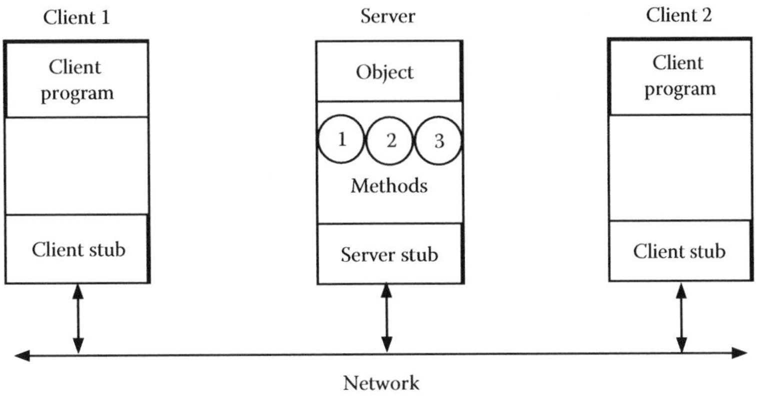

FIGURE 2.7 Remote object invocation scheme.

know if another client has already invoked a method for the remote object? Clearly, the implementation of serializability for remote objects requires additional coordination among clients.

Java RMI passes local and remote objects differently. All local objects are passed by value, while remote objects are passed by reference. While calling *by value*, the states of objects are explicitly copied and included in the body of the method call. An example of a *call by reference* is as follows: Each proxy contains (1) the network address of the server S1 containing a remote object X and (2) the name of the object X in that server. As a part of an RMI in another server S2, the client passes the proxy (containing the reference to X) to the server S2. Now S2 can access X on S1. Since each process runs on the same Java virtual machine, no further work is required following the unmarshaling of the parameters despite the heterogeneity of the hardware platforms.

2.6 MESSAGES

Most distributed applications are implemented using message passing. The messaging layer is logically located just above the TCP/IP or the UDP/IP layer, but below the application layer. The implementation of sockets at the TCP or the UDP layer helps processes address one another using specific socket addresses.

2.6.1 Transient and Persistent Messages

Messages can be transient or persistent. In *transient communication*, a message is lost unless the receiver is active at the time of the message delivery and retrieves it during the life of the application. An example is the interprocess communication via message buffers in a shared memory multiprocessor. In *persistent communication*, messages are not lost, but saved in a buffer for possible future retrieval. Recipients eventually receive the messages even if they were passive at the time of message delivery. An example is email communication. Messages are sent from the source to the destination via a sequence of routers, each of which manages a *message queue*.

2.6.2 Streams

Consider sending a video clip from one user to another. Such a video clip is a stream of frames. In general, *streams* are sequences of data items. Communication using streams requires a connection to be established between the sender and the receiver. In streams for multimedia applications, the QoS is based on the *temporal relationship* (like the number of frames per second or the propagation delay) among items in the stream, and its implementation depends on the available network bandwidth, buffer space in the routers, and processing speeds of the end machines. One way to guarantee the QoS is to reserve appropriate resources before communication starts. RSVP is a transport-level protocol for reserving resources at the routers.

2.7 WEB SERVICES

Web services provide a different way of using the Internet for a variety of applications based on machines communicating with one another over the WWW. One can design an application on a machine by using data from the website hosted by a different machine. For example, a user can implement a service on her smartphone using a city transit's website to obtain information about the latest bus schedule and set an alarm to remind about the departure time for catching the bus at a particular stop. Another user can design an application by pulling data from the website of a popular restaurant, so that whenever they will have her special dish for dinner on the menu, she will receive a text message. Web services are a form of middleware that helps to integrate applications. They facilitate applications talking to one another.

Historically, Microsoft pioneered web services as a part of their .NET initiative. Today, most web services are based on XML that is widely used for cross-platform data communication—these include simple object access protocol (SOAP); web service description language (WSDL) and universal description, discovery, and integration specification (UDDI); and Java web services.

WSDL describes the public interface to the web service. It is an XML-based service description that describes how a client should communicate using the web service. These include protocol bindings and message formats required to interact with the web services listed in its directory. A client can find out what functions are available on the web service and use SOAP to make the function call.

SOAP allows a one-way message containing a structured data to be sent from one process to another using any transport protocol like TCP or HTTP or SMTP. The message exchange is completed when the recipient sends a reply back to the sender. One can use a Java API for an XML-based RPC implementation using SOAP to make RPCs on the server application. A standard way to map RPC calls to SOAP messages allows the infrastructure to automatically translate between method invocations and SOAP messages at run time, without redesigning the code around the web service platform.

XML is virtually a ubiquitous standard. The attention is now focused on the next level of standards derived from XML—UDDI, WSDL, and SOAP. The rapid growth of web services to access the information stored in diverse databases makes some people wary about privacy questions.

2.8 EVENT NOTIFICATION

Event notification systems help establish a form of asynchronous communication among distributed objects on heterogeneous platforms and have numerous applications. An example is the *publish–subscribe* middleware. Consider the airfares that are regularly published by the different airlines on the WWW. You are planning a vacation in Hawaii, so you may want to be notified of an *event* when the round-trip airfare from your nearest airport to Hawaii drops below $400. This illustrates the nature of publish–subscribe communication. Here, you are the subscriber of the event. Neither publishers nor subscribers are required to know anything about one another, but communication is possible via a brokering arrangement. Such event notification schemes are similar to interrupts or exceptions in a centralized environment. By definition, they are asynchronous.

Here are a few other examples. A *smart home* may send a phone call to its owner away from home whenever the garage door is open, or there is a running faucet, or there is a power outage. In a collaborative work environment, processes can resume the next phase of work, when everyone has completed the current phase of the work—these can be notified as events. In an intensive care unit of a medical facility, physicians can define events for which they need notification. Holders of stocks may want to be notified whenever the price of their favorite stock goes up by more than 5%. An airline passenger would like to be notified via an app in her smartphone if the time of the connecting flight changes.

Apache River (originally *Jini* developed by Sun Microsystems) provides event notification service for Java-based platforms. It allows subscribers in one Java virtual machine (JVM) to receive notification of events of interest from another JVM. The essential components are as follows:

1. An *event generator* interface, where users may register their events of interest.

2. A *remote event listener* interface that provides notification to the subscribers by invoking the *notify* method. Each notification is an instance of the *remote event* class. It is passed as an argument to the *notify* method.

3. *Third-party agents* play the role of observers and help coordinate the delivery of similar events to a group of subscribers.

Apples's iOS running on iPhones or iPads uses the *push notification service* for sending notification of events to users via a short text message or a sound byte.

2.9 VIRTUALIZATION: CLOUD COMPUTING

Cloud computing is an environment in which your data and/or programs are not stored or managed by your own machine, but done by a third party: the *cloud*, who offers this as a service. You can access your data or results of your computation through a browser that is a part of every computer system. Cloud computing is the convergence of several technologies, like distributed computing, cluster computing, and web services.

Even before the term *cloud computing* was coined, the concept was getting momentum under various names. All the web applications are now included under cloud computing.

Many of us use Google mail or Yahoo mail to store and access email messages. These emails are stored in the servers managed by the respective companies. Google Doc allows users to upload, edit, and retrieve documents to Google's data servers for the purpose of sharing. YouTube hosts millions of user-uploaded videos and allows us to store and share our videos with others without any cost. Many of us routinely use Flickr and Picasa to upload and share millions of our personal digital photographs. Social networking sites like Facebook allow members to post pictures and videos that are stored in the site's servers.

Not only for the dedicated cloud enthusiasts but also for the average users, Amazon's Elastic Computing Cloud (EC2) virtual computing environment might be the answer to all the needs. Rather than purchasing servers, software, network equipment, and so on, users would buy into a fully outsourced set of online services instead. The users need not care about where these servers are geographically located—they are all part of the cloud, residing in a single data center or distributed across a number of data centers and controlled by a single organization.* This not only subscribes to the concept of *virtualization* but also reinforces the concept of pay-as-you-use or *utility computing*, where the users pay a metered price in exchange for the service. There is no upfront investment and it allows flexible scaling of resources. Above all, the current prices for such services (as of December 2013) are extremely reasonable—for an average student user job running for an hour or so, it costs pennies only.

To use a cloud service like Amazon's EC2, note that Amazon web service (AWS) is the umbrella of all services provided by Amazon, including EC2. So once a user signs up for AWS, she can go to the EC2 link to create her virtual server and access it using her remote desktop connection.

2.9.1 Classification of Cloud Services

Some cloud platforms offer a mix of proprietary development and application packages, while others are simply metered services. To distinguish between the various forms, cloud services have been classified into *software-as-a-service* (SaaS), *platform-as-a-service* (PaaS), and *infrastructure-as-a-service* (IaaS), respectively:

SaaS: It is the most basic among these services. Think of Google Docs. Creating a Google account is free. Once you have an account, all you have to do is log in to *google.com/docs* and you instantly have access to a powerful word processor, spreadsheet application, and presentation creator. Google makes all these services available to you, and you can manage everything from the web browser without installing anything else on your machine. Dropbox, Salesforce.com, etc., are applications that qualify as SaaS and the basic services are free.

PaaS: PaaS facilitates the development of an application that will run in a specific cloud environment. One can utilize the available infrastructure and tools of an established environment. This also gives the developer the ability to quickly make her app available to a

---

* These are different from grids that are a less tightly coupled federation of heterogeneous resources under the control different organizations. As a result, grid computing must deal with verifying credentials across multiple administrative domains.

wide audience. For example, on the Facebook platform, developers can write new applications and make them available to other users of that platform. Google Apps provides APIs and functionalities that facilitate developers to build web applications by leveraging its different services such as maps and calendar. These are very handy for developing lightweight web applications.

IaaS: This is the most comprehensive cloud platform and is mainly used by full-time developers or large-scale enterprise customers or by occasional users who have the IT skills but don't have the infrastructure. While SaaS allows usage of cloud apps, and PaaS allows you to develop apps, IaaS gives you infrastructure for developing, running, and storing your apps in cloud environments. The benefit of IaaS is the virtually limitless storage and computing power available to the developers without having any physical hardware on site. Amazon EC2 is a great example of IaaS. From the smallest application to full-scale websites needed for launching a business, EC2 provides the necessary cloud infrastructure. Users have the freedom to develop using an assortment of tools, from MySQL to Ruby on Rails, and can choose from several Linux or Windows environments.

With the growth of cloud computing, the number of data centers is also growing at a rapid pace. Each data center is equipped with thousands of processors whose collective computing speed adds up to the petaflop range, virtually unlimited storage, and high-speed networking facility. Many of these data centers are automatically and remotely managed without any human operator being around. Energy is a big budget item for these data centers, as well as water usage that is needed for cooling. The enormous volume of water required to cool high-density server farms is making water management a growing priority for data center operators. A 15 MW data center can use up to 360,000 gal of water a day [M09].

2.9.2 MapReduce

One significant feature of modern day distributed systems is the large volume of data that they are expected to handle. Photos and audio and video recordings are continuously created and passed to the social network sites. Amazon and eBay routinely deal with enormous volumes of data from all over the world. The search engines Google, Yahoo, and Bing routinely crunch enormous amount of data to process user queries from all over the world. Massively parallel data processing engines have become indispensable for generating prompt responses in these scenarios. To crunch massive volumes of data, Google invented *MapReduce* in 2004, and in 2007, Yahoo made it into the open-source project *Hadoop* that is essentially an operating system to enable MapReduce programs to run on computing clusters.

MapReduce algorithms efficiently harness the built-in parallelism exhibited by many large-scale or data-intensive problems. To design such algorithms, one has to start with the two basic operations: *map* and *reduce*. The function *map* applies a specific function to all the elements of a list, and their execution is meant to run in parallel. Each process or thread is assigned one such task. The function *reduce* is an aggregation operation that takes the output of the map operation and aggregates to a final result.

The basic data structure in MapReduce is the *<key, value>* pairs. If $x = 5$, then the *key* is the integer variable x, and its *value* is 5. For a graph, the key may be the identifier of a node and its value may be the list of its neighbors. For a web page, the key is its URL, and the value is the HTML code for that page. In the succeeding text, we illustrate a typical application of MapReduce via an example: A directory D has a number of text files, each containing some text—our goal is to capture and count every occurrence of the word *dog* and every occurrence of the word *cat* in these two files and produce the cumulative result for each.

```
{Pseudocode for the word count problem}
1: Mapper (directory D; file f)
2: for every term "dog" in file f do
3: Output ("dog", value 1)

1: Reducer (term "dog"; set of values)
2: total: = 0
3: for each t in set of values do
4: total: = total + t
5: Output ("dog", total)
```

Several instances of mapper can run on different machines, and there can be another mapping task for the key *cat*. The execution framework of MapReduce guarantees that all values associated with the same key are brought together in the reducer via a shuffle stage. As a result, the word count algorithm adds all counts associated with each key, which is also a key–value pair. Thus, the basic framework of a MapReduce job can be summarized as follows:

Map:　　　$\langle key, value \rangle \rightarrow$ list of $\langle key, value \rangle$
　　　　　　{This is the intermediate $\langle key, value \rangle$ pair}
Reduce:　　$\langle key, \text{list of } values \rangle \rightarrow$ list of $\langle key, value \rangle$

In the earlier example, the output of the mapper for counting the occurrence of *dog* is a list, $\langle dog, 1 \rangle$, $\langle dog, 1 \rangle$, $\langle dog, 1 \rangle$..., and the output of the mapper for counting the occurrence of *cat* is a list, $\langle cat, 1 \rangle$, $\langle cat, 1 \rangle$, $\langle cat, 1 \rangle$.... The program will run across all files in a directory. After the completion of the mapping phase, the intermediate $\langle key, value \rangle$ pairs are exchanged between machines to send all values with the same key to a single reducer (shuffling). The reducer produces an output $\langle dog, 524 \rangle$, $\langle cat, 316 \rangle$, $\langle cow, 23 \rangle$.... Figure 2.8 illustrates this scheme.

It is important to realize that although MapReduce can be directly used for a large class of problems that exhibit embarrassingly parallel feature, many algorithms cannot be easily expressed as a single MapReduce job. Complex algorithms have to be decomposed into a sequence of jobs, and data routing has to be orchestrated so that the output of one job becomes the input to another job. These include iterative tasks that need to be repeated before convergence is reached.

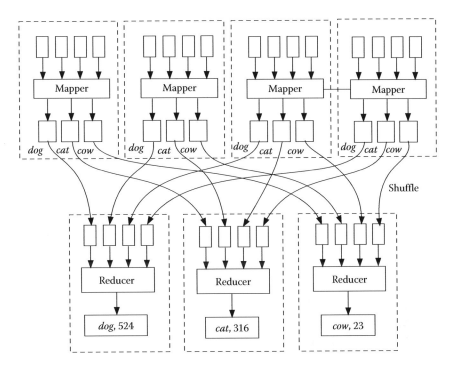

FIGURE 2.8   The architecture of MapReduce: the mappers receive their inputs from the files that store data. Each box is physically mapped to a computing node.

2.9.3 Hadoop

In the world of cluster-level parallelism, libraries implementing message-passing interface (MPI), or language extensions like OpenMP for shared memory parallelism, are well known. These logical abstractions hide many details of synchronization and communications at the process level. However, for a smooth implementation of such abstractions, developers have to keep track of these aspects, as well as how resources are made available to workers. This applies to MapReduce too. Apache Hadoop is an open-source implementation of the infrastructure that MapReduce needs.

Hadoop uses the master–slave architecture. A master node (also called a NameNode) receives the job from the user. The input data are broken into pieces and stored in the Hadoop distributed file system (HDFS). Physically, these data pieces are scattered over a large number of nodes in the system. The master node assigns the components of the job (*map* processes) to the various slave nodes (also called DataNodes). Each slave contributes some local storage and computational power to solve the problem. When the slave nodes complete their tasks, they return the results to the master. The master runs the *reduce* process to aggregate the pieces and generate the result. If necessary, the slave nodes may further subdivide the task and assign them to other slave nodes. Collectively, such an implementation involves thousands of computing nodes and can handle petabytes of data. DataNode failures are automatically handled in this framework. Hadoop is written in Java.

2.10 MOBILE AGENTS

A different mode of communication between processes is possible via *mobile agents*. A mobile agent is a piece of code that migrates from one machine to another. The code, which is an executable program, is called a *script*. In addition, agents carry data values or procedure arguments or interim states that need to be transported across machines. The use of an interpretable language like Tcl makes it easy to support mobile agent–based communication on heterogeneous platforms. Compared with messages that are passive, agents are active and can be viewed as *messengers*. Agent migration is possible via the following steps:

1. The sender writes the script and calls a procedure *submit* with the parameters (name, parameters, target). The state of the agent and the parameters are marshaled and the agent is sent to the target machine. The sender either blocks itself or continues with an unrelated activity at its own site.

2. The agent reaches the destination, where an agent server handles it. The server authenticates the agent, unmarshals its parameters, creates a separate process or thread, and schedules it for the execution of the agent script.

3. When the script completes its execution, the server terminates the agent process, marshals the state of the agent as well as the results of the computation if any, and forwards it to the next destination, which can be a new machine, or the sender. The choice of the next destination follows from the script.

Unlike an RPC that always returns to the sender, agents can follow a predefined itinerary or make autonomous routing decisions. Such decisions are useful when the next process to visit has crashed and the agent runs the risk of being trapped in a black hole. Mobile agents thus overcome the limitations of the traditional client–server architecture.

2.11 BASIC GROUP COMMUNICATION SERVICES

With the rapid growth of social networking and electronic commerce on the web, group-oriented activities have substantially increased in recent years. Examples of groups are the following:

- The batch of students graduating from a high school in a given year

- The friends of a person on Facebook or Twitter

- The clients of a video distribution service

- A set of replicated servers forming a highly available service

Group communication services include (1) a membership service that maintains a list of current members by keeping track of who joined the group, and which members left the group (or crashed), and (2) various types of multicasts within the group. Group members can use such multicasts as primitives for implementing different kinds of services. One example is *atomic multicast*, which guarantees that regardless of failures, either all nonfaulty members receive the message or no one receives the message. Another example is an ordered multicast, where in addition to the atomicity property, all nonfaulty members are required to receive the messages in a specific order. Each of these multicasts has applications in the implementation of specific group services. We will elaborate these in Chapter 15.

2.12 CONCLUDING REMARKS

The ISO model is a general framework for the development of network protocols. Real-life protocol suites, however, do not always follow this rigid structure, but often, they can be mapped into the ISO framework. TCP/IP is by far the most widely used networking protocol today.

Middleware services help create an abstraction over the networking layers and relieve users of many intricacies of network programming on wide area networks (WANs). Without these services, the development of applications would have been slow and error prone. The open nature of the request brokering services like CORBA adds to the interoperability of the distributed applications. Future implementations will increase the efficiency of the services with no need to rewrite the applications, as long as the interfaces remain unchanged.

2.13 BIBLIOGRAPHIC NOTES

An in-depth treatment of networking is available in Peterson and Davie's book [PD96]. Gray's book [G97] provides a detailed description of interprocess communication tools in Unix. Needham's article [N93] is a great introduction to *naming*. Albitz and Liu's book [AB01] describes the implementation of DNS. Birrell and Nelson [BN84] introduced RPC. Several tutorials on Java RMI can be found in the Oracle's website (Oracle acquired Sun Microsystems in 2010). Waldo [W98] presented a comparison between RPCs and RMIs. Arnold et al. [AOSW+99] described Jini (now called Apache River) event service. Gray [G96] and subsequently Kotz et al. [KGNT+97] introduced agent Tcl. The growth in web services and later cloud computing started with web 2.0 that includes blogs and social networking sites. Since the late 1990s, network bandwidth witnessed significant growth, and it opened up a window of opportunities. Salesforce.com led the idea of delivering enterprise solutions via a website. In 2002, Amazon took it to the next step by providing a suite of web services that makes storage and computation available to users through their browsers. In 2006, Amazon's EC2 enabled entrepreneurs to rent computers as utilities and run their applications through browser-based interfaces. Dean and Ghemawat [DG04] introduced MapReduce as a divide-and-conquer-based method for tackling large data sets on clusters.

EXERCISES

The following exercises ask you to study outside materials and investigate how communication takes place in the real world. The materials in this chapter provide some pointers and serve as a skeleton.

2.1 Study the domain hierarchy of your organization. Show how it has been divided into the various servers and explain the responsibility of each server.

2.2 You do *net banking* from your home for paying bills to the utility companies. Study and explain how the following communications take place.

2.3 Two processes P and Q communicate with one another. P sends a sequence of characters to Q. For each character sent by P, Q sends an ack back to P. When P receives the ack, it sends the next character. Show an implementation of the earlier interprocess communication using sockets.

2.4 Explain with an example why CSMA/CD cannot resolve media access contention in wireless networks and how the RTS–CTS signals used optionally with CSMA/CA resolve contention in the MAC layer.

2.5 Create an account with AWSs and learn how to use Amazon's EC2. Then implement the following distributed system using EC2: A writer process W writes a stream of characters into a buffer of size 128, and a reader process R reads those characters from the buffer. There are two constraints—the writer W should be blocked when the buffer is full, and the reader R should be blocked when the buffer is empty. You can assume that only one of the two processes is active at any time.

2.6 *Instant messaging* is a popular tool on your laptop or smartphone for keeping up with your buddies. Explore how instant messaging works.

2.7 Check if your city transit system has a website. Develop an application that uses web services to find a connection from point A to point B at a given time.

II

Foundational Topics

CHAPTER 3

Models for Communication

3.1 NEED FOR A MODEL

A distributed computation involves a number of processes communicating with one another. We observed in Chapter 2 that interprocess communication mechanism is fairly complex. If we want to develop algorithms or build applications for a distributed system, then the details of interprocess communication can be quite overwhelming. In general, there are many dimensions of variability in distributed systems. These include network topology, interprocess communication mechanisms, failure classes, and security mechanisms. Models are simple abstractions that help understand the variability—abstractions that preserve the essential features, but hide the implementation details from observers who view the system at a higher level. Obviously, there can be many different models covering many different aspects. For example, a network of static topology does not allow the deletion or addition of nodes or links, but a dynamic topology allows such changes. Depending on how models are implemented in the real world, results derived from an abstract model can be applied to a wide range of platforms. Thus, a routing algorithm developed on an abstract model can be applied to both ad hoc wireless LAN and sensor network without much extra work. A model is acceptable, as long as its features or specifications can be implemented, and these adequately reflect the events in the real world.

3.2 MESSAGE-PASSING MODEL FOR INTERPROCESS COMMUNICATION

Interprocess communication is one dimension of variability in a distributed system. Two primary models that capture the essence of interprocess communication are the *message-passing* model and the *shared-memory* model. In this section, we highlight the properties of a message-passing model.

3.2.1 Process Actions

Represent a distributed system by a graph $G = (V, E)$, where V is a set of nodes and E is a set of edges joining pairs of nodes. Each node is a sequential process, and each edge corresponds to a communication channel between a pair of processes. Unless otherwise stated, we assume the graph to be directed—an edge from node i to node j will be represented by the ordered pair (i, j). An undirected edge between a pair of processes i, j is equivalent to a pair of directed edges, one from i to j and the other from j to i. The *actions* by a node can be divided into the following four classes:

1. *Internal action*: An action is an *internal action* when a process performs computations in its own address space resulting in the modification of one or more of its local variables.

2. *Communication action*: An action is a *communication action* when a process sends a message to another process or receives a message from another process.

3. *Input action*: An action is an *input action* when a process reads data from sources *external* to the system. For example, in a process control system, one or more processes can input the values of environmental parameters monitored by sensors. These data can potentially influence the operation of the system under consideration.

4. *Output action*: An action is an *output action* when it controls operations that are *external* to the system. An example is the setting of a flag or raising an alarm. For a given system, the part of the universe external to it is called its *environment*.

Messages propagate along directed edges called *channels*. Communications are assumed to be point to point—a multicast is a set of point-to-point messages originating from a designated sender process. Channels may be reliable or unreliable. In a reliable channel, the loss or corruption of messages is ruled out. Unreliable channels will be considered in Chapter 12. In the rest of this chapter, we assume reliable channels only.

3.2.2 Channels

The following axioms form a sample specification for a class of reliable channels:

Axiom 1

Every message sent by a sender is received by the receiver, and every message received by a receiver is sent by some sender in the system.

Axiom 2

Each message has an arbitrary but finite, nonzero propagation delay.

Axiom 3

Each channel is a FIFO channel. Thus, if x and y are two messages sent by one process P to another process Q and x is sent before y, then x is also received by Q before y.

We make no assumption about the upper bound of the propagation delay. Let c be a channel (Figure 3.1) from process P to process Q, $s(c)$ be the sequence of messages sent by process P to process Q, and $r(c)$ be the sequence of messages received by Q. Then it follows from the earlier axioms that at any moment, $r(c)$ is a prefix of $s(c)$.

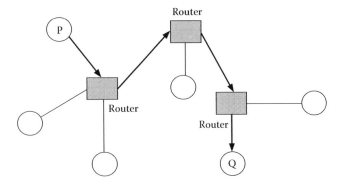

FIGURE 3.1   Channel from a process P to a process Q.

The axioms presented earlier are true for *our* model only, but they may not be true for all distributed systems. Some clarification about these axioms is given in the following. Axiom 1 rules out loss of messages as well as reception of spurious messages. When channels are unreliable, Axiom 1 may be violated at the data link or the transport layer. Recovery of lost messages is an important aspect of link and transport layer protocols.

In Axiom 2, the absence of a predefined upper bound for the propagation delay is an important characteristic of *asynchronous* channels. It also *weakens* the computational model. In reality, depending on the nature of the channel and the distance between the sender and the receiver, it is sometimes possible to specify an upper bound on the propagation delay. For example, consider a system housed inside a small room, and let signals directly propagate along electrical wires from one process to another. Since electrical signals travel approximately 1 ft/ns, the propagation delay across a 20 ft link inside the room will *apparently* barely exceed 20 ns.* However, if the correct operation of a system depends on the upper bound of the propagation delay, then the correctness of the same system may be jeopardized when the length of the link is increased to 300 ft. The advantage of weakening the model is that a system designed for channels with arbitrarily large but finite delay continues to behave correctly regardless of the actual value of the propagation delay. Delay insensitivity thus adds to the robustness and the universal applicability of the system.

Axiom 3 is not necessarily satisfied by a datagram service—packets may arrive out of order at the receiving end. In order that our model becomes applicable, it is necessary to assume the existence of a layer of service that *resequences* the packets before these are delivered to the receiver process. We want to carefully separate the intrinsic properties of our model from its implementation in an actual application environment.

How many messages can a channel hold? There are two possible ways to abstract this: the capacity of a channel may be either an infinite or finite. With a channel of infinite capacity, the sender process can send messages as frequently as it wants—the channel is never blocked (or it never drops messages due to storage limitations) regardless of the slowness of the receiver process. With a finite-capacity channel, however, the channel may sometimes be

---

* In *most cases* of interprocess communication, the major part of the delay is not due to the propagation time along the wire, but due to the handling of the *send* and the *receive* operations by the various layers of the network protocols, as well as delay caused by the routers.

full, and attempts to send messages may either block the sender or return an error message or cause messages to be dropped. Although any real channel has a finite capacity, this capacity may often be so large that the sender is rarely affected. From this perspective, arbitrarily large capacity of channels is a useful simplification, and not an unreasonable abstraction.

A channel is an interesting type of shared data object. It differs from a memory cell in many ways. For example, it is not possible to *write* anything (in the sense that a traditional write operation erases the old contents of a memory cell) into the channel—one can only *append* something to the existing contents of the channel. An immediate consequence of this limitation is that it is not possible to *unsend* a message that has already been sent along a channel. Similarly, one cannot *read* the contents (traditional reads are nondestructive) of a channel—one can only *read and delete* the header element from a channel. Other than *append* and *delete*, it is tempting to assume the existence of a Boolean function *empty(c)* that returns *true* if channel *c* is empty. However, it is far from trivial to determine if a channel is empty at a given moment.

3.2.3 Synchronous versus Asynchronous Systems

Another dimension of variability in distributed systems is *synchrony* and *asynchrony*. The broad notion of synchrony is based on senders and receivers maintaining synchronized clocks and executing actions with a rigid temporal relationship. However, a closer view reveals many aspects of synchrony, and the transition from a fully asynchronous to a fully synchronous model is a gradual one. Some of the behaviors characterizing a synchronous system are as follows:

Synchronous clocks: In a system with synchronous clocks, the local clocks of every processor show the same time. The readings of a set of independent clocks tend to drift, and the difference grows over time. Even with atomic clocks, drifts are possible, although the extent of this drift is much smaller than that between clocks designed with ordinary electrical components. For less stringent applications, the domestic power supply companies closely mimic this standard, where a second is equal to the time for 60 oscillations of the alternating voltage (or 50 oscillations according to the European standard and also in Australian and Asian countries) entering our premises.* Since clocks can never be perfectly synchronized, a weaker notion of synchronized clocks is that the drift rate of local clocks from real time has a known upper bound.

Synchronous processes: A system of synchronous processes takes actions in lockstep synchrony, that is, in each step, all processes execute an eligible action. In real life, however, every process running on a processor frequently stumbles because of interrupts. As a result, interrupt service routines introduce arbitrary amounts of delay between the executions of two consecutive instructions, making it appear to the outside world as if instruction execution speeds are unpredictable with no obvious lower bound. Using a somewhat different characterization, a process is sometimes called synchronous, when there is a known lower bound of its instruction execution speed.

Even when processes are asynchronous, computations sometimes progress in *phases* or in *rounds*—in each phase or round, every process does a predefined amount of work, and no process starts the $(i + 1)$th phase until all processes have completed their ith phases.

---

* These clock pulses are used to drive many of our older desktop and wall clocks, but now, it is obsolete.

The implementation of such a *phase-synchronous* or *round-synchronous* behavior requires the use of an appropriate phase synchronization protocol.

Synchronous channels: A channel is called synchronous when there is a known upper bound on the message propagation delay along that channel. Such channels are also known as *bounded-delay* channels.

Synchronous message order: The message order is synchronous when the receiver process receives messages in the same order in which sender process sent them (Axiom 3 in Section 3.2.2 satisfies this).

Synchronous communication: In synchronous communication, a sender sends a message only when the receiver is ready to receive it and vice versa. When the communication is asynchronous, there is no coordination between the sender and the receiver. The sender of message number i does not care whether the previous message $(i - 1)$ sent by it has already been received. This type of send operation is also known as a *nonblocking send* operation. In a *blocking send*, a message is sent only when the receiver signals its readiness to receive the message. Like send, receive operations can also be blocking or nonblocking. In *blocking receive*, a process waits indefinitely long to receive a message that it is expecting from a sender. If the receive operation is *nonblocking*, then a process moves to the next task in case the expected message does not (yet) arrive and attempts to receive it later.

Synchronous communication involves a form of handshaking between the sender and the receiver processes. In real life, postal or email communication is a form of asynchronous communication, whereas a telephone conversation is a great example of a synchronous communication.

In [H78], Tony Hoare introduced the communicating sequential process (CSP) model—it adopts a version of synchronous communication, where a pair of neighboring processes communicates through a channel of zero capacity. A sender process P executes an instruction $Q!x$ to output the value of its local variable x to the receiver process Q. The receiver process Q executes the instruction $P?y$ to receive the value from P and assign it to its local variable y. The execution of the instructions $Q!x$ and $P?y$ is synchronized, inasmuch as the execution of anyone of these two is blocked, until the other process is ready to execute the other instruction. More recent examples of synchronous communication are Ada rendezvous, RPCs [BN84], and RMI.

Note that of these five features, our message-passing model introduced at the beginning of this chapter assumed synchronous message order only.

3.2.4 Real-Time Systems

Real-time systems form a special class of distributed systems that are required to respond to inputs in a timely and predictable way. Timeliness is a crucial issue here, and usually, the cost of missing deadlines is high. Depending on the seriousness of the deadline, real-time systems are classified as *hard* or *soft*. The air-traffic control system is an example of a hard real-time system where missed deadlines are likely to cost human lives. A vending machine is an example of a soft real-time system, where a delay in receiving an item after inserting the proper amount in coins can cause some annoyance to the customer, but nothing serious will happen.

3.3 SHARED VARIABLES

In the message-passing model, each process has a private address space. In an alternative model of computation, the address spaces of subsets of processes overlap, and the overlapped portion of the address space is used for interprocess communication. This model is known as the *shared-memory* model. The shared-memory model has a natural implementation on tightly coupled multiprocessors. Traditionally, concurrent operations on shared variables are serialized, and the correctness of many shared-memory algorithms relies on this serialization property.

There are important and subtle differences between the message-passing and the shared-memory models of computation. One difference is that in a shared-memory model, a single copy of a variable or a program code is shared by more than one process, whereas to share that variable in the message-passing model, each process must have an exclusive copy of it. Consequently, consistent serialization of updates is a nontrivial task.

Due to the popularity of shared variables, many computing clusters support *DSM*, an abstraction for sharing a virtual address space between computers that do not share physical memory. The underlying hardware is a multicomputer, and the computers communicate with one another using message passing. The primary utility of DSM is to relieve the users of the intricate details of message passing and let them use the richer programming tools of parallel programming available for shared-memory systems.

Two variations of the shared-memory model used in distributed algorithms are (1) the *state-reading* model (also known as the *locally shared variable model*) and (2) the *link register* model. In the state-reading model, any process can read, in addition to its own state, the state of each of its neighbors from which a channel is incident on it. However, such a process can update only its own state. Figure 3.2a illustrates the state-reading model, where processes 1 and 3 can read the state of process 2 and process 0 can read the state of process 1.

In the *link register* model, each link or channel is a *single-reader single-writer* register (Figure 3.2b). The sender writes into this register, and the receiver reads from that register. To avoid additional complications, the link register model also assumes that all read and write operations on a link register are serialized, that is, write operations never overlap with read operations.* A bidirectional link (represented by an undirected edge) consists of a pair of link registers.

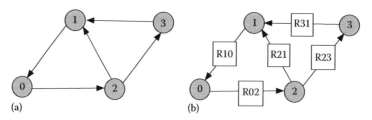

(a) (b)

FIGURE 3.2 (a) The state-reading model: a directed edge (2, 3) means that 3 can read the state of 2. (b) The link register model: 1 and 3 access the information from the link registers R21 and R23, respectively, which are updated by 2. The contents of these registers are limited by how much 2 wants to share with them.

---

* The behavior of registers under overlapping read or write operations is beyond the scope of this chapter.

One difference between the earlier two models is that in Figure 3.2b, neighbors 1 and 3 access the information from the link registers R21 and R23, respectively, and their contents are limited by how much process 2 wants to share with them. These may be different from the state of process 1. It also depends on *when* these register values are updated.

3.3.1 Linda

David Gelernter at Yale University developed Linda in 1985 [G85]. It is a concurrent programming model using the concept of a shared *tuple space*, which is essentially a shared communication mechanism. The general principle is similar to *blackboard systems* used in artificial intelligence. Processes collaborate, by depositing and withdrawing tuples. Processes may not know each other or at least they do not directly communicate with each other. Communication is based on pattern matching, that is, a process may check for a needed tuple very much like the query-by-example paradigm in database systems and retrieve one or more tuples that satisfy the pattern. Depositing tuples is asynchronous, while a querying process may choose to block itself until such time when a tuple is matched.

Linda tuples are unordered and accessed by six primitives. The primitive OUT and IN are used to deposit a tuple into the tuple space and extract a tuple from the tuple space—these simulate the *send* and the *receive* operations, respectively. The primitive RD also receives a tuple, but unlike IN, it does not delete the tuple from the space. INP and RDP are nonblocking versions of IN and RD—these work like IN and RD when there are matching tuples, but return a *false* otherwise. Finally, EVAL creates new processes and can be compared with the Unix *fork* command.

The following is an example of a Linda program written as an extension of a C program. A master process delegates tasks to *n* slave processes. When all slaves finish their tasks, they output "done" into the tuple space. The master inputs this from the tuple space and then prints a message that all tasks have been completed.

```
/** main program**/
real_main(argc,argv)
int argc;
char *argv[];
{
        int nslave, j, hello();
        nslave = atoi (argv[1]);

        for (j = 0; j < nslave; j++)
        EVAL ("slave", hello(j));
        for(j = 0; j < nslave; j++)
        IN("done");

        printf("Task completed.\n");
}
```

```
/** subroutine hello **/
        int hello (i)
        int i;
{
        printf("Task from number %d.\n",i);
        OUT("done");
        return(0);
}
```

JavaSpace is an object coordination system that is based on Linda. Tuple space can be implemented as a distributed data structures, and parts of the space can be physically mapped on the local memory of the different processes in a network. Further details can be found in [CG89].

3.4 MODELING MOBILE AGENTS

A mobile agent is a *program code* that migrates from one process to another. Unlike a message that is *passive*, a mobile agent is an *active* entity that can be compared with a messenger. The agent code is executed at the host machine where the agent can interact with the variables of programs running on the host machine, use its resources, and take autonomous routing decisions. During migration, the process in execution transports its state from one machine to another while keeping its data intact. Two major categories of mobile agents are in use: some support *strong mobility*, and others don't. With strong mobility, agents are able to transfer its control state from one machine to another—thus after executing instruction k in a machine A, the mobile agent can execute instruction $(k + 1)$ in the next machine B. With *weak mobility*, the control state is not transferred—so at each host, the code executes without this information.

Mobile agents complement the existing technologies of interprocess communication. In applications involving very large databases, network bandwidth can be saved when a client sends a mobile agent to the database server with a few queries, instead of pulling huge volumes of data from the server to the client site. Mobile agents also facilitate disconnected modes of operation. Messages and mobile agents can coexist: while certain tasks use message passing, a mobile agent can carry out the task of coordinating an activity across the entire network or in a fraction of it. Dartmouth's D'Agents and IBM's Aglets are two well-known mobile agent systems.

A mobile agent with weak mobility can be modeled as follows: call the initiator of the agent its *home*. Each mobile agent is designated by (at least) three components (I, P, B). The first component I is the agent identifier and is unique for every agent. The second component P designates the agent program that is executed at every process visited by it. The third component B is the *briefcase* and represents the data variables to be used by the agent. Two additional variables *current* and *next* keep track of the current location of the agent and the next process to visit. Here is an example of a computation using mobile agents: Consider the task of computing the lowest price of an item that is being sold in *n* different stores. Let *price(i)* denote the price of the item in store *i*, and let the briefcase variable *best* denote the lowest price of the item among the stores visited by the agent so far. Treat each store as a process. To compute *best*, an initiator process sends a mobile agent to a neighboring process. Thereafter, the agent executes the following program *P* at each site before returning home:

```
initially    current = home; best = price(home),
             visit next; {next depends on a traversal algorithm}
             {after reaching a new host}
while        current ≠ home do
             if price(i) < best then best := price(i) else skip end if;
             visit next;
end while
```

It will take another traversal to disseminate the value of *best* among all the processes. Readers may compare the complexity of this solution with the corresponding solutions on the message-passing or the shared-memory model.

Agents can be *itinerant* or *autonomous*. In the itinerant agent model, the initiator loads the agent with a fixed itinerary that the agent is supposed to follow. For an autonomous agent, there is no fixed itinerary—at each step, the agent is required to determine which process it should visit next to get the job done.

In network management, a class of primitive agents mimicking biological entities like *ants* has been used to solve problems like shortest path computation and congestion control. The individual agents do not have any explicit problem-solving knowledge, but intelligent action emerges from the collective action by ants. White [WP98] describes the operation of ant-based algorithms.

3.5 RELATIONSHIP AMONG MODELS

While all models discussed so far are related to interprocess communication only, other models deal with other dimensions of variability. For example, failure models abstract the behavior of faulty processes and are discussed in Chapter 13. Models are important to programmers and algorithm designers. Before devising the solution to a problem, it is important to know the rules of the game. Two models A and B are equivalent, when there exists a 1-1 correspondence between the objects and operations of A and those of B.

3.5.1 Strong and Weak Models

Informally, one model A is considered *stronger* than another model B, when to implement an object (or operation) in A, one requires more than one object (or operation) in B. For example, a model that supports multicast is stronger than a model that supports point-to-point communications, since ordinarily a single *multicast* is implemented using several *point-to-point* communications.* The term *stronger* is also attributed to a model that has more constraints compared to those in another model (called the *weaker* model). In this sense, message-passing models with bounded-delay channels are stronger than message-passing models with unbounded-delay channels. Using this view, synchronous models are stronger than asynchronous models. Remember that *strong* and *weak* are not absolute, but relative attributes, and there is no measure to quantify them. It is also true that sometimes two models cannot be objectively compared, so one cannot be branded as stronger than the other.

Which model would you adopt for designing a distributed application? There is no unique answer to this question. On one side, the choice of a strong model simplifies algorithm design, since many constraints relevant to the application are built into the model. This simplifies correctness proofs too. On the other side, when an application is implemented, a crucial issue is the support provided by the underlying hardware and the operating system. Occasionally, you may want to port an available solution from an alternate model to a target platform—for example, an elegant leader election algorithm running on the link register model is described in a textbook, and you may want to implement it on a message-passing architecture with bounded process delays and bounded channel capacities, because that is what the hardware architecture of the target platform supports. This may add to the complexity.

Such implementations are exercises in simulation and are of interest not only to practitioners but also to theoreticians. Can model A be simulated using model B? What are the time and space complexities of such simulations? Questions like these are intellectually challenging and stimulating, and building such *layers of abstraction* has been one of the major activities of computer scientists. In general, it is simpler to implement a weaker model from a stronger one, but the implementation of a stronger model using a weaker one may take considerable effort. In the remainder of this section, we outline a few such implementations.

3.5.2 Implementing a FIFO Channel Using a Non-FIFO Channel

Let c be a non-FIFO channel from process P to process Q. Assume that the message delay along c is arbitrary but finite. Consider a sequence of messages $m[0], m[1], m[2], ..., m[k]$ sent by process P. Here, i denotes the *sequence number* of the message $m[i]$. Since the channel is not FIFO, it is possible for $m[j]$ to reach Q after $m[i]$ even if $i > j$. To simulate FIFO behavior, process Q operates in two phases: *store* and *deliver*. The roles of these phases are explained as follows:

Store: Whenever Q receives a message $m[j]$, it stores it in its local buffer.

Deliver: Process Q delivers the message to the application, only after message $m[j - 1]$ has already been delivered.

---

* The claim becomes debatable for those systems where physical system supports broadcasts.

The implementation, referred to as a *resequencing protocol*, is described in the following. It assumes that every message has a sequence number that grows monotonically.

```
{Sender process P}                {Receiver process Q}
var i : integer {initially 0}     var k : integer {initially 0}
                                      buffer : buffer [0..∞] of
                                        message
repeat                                {initially for all k:
                                        buffer[k] = null}
                                  repeat {store}
        send m[i],i to Q;             receive m[i],i from P;
        i := i+1;                     store m[i] into buffer[i];
forever                               {deliver}
                                      while buffer[k] ≠ null do
                                      begin
                                        deliver the content of
                                          buffer k];
                                        buffer [k] := null; k := k+1;
                                      end
                                  forever
```

The solution is somewhat simplistic for two reasons: (1) the sequence numbers can become arbitrarily large, making it impossible to fit in a packet of bounded size, and (2) the size of the buffer required at the receiving end has to be arbitrarily large.

Now, change the model. Assume that there exists a known upper bound of T seconds on the message propagation delay along channel c, and messages are sent out at a uniform rate of r messages per second by process P. It is easy to observe that the receiving process Q will not need a buffer of size larger than $r \cdot T$, and it is feasible to use a sequence number of bounded size from the range $[0.. r \cdot T)$. Synchrony helps! Of course, this requires process Q to receive the messages at a rate faster than r.

How can we implement the resequencing protocol with bounded sequence numbers on a system with unbounded message propagation delay? A simple solution is to use acks. Let the receiving process have a buffer of size w to store all messages that have been received, but not yet been delivered. The sender will send messages $m[0], m[1], \ldots, m[w-1]$ and wait for an ack from the receiver. The receiver will empty the buffer and deliver them to the application and send an ack to the sender. Thereafter, the sender can recycle the sequence numbers $0..(w-1)$.

The price paid for saving the buffer space is a reduction in the message throughput rate. The exact throughput will depend on the value of w, as well as the time elapsed between the sending of $m[w-1]$ and the receipt of the ack. The larger is the value of w, the bigger is the buffer size, the fewer is the number of acks, and the better is the throughput.

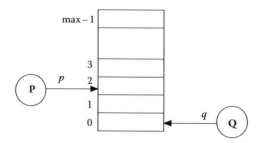

FIGURE 3.3   Implementation of a channel of capacity max from **P** to **Q**.

3.5.3 Implementing Message Passing Using Shared Memory

A relatively easy task is to implement message passing on a shared-memory multiprocessor. Such a simulation must satisfy the channel axioms. The implementation of a channel of capacity (max − 1) between a pair of processes uses a circular message buffer of size max (Figure 3.3).

Let $s[i]$ denote the ith message sent by the sender and $r[j]$ denote the jth message received by the receiver. The implementation is described in the following. Observe that the sender is blocked when the channel is full and the receiver is blocked when the channel is empty.

```
shared var p, q: integer {initially p = q}
buffer: array [0..max-1] of message
{Sender process P}
var s: array of messages sent by P, i : integer {initially 0}
repeat
     if p ≠ q - 1 mod max then
     begin
             buffer[p] := s[i]; i := i + 1; p := p + 1 mod max
     end
forever


{Receiver process Q}
var r: array of messages received by Q, j : integer {initially 0}
repeat
     if q ≠ p mod max then
     begin
             r[j] := buffer[q]; j := j + 1; q := q + 1 mod max
     end
forever
```

3.5.4 Implementing Shared Memory Using Message Passing

We now look into the implementation of a globally shared variable X in a system of n processes 0, 1, 2, …, $n - 1$ using message passing. For this, each process i maintains a local

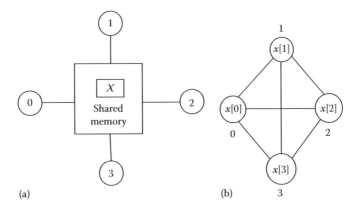

FIGURE 3.4 (a) A shared-memory location X; (b) its equivalent representation in message-passing model.

copy $x[i]$ of X (Figure 3.4). The important requirement here is that whenever a process wants to read X, its local copy must equal the latest updated value of X. A *first step* towards implementing the read and write operations on X is described as follows:

```
{Implementing shared memory by message passing: first attempt}
{read X by process i}
read x[i]  x[i]  := v

{write X := v by process i}
x[i]  := v
Multicast v to every other process j (j ≠ i) in the system;
Process j (j ≠ i), after receiving the multicast, sets x[j] to v.
```

The apparent goal is to make each local copy of the shared variable identical to one another after each update operation. However, there are problems. Assume that two processes 0 and 1 are trying to update the shared variable X simultaneously with different values. It is possible for these updates to reach processes 2 and 3 in different order, making $x[2] \neq x[3]$. It may appear that the problem can be resolved by making the multicast operation *write* X an *indivisible operation* (also known as an *atomic operation*), but actually, it does not resolve the problem, since it does not regulate the order of arrival of the updates at their destination processes. Thus, the solution described here is simplistic, flawed, and incomplete. Furthermore, implementing an atomic operation is far from trivial and has its own overhead.[*] What is needed is a guarantee that all processes will always receive the updates in identical order. This is known as *total order multicast*. Its implementation depends on how multicasts are performed and whether messages can be lost and whether processes are prone to failures. We will address the feasibility of this in Chapter 16.

---

[*] Implementation of atomic broadcasts will be addressed in Chapter 16.

3.5.5 Impossibility Result with Channels

Let us revisit the message-passing model and examine the problem of detecting whether a channel is empty. A familiar scenario is as follows: there are two processes i and j and two channels (i, j) and (j, i). During a cold start, these channels may contain an arbitrary sequence of messages. To properly initialize the channel, process j, before starting the execution of a program, wants to make sure that the channel (i, j) is empty, so that whatever message it receives was actually sent by process i. How can process j detect this condition?

If the clocks are synchronized and there exists an upper bound T on the message propagation delay along channel (i, j), then the problem is simple—both processes i and j pause for at least T seconds—this flushes both channels and each process rejects all the messages arriving during this time. After T seconds, process j knows that channel (i, j) is empty.

What if there is no known upper bound on the channel propagation delay? It is impossible for process j to wait for a bounded time and declare that channel (i, j) is empty. An alternative *attempt* is for process j to send a message to process i, requesting it to echo back the special message * along channel (i, j). If the channel is FIFO and process j receives the special message * before receiving any other message, then j might be tempted to conclude that the channel must have been empty. However, there is a fly in the ointment—it assumes that initially the channel (i, j) did not contain the special message *. This contradicts the assumption that initially the channel can contain arbitrary messages.

Although this is not a formal impossibility proof, it turns out that without an upper bound on the message propagation delay, it is not possible to detect whether a channel is empty even if it is known to be FIFO.

3.6 CLASSIFICATION BASED ON SPECIAL PROPERTIES

Distributed systems are also sometimes classified based on special properties or special features. Here are a few classifications.

3.6.1 Reactive versus Transformational Systems

A distributed system is called *reactive* when one or more processes constantly *react* to environmental changes or user requests. An example of a reactive system is a server. Ordinarily, the servers never sleep—whenever client processes send out requests for service, the servers provide the desired service. Another example of a reactive system is a token ring network. A process requesting service waits to grab the token and, after completing the send or receive operation, releases the token that is eventually passed on to another waiting process. This goes on forever.

The goal of nonreactive systems is to transform the initial state into a final state via actions of the component processes and reach a terminal point. An example is the computation of the routing table in a network of processes. When the computation terminates, or reaches a fixed point, every process has its routing table correctly configured. Unless there is a failure or a change in topology, there is no need to recompute the routing tables. Nonreactive systems are also known as *transformational systems*.

3.6.2 Named versus Anonymous Systems

A distributed system is called *anonymous* when the algorithms do not take into consideration the names or the identifiers of the processes. Otherwise, it is a *named* system. Most real systems are named systems. However, anonymity is an esthetically pleasing property that allows a computation to run unhindered even when the processes change their names or a newly created process takes up the task of an old process. From the space complexity point of view, each process needs at least $\log_2 n$ bits to store its name where n is the number of processes. This becomes unnecessary in anonymous systems.

Anonymous systems pose a different kind of challenge to algorithm designers. Without identifiers or names, processes become indistinguishable. This symmetry creates problems for those applications in which the outcome is required to be asymmetric. As an example, consider the task of electing a leader in a network of n ($n > 1$) processes. Since by definition there can be only one leader, the outcome is clearly asymmetric. However, since every process will start from the same initial state and will execute identical instructions at every step, there is no obvious guarantee that the outcome will be asymmetric, at least using deterministic means. In such cases, probabilistic techniques become useful for breaking symmetry.

3.7 COMPLEXITY MEASURES

The cost or complexity of a distributed algorithm depends on the algorithm, as well as on the model. Two well-known measures of complexity are the *space complexity* (per process) and the *time complexity*.

The space complexity of an algorithm is the amount of memory space required to solve an instance of the algorithm as a function of the size of the system. One may wonder if we should care about space complexity, when the cost of memory has come down drastically. In the context of present-day technology, the absolute measure may not be very significant for most applications, but the *scale of growth* as a function of the *number of nodes* in the network (or the *diameter* of the network) may be a significant issue. A constant space complexity, as represented by $O(1)$ using the big-O notation, is clearly the best, since the space requirement for each process is immune to network growth. Also, many applications require processes to send the value of their current state to remote processes. The benefit of constant space is that message sizes remain unchanged regardless of the size or the topology of the network.

For time complexity, numerous measures are available. Some of these measures evolved from the fuzzy notion of time across the entire system, as well as the nondeterministic nature of distributed computations. An accepted metric is the total *number of steps* needed by all the processes in the worst case from start to end, each step accounting for a basic action. The execution of an algorithm may halt from time to time due to interruptions generated by the operating systems, but the time complexity measure is immune to these unpredictable interruptions.

With today's technology, processor clocks tick at rates greater than 1 GHz, but message propagation delays still range from a few microseconds to a few milliseconds.

Communication cost is the major overhead of the time needed to execute a distributed algorithm. Accordingly, a more relevant metric is the *message complexity*, which is the number of messages exchanged during an instance of the algorithm as a function of the size of the system. Message complexity fills the void left in the traditional measure of time complexity.

An argument against the use of the number of messages as a fair measure of message complexity is as follows. Messages do not have constant sizes—the size of a message may range from 64 or 128 bits to several million bits. If message sizes are taken into account, then sometimes the cost of sending a large number of short messages can be lower than the cost of sending a much smaller number of large messages. This suggests that the *total number of bits exchanged* (i.e., the bit complexity) during the execution of an algorithm should be a more appropriate measure of the communication cost. Another form of classification in message complexity is based on the LOCAL and the CONGEST models [Peleg00]. The LOCAL model is simplistic—it disregards the size of the messages (and the variability in the time needed to process them) and assumes that all enabled processes execute their actions simultaneously. The CONGEST model takes into account the volume of communication and enforces a limit of $O(\log n)$ on the basic message size, where n is the size of the system. The CONGEST model however allows process actions to be synchronous or asynchronous.

In a purely asynchronous message-passing model with arbitrary message propagation delays, absolute time plays no role. However, in models with bounded channel delays and approximately synchronized clocks, a useful alternative metric is the *total time* required to execute an instance of the algorithm. One can separately estimate the average and the worst-case complexities.

The shared-memory models handle shared communication costs in a simplistic way. The program of each process is a sequence of discrete steps that include the communication overhead via reading of the state variables of a neighboring process (thus, it requires one step to read the state of a neighboring process). The size or grain of these discrete steps is called the *atomicity* of the computation, and the entire computation is an *interleaving* of the sequence of atomic steps. The time complexity of an algorithm is the total number of steps taken by all the processes during its execution as a function of the size of the system. As with space complexity, here also, the key issue is the growth of time complexity with the growth in network size or the network diameter.

Example 3.1: Multicasting in a Hypercube

Consider a k-cube with $n = 2^k$ processes. Each vertex represents a process and each edge represents a bidirectional FIFO channel. Process 0 is the initiator of a multicast— it periodically multicasts a value that updates a local variable x in each process in the n-cube. Let $x[i]$ denote the value of x in process i. The initial values of $x[i]$ can be arbitrary. The case of $k = 3$ is illustrated in Figure 3.5.

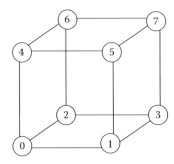

FIGURE 3.5   Multicasting in a 3-cube: process 0 is the initiator.

Let $N(i)$ denote the set of neighbors of process i. Process 0 starts the multicast by sending the value to each node in $N(0)$. Thereafter, every process i executes the following program:

```
{Program for process i > 0}
receive message m {m contains the value};
if m is received for the first time
      then   x[i] := m.value;
             send x[j] to each node in {j ∈ N(i): j > i}
      else   discard m
end if
```

The multicast terminates when every process j has received a message from each neighbor $i < j$ and sent a message to every neighbor $k > j$ (whenever there exists one).

Since messages traverse every edge exactly once, the message complexity of the proposed algorithm is $|E|$, where E is the set of edges. It is easy to observe that $|E| = (1/2)n \cdot \log_2 n$.

Example 3.2

We now solve the same problem on the state-reading model, where each process can read the states of all of its neighbors. As before, processes start from an arbitrary initial state. Process 0 first executes $x[0] := v$ (where v is the value to be multicast), and thereafter, process 0 remains idle. Every other process executes the following program:

```
{Program for process i>0}
while ∃j ∈N(i): (j < i) ∧ (x(i) ≠ x(j)) do x(i) := x(j) end while
```

When the multicast is complete, $\forall i, j: x(i) = x(j)$.

In the state-reading model, the sender is passive—it is the receiver's responsibility to *pull* the appropriate value from the sender (as opposed to *pushing* the value in the message-passing model). Accordingly, when a process pulls a different value of x from a lower-numbered process, it has no way of knowing if it is the value chosen by the source or it is the unknown initial value of that node. The time complexity, as measured by the *maximum number* of assignment statements executed by all the processes, can be arbitrarily large. To understand why, assume that in Figure 3.5, the initial states of the nodes 3, 5, 6, and 7 are different from one another. Here, node 7 may continue to copy the states of these three neighbors one after another for an indefinitely long period. However, eventually, nodes 3, 5, 6 will set their states to that of node 0, and after this, node 7 will require one more step to set $x[7]$ to the value multicast from node 0. This is why the upper bound of the time complexity is finite, but arbitrarily large.

If however the initial values of $x[i]$ are not arbitrary (which is the case after the first multicast is over), then the message complexity will be the same as that in the message-passing model. Even when the initial values of x are arbitrary, we can devise an alternative solution with bounded time complexity as follows: Allocate an additional buffer space of size $\log n$ per process, and ask every process to read and memorize the states of *all* of its neighbors that have a lower id before modifying its own state. A process will update its state only when the states of all of its lower-numbered neighbors are identical and it is different from its own state. The modified rule for process i will be as follows:

```
while ∀j, k ∈ N(i):  j < i ∧ k < i,  x(j) = x(k) ∧ x[i] ≠ x[j]
   do x(i) := x(j) end while
```

The multicast terminates when the predicate is false for all processes. In a k-cube, the maximum number of steps required by all the processes to complete the multicast can be calculated as follows. We focus on the maximum number of steps needed by the process that is farthest from the source to acquire the value from the source, since it is easy to argue that by that time, the multicast will reach all other processes. After one step, the process that is farthest from process 0 (i.e., at distance k) copies the state of its distance-1 neighbors; after an additional $(k_{C_1} + 1)$ steps, it copies the state of its distance-2 neighbors; after an additional $(k_{C_2} + k_{C_1} + 1)$ steps, it copies the state of its distance-3 neighbors; and so on. The total number of steps to complete the multicast is the sum of all these. Since the number of steps required to copy a value from the farthest process is $(k_{C_k} + k_{C_{k-1}} + \cdots + k_{C_2} + k_{C_1} + 1) \leq 2^k$, and the time to copy the value from any interim distance is bounded by this upper limit, the maximum number of steps needed to complete the multicast is $\leq k \cdot 2^k$, that is, $O(n \log n)$.

Round complexity: Another measure of time complexity in asynchronous systems is based on *rounds*. Historically, a *round* involves synchronous processes that execute their actions in lockstep synchrony, and round complexity is the number of steps taken by each process as a function of the size of the input. In asynchronous systems, a round is an execution sequence in which the *slowest enabled process* executes one step.

Naturally, during this period, the faster processes may have taken one or more steps. Thus, the following execution in a system of four processes 0, 1, 2, 3 constitutes two rounds:

$$\underline{1\ 2\ 0\ 2\ 1\ 3}\ \underline{2\ 1\ 0\ 1\ 3}$$

The notion of rounds gives us a measure of the number of steps that will be necessary, if processes execute their actions in lockstep synchrony. The algorithm in the *modified* version of Example 2 will take $\log_2 n$ rounds to complete, $\log_2 n$ being the diameter of the network.

The examples illustrated here use deterministic actions only. There is a rich set of algorithms in distributed computing that uses probabilistic or randomized actions, where the next state of a process or the choice of a neighbor is decided using a coin flip. Since the time complexity can vary from one run to another, the time complexity in such cases is determined either by the *expected number* of steps or by the number of required steps *with high probability* (abbreviated as *w.h.p.*), which is synonymous with the probability $\geq \left(1 - \left(1/n^c\right)\right)$, where $c > 1$. We will study a couple of probabilistic algorithms in Chapter 10.

3.8 CONCLUDING REMARKS

A model is an abstraction of a real system. When the real system is complex, reasoning about correctness becomes complicated. In such cases, various mechanisms of abstraction are used to hide certain details irrelevant to the main issues. Weaker models have fewer constraints. Systems functioning correctly on weaker models usually function correctly on stronger models without additional work, but the converse is not true. This chapter presents a partial view—it only focuses on some of the broader features of the various models, but ignores more difficult and subtle issues related to scheduling of actions or grains of computation. These will be addressed in Chapter 5 where we discuss program correctness. The implementation of one kind of model using another kind of model is intellectually challenging, particularly when we explore the limits of space and time complexities. Results related to impossibilities, upper and lower bounds of space, time, or message requirements, form the foundations of the theory of distributed computing.

3.9 BIBLIOGRAPHIC NOTES

Brinch Hansen's RC4000 operating system [BH73] is one of the first practical systems based on the message-passing model. The original nucleus supported four primitives to enable client–server communication using a shared pool of buffers. Cynthia Dwork first presented the taxonomy of synchronous behavior. In a classic paper (CSPs), Hoare introduced synchronous message passing [H78] as a form of communication via handshaking. David May of INMOS implemented it in Occam [M83]. Other well-known examples of this model are Ada Rendezvous and RPC. David Gelernter [G85] developed Linda. Since then, the Linda primitives have been added to several languages like C or FORTRAN. Two prominent and contemporary uses of message passing are MPI (introduced in the MPI forum in 1994— see the article by Marc Snir et al. [DOSW96]) and parallel virtual machine (PVM) by Geist et al. [GBDW+94]. Various middlewares introduced in Chapter 2 highlight the importance of the message-passing model. Two well-known tools for mobile agent implementations are Dartmouth's D'Agent [GKC+98] and IBM's Aglets [LO98], and various abstraction of mobile agents can be found in [AAKK+00] and [G00].

Shared variables have received attention from the early days of multiprocessors and have been extensively studied in the context of various synchronization primitives. In the modern context, the importance of shared variables lies in the fact that many programming languages favor the shared variable abstraction regardless of how they are implemented. DSM creates the illusion of shared memory on an arbitrary multicomputer substrate. In client–server computing, clients and servers communicate with one another using messages, but interclient communication uses shared objects maintained by the server.

For discussions on algorithmic complexities, read the classic book by Cormen et al. [CLR+01]. The LOCAL and the CONGEST models were introduced by Peleg [Peleg00].

EXERCISES

3.1 Distinguish between the link register model and the message-passing model in which each channel has unit capacity. Implement the link register model using the message-passing model.

3.2 A keyboard process K sends an arbitrarily large stream of key codes to a process P, which is supposed to store them in its local memory. In addition to receiving the inputs from K, P also executes an infinite computation. The channel (K, P) has a finite capacity. Explain how K and P will communicate with each other in the following two cases:

a. K uses blocking send, and P uses nonblocking receive.

b. K uses blocking send, and P uses blocking receive.

Comment on the progress of the computation in each case.

3.3 The CSP language proposed by Hoare [H78] uses a form of synchronous message passing: a process sending a message is delayed or blocked until the receiver is ready to receive the message, and vice versa. The symbols ! and ? are used to designate output and input actions, respectively. To send a value e to a process Q, the sending process P executes the statement $Q!e$, and to receive this value from P and assign it to a local variable x, process Q executes the action $P?x$. Write a program with three processes P, Q, and move, so that process move receives the values of an array x of known size from P one after another and sends them to Q, which assigns these values to a local array y.

3.4 n client processes $0..(n - 1)$ share a resource managed by a resource server. The resources are to be used in a mutually exclusive manner. The clients send requests for using these resources, and the server guarantees to allocate the resource to the clients in a fair manner (you are free to invent your own notion of fairness here and try different definitions), and the clients guarantee to return the resources in a finite time. Write down a program to illustrate the client–server communication using (a) message passing and (b) shared memory.

3.5 A wireless sensor network is being used to monitor the maximum temperature in a region. Each node monitors the temperature of a specific point in the region. Propose an algorithm for computing the maximum temperature and broadcasting the maximum value to every sensor node. Assume that communication is by broadcasting only, and the broadcasts are interleaved (i.e., they do not overlap), so two sensor nodes never send or receive data at the same time.

3.6 Consider an anonymous distributed system consisting of n processes. The topology is a completely connected network, and the links are bidirectional. Propose an algorithm using processes that can acquire unique identifiers.

(Hint: Use coin flipping, and organize the computation in rounds.) Justify why your algorithm will work.

3.7 In a network of mobile nodes, each node supports wireless broadcast only. These broadcasts have a limited range, so every node may not receive all the messages, and thus the network may not be connected. How will any given node P find out if it can directly or indirectly communicate with another node Q? What kind of computation models are you using for your solution?

3.8 Alice and Bob enter into an agreement: whenever one falls sick, she or he will call the other person. Since making the agreement, no one called the other person, so both concluded that they are in good health. Assume that the clocks are synchronized, communication links are perfect, and a telephone call requires zero time to reach. What kind of interprocess communication model is this?

3.9 Alice decides to communicate her secret to Bob as follows: the secret is an integer or can be reduced to an integer. The clocks are synchronized, and the message propagation is negligibly small. Alice first sends a 0 at time t, and then sends a 1 at time $t + K$, K being the secret. Bob deciphers the secret by recording the time interval between the two signals.

What kind of interprocess communication model is this? Will this communication be possible on a partially synchronous model where the clocks are approximately synchronized and the upper bound of the difference between the two clocks is known? Explain your answer.

3.10 Using the mobile agent model, design an algorithm to detect biconnectivity between a pair of processes i and j in a network of processes. Assume that both i and j can send out mobile agents.

3.11 A combinational digital circuit is to be simulated using an acyclic network of processes. Each process in the simulated model will represent a gate. Model the hardware communication using (1) CSP (explained in Exercise 3.3) and (2) link register. Provide an example in each case.

3.12 Figure 3.6 illustrates a pipeline with n processes 1 through n. Streams of tasks are sent to the pipeline through the entry point. Each process completes $1/n$ fraction of a total task and passes it on to the next process. The finished task exits the pipeline at the exit point. Process i accepts an item for processing, when (1) it has completed its own part, (2) it sends an ack to its predecessor $(i - 1)$, and (3) the predecessor $(i - 1)$ delivers the item to it. Initially, processes 2 through n have sent acks to their predecessors. Write a program for process i $(1 < i < n)$ using the *state-reading* model of interprocess communication.

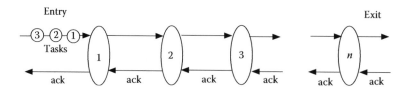

FIGURE 3.6   Pipeline of n processes 1 through n.

3.13 Assuming that message propagation delays have known upper bounds, show an implementation of the state-reading model using a message-passing model.

3.14 Consider the problem of *phase synchronization*, where a set of processes executes their actions in phases and no process executes phase $(k + 1)$ until every process has completed its phase $k(k \geq 0)$. This is useful in parallelizing loop computations. Implement phase synchronization using the shared-memory model.

3.15 In synchronous communication (also known as synchronous message passing), a message m is sent by a process **P** only when the receiving process **Q** is ready to receive it and vice versa. Assuming that **P** and **Q** are connected with each other by a pair of unidirectional unit capacity links, outline the implementation of *synchronous message passing* using *asynchronous message passing*.

3.16 A process P sends messages to another process Q infinitely often via a unidirectional channel. The communication channel c is not FIFO. Assume now that there exists a known upper bound of T seconds on the message propagation delay along channel c, messages are sent out at a uniform rate of r messages per second by process P, and process Q is faster than process P. What is the smallest size buffer that process Q is required to maintain, if it wants to accept the messages in the same order in which P sent them?

3.17 Processes in a *named* system rely on process identifiers for taking certain actions. Can you design a distributed algorithm (of your choice) that has a space complexity $O(1)$ on a named system? Explain your answer.

3.18 Consider a population of size n in a town and the task of multicasting a message m to every resident of that town. Residents communicate using Twitter, and communication is round based: in each round, a person can tweet a message to exactly one other person. The goal is to complete the multicast as quickly as possible.

> *Part 1.* Assume that each sender has full knowledge of who has not currently received the message. Propose an algorithm using the multicast that is completed in the *fewest number of rounds* (only the main idea using pseudocodes is needed here). Calculate the time complexity in rounds.

> *Part 2.* Now assume that senders have no knowledge of who has already received the message (senders are lazy and no one maintains the list of the residents to whom she or he has already tweeted the message in the previous rounds). So each sender randomly picks a resident and tweets him or her. Calculate the expected number of rounds needed for the message to reach every resident.

CHAPTER 4

Representing Distributed Algorithms

Syntax and Semantics

4.1 INTRODUCTION

This chapter introduces a minimal set of notations to represent distributed algorithms. These notations do not always conform to the syntax of popular programming languages like C or Java. They are only useful to appropriately specify certain key issues of atomicity, nondeterminism, and scheduling in a succinct way. The notations have enough flexibility to accommodate occasional use of even word specifications for representing actions. These specifications are only meant for a human user who is trying to implement the system or reason about its correctness. In [D76], Dijkstra argued about the importance of such a language to *influence our thinking habits*. This is the motivation behind introducing these simple notations for representing distributed algorithms.

4.2 GUARDED ACTIONS

A sequential process consists of a *sequence of actions*. Here, each action or statement corresponds to either an internal action or a communication action. The following notation represents a sequential process consisting of $(n + 1)$ actions S_0 through S_n:

$$S_0; S_1; S_2; \ldots; S_n;$$

We structure programs as follows. To name a particular program, we will use **program** <name> in the opening line. The variables and constants of a program will be introduced in the **define** section, and any initial values will be separately declared in the **initially** section. This will be followed by the program statements. We will designate a message as a variable of type **message**. If the structure of the message is important, then we will optionally define it as a *record*. Thus, a message m with three components a, b, c will be denoted as follows:

```
type message = record
                a: integer
                b: integer
                c: boolean
              end
define m: message
```

The individual components of m will be designated by $m.a$, $m.b$, and $m.c$. For representing other forms of structured data, we will freely use Pascal-like notations.

A *simple assignment* is of the form $x := E$, where x is a local variable and E is an expression. A *compound assignment* assigns values to more than one variable in one indivisible action. Thus, $x, y := m.a, 2$ is a single action that assigns the value of $m.a$ to the variable x and the value 2 to the variable y. The compound assignment $x, y := y, x$ swaps the values of the variables x and y.

On many occasions, an action S takes place only when some condition G holds. Such a condition is often called a *guard*. We will represent a *guarded action* by $G \rightarrow S$. Thus, $x = y \rightarrow x := y + 1$ is a guarded action. A guarded action will also be called a *statement*.

A trivial extension is the *alternative construct* that contains more than one guarded action. Consider the following construct:

```
If     G₀  →  S₀
[]     G₁  →  S₁
[]     G₂  →  S₂
.

.

[]     Gₙ  →  Sₙ
fi
```

It will mean that the action S_i takes place only when guard G_i is true. When no guard is true, this reduces to the statement *skip* (do nothing). In case more than one guard is true, the choice of the action that will be executed is *completely arbitrary*, unless specified otherwise by the scheduler. The previous notations are adequate to represent possible *nondeterministic* behaviors of programs.

Since we will deal only with abstract algorithms and not executable codes, program declarations or statements may occasionally be relaxed, particularly when their meanings are either obvious or self-explanatory. As an illustration, let process i send a message to process j—process j, upon receiving a message from process i, will initiate some action. The complete syntax for the action by process j will be

```
if ¬empty(i,j)  →   receive message m;
                    if    m = hello  →  ...
                    []    m = hi  →  ...
                    fi
                    ...   ...   ...
fi
```

However, if (i, j) is the only channel incident on process j, or the identification of the channel through which the message is received is not relevant to our discussion, then with little loss of clarity, we can represent the same program as

```
if      message = hello → ...
[]      message = hi →

        ... ... ...
fi
```

It is assumed that the message is consumed from the message buffer when it is received and examined. Finally, we describe the *repetitive construct*. The notation

```
do      G₀ → S₀
[]      G₁ → S₁
[]      G₂ → S₂
 .

 .

[]      Gₙ → Sₙ
od
```

will represent a loop that is executed as long as *at least one* of the guards $G_0...G_n$ is true. When two or more guards are true, any one of the corresponding actions can be chosen for execution, and the choice is arbitrary. The execution of the loop terminates when *all* the guards are false. As an example, consider the following program:

```
program uncertain;
define x : integer;
initially x = 0
do      x < 4 → x := x + 1
[]      x = 3 → x := 0
od
```

For the first three steps in the execution of this program, only the first guard is true, so the action $x := x + 1$ is executed, and the value of x becomes 3. But what happens after the third step? Note that both the guards are true now, so we will allow only one of the corresponding actions to take place, the choice being completely arbitrary. If the second action is chosen, then the value of x again becomes 0 and the first statement has to be executed during the next three steps. If however the first action is chosen, then the value of x becomes 4, and the loop terminates since all the guards are false. The corresponding state diagram is shown in Figure 4.1.

Can we predict if the second action will at all be chosen when $x = 3$? No. This is because the choice of an action is completely determined by the *fairness* of the scheduler. A fair scheduler will eventually select the first action when $x = 3$, and so the program will terminate. However, an unfair scheduler may not do so, and, therefore, termination is not guaranteed. We will address fairness issues in Section 4.5. Readers are encouraged to explore how these execution semantics can be specified using a well-known programming language of their choice.

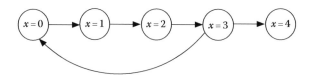

FIGURE 4.1 The state transitions in program uncertain.

A major issue in a distributed computation is *global termination*, which corresponds to reaching a state in which (1) the execution of the program for each process has terminated, that is, all guards are false, and (2) there is no message in transit along any of the channels. This is useful when a computation is structured into a sequence of phases, each process needs to detect whether the computation of the current phase has terminated before beginning the next phase. We will address termination detection in Chapter 9.

4.3 NONDETERMINISM

Guarded actions representing alternative and repetitive constructs involve *nondeterminism*: whenever two or more guards are simultaneously enabled, the choice of which action will be scheduled is at the discretion of the scheduler. Define the global state A of a distributed system as the set of all local states and channel states, and define the *behavior* of a computation as a sequence of global states $A_0 \rightarrow A_1 \rightarrow A_2 \rightarrow \dots \rightarrow A_k \rightarrow A_{k+1} \rightarrow \dots$, where A_0 is the initial state. Each state transition is due to an action by some process in the system. In a deterministic computation, the behavior remains the same during every run of the program. However, in a nondeterministic computation, starting from the same initial state, the behavior of a program may be different during different runs, since the scheduler has discretionary choice about alternative actions.

This apparently complicates matters, and nondeterminism may appear to be an unnecessary digression. However, in distributed systems, nondeterminism is quite natural. Operating system needs to guarantee that even if device interrupts are received at unpredictable moments, every execution of a well-behaved program *consistently* produces the same output. Similarly, network delays are arbitrary, so in different runs of a distributed algorithm, a process may receive the same set of messages in different order. This makes the case of nondeterministic behavior a rule in distributed systems and determinism a special case.

Deterministic scheduling of actions is sometimes inadequate to generate all the possible outcomes of a nondeterministic distributed computation. Consider a server process with k input channels $c_0, c_1, c_2, c_3, \dots, c_{k-1}$: when a client sends a message through one of these channels, the server sends an acknowledgement to the sender:

```
define x : array [0..k - 1]of boolean
initially for all channels are empty
do ¬empty(c₀)    →    send acknowledgement along c₀
[] ¬empty(c₁)    →    send acknowledgement along c₁
   ...  ...  ...
[] ¬empty(cₖ₋₁) →    send acknowledgement along cₖ₋₁
od
```

A deterministic scheduler will poll the channels *in a fixed order*. Thus, if three messages arrive via c_0, c_1, and c_2 at the same time, then the server will only send acknowledgments in the order $c_0 c_1 c_2$ and never acknowledge these in the order $c_0 c_2 c_1$, although this is feasible if the channels are polled in a nondeterministic order. The semantics of deterministic choices using **if... then... else...** produce a *subset* of the set of behaviors that are possible using nondeterministic choice. A system that is proven correct with nondeterministic choice is guaranteed to behave correctly under a deterministic scheduler.

4.4 ATOMIC OPERATIONS

Consider a process P with two input channels: *red* and *blue*. Suppose that two infinite streams of messages arrive through these two channels. Also, let x be a local variable of process P. What happens to x when the following program is executed?

```
do     ¬empty(red)  → x := 0    {red action}
[]     ¬empty(blue) → x := 15   {blue action}
od
```

Regardless of how nondeterminism is handled, we would expect the value of x to be an arbitrary sequence of 0's and 15's. However, there are some subtle issues that require closer examination.

If each assignment is an *indivisible action*, then the previous conclusion is definitely true. However, this has not been specified in the program! To realize how such a specification can make a difference, assume that x is a four-bit integer $x_3 x_2 x_1 x_0$ and assume that each assignment is executed in four steps—each step updating only one bit of x. Thus, $x := 0$ is translated to $x_3 := 0; x_2 := 0; x_1 := 0; x_0 = 0$ and $x := 15$ is translated to $x_3 := 1; x_2 := 1; x_1 := 1; x_0 = 1$. With both guards constantly enabled, the scheduler may choose to interleave the actions in the following sequence:

$$x_3 := 0; \quad \text{\{red action\}}$$

$$x_3 := 1; \quad \text{\{blue action\}}$$

$$x_2 := 1; \quad \text{\{blue action\}}$$

$$x_2 := 0; \quad \text{\{red action\}}$$

$$x_1 := 0; \quad \text{\{red action\}}$$

$$x_1 := 1; \quad \text{\{blue action\}}$$

$$x_0 := 1; \quad \text{\{blue action\}}$$

$$x_0 := 0; \quad \text{\{red action\}}$$

With this scenario, $x_3 = 1$, $x_2 = 0$, $x_1 = 1$, $x_0 = 0$ (i.e., $x = 10$ in decimal) will be the value of x. In fact, depending on the pattern of interleaving, x can assume any value between 0 and 15!

The previous example shows that the result of a computation can be influenced by what operations are considered indivisible. Such an indivisible action is called an *atomic action* or *atomic operation*. A distributed computation is an interleaved sequence of operations, and *atomicity* (also called *granularity*) determines the types of permissible interleaving in a distributed computation. If the actions $x := 0$ and $x := 15$ were atomic, then at any time, x would have assumed one of these two values only.

As another illustration of the significance of atomic operations, consider a system of three processes P, Q, R. Let process P execute the program:

```
define      b : boolean
initially   b = true
do    b →  send message to process Q
[]    ¬empty(R, P)  →  receive message from process R; b := false
od
```

If the send operation takes a long time, but a message from R arrives before the send operation is complete, should the arrival of the message interrupt the send operation? No, as long as the send operation is treated as an atomic operation. However, if P recognizes the arrival of the message from R *before* the send operation is scheduled and schedules the receive operation, then it will execute the statement $b := false$, and the send operation will not be scheduled at all.

The *grain* of an atomic operation is determined by what is considered indivisible. An example of fine-grain atomicity is *read–write atomicity*, where only read and write operations on single variable are considered indivisible. Consider a completely connected network of n ($n > 2$) processes $0..n - 1$, and each process i executes a program **do** $G_i \to S_i$ **od.** Assume that to evaluate the guard G_i, process i needs to read the states of all the remaining $n - 1$ processes in the system. With coarse-grained atomicity, the evaluation of each G_i is atomic action. In the *read–write atomicity* model, however, between two consecutive read operations by one process, another process can change its state by executing an action. Consequently, program behaviors and outcome can change if read–write atomicity model is assumed.

Atomic actions have the *all-or-nothing property*. If a process performs an *atomic multicast,* then either every receiver receives the message, or none of them receives it. The duration of an atomic action is no more relevant—and for the purpose of reasoning, we can reduce the duration to a point on the time axis.

How does a system implement the granularity or the atomicity of an action? Some levels of atomicity are guaranteed by the processor hardware—whereas others have to be implemented by software. In a single processor that does not use instruction pipelining, the execution of every instruction is atomic—external interrupts are not recognized until the execution of the current instruction is complete. In a shared-memory bus-based multiprocessor, the bus controller implements atomic *memory read* and *memory write* operations by serializing concurrent accesses to the shared memory. It thus provides natural support to read–write atomicity. In many shared-memory multiprocessors, special atomic instruction like *test-and-set (TS)* or *compare-and-swap* locks the memory bus (or switch) for two consecutive memory cycles, enabling the program to perform an indivisible read–modify–write (RMW) operation. Such operations are effective tools for implementing various types of locks and can be used to efficiently implement critical sections (CSs), which are atomic units of arbitrarily large size.

In our computation model, unless otherwise stated, we will assume that every guarded action is an atomic operation. Thus, once a guard G_i in the guarded action $G_i \rightarrow S_i$ is *true* and the scheduler initiates the execution of the action S_i, its execution must be allowed to complete (regardless of how much time it takes) before the guards are reevaluated and another action is scheduled. Consider the following program:

```
program      switch
define       a, flag: boolean
initially    a = true, flag = false
do           a → flag := true;
                 flag := false
[]    flag ∧ a →    a := false
od
```

Since the first action is atomic, every time the guards are evaluated, *flag* is found to be *false* and the second statement is never executed! So the program does not terminate. For a more complete understanding of such issues, we introduce the notion of *fairness*.

4.5 FAIRNESS

In a nondeterministic program, whenever multiple guards are true, there is more than one action to choose from. The choice of these alternatives is determined by the notion of *fairness*. Fairness is a property of the *scheduler*, and it can affect the behavior of a program.

To understand what fairness is, consider the set of all possible schedules (i.e., sequences of actions) in a given program. Any criterion that discards some of these schedules is a fairness criterion. With such a general definition of fairness, it is possible to define numerous types of fairness criteria. Of these, the following three types of fairness have received wide attention:

1. Unconditional fairness

2. Weak fairness

3. Strong fairness

A scheduler that allows all possible schedules characterizes an *unfair scheduler*. Consider the following program:

```
program  test
define   x: integer {initial value undefined}
do    true  →  x := 0
[]    x = 0 →  x := 1
[]    x = 1 →  x := 2
od
```

If the scheduler is unfair, then it is not impossible for it to always schedule the first action even if the guard of the second action remains true. As a result, the value of x may never be 1 or 2.

Unconditionally fair scheduler. A scheduler is *unconditionally fair* when each statement is *eventually* scheduled*, regardless of the value of its guard.

This version of fairness deals with scheduling at the statement level. All *unconditional statements* will eventually be scheduled for execution. However, an action that is scheduled is executed only if its guard is true at that time; otherwise, it is ignored. An unconditionally fair scheduler is a *primitive version of a fair scheduler* that intends to give every action a fair chance to execute. Thus, in the program *test*, an unconditionally fair scheduler may schedule the guarded action $x = 1 \rightarrow x := 2$ at a time when $x = 0$, but clearly, the action will not be executed. However, the action $x = 0 \rightarrow x := 1$ will definitely be executed when it is scheduled.

As an example, consider the scheduling of n processes in a multiprogrammed uniprocessor. Traditional scheduling policies reflect unconditional fairness, which guarantees that processor time is allocated to each of the n processes infinitely often. More refined version of schedulers pay attention to the guards while scheduling an action. This leads to the concepts of *weakly fair* and *strongly fair* schedulers.

Weakly fair scheduler. A scheduler is *weakly fair* when it eventually schedules every guarded action whose guard becomes true and remains true thereafter.

Consider the program *test* again. Initially, only the first action is guaranteed to execute. After this, the condition $x = 0$ holds, so the guards of the first two actions remain enabled. The weakly fair scheduler will eventually schedule each of these two actions. Once the second action is executed, the condition $x = 1$ holds, which asserts the guard of the third action while the first guard remains enabled. So the scheduler has to choose between the first and the third actions. If the scheduler chooses the first action, then the guard of the third action again becomes disabled. In a valid execution, the weakly fair scheduler may schedule the first two actions infinitely often, but may *never* schedule the third action, since its guard does not *remain* enabled.

Finally, consider the following program *fair* and examine if the program will terminate under a weakly fair scheduler:

```
program     fair
define      x,b : boolean
initially   b = true
do   b   →   x := true
[]   b   →   x := false
[]   x   →   b := false
[]   x   →   x := ¬x
od
```

Termination of the program *fair* is assured only if the scheduler chooses the *third and fourth actions*. But will these actions be chosen at all? With a weakly fair scheduler, there is

---

* Note that the term *eventually* corresponds to the inevitability of an action. It does not specify when or after how many steps the action will take place, but it guarantees that the delay is finite.

no guarantee that this will happen, since the guard x never stabilizes to the value *true* from the first two actions. However, program *fair* will terminate if the scheduler is *strongly fair*.

Strongly fair scheduler. A scheduler is strongly fair if it eventually schedules every guarded action whose guard is true *infinitely often.*

Note the difference between the two types of schedulers. The guard x in program *fair* does not remain enabled, but is enabled infinitely often—so a strongly fair scheduler must eventually execute the third action $b := false$, which negates the first two guards. On the same ground, it also eventually executes the fourth action, after which the program terminates.

In all these examples, all guarded actions belong to the same process—however, this is not necessary. The same definitions of fairness will apply when the guarded actions belong to distinct processes in a network or the guards include variables from a pair of neighboring processes.

Note that for a given program with a predefined initial state, the set of possible schedule of actions (also called *executions* or *behaviors*) under a weakly fair scheduler is a subset of the set of possible schedule of actions under a strongly fair scheduler. On one side, this implies that from a given initial configuration A_0, any state A_k that is reachable under a weakly fair scheduler is also reachable under a strongly fair scheduler. However, on the other side, the additional feasible executions under a strongly fair scheduler may sometimes affect program termination, when these executions have the potential to lead the system through a cycle of nonterminal configurations.

It is possible to come across many other types of schedulers in common applications. An example is a FIFO scheduler. A FIFO scheduler guarantees that between two consecutive executions of one action, every other action with an enabled guard gets the opportunity to execute once. If there is a mechanism to order the instants at which the guards were enabled, then FIFO scheduling guarantees that the corresponding actions are executed in that order. In real-time systems, fairness can be specified using physical clocks—for example, "engine number 2 must start 12 seconds after the blast-off of the spacecraft" specifies a stringent fairness property.

4.6 CENTRAL VERSUS DISTRIBUTED SCHEDULERS

The examples illustrated in the previous section dealt with a single process, whose program consists of one or more guarded actions. We will now examine the various possibilities of scheduling actions in a *network* of processes.

Since each individual process has a local scheduler, one possibility is to leave the scheduling decisions to these individual schedulers, without attempting any kind of global coordination. This is most natural and characterizes *distributed schedulers*.

Consider the system of two processes in Figure 4.2. This system uses the state-reading model. Each process i has a Boolean variable $x[i]$ and their initial values are arbitrary. The goal is to lead the system to a configuration in which the condition $x[0] = x[1]$ holds. To meet this goal, assume that process i executes the following program:

```
do
      x[i + 1 mod 2] ≠ x[i]  →  x[i]  :=  ¬x[i]
od
```

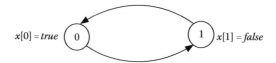

FIGURE 4.2   A system of two processes.

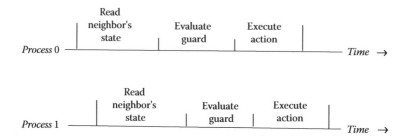

FIGURE 4.3   Overlapped actions with distributed schedulers.

With the initial values of x as shown in Figure 4.2, both processes have enabled guards. Using the *distributed scheduler model,* each process can take independent scheduling decisions as shown in Figure 4.3. Here, it is not impossible for each process i to concurrently read the state of the other process, detect that the guard is true, and eventually complement $x[i]$. As a result, the computation may potentially run forever and the goal is never reached.

The other scheduling model is based on the interleaving of actions. It assumes the presence of an invisible *demon** that coordinates actions on a global basis. In particular, this demon finds out *all* the guards that are enabled, arbitrarily picks *any one* of these guards, schedules the corresponding action, and waits for the completion of this action before reevaluating the guards. This is the model of a *central scheduler* or a *serial scheduler*. With reference to Figure 4.2 again, if the central scheduler chooses process 0, then until $x[0]$ has changed from *true* to *false*, process 1 cannot evaluate the guard or execute an action. So the computation terminates and the goal is reached in one step. This illustrates that the choice of the type of scheduler can make a difference in the behavior of a distributed system.

To simulate the distributed scheduling model under *fine-grained atomicity,* each process must maintain a private copy of the state of each of its neighbors. Let $k \in N(i)$. Let $s[i, k]$ designate the local copy of the state $x[k]$ of process k as maintained by process i. The evaluation of the guard by process i is a two-phase operation: in the first phase, process i copies the state of *each neighbor k*, that is, $\forall k \in N(i)$, $s[i, k] := x[k]$. In the second phase, each process evaluates its guard(s) using the local copies of its neighbors' states and decides if an action will be scheduled. The number of steps required to copy the neighbors' states will depend on the grain of atomicity and the size of the state space. For example, when *read–write atomicity* is assumed, only one variable of a neighbor is copied at a time, and all read

---

* A demon is an alternative name for a scheduler.

and write operations on $x[k]$ are interleaved. However, the *coarse-grained atomicity* model allows processes to read the states of all the neighbors in a single step.

The central or serial scheduler is a convenient abstraction that simplifies the reasoning about program correctness, but its implementation requires additional effort. Based on whatever hardware support is available, a central scheduler can be implemented by creating a *single token* in the entire system and circulating that token infinitely often among the processes in the system, as in a token ring network. Any process that receives the token is entitled to execute a guarded action before relinquishing the token to another process. The implementation of a single token requires an appropriate mutual exclusion protocol.

Central scheduling exhibits poor parallelism and poor scalability. Also its implementation in a distributed setting is nontrivial. This leads to the obvious question: Why should we care about central schedulers, when a strictly serial schedule often appears contrived? The primary reason is the relative ease of correctness proofs, as we will observe in Chapter 5. No system will correctly work under a distributed scheduler if it does not correctly work with a central scheduler. In restricted cases, correct behavior with a central scheduler guarantees correct behavior with a distributed scheduler. The following theorem represents one such case.

Theorem 4.1

If a distributed system works correctly with a central scheduler, and no enabled guard of a process is disabled by the actions of their neighbors, then the system is also correct with a distributed scheduler.

Proof: Assume that i and j are neighboring processes executing the guarded actions $G_i \rightarrow S_i$ and $G_j \rightarrow S_j$, respectively. Consider the following four events: (1) the evaluation of G_i as true, (2) the execution of S_i, (3) the evaluation of G_j as true, and (4) the execution of S_j. With a distributed scheduler, these can be scheduled in any order subject to the constraint that (1) happens before (2) and (3) happens before (4). Without loss of generality, assume that the scheduler evaluates G_i first. Then, distributed schedulers allow the following three schedules:

(1) (2) (3) (4)

(1) (3) (4) (2)

(1) (3) (2) (4)

However, by assumption, (4) does not affect (1), and (2) does not affect (3)—so these are causally independent events. Also, (2) and (4) are causally independent for obvious reasons. Since the scheduling of causally independent events has no impact on the final outcome, they can be swapped at will, which means

(1) (3) (4) (2) ≡ (1) (3) (2) (4) ≡ (1) (2) (3) (4)

that is, the second and the third schedules can be reduced to the first one by the appropriate swapping of events. But the first schedule corresponds to that of a central scheduler. ▪

4.7 CONCLUDING REMARKS

The semantics of a distributed computation depend on specific assumptions about atomicity and scheduling policies. In the absence of a complete specification, the weakest possible assumptions hold.

To illustrate the importance of atomicity and scheduling policies, let us take a second look at program *switch* in Section 4.3. Will this program terminate with a strongly fair scheduler, since the variable *flag* becomes true infinitely often? With the assumption about the atomicity of each guarded action, the answer is no. In fact, a weakly fair scheduler also does not guarantee termination. This is because an atomic statement is indivisible by definition, so each time the value of *flag* is monitored, it is found to be false.

Termination of program *switch* (Section 4.3) is possible with a strongly fair scheduler, when we split the first guarded statement as follows:

```
do     a → flag := true;
[]     a → flag := false
       ... ... ...
od
```

Such a split reduces the grain of atomicity. However, even with this modification, a weakly fair scheduler cannot guarantee termination.

Finally, a note for programmers is that the language presented in this section is a *specification language* only. The goal is to correctly represent the permissible overlapping or interleaving of various types of actions as well as various possible scheduling policies that might influence the behavior of a distributed system. For the sake of simplicity, the use of additional notations have been reduced to a minimum and often substituted by unambiguous sentences in English. Elsewhere, readers may find the use $\langle S \rangle$ to represent the atomic execution of S or different variations of [] to distinguish between strong and weak fairness in scheduling policies—but we decided to get rid of such additional symbols, since the language is for *human* interpretation and not for *machine* interpretation. For an exact implementation of any of these programs in an existing programming language like Java or C++, one should not only translate the guards and the actions but also implement the intended grain of atomicity, nondeterminism, and the appropriate fairness of the scheduler. This may not always be a trivial task.

4.8 BIBLIOGRAPHIC NOTES

Dijkstra introduced guarded actions (commands) in [D75]. The same article showed the importance of nondeterminism and techniques for handling it. No one should miss Dijkstra's landmark book *A Discipline of Programming* [D76] for an in-depth look at program derivation and reasoning about its correctness. In [D68], Dijkstra illustrated the importance of atomic actions when he introduced the P and V operators for solving the CS problem. The database community is well versed with the concept of atomic actions in the specification of the ACID* properties of transactions. Lamport extensively stud-

---

* ACID is the acronym for *a*tomicity, *c*onsistency, *i*solation, and *d*urability.

ied the role of atomicity in the correctness proof of concurrent programs (see [L77] and [L79]). In [L74], Lamport presented his *bakery algorithm* that showed how to implement a coarse-grained atomic action (like a CS) without the support of read–write atomicity at the hardware level. Francez presented a comprehensive study of fairness in his book [F86]—we have chosen only three important types of fairness here. Central and distributed schedulers are two prominent scheduling models in a distributed computation. It is unclear who introduced these first and seems to be folklore—but widely used in correctness proofs.

EXERCISES

4.1 Consider a completely connected network of n processes. Each process has an integer variable called *phase* that is initialized to 0. The processes operate in rounds—at the end of each round, every process sends the value of its *phase* to every other process and then waits until it receives the value of *phase* from every other process, after which the value of phase is incremented by 1, and the next round begins.

Specify the program of each process using guarded actions in the state-reading model.

4.2 Consider a strongly connected network of n processes 0, 1, 2, ..., $n - 1$. Any process i (called the source) can send a *message* to any other process j (called the destination) in the network. Each process i has a local variable v_i. A message sent out by a source process i consists of (1) the sequence number of the message, (2) the source id, (3) the destination id, and (4) the value of v_i. We focus on the routing of the messages.

Any message has to be routed through zero or more processes. A process j receiving a message accepts it only if it is the destination and executes the assignment $v_j :=$ v_i—otherwise, it forwards the message along its outgoing edges. The communication is considered to be complete, when at least one copy of the message reaches the destination, the value of the local variable is appropriately updated, and there are no other circulating messages or pending actions.

Specify the communication from i to j using guarded actions in the message-passing model.

4.3 A soda machine accepts only quarters and dimes. A customer has to pay 50 cents to buy a can of soda from the machine. Specify the behavior of the machine using guarded actions. Assume that the machine rejects any coin other than quarters and dimes, since it cannot recognize those coins. Also, assume that the machine will not issue a refund if a customer deposits excess money.

4.4 In a graph $G = (V, E)$, each node $i \in V$ represents a process whose state is denoted by a nonnegative integer $c(i)$. The program for each process i is described in the following:

```
do ∃j ∈ N(i):c(i) = c(j) → c(i) := c(i)+ 1 od
```

Transform the program from the state-reading model to the message-passing model using a round-based computation.

4.5 Consider the following program:

```
program      luck
define       x,z : boolean
initially    x = true
do      x → z := ¬z
[]      z → (x,z) := (false,false)
od
```

Will the previous program terminate if the scheduler is (1) weakly fair or (2) strongly fair?

4.6 There are n distinct points 0, 1, 2, ..., $n - 1$ on a 2D plane. Each point i represents a mobile robot that is controlled by a process P_i. P_i can read the position of the points $(i - 1)$ and $(i + 1)$ in addition to its own and execute an action to relocate itself to a new position.

The goal of the following algorithm is to make the n points collinear (i.e., they fall on the same straight line) from arbitrary initial positions. Process P_0 and P_{n-1} do not do anything—every other process $P_i (0 < i < n - 1)$ executes the following program:

```
program align {for i}
do (i - 1,i,i + 1) are not collinear → move i so that (i - 1,i,i + 1)
   are aligned od
```

To make the three points collinear, assume that node i first modifies its y-value, and then modifies its x-value only if it becomes necessary.

Will the points be aligned if there is a central scheduler that allows only one process to move at a time? What happens with distributed schedulers and fine-grained atomicity? Justify your answer.

4.7 A round-robin scheduler guarantees that between two consecutive actions by the same process, every other process with an enabled guard executes its action exactly once.

Consider a completely connected network of n processes. Assuming that a central scheduler is available, implement a round-robin scheduler using the locally shared-memory model. Provide brief arguments in support of your implementation.

4.8 Alice, Bob, and Carol are the members of a library that contains, among many other books, single copies of the books A, B. Each member can check out at most two books, and the members promise to return the books within a bounded period. Alice periodically checks out the book A, and Bob periodically checks out the book B. Carol wants to use both books *at the same time*. How will the librarian ensure that Carol receives her books? Write the program for Alice, Bob, Carol, and the librarian. What kind of fairness is needed in your solution? Can you solve the problem using a weakly fair scheduler?

4.9 Consider the problem in Exercise 4.8, and now assume that there are four members Alice, Bob, Carol, and David, who are trying to share single copies of four books A, B, C, D from the library. Periodically, Alice wants both A and B, Bob wants both B and C, Carol wants both C and D, and David wants both D and A, and each member promises to return the books in a bounded time. Write the program for the four members and librarian so that every member eventually receives their preferred books.

4.10 A distributed system consists of a completely connected network of three processes. Each process wants to pick a *unique identifier* from the domain $Z = \{0, 1, 2\}$. Let *name*[i] represent the identifier chosen by process i. The initial values of identifiers are arbitrary. A suggested solution with coarse-grained atomicity is as follows:

```
program pick-a-name {for process i}
define name: integer ∈ {0, 1, 2}
do ∃j ≠ i:name[j] = name[i] ∧ x ∈ Z\{name[k]:k ≠ i} → name[i] := x
od
```

a. Verify if the solution is correct, and if not, then fix it.

b. Suggest a solution using read–write atomicity. Briefly justify why your solution will work.

4.11 Consider a ring of $n(n > 2)$ identical processes. The processes are sympathetic to one another, so that when any process i executes an action, every process $j \neq i$ executes the same action. Represent the operation of the processes using guarded actions using the message-passing model.

(Hint: This is a consensus problem, where the processes have to agree about the next action to be executed.)

4.12 (Studious philosophers) Three philosophers 0, 1, and 2 are sitting around a table. Each philosopher's life alternates between reading and writing. There are three books on the table B_0, B_1, and B_2—each book is placed between a pair of philosophers (Figure 4.4). While reading, a philosopher grabs two books—one from the right and one from the left. Then he or she reads them, takes notes, and puts the books back on the table.

a. Propose a solution to the previous resource-sharing problem by describing the life of a philosopher. A correct solution implies that each philosopher can eventually grab both books and complete the write operation. Your solution must work

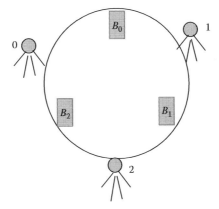

FIGURE 4.4   Three studious philosophers 0, 1, 2.

with a strongly fair scheduler, which means that if a book appears on the table infinitely often, the philosopher will eventually grab that book. (Your solution may not work with a weakly fair scheduler.)

b. Next, propose a solution (once again by describing the life of a philosopher) that will work with a weakly fair scheduler too. Provide brief arguments why your solution will work.

4.13 Three PhD candidates are trying to concurrently schedule their PhD defenses. In each committee, there are five members. No student has a prior knowledge of the schedule of any faculty member, so they ask each faculty member when they will be available. Once a member makes a commitment, he or she cannot back out unless the student requests to cancel the appointment. Suggest an algorithm for committee formation that leads to a feasible schedule, assuming that such a schedule exists.

CHAPTER 5

Program Correctness

5.1 INTRODUCTION

The designer of a distributed system has the responsibility of certifying the correctness of the system before the users start using it. This guarantee must hold as long as every hardware and software component works according to its specification. A system may function incorrectly when its components fail, or the process states are corrupted by external perturbations, and there is no provision for fault tolerance. This chapter explains what correctness criteria are considered important for distributed systems and how to prove the correctness properties.

Consider a distributed system consisting of n processes 0, 1, 2, ..., $n - 1$. Let s_i denote the local state of process i. The global state (also called the *configuration*) S of the distributed system consists of the local states of all the processes and is defined as $S = s_0 \times s_1 \times s_2 \times \cdots \times s_{n-1}$. While this is adequate for systems that use shared memory for interprocess communication, for message-passing models, the global state also includes the states of the channels. The global state of a distributed system is also called its *configuration*.

From any global state s_i, the execution of an eligible action takes the system to the next state S_{i+1}. The central concept is that of a *transition system*. A *computation* is a sequence of atomic actions $S_0 \rightarrow S_1 \rightarrow S_2 \rightarrow \cdots \rightarrow S_f$ that transforms a given *initial state S_0* to a *final state S_f*. A sequence of states and state transitions is also called a *behavior* of the system. With partial ordering of events and nondeterministic scheduling of actions, such sequences are not always unique—depending on the system characteristics and implementation policies, the sequence of actions from any given configuration can vary from one run to another. Yet, from the perspective of a system designer, it is important to certify that the system operates *correctly* for every possible run.

Figure 5.1 represents the *history* of a computation that begins from the initial global state A and ends in the final global state L. Each arc corresponds to an atomic action that causes a state transition. Note that in each of the states B and G, there are two possible actions: this corresponds to either data-dependent actions or nondeterministic choices made by the scheduler(s). The *history* can be represented as the set of the following three

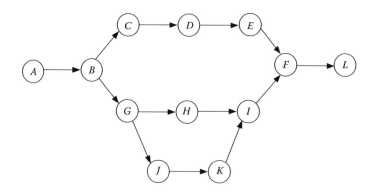

FIGURE 5.1 The history of a distributed system: the circles represent states and the arcs represent actions causing state transitions.

state sequences: {*ABCDEFL, ABGHIFL, ABGJKIFL*}. If a computation does not terminate, then some of the behaviors can be infinite.

Regardless of what properties are considered to judge correctness, a handful of test runs of the system can never guarantee that the system will behave correctly under all possible circumstances. This is because such test runs may at best certify the correctness for some specific behaviors but can rarely capture all possible behaviors. To paraphrase Dijkstra, "test runs can at best reveal the presence of bugs, but not their absence."

It is tempting to prove correctness by enumerating all possible interleavings of atomic actions and reasoning about each of these behaviors. However, there is a scalability cliff— due to the explosive growth in the number of such behaviors, this approach soon turns out to be impractical, at least for nontrivial distributed systems. For example, with n processes each executing a sequence of m atomic actions, the total number of possible interleavings is

$$\frac{(n \cdot m)!}{(m!)^n}$$

Even for modest values of m and n, this is a very large number.* Therefore, to exhaustively test even a small system within a reasonable time, the computing capacity available with today's largest and fastest computers becomes inadequate.

5.2 CORRECTNESS CRITERIA

Most of the useful properties of a system can be classified as either *liveness* or *safety* properties.

5.2.1 Safety Properties

A safety property intuitively implies that "*bad things* never happen." Different systems have different notions of what can be termed as a bad thing. Consider the history shown in Figure 5.1 and let a safety property be specified by the following statement: "the value of a certain integer

---

* For $n = 10$, $m = 4$, this number $>10^{34}$.

variable *temperature* should never exceed **100**." If this safety property has to hold for a system, then it must hold for every state of the system. Thus, if we find that in state **G** *temperature* = **107**, then we immediately conclude that the safety property is violated—we need not wait for what will happen to *temperature* after state **G**. To demonstrate that a safety property is violated, it is sufficient to demonstrate that it does not hold during an initial prefix of a behavior. Many safety properties can be specified as an *invariant* over the global state of the system. What follows are some examples of safety properties in well-known synchronization problems.

Mutual exclusion: Consider a number of processes trying to periodically enter a critical section. Once a process successfully enters the critical section, it is expected to do some work, exit the critical section, and then try for a reentry later. The program for a typical process has the following structure:

```
do true →
      entry protocol;
      critical section;
      exit protocol
od
```

Here, a safety property is that at most one process can be inside its critical section. Accordingly, the safety invariant can be written as $N_{cs} \leq 1$ where N_{cs} is the number of processes in the critical section at any time. A bad thing corresponds to a situation in which two or more processes are in the critical section at the same time.

Bounded capacity channel: A transmitter process P and a receiver process Q are communicating through a channel of bounded capacity B. The usual conditions of this communication are as follows: (1) The transmitter should not send messages when the channel is full, and (2) the receiver should not receive messages when the channel is empty. The following invariant represents a safety property that must be satisfied in every state of the system:

$$nC \leq nP \leq nC + B$$

where
    nP is the number of items produced by the transmitter process
    nC is the number of items consumed by the receiver process
    B is the channel capacity

Let $B = 20$. A bad thing happens when $nP = 45$, $nC = 25$, and the producer produces one more item and puts it into B.

Readers and writers' problem: Assume that nR *reader* processes simultaneously read a shared file that is updated by nW *writer* processes. To prevent the content of the file from being garbled and to help the readers read out meaningful copies, (1) the writers must get exclusive access to the file, and (2) the readers must access the file only when no writer is writing. This safety property can be expressed by the invariant

$$(nW \leq 1) \wedge (nR = 0) \vee (nW = 0) \wedge (nR \geq 0)$$

A bad thing will happen if a writer process is granted write access when a reader is reading the file.

Absence of deadlock: A system is deadlocked when it has not reached the final configuration, but no process has an eligible action, making further progress impossible. Clearly, deadlock is a bad thing for any distributed system. Consider a computation that starts from a configuration that satisfies the precondition P and is expected to satisfy the postcondition Q upon termination. Let GG be the disjunction of all the guards of all the processes. Then the desired safety property can be expressed by the invariant $Q \vee GG$.

Partial correctness: An important type of safety property is *partial correctness*. Partial correctness of a program asserts that if the program terminates, then the resulting state is the final state satisfying the desired postcondition. The bad thing here is the possibility of the program terminating with a wrong answer or entering into a deadlock. Using the example from the previous paragraph, a program is partially correct when $\neg GG \Rightarrow Q$, so the same safety invariant $Q \vee GG$ applies to partial correctness also. Partial correctness does not, however, say anything about whether the given program will terminate—that is a different and often a deeper issue.

The absence of safety can be established by proving the existence of a state that is reachable from the initial state and violates the safety criterion. To prove safety, it is thus necessary to assert that in every state that is reachable from the initial state, the safety property holds.

5.2.2 Liveness Properties

The essence of a liveness property is that *"good things* eventually happen." *Eventuality* is a tricky issue—it simply implies that the event happens after a finite number of actions, but no expected upper bound for the number of actions is implied in the statement.* Consider the statement:

Every criminal will eventually be brought to justice.

Suppose that the crime was committed on January 1, 1990, but the criminal is still at large. Can we say that the statement is false? No—since who knows, the criminal may be captured tomorrow! It is impossible to prove the falsehood of a liveness property by examining a finite prefix of the behavior. Of course, if the accused person is taken to court today and proven guilty, then the liveness property is trivially proved. But this may be a matter of luck or may depend on many other things—the definition does not specify how long we have to wait for a liveness property to hold, as long as the waiting period is finite. Here are some examples of well-known liveness properties:

Progress: Let us revisit the classical mutual exclusion problem, where a number of processes try to enter their critical sections. A desirable feature here is that once a process executes its entry protocol to declare its intention to enter its critical section, it must make *progress*

---

* In probabilistic systems where the course of actions is decided by flipping a coin, it is sufficient to guarantee that the events happen with probability 1.

toward the goal and *eventually* enter the critical section. Thus, progress toward the critical section is a liveness property. Even if there is no deadlock, the progress is violated if there exists at least one infinite behavior, in which a process remains outside its critical section. The absence of guaranteed progress is known as *livelock* or *starvation*.

Fairness: *Fairness* is a liveness property, since it determines whether the scheduler will schedule an action in a finite time. Like most progress properties, fairness does not ordinarily specify when or after how many steps the action is scheduled.

Reachability: Reachability addresses the following question: Given a distributed system with an initial state S_0, does there exist a finite behavior that changes the system state to S_k? If so, then S_k is said to be *reachable* from S_0. Reachability is a liveness property.

Network protocol designers sometimes run simulation programs to test protocols. They explore the possible states that the protocol could lead the system into and check if any of these is a bad or undesirable state. However, even for small protocols with a few lines of code, the number of states can sometimes be so large that most simulations succeed in reaching a fraction of the set of possible states within a reasonable time. Many protocols are certified using this type of testing. The testing of reachability through simulation is rarely foolproof and takes a heavy toll of system resources, often leading to the so-called *state-explosion* problem. Testing a protocol is not an alternative to proving its correctness.

Termination: Program *termination* is a liveness property. It guarantees that starting from the initial state, every feasible behavior leads the system to a configuration in which all the guards are false and the terminal configuration is reached. Recall that partial correctness simply ensures that the desired postcondition holds *when all guards are false*. It does not tell us anything about whether the terminal state is reachable via all admissible behaviors. Thus, total correctness of a program is the *combination of partial correctness* and *termination*. Here is an example:

Example 5.1

Consider a system of four processes P_0 through P_3 as shown in Figure 5.2. Each process has a color c represented by an integer from the set {0,1,2,3}. Let $c[i]$ represent the color of process P_i. The objective is to devise an algorithm, so that regardless of the initial colors of the different processes, the system eventually reaches a configuration where no two *adjacent* processes have the same color.

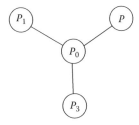

FIGURE 5.2   A system of four processes: each process tries to acquire a color that is different from the colors of its neighbors.

Let $N(i)$ denote the set of neighbors of process P_i. We propose the following program for every process P_i to get the job done:

```
program    colorme{for process Pᵢ}
do ∃Pⱼ ∈ N(i):c(i) = c[j] → c[i] = c[i] + 2 mod 4 od
```

Is the program partially correct? By checking the guards, we conclude that *if* the program terminates, that is, if all the guards are false, then the following condition holds:

$$\forall P_j \in N(i): c[i] \neq c[j] \tag{5.1}$$

By definition, this is the desired postcondition. So the system is partially correct.

However, it is easy to find out that the program *may not* terminate. Consider the initial state A $c[0] = 0$, $c[1] = 0$, $c[2] = 2$, $c[3] = 2$. Figure 5.3 shows a possible sequence of actions ABCDEFGHIJKLA in which the system returns to the starting state A without ever satisfying the desired postcondition (5.1). This cyclic behavior demonstrates that it is possible for the program to run forever. Therefore, the program is partially correct, but not totally correct.

Note that it is possible for this program to reach one of the terminal states X or Y if the schedulers choose an alternate sequence of actions. For example, if in state A, process P_1 makes a move, then the state $c[0] = 0$, $c[1] = 2$, $c[2] = 2$, $c[3] = 2$ is reached and condition (5.1) is satisfied! However, termination is not *guaranteed* as long as there exists a single infinite behavior where the conditions of the goal state are not satisfied. This makes

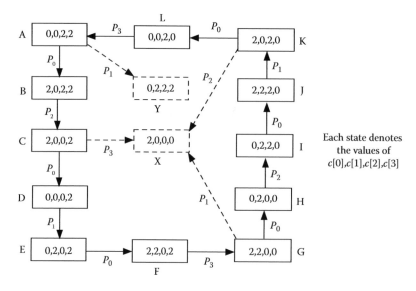

FIGURE 5.3   A partial history of the system in Figure 5.2 where the edges are labeled with the identifiers of the processes causing that transition: it shows an infinite behavior ABCDEFGHIJKLA. Note that X, Y are terminal states and are reachable, but there is no guarantee that the adversary will choose the transitions leading to those states.

the protocol incorrect. In this context, it is important to understand the role of the *adversary* (or the *demon* or the *scheduler*). For every distributed algorithm, think of an invisible adversary that is trying to challenge your design with the worst possible schedule. When you feel assured that the terminal state is reachable, the omniscient adversary explores if there exists a single behavior that can prevent the algorithm for reaching the terminal state within a bounded time. If such a behavior is found, then the adversary wins, and you lose. The adversary here represents the real world. To win against the adversary and guarantee termination, one must ascertain that there is no behavior that prevents the system from reaching the terminal configuration in a bounded number of steps.

Although most useful properties of a distributed system can be classified either as liveness or a safety property, there are properties that belong to neither of these two classes. Consider the statement, "there is a 90% probability that an earthquake of magnitude greater than 9.5 on the Richter scale will hit California before the year 2025." This is neither a liveness nor a safety property.

An implicit assumption made in this chapter is that all well-behaved programs eventually terminate. This may not always be the case—particularly for open or dynamic systems. An *open system* (also called a *reactive system*) responds to changes in the environment. Many real-time systems like the telephone network or the air-traffic control network are open systems. A system that assumes the environment to be fixed is a *closed system*.

Correctness also depends on assumptions made about the underlying model. Such assumptions include program semantics, the choice of the scheduler, or the grain of atomicity. A given property may hold if we assume strong fairness, but may not hold if we assume weak fairness. Another property may be true only if we choose a coarse-grain atomicity but may cease to hold with fine-grain atomicity.

5.3 CORRECTNESS PROOFS

The set of possible behaviors of a distributed system can be very large, and testing is not a feasible way of demonstrating the correctness of nontrivial system. What is required is some form of mathematical reasoning. Established methods like *proof by induction* or *proof by contradiction* are widely applicable. However, mathematical tools used to prove correctness often depend on what properties are being investigated. The techniques for proving safety properties are thus different from the techniques for proving liveness properties. In this chapter, we will review some of the well-known methods for proving correctness, as well as a few formal systems and transformation techniques that lead to a better understanding of the semantics of distributed computation. We particularly focus on the following four topics:

1. Assertional methods of proving safety properties

2. Use of well-founded sets for proving liveness properties

3. Programming logic

4. Predicate transformers

Most of these methods require a good understanding of propositional logic and predicate logic. We therefore begin with a brief review of propositional and predicate logic.

5.3.1 Quick Review of Propositional Logic

A proposition is a statement that is either *true* or *false*. Thus, *Alice earns $2000 a month* is a proposition, but *x is very large* is not a proposition. The axioms and expressions of propositional logic use the following symbols:

- The propositional constants *true* and *false* represent universal truth and universal falsehood, respectively.

- The propositional variables *P, Q, R*, etc., can have a value *true* or *false*.

- The propositional operators ¬∧∨= capture the notions of *not, and, or, implies*, and *equal to*, respectively, in commonsense reasoning.

The basic axioms of propositional logic are shown in Figure 5.4. Readers are encouraged to reason each of these axioms using commonsense. These axioms can be used to prove every assertion in propositional logic. Consider the following example.

Example

Prove that $P \Rightarrow P \lor Q$.

Proof $P \Rightarrow P \lor Q$
$$= \neg P \lor (P \lor Q) \quad \{\text{Axiom 5}\}$$
$$= (\neg P \lor P) \lor Q \quad \{\text{Axiom 8a}\}$$
$$= (\neg P \lor P) \lor Q \quad \{\text{Axiom 1a}\}$$
$$= true \lor Q \quad \{\text{Axiom 7a}\}$$
$$= true \quad \{\text{Axiom 2a}\}$$

0. $\neg(\neg P)$		
1(a). $P \lor \neg P = true$	1(b). $P \land \neg P = false$	
2(a). $P \lor true = true$	2(b). $P \land false = false$	
3(a). $P \lor false = P$	3(b). $P \land true = P$	
4(a). $P \lor P = P$	4(b). $P \land P = P$	
5. $(P \Rightarrow Q) = \neg P \lor Q$		
6. $(P = Q) = (P \Rightarrow Q) \land (Q \Rightarrow P)$		
7(a). $P \lor Q = Q \lor P$	7(b). $P \land Q = Q \land P$	
8(a). $P \lor (Q \lor R) = (P \lor Q) \lor R$	8(b). $P \land (Q \land R) = (P \land Q) \land R$	
9(a). $P \land (Q \lor R) = (P \land Q) \lor (P \land R)$	9(b). $P \lor (Q \land R) = (P \lor Q) \land (P \lor R)$	
10(a). $\neg(P \land Q) = \neg P \lor \neg Q$	10(b). $\neg(P \lor Q) = \neg P \neg Q$	

FIGURE 5.4   The basic axioms of propositional logic.

Pure propositional logic is not adequate for proving the properties of a program, since propositions cannot be related to program variables or program states. This is, however, possible using predicate logic, which is an extension of propositional logic.

5.3.2 Brief Overview of Predicate Logic

Propositional variables, which are either *true* or *false*, are too restrictive for real applications. In predicate logic, predicates are used in place of propositional variables. A predicate specifies the property of an object or a relationship among objects. Consider a program variable x. Then the relation $x < 1000$ is a predicate. A predicate is associated with a set, whose properties are often represented using the *universal quantifier* $\forall$ (for all) and the *existential quantifier* $\exists$ (there exists). A predicate using quantifiers takes the following form:

<Quantifier> <bound variable(s)> : <range> :: <property>

Examples of predicates using quantifiers are as follows:

- $(\forall x, y : x, y$ are positive integers $:: x \cdot y = 63)$ designates the set of values $\{(1, 63), (3, 21), (7, 9), (9, 7), (21, 3), (63, 1)\}$ for the pair of bound variables (x, y).

- A system contains $n(n > 2)$ processes 0, 1, 2,...$n - 1$. The state of each process is either 0 or 1. Then the predicate $(\forall i, j : 0 \leq i, j \leq n - 1::$ state of process $i =$ state of process $j)$ characterizes a property that holds for the set of processes $\{0, 1, 2, ... n - 1\}$. Some widely used axioms with quantified expressions are shown in Figure 5.5.

The axioms of predicate logic are meant for formal use. However, for the sake of developing familiarity with these axioms, we encourage the reader to develop an intuitive understanding of these axioms. As an example, consider the infinite set $S = \{2, 4, 6, 8, ...\}$. Here, "for all x in the set S, x is positive and even" is true. Application of Axiom 5 in Figure 5.5 and Axiom 10a in Figure 5.4 leads to the equivalent statement: "there does not exist an element x in the set S, such that x is not even or not positive"—a fact that can be understood without any difficulty.

---

1. $\forall x : R :: A \vee B = (\forall x : R :: A) \vee (\forall x : R :: B)$

2. $\forall x : R :: A \wedge B = (\forall x : R :: A) \wedge (\forall x : R :: B)$

3. $\exists x : R :: A \vee B = (\exists x : R :: A) \vee (\exists x : R :: B)$

4. $\exists x : R :: A \wedge B = (\exists x : R :: A) \wedge (\exists x : R :: B)$

5. $\forall x : R :: A = \neg(\exists x : R :: \neg A)$

6. $\exists x : R :: A = \neg(\forall x : R :: \neg A)$

FIGURE 5.5   Some widely used axioms in predicate logic.

5.4 ASSERTIONAL REASONING: PROVING SAFETY PROPERTIES

Assertional methods have been extensively used to prove the correctness of sequential programs. In distributed systems, assertional reasoning is an important tool for proving safety properties. Let *P* be an *invariant* representing a safety property. To demonstrate that *P* holds for every state of the system, we will use the method of induction. We will show that (1) *P* holds in the initial state, and (2) if *P* holds at a certain state, then the execution of *every* action enabled at that state preserves the truth of *P*. A simple example is given in the following:

Example 5.2

Consider the system of Figure 5.6 where a pair of processes *T* and *R* communicates with each other by sending messages along the channels *c1* and *c2*. Process *T* has a local variable *t*, and process *R* has a local variable *r*. The program for *T* and *R* are described in the following paragraph. We will demonstrate that the safety property *P*, "The total number of messages in both channels is ≤10," is an invariant for this system.

```
(Communication between two processes T and R)
define      c1,c2: channel;
initially   c1 = Ø, c2 = Ø;
{program for T}
define t: integer {initially t = 5}
1       do    t > 0           →    send a message along c1; t := t - 1
2       []    ¬empty(c2)      →    receive a message from c2; t := t + 1
        od
{program for R}
define r: integer {initially r = 5}
3       do    ¬empty(c1)      →    receive a message from c1; r := r + 1
4       []    r > 0           →    send a message along c2; r := t - 1
        od
```

Let *n1* and *n2* denote the number of messages in the channels *c1* and *c2*, respectively. To prove the safety property *P*, we will establish the following invariant:

$$I \equiv (t \geq 0) \wedge (r \geq 0) \wedge (n1 + t + n2 + r = 10)$$

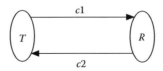

FIGURE 5.6   A two-process system.

It trivially follows that $I \Rightarrow P$. We now prove by induction that I holds at every state of the system.

Basis: Initially, $n1 = 0$, $n2 = 0$, $t = 5$, $r = 5$ so I holds.

Inductive step: Assume that I holds in the current state. We need to show that I will hold after the execution of every eligible guarded action in the program.

> {After action 1} The values of $(t + n1)$, $n2$, and r remain unchanged. Also, since the guard is true when $t > 0$, and the action decrements t by 1, the condition $t \geq 0$ continues to hold. Therefore, I holds.
>
> {After action 2} The values of $(t + n2)$, $n1$, and r remain unchanged. Also, the value of t can only be incremented, so $t \geq 0$ holds. Therefore, I continues to hold.
>
> {After action 3} The values of $(r + n1)$, $n2$, and t remain unchanged. Also, the value of r can only be incremented, so $r \geq 0$ holds. Therefore, I continues to hold.
>
> {After action 4} The values of $(r + n2)$, $n1$, and t remain unchanged. Also, since the guard is true when $r > 0$ and the action decrements r by 1, the condition $r \geq 0$ continues to hold. Therefore, I holds.

To summarize, I is true in the initial state. Also, if I holds at a certain state, then I holds at the following state. Therefore, I holds in every state of the system. Since $I \Rightarrow P$, P holds.

5.5 PROVING LIVENESS PROPERTIES USING WELL-FOUNDED SETS

A classical method of proving a liveness property is to discover a mapping function f: $S \rightarrow WF$, where S is the set of global states of the system and $WF = \{w1, w2, w3, \ldots\}$ is a well-founded set. Among the elements of the well-founded set, there should be a total order $\gg$, such that the following two properties hold:

- There does not exist any infinite chain $w1 \gg w2 \gg w3\ldots$ in WF.

- If an action changes the system state from $s1$ to $s2$, and $w1 = f(s1)$, $w2 = f(s2)$, then $w1 \gg w2$.

Eventual convergence to a goal state is guaranteed by the fact that if there exists an infinite behavior of the system, then it must violate the first property. The function f is called a *measure function* (also called a *variant* function), since its value is a measure of the progress of computation toward its goal.

The important issue in this type of proof is to discover the right WF and f for a given computation. A convenient (but not the only possible) choice of WF is the set of nonnegative integers, with $\gg$ representing the $>$ (greater than) relationship. In this framework, the initial state maps some positive integer in WF and the goal state often corresponds to the integer 0. The proof obligation reduces to finding an appropriate measure function f, so that every eligible action from a state S reduces the value of $f(s)$. Another example of WF is a set of tuples with $\gg$ denoting the *lexicographic* order. The next example illustrates this proof technique.

Example 5.3: Phase Synchronization Problem

Consider an array of clocks $0, 1, 2, \ldots, n-1$ as shown in Figure 5.7. Each clock has three values $0, 1, 2$ called its *phase*. These clocks tick at the same rate in lock-step synchrony. Under normal conditions, every clock displays the same phase. This means that if every clock displays x at the current moment, then after the next step (i.e., clock tick), all clocks will display $(x+1) \bmod 3$. When the clocks exhibit this behavior, we say that their phases are synchronized.

Now assume that due to unknown reasons, the clocks are out of phase. What program should the clocks follow so that eventually their phases are synchronized and remain synchronized thereafter?

Let $c[i]$ represent the phase of clock i and $N(i)$ denote the set of neighbors of clock i. We choose the model of locally shared variables, where each clock reads the phases of all of its neighbors in one atomic step but updates only its own phase. We propose the following program for every clock i:

```
{program synch: program for clock i}
do  ∃j ∈ N(i):c[j] = c[i] + 1 mod 3   →  c[i]  :=  c[i] + 2 mod 3
[]  ∀j ∈ N(i):c[j] ≠ c[j] + 1 mod 3   →  c[i]  :=  c[i] + 1 mod 3
od
```

Before demonstrating the proof of convergence (which is a liveness property), we encourage the readers to try out a few cases and watch how the phases are synchronized after a finite number of clock ticks. Observe that once the clock phases are synchronized, they remain so forever, since only the second action is chosen.

To prove convergence to a *good state* using a well-founded set, first consider a pair of *neighboring* clocks i and $(i + 1)$. If $c[i + 1] = c[i] + 1 \bmod 3$, then draw an arrow ($\leftarrow$) from clock $(i + 1)$ to i; else if $c[i] = c[i + 1] + 1 \bmod 3$, then draw an arrow ($\rightarrow$) from clock i to $(i + 1)$. There is no arrow between i and $(i + 1)$ when $c[i] = c[i + 1]$. From an arbitrary initial state (which may be reached via a failure or a perturbation), observe the following facts about the proposed protocol:

Observation 1: If a clock $i(0 < i < n - 1)$ has only a $\rightarrow$ but no $\leftarrow$ pointing toward it, then after one step, the $\rightarrow$ will shift to clock $(i + 1)$. In the case of $i = n - 1$, the $\rightarrow$ will disappear after one step. There can be no arrow pointing toward clock 0.

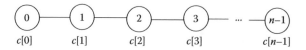

FIGURE 5.7   An array of three-phase clocks: every clock ticks as $0, 1, 2, 0, 1, 2, \ldots$.

Observation 2: If a clock i ($0 < i < n − 1$) has only a ← but no → pointing toward it, then after one step, the ← will shift to clock $i − 1$. In the case of $i = 0$, this arrow will disappear after one step. There can be no arrow pointing towards clock $n − 1$.

Observation 3: If a clock i ($0 < i < n − 1$) has both a ← and a → pointing toward it, then after one step, both arrows will disappear. This possibility is ruled out for both clock 0 and clock $n − 1$.

To prove that the clocks will eventually synchronize, define a function D that maps the set of global states of the system to a set of nonnegative integers:

$$D = d[0] + d[1] + d[2] + \cdots + d[n−1]$$

where

$$
\begin{aligned}
d[i] &= 0 && \text{if there is no arrow pointing toward clock } i \\
&= i + 1 && \text{if there is a ← pointing toward clock } i \\
&= n − 1 && \text{if there is a → pointing toward clock } i \\
&= 1 && \text{if there are both a ← and a → pointing toward clock } i
\end{aligned}
$$

By definition, $\forall i: d(i) \geq 0$, so $D \geq 0$. Based on observations 1–3, if $D > 0$, then D *will decrease* after each step of the proposed algorithm. Therefore, regardless of the size of the system, in a bounded number of moves, the value of D will be reduced to 0, after which the first guard of the proposed program will no longer be enabled, and the phases of all the clocks will be synchronized. Once synchronized, they will remain so thereafter (action 2).

Such counting arguments help compute the upper bound on the number of steps required for convergence to the target configuration. Each arrow can take at most $n − 1$ steps to disappear, and the arrows move synchronously. Therefore, it requires at most $n − 1$ clock ticks for the phases to be synchronized.

Nondeterminism and *fairness* models sometimes make it difficult to compute an upper bound for time complexity. Consider the following example:

```
program     step
define      m, n: integer
initially   m = 1, n = 0
do    m ≠ 0 →      m := 0
[]    m ≠ 0 →      n := n + 1
od
```

If the scheduler is unfair, then termination of the aforementioned program is not guaranteed. With a weakly fair scheduler, termination is guaranteed, but it is impossible to determine as such after how many steps this program will terminate, since there is no clue about when the first action will be scheduled. Much will depend on how the nondeterminism and fairness models are implemented.

5.6 PROGRAMMING LOGIC

Programming logic, first introduced by Hoare [H69, H72], is a formal system that manipulates predicates consisting of program states and relations characterizing the effects of program execution. One can reason about many aspects of program correctness using the rules of programming logic.

A program S transforms a *precondition* P that holds before the execution of the program into a *postcondition* Q that holds after the execution of the program. In programming logic, this is represented using the triple:

$$\{P\}\ S\ \{Q\}$$

Here, P and Q are predicates or assertions. An example of a triple is

$$\{x = 5\}\ x := x + 1\ \{x = 6\}$$

Every valid triple is called a *theorem* of programming logic. A theorem can be derived from more basic theorems (called *axioms*) using the transformation rules of programming logic. Some important axioms of programming logic are presented in the following:

Axiom 5.1

$\{P\}$ *skip* $\{P\}$ (*skip* means *do nothing*)

Axiom 5.2

$\{Q[x \leftarrow E]\}\ x{:} = E\{Q\}$

Here, E denotes an expression or a value, $x \leftarrow E$ is an assignment, and $Q[x \leftarrow E]$ denotes the condition derived from Q by substituting every occurrence of the variable x by the corresponding expression E. Consider the following examples, where ? denotes the unknown precondition of the triple $\{?\}\ S\ \{Q\}$:

Example 5.4

$$\{?\}\ x{:} = 1\ \{x = 1\}$$
$$? = (1 = 1) = true$$

Thus, $\{true\}x{:}= 1\{x = 1\}$ is a theorem.

Example 5.5

$$\{?\}\ x{:} = 2x + 1\ \{x > 99\}$$
$$? = (2x + 1 > 99) = (2x > 99) = (x > 49)$$

Thus, $\{x > 49\}x{:} = 2x + 1\{x > 99\}$ is a theorem.

Example 5.6

$$\{?\}\ x{:} = 100\ \{x = 0\}$$
$$? = (100 = 0) = false$$

Thus, $\{false\}\ x{:} = 100\ \{x = 0\}$ is a theorem.

Axiom 5.3

$$\{Q[x \leftarrow y, y \leftarrow x]\}\ (x, y){:} = (y, x)\ \{Q\}$$

Axiom 5.3 tells us how to compute the precondition in case of a swap. Here, $Q[x \leftarrow y, y \leftarrow x]$ implies simultaneous substitution of x by y and y by x in Q.

Appropriate *inference rules* extend the applicability of the axioms of programming logic. Let the notation H/C designate the fact that the *hypothesis* H leads to the *conclusion* C. To illustrate the use of these inference rules, consider the triple:

$$\{x = 100\}\ x := 2x + 1\ \{x > 99\}$$

Intuitively, this is a correct triple. But how do we show that it is a theorem in programming logic? Example 5.5 computes the corresponding precondition as $(x > 49)$, which is different from $(x > 100)$! However, it follows from simple predicate logic that $(x = 100) \Rightarrow (x > 49)$, so the triple $\{x = 100\}x := 2x + 1\{x > 99\}$ should be a theorem. Similarly, $\{x = 100\}$ $x := 2x + 1\{x > 75\}$ should also be a theorem, since $(x = 99) \Rightarrow (x > 75)$. The inference rule can now be represented by Axiom 5.4:

Axiom 5.4

$$\frac{(P' \Rightarrow P, \{P\}\ S\ \{Q\}, Q \Rightarrow Q')}{\{P'\}S\{Q'\}}$$

The essence of the aforementioned axiom is that the *strengthening* of the precondition and the *weakening* of the postcondition have no impact on the validity of a triple.

Axiom 5.5 illustrates how inference rules can help specify the semantics of the sequential composition operator ";."

Axiom 5.5

$$\frac{(\{P\}\ S_1\ \{R\}, \{R\}\ S_2\ \{Q\}}{\{P\}S_1; S_2\{Q\}}$$

Example 5.7

Prove that $\{y > 48\}$ $x := y + 1; x := 2x + 1$ $\{x > 99\}$ is a theorem.

$$\{x > 49\}x := 2x + 1\{x > 99\} \qquad \text{(Example 5.5)} \qquad (5.2)$$

Also, $\{y + 1 > 49\}$ $x := y + 1$ $\{x > 49\}$ (Axiom 5.2).
    This implies

$$\{y > 48\} \ x := y + 1\{x > 49\} \qquad\qquad (5.3)$$

Therefore, $\{y + 1 > 49\}$ $x := y + 1\{x > 99\}$ {using (5.2), (5.3), and Axiom 5.5}.

Now consider the alternative construct *IF* in a triple $\{P\}IF\{Q\}$:

```
if   G₁  →  S₁
[]   G₂  →  S₂
.

.
[]   Gₙ  →  Sₙ
fi
```

Assume that the guards are disjoint and let $GG = G_1 \lor G_2 \lor \cdots \lor G_n$. When $\neg GG$ holds (which means all guards are false), *IF* reduces to a *skip* statement. From Axiom 5.1, it follows that $(\neg GG \land P) \Rightarrow Q$. However, if some guard G_i is true, then the execution of S_i leads to the postcondition Q, so $\{P \land G_i\}S_i\{Q\}$ is a theorem. These interpretations lead to the following semantics of the alternative construct:

Axiom 5.6

$$\frac{(\neg GG \land P \Rightarrow Q, \{P \land G_i\} \ S_i \ \{Q\} : 1 \le i \le n)}{\{P\}IF\{Q\}}$$

Finally, consider the iterative construct *DO* in the triple $\{P\}DO\{Q\}$:

```
do   G₁  →  S₁
[]   G₂  →  S₂
.

.
[]   Gₙ  →  Sₙ
od
```

The semantics of this iterative construct can be represented in terms of a loop invariant. A loop invariant I is a predicate that is true at every stage from the beginning to the end of a loop. Initially, $P \equiv I$. Also, in order that the postcondition Q holds, the loop must terminate—so $\neg GG$ must eventually hold. Therefore, $Q = \neg GG \wedge I$. Furthermore, since the execution of each S_i preserves the invariance of I, $\{I \wedge G_i\} S_i \{I\}$ holds. Combining these facts, we obtain the following axiom:

Axiom 5.7

$$\frac{(\{I \wedge G_i\} S_i \{I\} : 1 \le i \le n)}{\{I\} DO \{\neg GG \wedge I\}}$$

Note that in *DO* loops, programming logic cannot determine if the loop will terminate—it can only decide that *if* the program terminates, *then* the postcondition holds.

The axioms of predicate logic and programming logic have useful applications in correctness proofs. An example of application is the assertional proof of sequential programs. Starting from the given precondition P for a program $S = S_1; S_2; S_3; \ldots; S_n$, the assertion Q_i is computed after the execution of each statement S_i. The program is proven to be correct when $Q_n \Rightarrow Q$, the desired postcondition. An annotated program showing the intermediate assertions is called a *proof outline*.

5.7 PREDICATE TRANSFORMERS

Consider the question: "What is the *largest set* of initial states, such that the execution of a program S starting from *any* of these states (1) is guaranteed to terminate and (2) results in a postcondition Q?" This question is of fundamental importance in the field of program derivation, and we will briefly address it here. The set of *all* initial states satisfying the aforementioned two conditions is known as the *weakest precondition wp(S, Q)*. Since *wp* maps the predicate Q into the predicate $wp(S, Q)$, it is also called a *predicate transformer*. If $P \Rightarrow wp(S, Q)$, then $\{P\} S \{Q\}$ is a theorem in programming logic. Note that a theorem in programming logic does not require termination, whereas predicate transformers imply properly terminating behavior. Some useful axioms with predicate transformers [D76] are given in the following:

Axiom 5.8

$wp(S, false) = false$ (law of excluded miracle)

Proof Assume that for some program S, $wp(S, false) \ne false$. Then there exists an initial state from which the execution of S terminates and results in a final state that satisfies *false*. But no state satisfies the predicate *false*. ■

Axiom 5.9

If $Q \Rightarrow R$, then $wp(S, Q) \Rightarrow wp(S, R)$ (property of monotonicity).

Axiom 5.10

$wp(S, Q) \land wp(S, R) \Rightarrow wp(S, Q \land R)$

Axiom 5.11

$wp(S, Q) \lor wp(S, R) \Rightarrow wp(S, Q \lor R)$

Proof

$wp(S, Q) \Rightarrow wp(S, Q \lor R)$ (Axiom 5.9)
Similarly, $wp(S, R) \Rightarrow wp(S, Q \lor R)$ (Axiom 5.9).
Therefore, $wp(S, Q) \lor wp(S, R) \Rightarrow wp(S, Q \lor R)$ (propositional logic).   ■

Note: For a deterministic computation, the $\Rightarrow$ in Axiom 5.11 can be replaced by =. As an illustration, consider the following program, where nondeterminism plays a crucial role in deciding the value of x after every step:

```
Program toss;
define x : integer;
if     true → x := 0
[]     true → x := 1
fi
```

Here, $wp(toss, x = 0) = false$, and $wp(toss, x = 1) = false$
(since no initial state can guarantee that the final value of x will be 0 or 1).
However, $wp(toss, x = 0 \lor x = 1) = true$
(since from every initial state, the final value of x must be either 0 or 1).

This is in agreement with the statement of Axiom 5.11, but it falls apart when the implication $\Rightarrow$ is replaced by the stronger relation =. Now consider the next program *mod*, which is different from *toss*.

```
Program mod
define x: integer
if     x = even → x := 0
[]     x = odd  → x := 1
fi
```

Here, unlike program *toss*, regardless of the initial value of *x*, if $wp(\text{mod}, x = 1) = false$, then $wp(\text{mod}, x = 1) = true$ and vice versa. Since $wp(\text{mod}, x = 0 \lor x = 1) = true$, it is in agreement with the note following Axiom 5.11. Of course, it is only an example and not a proof.

Axiom 5.12

$$wp\big((S_1;S_2),Q\big) = wp\big(S_1, wp(S_2,Q)\big)$$

Axiom 5.13

$$wp(IF, Q) = (\neg GG \Rightarrow Q) \land (\forall i: 1 \leq i \leq n:: G_i \Rightarrow wp(S_i, Q))$$

The most significant difference from programming logic is in the semantics of the *DO* loop, since termination is a major issue. The predicate transformer of a *DO* statement requires the loop to terminate in *k* or fewer iterations (i.e., *k* being an upper bound). Let $H_k(Q)$ represent the *largest set of states* starting from which the execution of the *DO* loop terminates in *k* or fewer iterations. The following two conclusions immediately follow:

- $H_0(Q) = \neg GG \Rightarrow Q$
- $H_k(Q) = H_0(Q) \lor wp(IF, H_{k-1}(Q))$

Using these, the semantics of the *DO* loop can be specified as follows:

Axiom 5.14

$$wp(DO, Q) = \exists k \geq 0: H_k(Q)$$

An extensive discussion of program derivation using predicate transformers can be found in [D76] and [G81].

5.8 CONCLUDING REMARKS

Formal treatments sharpen our reasoning skills about why a program should work or fail. While such treatments can be convincingly demonstrated for toy examples, formal reasoning of nontrivial examples is often unmanageably complex. This obviously encourages the study of automated reasoning methods. An alternative is program synthesis. A program *S* composed of a set of subprograms using a set of composition rules is guaranteed to work correctly, if each subprogram works correctly, and the composition rules are certified by the axioms of a formal program derivation system.

A powerful tool for reasoning about the dynamic behavior of programs and their properties is *temporal logic*. It provides a succinct expression for many useful program properties using a set of temporal operators. Note that predicate logic and propositional logic cater to timeless properties. To take a peek at temporal logic, let us consider the two important operators $\Diamond$ and $\square$. If P is a property, then consider the following:

- $\square P$ implies that P is *always* true. This is useful for expressing invariants in safety properties.

- $\Diamond P$ implies that P will *eventually* become true. This is useful for expressing liveness properties.

The two operators are related as $\Diamond P = \neg\square(\neg P)$. This can be intuitively reasoned as follows: if P represents the property of termination of a program S, then the statement "the program S will eventually terminate" can also be stated as "it is not always true that program S will not terminate."

A *formula* is an assertion about a behavior. Every property can be expressed by a formula, which is built from elementary formulas using the operators of propositional logic and the operators $\Diamond$ and $\square$. The following are useful in dealing with temporal properties and can be reasoned with intuition:

$$\square P \wedge \square Q = \square(P \wedge Q)$$

$$\square P \vee \square Q \Rightarrow \square(P \vee Q)$$

(Note that $\square P \vee \square Q \neq \square(P \vee Q)$, since if P and Q change with time, and $Q = \neg P$, then the left side may be false, but the right side is true.)

$$\Diamond P \vee \Diamond Q = \Diamond(P \vee Q)$$

$$\square\Diamond(F \vee G) = (\square\Diamond F) \vee (\square\Diamond G)$$

(Note that $\square\Diamond P$ (always eventually true) is true for a behavior if and only if $\Diamond P$ is true at all times during that behavior. This is synonymous with "P is true infinitely often."). Finally,

$$\Diamond\square P \Rightarrow \square\Diamond P$$

Temporal logic formulas can be used to specify various fairness models introduced in Chapter 4. For example, consider the construct **do** $G_i \rightarrow S_i \; []G_j \rightarrow S_j$ **od**. Under a weakly fair scheduler, action S_j will be scheduled for execution if $\Diamond\square G_j$ holds, which means there is a time t after which G_j will be *always true*. Under a strongly fair scheduler, action S_j will be scheduled for execution if $\square\Diamond G_j$ holds, that is, G_j is true infinitely often.

To learn more about temporal logic, read Manna and Pnueli's book [MP92].

5.9 BIBLIOGRAPHIC NOTES

The original work on proving the correctness of sequential programs was done by Floyd [F67]. In [H69], Hoare developed the framework of programming logic. Ashcroft and Manna [AM71] were the first to prove properties of about concurrent programs. In [H72], Hoare extended his partial correctness proof of sequential programs to include concurrency. A more complete treatment of the subject is available in Susan Owicki's dissertation, a summary of which can be found in [OG76]. Here, Owicki and Gries proposed how to prove the partial correctness of concurrent programs where processes communicate through a shared memory.

Lamport [L77] introduced the terms *safety* and *liveness*. These were originally taken from the Petri net community who used the term *safety* to designate the condition that no place contains more than one token and the term *liveness* to designate the absence of deadlock. The book by David Gries [G81] is a comprehensive text on correctness proofs that no beginner should miss. Dijkstra [D76] introduced predicate transformers and demonstrated how they can be used to derive programs from scratch. Amir Pnueli introduced temporal logic [P77]. Manna and Pnueli's book [MP92] contains a complete treatment of the specification and verification of concurrent systems using temporal logic. Owicki and Lamport [OL82] demonstrated the use of temporal logic to prove liveness properties of concurrent programs. In [L94], Lamport developed a complete proof system called TLA based on temporal logic. Chandy and Misra developed an alternative proof system called UNITY—a comprehensive treatment of their work can be found in their book [CM88].

EXERCISES

5.1 Use predicate logic to represent the following:

a. A set B of n balls ($n > 2$) of which *at least* two balls are red and the remaining balls are white.

b. Let V define a set of points on a 2D plane. Represent the distance D between a pair of points (u, v) belonging to V such that the distance between them is the smallest of the distances between all pairs of points in V.

5.2 Consider the program for clock phase synchronization in Example 5.3. If the topology is a cycle instead of a linear array, then will the clock phases be synchronized? Briefly justify your answer. Also discuss how you can generalize the solution for k-phase clocks ($k > 2$).

5.3 The following program is designed to search an element t in the integer array X:

```
define X: array [0..n - 1] of integers
       i, t: integer
initially i = 0
do X[i] ≠ t → i := i + 1 od
```

Assuming that t is indeed the value of one of the elements in the array X, define a well-founded set and use it to prove that the program terminates in a bounded number of steps.

5.4 Two processes P and Q communicate with each other using *locally shared variables* p and q. Their programs are as follows:

```
Program P              program Q
define p : boolean     define q: boolean
do p = q → p := ¬p od  do p ≠ q → q := ¬q od
```

Prove that the program does not terminate.

5.5 Consider a connected network (V, E), where each vertex $v \in V$ is an *unbounded clock*. All clocks are ticking at the same rate and displaying the same value. Due to electrical disturbances, one or more of these clock values might occasionally be perturbed. The following program synchronizes the clock values in a bounded number of steps following a perturbation:

```
{program for clock i}
Define c[i]: integer {non-negative integer representing value
   of clock i}
        {N(i) denotes the set of neighbors of clock i}
do true → c[i] := 1 + max{c[j]:j ∈ N(i) ∪ i} od
```

Assume a synchronous model where all clocks simultaneously execute the aforementioned action with each clock tick and the action takes zero time to complete. Prove using a well-founded set and an appropriate variant function that the clocks will start displaying the same value in a bounded number of steps. Also, what is the round complexity of the algorithm?

5.6 If a distributed computation does not terminate with a strongly fair scheduler, then can it terminate with a weakly fair scheduler? What about the converse? Provide justification (or example) in support of your answer.

5.7 In a coffee jar, there are black and white beans of an unknown quantity. You dip your hands into the jar and randomly pick two beans. Then play the following game until the jar contains only one bean:

a. If both beans have the same color, then throw them away and put one black bean back into the jar (assume that by the side of the jar, there is an adequate supply of black beans).

b. If the two beans have different colors, then throw the black bean and return the white bean into the jar.

What is the color of the *last bean*, and how is it related to the initial content in the coffee jar? Furnish a proof in support of your answer.

5.8 There are two processes P and Q. P has a set of integers A, and Q has another set of integers B. Using message-passing model, develop a program by which processes P and Q exchange integers, so that eventually every element of A is greater than every element of B. Present a correctness proof of your program.

5.9 Consider a bag of numbers and play the following game as long as possible. Pick any two numbers from this bag. If these numbers are unequal, then do nothing—otherwise, increment one, decrement the other, and put them back in the bag. The claim is that in a bounded number of steps, no two numbers in the bag will be equal to one another. Can you prove this?

5.10 Consider a *tree* (V, E), where each node $i \in V$ represents a process. Each node i has a color $c[i] \in \{0,1\}$. Starting from an arbitrary initial configuration, the nodes have to acquire a color such that no two neighboring nodes have the same color. We propose the following algorithm for each process i:

```
program   twocolor
define c[i]: color of process i {c = 0 or 1}
do j ∈ N(i):c[i] = c[j] → c[i] := 1 - c[i] od
```

Assume that the scheduler is weakly fair. Will the algorithm terminate? If not, then explain why. Otherwise, give a proof of termination.

5.11 This is a logical continuation of the previous question. Consider that you have a *rooted tree* with a designated root. Each node (except the root) has a neighbor that is designated as its parent node. If j is the parent of i, then i cannot be the parent of j. Now try the same problem (of coloring) as in Question 10, but this time, use a different algorithm:

```
program   treecolor (for process i}
define    c[i]: color of process i
          p[i]: parent of process i

do c[i] = c[p[i]] → c[i] := 1 - c[i] od
```

Assume that the scheduler is weakly fair. Will the algorithm terminate? If not, then explain why. Otherwise, give a proof of termination.

5.12 Consider an array of $n(n > 3)$ processes. Starting from a terminal process, mark the processes alternately as even and odd. Assume that the even processes have states $\in\{0, 2\}$, and the odd processes have states $\in\{1, 3\}$. The system uses the state-reading model and distributed scheduling of actions. From an unknown starting state, each process executes the following program:

```
program   alternator {for process i}
define    s ∈{0,1,2,3}{state of a process}
do ∀j ∈ N(i):s[j] = s[i] + 1 mod 4 → s[i] := s[i] + 2 mod 4
od
```

The program will enable the processes to settle down to a steady behavior. Observe and summarize the *steady-state behavior* of the aforementioned system of processes. What is the *maximum number* of processes that can eventually execute their actions concurrently in the steady state? Briefly justify your answer.

5.13 The following computation runs on a unidirectional ring of n processes 0, 1, 2, ..., $n - 1$ ($n > 3$). Processes 0 and $n - 1$ are neighbors. Each process i has a local integer variable $x[i]$ whose value is in the range $0.. k - 1$ ($k > 1$).

```
{process 0}      do x[0] ≠ x[n - 1] → x[0] := x[0] + 1 mod n od
{process i > 0} do x[i] ≠ x[i - 1] → x[i] := x[j - 1] od
```

Prove that the aforementioned computation will not deadlock.

5.14 In a completely connected network of processes, each process i has an integer variable $x[i]$. Initially, $\forall i : x[i] = 0$. Each process i executes the following program:

```
do    ∀j ∈ N(i):x[i] ≤ x[j] → x[i] := x[i] + 1
[]    ∃j ∈ N(i):x[i] > x[j] → skip
od
```

Prove that the safety property $\forall i,j : |x[i] - x[j]| \leq 1$ holds.

CHAPTER 6

Time in a Distributed System

6.1 INTRODUCTION

Time is an important parameter in a distributed system. Consistency maintenance among replicated data relies on which update is the most recent one. Real-time systems like air-traffic control must have accurate knowledge of time to provide useful service and avoid catastrophe. Important authentication services (like Kerberos) rely on synchronized clocks. Wireless sensor networks rely on accurately synchronized clocks to compute the trajectory of fast-moving objects. *High-frequency trading* [D09], where fast computers scan millions of accounts in a second and run complex algorithms for implementing various trading strategies in a matter of milliseconds, relies heavily on the accuracy of time synchronization. Before addressing the issues related to time in distributed systems, let us briefly review the prevalent standards of physical time.

6.1.1 Physical Time

The notion of time and its relation to space have intrigued scientists and philosophers since the ancient days. According to the laws of physics and astronomy, real time is defined in terms of the rotation of Earth in the solar system. A solar second equals 1/86,400th part of a solar day, which is the amount of time that the Earth takes to complete one revolution around its own axis. This measure of time is called the real time (also known as Newtonian time) and is the primary standard of time. Our watches or other timekeeping devices are secondary standards that have to be calibrated with respect to the primary standard.

Modern timekeepers use atomic clocks as a de facto primary standard of time. Per this standard, a second is precisely the time for *9,192,631,770* orbital transitions of the *cesium 133* atom. In actual practice, there is a slight discrepancy—86,400 atomic seconds is close to 3 ms less than a solar day, so when the discrepancy grows to about 1 s, a *leap second* is added to the atomic clock.

International Atomic Time (TAI) is an accurate time scale that reflects the weighted average of the readings of nearly 300 atomic clocks in over 50 national laboratories worldwide. It has been available since 1955 and became the international standard on which

UTC is based. UTC was introduced on January 1, 1972, following a decision taken by the 14th General Conference on Weights and Measures (CGPM). The International Bureau of Weights and Measures is in charge of the realization of TAI.

UTC, popularly known as GMT (Greenwich Mean Time) or *Zulu* time, differs from the local time by the number of hours of your time zone. The use of a central server receiving the WWV shortwave signals from Fort Collins, Colorado, and periodically broadcasting the UTC-based local time to other timekeepers is quite common. In fact, inexpensive clocks driven by these signals are now commercially available.

Another source of precise time is GPS. A system of 32 satellites deployed in the Earth's orbit maintains accurate spatial coordinates and provides precise time reference almost everywhere on Earth where GPS signals can be received. Each satellite broadcasts the value of an onboard atomic clock. To use the GPS, a receiver must be able to receive signals from at least four different satellites. While the clock values from the different satellites help obtain the precise time, the spatial coordinates (latitude, longitude, and the elevation of the receiver) are computed from the distances of the satellites estimated by the propagation delay of the signals. The clocks on the satellites are physically moving at a fast pace, and per the theory of relativity, this causes the onboard clocks to run at a slightly slower rate than the corresponding clocks on the Earth. The cumulative delay per day is approximately 38 ms, which is compensated using additional circuits. The atomic clocks that define GPS time record the number of seconds elapsed since January 6, 1980. At present (i.e., in 2013), the GPS time is nearly 16 s ahead of UTC, because it does not use the leap second correction. Receivers thus apply a clock-correction offset (which is periodically transmitted along with the other data) in order to display UTC correctly and optionally adjust for a local time zone.

6.1.2 Sequential and Concurrent Events

Despite technological advances, the clocks commonly available at the processors distributed across a system do not exactly show the same time. Built-in atomic clocks are not yet cost-effective—for example, wireless sensor networks cannot (yet) afford an atomic clock at each sensor node, although accurate timekeeping is crucial to detecting and tracking fast-moving objects. Certain regions in the world cannot receive such time broadcasts from reliable timekeeping sources. GPS signals are difficult to receive inside a building. The lack of a consistent notion of system-wide global time leads to several difficulties. One difficulty is the computation of the global state of a distributed system (defined as the set of local states of all processes at a given time). The special theory of relativity tells us that simultaneity has no absolute meaning—it is relative to the location of the observers. If the time stamps of a pair of events are nanoseconds apart, then we cannot convincingly say which one happened earlier, although this knowledge may be useful to establish a cause–effect relationship between the events. However, causality is a basic issue in distributed computing—the ordering of events based on causality is more fundamental than that obtained using physical clocks. Causal order is the basis of *logical clocks*, introduced by Lamport [L78]. In this chapter, we will address how distributed systems cope with uncertainties in physical time.

6.2 LOGICAL CLOCKS

An event corresponds to the occurrence of an action. A set of events $(a, b, c, \ldots)$ in a single process is called *sequential*, and their occurrences can be totally ordered in time using the clock at that process. For example, if Bob returns home at 5:40 p.m., answers the phone at 5:50 p.m., and eats dinner at 6:00 p.m., then the events (*return home, answer phone, eat dinner*) define an ascending sequential order. This total order is based on a single and consistent notion of time that Bob believes in accordance with his clock.

In the absence of perfectly reliable timekeepers, two different physical clocks at two different locations will always drift. Even if they are periodically resynchronized, some inaccuracy in the interim period is unavoidable. The accuracy of synchronization depends on clock drift as well as on the resynchronization interval. Thus, 6:00 p.m. for Bob is not necessarily exactly 6:00 p.m. for Alice at a different location, even if they live in the same time zone. Events at a single point can easily be totally ordered on the basis of their times of occurrences at that point. But how do we decide if an event with Bob happened *before* another event with Alice? How do we decide if two events are concurrent?

To settle such issues, we depend on an obvious law of nature: no message can be received *before* it is sent. This is an example of *causality*. Thus, if Bobs ends a message to Alice, then the event of sending the message must have happened *before* the event of receiving that message regardless of the clock readings. The importance of causal relationship can be traced in many applications. For example, during a chat, let Bob send a message M to Carol and Alice and Alice post a reply *Re:M* back to Bob and Carol. Clearly M happened before *Re:M*. To make any sense of the chat, Carol should always receive M before *Re:M*. If two events are not causally related, then we do not care about their relative orders and call them *concurrent*.

The aforementioned observations lead to three basic rules about the causal ordering of events, and they collectively define the *happened before* (or the *causally ordered before*) relationship $\prec$ in a distributed system:

Rule 1: Let each process have a physical clock whose value is monotonically increasing. If a, b are two events within a single process P, and the time of occurrence of a is earlier than the time of occurrence of b, then $a \prec b$.

Rule 2: If a is the event of sending a message by process P, and b is the event of receiving the same message by another process Q, then $a \prec b$.

Rule 3: $(a \prec b) \wedge (b \prec c) \Rightarrow (a \prec c)$.

Figure 6.1 illustrates these rules using a space–time diagram. Here, P, Q, and R are three different sequential processes at three different sites. At each site, there is a separate physical clock, and these clocks tick at an unknown pace. The horizontal lines at each site indicate the passage of time. Based on the rules mentioned previously, the following results hold:

$$b \prec h \quad \text{since} \quad (b \prec c) \wedge (c \prec g) \wedge (g \prec h)$$

$$a \prec d \quad \text{since} \quad (a \prec b) \wedge (b \prec c) \wedge (c \prec d)$$

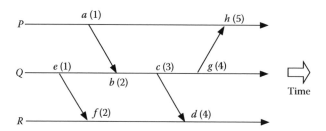

FIGURE 6.1   A space–time view of events in a distributed system consisting of three processes *P*, *Q*, *R*: the horizontal lines indicate the timelines of the individual processes, and the diagonal lines represent the flow of messages between processes. Each event is tagged with its logical clock value.

However, it is impossible to determine any causal ordering between the events (*a*, *e*)—neither *a* ≺ *e* holds, nor *e* ≺ *a* holds. The same thing applies to (*c*, *f*) and (*b*, *h*). In such a case, we call the two events *concurrent* (*a* ∥ *b*). It must be clear from this example that the events in a distributed system cannot always be totally ordered. The happened before relationship defines a *partial order*, and concurrency corresponds to the absence of causal ordering.

A *logical clock* is an event counter that respects causal ordering. Consider the sequence of events in a single sequential process. Each process has a counter *LC* that represents its logical clock. Initially, for every process, *LC* = 0. The occurrences of events correspond to the *ticks* of the logical clock local to that process. Every time an event takes place, *LC* is incremented. Logical clocks can be implemented using three simple rules:

LC1: Each time a local event takes place, increment *LC* by 1.

LC2: When sending a message, append the value of *LC* to the message.

LC3: When receiving a message, set the value of *LC* to 1+ max (*local LC*, *message LC*), where *local LC* is the local value of *LC* and *message LC* is the *LC* value appended with the incoming message.

The aforementioned implementation of logical clocks provides the following limited guarantee for a pair of events *a* and *b*:

$$a \prec b \Rightarrow LC(a) < LC(b)$$

However, the converse is not true. In Figure 6.1, $LC(f) = 2$ and $LC(h) = 5$, but there is no causal order between *f* and *h*. This is a limitation of logical clocks.

Although causality induces a partial order, in many applications, it is important to define a total order among events. For example, consider the server of an airline handling requests for reservation from geographically dispersed customers. A fair policy for the server will be to allocate the next seat to the customer who sent the request ahead of others. If the physical clocks are perfectly synchronized and the message propagation delay is zero, then it is trivial to determine the order of the requests using physical clocks. However, if the physical clocks are not synchronized and the message propagation delays are arbitrary, then determining a total order among the incoming requests becomes a challenge.

One way to evolve a consistent notion of *total ordering* across an entire distributed system is to *strengthen* the notion of logical clocks. If a and b are two events in processes i and j (not necessarily distinct), respectively, then define *total ordering* ($\ll$) as follows:

$$a \ll b \quad \text{iff} \quad \text{either} \quad LC(a) < LC(b)$$

$$\text{or} \quad LC(a) = LC(b) \quad \text{and} \quad i < j$$

where $i < j$ is determined either by the relative values of the numeric process identifiers or by the lexicographic order of their names. Whenever the logical clock values of two distinct events are equal, their process numbers or names will be used to break the tie. The (id, LC) value associated with an event is called its *time stamp*.

It should be obvious that $a \prec b \Rightarrow a = b$. However, its converse is not necessarily true.

While the definition of causal order is quite intuitive, the definition of concurrency as the absence of causal order leads to tricky situations that may appear counterintuitive. For example, the concurrency relation is *not* transitive. Consider Figure 6.1 again. Here, f is concurrent with g, and g is concurrent with d, but f is *not* concurrent with d.

Even after the introduction of causal and total order, some operational aspects in message ordering remain unresolved. One such problem is the implementation of FIFO communication across a network. Let m, n be two messages sent successively by a process P to another process R (via arbitrary routes). Here, the FIFO property implies that if $send(m) \prec send(n)$ holds, then the recipient process R must receive message m before receiving message n.

In Figure 6.2, P first sends the first message m directly to R and *then* sends the next message n to Q, who forwards it to R. Although $send(m) \prec send(n)$ holds, and each channel individually exhibits FIFO behavior, there is no guarantee that m will reach R before n. Thus, $send(m) \prec send(n)$ does not necessarily imply $receive(m) \prec receive(n)$.

This anomalous behavior can sometimes lead to difficult situations. For example, even if a server uses the policy of servicing requests based on time stamps, it may not always be able to do so, because the request bearing a smaller time stamp may not have reached the server before another request with a larger time stamp. At the same time, no server is clairvoyant—that is, it is impossible for the server R to know whether a request with a lower time stamp will ever arrive *in future*. This is an important issue in distributed

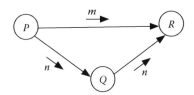

FIGURE 6.2   A network of three processes connected by FIFO channels.

simulation, which requires that the temporal order of events in the simulated environment be a true reflection of the corresponding order in the real system. We will address this in Chapter 18.

6.3 VECTOR CLOCKS

One major weakness of logical clocks is that the *LC* values of two events cannot reveal if they are causally ordered. *Vector clocks*, independently discovered by Fidge [F88] and Mattern [M88] overcome this weakness. The primary goal of vector clocks is to detect causality. Let *V* denote the set of all events in a distributed system of n processes $0\ldots n-1$ and *A* denote the set of nonnegative integer vectors of size n. Then vector clock is a mapping $VC: V \rightarrow A$.

Let $a, b \in V$. Denote the ith element of $VC(a)$ by $VC_i(a)$. Define a partial order < among the vector clock values as follows: $VC(a) < VC(b)$ if and only if the following two conditions hold:

1. $\forall i: 0 \leq i \leq n - 1: VC_i(a) \leq VC_i(b)$

2. $\exists j: 0 \leq j \leq n - 1: VC_j(a) < VC_j(b)$

For a pair of events a, b, if neither $VC(a) < VC(b)$ nor $VC(b) < VC(a)$ holds, then $VC(a)$: $VC(b)$, and the events are concurrent, that is, $a \parallel b$.

Since causality detection is one of the primary goals of vector clocks, its implementation is required to satisfy the following condition:

$$a \prec b \Leftrightarrow VC(a) < VC(b)$$

To implement a system of vector clocks, each process i initializes its vector clock $VC[i]$ to $0, 0, 0, \ldots, 0$ (n components). Subsequently, each process follows the following three rules:

Rule 1: Each local event at process i increments the ith component of its latest vector clock value by 1 (i.e., $VC_i[i]:=VC_i[i]+1$).

Rule 2: The sender of a message appends the vector clock value of the send event to every message that it sends.

Rule 3: When process j receives a message with a vector clock value T from another process, it first increments the jth component of its own latest vector clock by 1 (i.e., $VC_j[i]:=VC_j[i]+1$) and then updates its vector clock as follows:

$$\forall k: 0 \leq k \leq n-1 :: VC_k[j] := \max(T_k, VC_k[j])$$

An example is shown in Figure 6.3. The event with vector time stamp (2,1,0) is causally ordered before the event with the vector time stamp (2,1,4) but is concurrent with the event having time stamp (0,0,2).

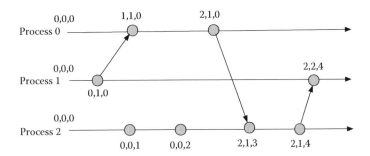

FIGURE 6.3   Example of vector time stamps.

Although vector clocks detect causal ordering or the lack of it, a problem is their poor scalability. As the size of the system increases, so does the size of the clock. Consequently, in dynamic systems, the addition of a process requires a reorganization of the state space across the entire system. Even if the topology is static, for large-scale systems, the communication bandwidth suffers when messages are stamped with the value of the vector clock. In Chapter 15, we will discuss the use of vector time stamps in solving the problem of causally ordered group communication.

6.4 PHYSICAL CLOCK SYNCHRONIZATION

6.4.1 Preliminary Definitions

Consider a system of n physical clocks $(0, 1, 2, …, n − 1)$ ticking approximately at the same rate. Such clocks may not accurately reflect real time, and despite great care taken in building these clocks, their readings slowly drift apart over time. Therefore, these clocks need to be periodically resynchronized to bring the discrepancies within acceptable bounds. The availability of synchronized clocks simplifies many problems in distributed systems. Air-traffic control systems rely on accurate timekeeping to monitor flight paths and avoid collisions. Some security mechanisms depend on the physical times of events, so a loss of synchronization may be a potential security lapse. Multiversion objects need accurate time stamps to recognize the most recent version.

Ordinary quartz-oscillator-based clocks maintain time with an accuracy of 0.2 s/day, but the accuracy is affected by temperature and aging. GPS receivers can provide time with an accuracy of 1 ms, but they work mostly outdoors for a clear sky view. The National Institute of Science and Technology (NIST) radio station WWV that broadcasts from Boulder, Colorado, or Washington, DC, provides time information with an accuracy of 4–10 ms depending in the distance of the receiver. The overhead of processing these signals by a computer adds some more slack to the accuracy of timekeeping. This is sometimes compounded by failures and occasional unpredictability in the signal propagation delays. For reasons of convenience and cost, many of these solutions are not ideal for every machine or every application. Accordingly, most machines try to adjust their times by periodically asking other machines. Three main problems have been studied in the area of physical clock synchronization:

External synchronization: The goal of external synchronization is to maintain the reading of each clock as close to the UTC as possible. A time server is a machine that provides accurate time information to be used as a reference by other machines. The NTP (Network Time Protocol) is an external synchronization protocol that runs on the Internet and coordinates a number of time servers. This enables a large number of computers connected to the Internet to synchronize their local clocks to within a few milliseconds from the UTC. NTP takes appropriate recovery measures against possible failures of one or more servers as well as the failure of links connecting the servers.

Internal synchronization: The goal of internal synchronization is to keep the readings of a system of *autonomous clocks* closely synchronized with one another, despite the failure or malfunction of one or more clocks. These clock readings may not have any connection with UTC or GPS time—mutual consistency is the primary goal. Of course external synchronization implies internal synchronization, but many applications do not need the extra work. For example, wireless sensor networks measure the distance between a source and a destination node by measuring the propagation delay of a reference signal. The internal clocks of the sensor nodes are ordinarily built from inexpensive oscillators. Accurate internal synchronization between the source and the destination clocks allows the application to accurately measure the propagation delay of the signal and calculate the distance by multiplying the propagation delay by the speed of signal. Here, external synchronization is unnecessary.

Phase synchronization: Many distributed computations run in phases: in a given phase, all processes execute some actions, which are followed by the next phase. A phase clock is an integer-valued variable that is incremented each time a phase completes. Each process has its own copy of the phase clock. In the clock phase synchronization problem, we assume a synchronous model where all phase clock variables are incremented in unison, as if all of them are driven by the same clock. Clearly, once all the phase variables are equal, they remain so forever, and synchronization becomes unnecessary. However, due to transient failures, phase clocks may occasionally differ, so that while all the nonfaulty clocks tick as 1,2,3,4,..., the faulty clock might tick as 6,7,8,9,... during the same time. A clock phase synchronization algorithm guarantees that starting from an arbitrary configuration, eventually the values of all the phase clocks become identical.

Bounded and unbounded clocks: A clock is *bounded*, when with every tick, its value c is incremented in a mod M field, $M > 1$. Such a clock has only a finite set of possible values $(0, 1, 2, ..., M - 1)$. After $M - 1$, the value of c rolls back to 0. The value of an *unbounded clock* on the other hand increases monotonically, and thus, such a clock can have an infinite number of possible values.

Due to the finite space available in physical systems, only bounded clock values can be recorded. It may appear that by appending additional information like year and month, a clock reading can look unbounded. Consider Figure 6.4 for an example of a clock reading.

Even this clock will overflow in the year 10,000. Anyone familiar with the Y2K problem [Y2K] knows the potential danger of bounded clocks: as the clock value increases beyond M, it changes to 0 instead of $M + 1$, and computer systems consider the corresponding event as an event from the past. As a result, many anticipated events may not be scheduled.

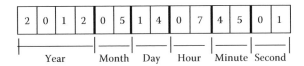

| 2 | 0 | 1 | 2 | 0 | 5 | 1 | 4 | 0 | 7 | 4 | 5 | 0 | 1 |

Year   Month   Day   Hour   Minute   Second

FIGURE 6.4   The clock reading when the drawing of this diagram was completed.

The solution to this problem by allocating additional space for storing the clock values is only temporary. All that we can guarantee is that clocks will not overflow in the foreseeable future. A 64-bit clock that is incremented every microsecond will not overflow for nearly 20 trillion years. However, when we discuss about fault tolerance, we will find out that a *faulty* clock can still overflow very easily and cause the old problem to resurface.

Drift rate: The maximum rate by which the value of a clock drifts from the ideal time (or the real time) is called the *drift rate* ρ. With ordinary crystal-controlled clocks, the drift rate is around 1 in 10^6. For the best atomic clocks, the drift rate is much smaller (around 1 in 10^{13}). A drift rate ρ guarantees that

$$(1-\rho) \le \frac{dC}{dt} \le (1+\rho)$$

where

    C is the clock time
    t represents the real time

Clock skew: The maximum difference δ between any two clocks that is allowed by an application is called the *clock skew* (Figure 6.5).

Resynchronization interval: Unless driven by the same external source, physical clocks drift with respect to one another as well as with respect to UTC. To keep the clock readings

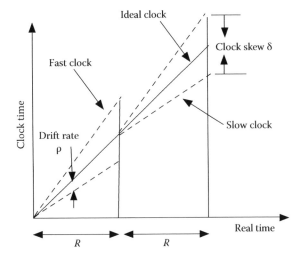

FIGURE 6.5   The cumulative drift between two clocks drifting apart at the rate *r* is brought closer after every resynchronization interval *R*.

close to one another, the clocks are periodically resynchronized. The maximum difference R between two consecutive synchronization actions is called the *resynchronization interval*, and it depends on the maximum permissible clock skew in the application.

6.4.2 Clock Reading Error

In physical clock synchronization, there are some unusual sources of error that are specific to time-dependent variables only. Having accurate clocks is not of much value unless those clocks can be accurately read. Some of the sources of error in reading clocks are as follows:

Propagation delay: If clock i sends its reading to clock j, then the accuracy of the received value must depend not only on the value that was sent but also on the message propagation delay. Note that this is not an issue when other types of values (which are not time sensitive) are sent across a channel. Even when the propagation delay due to the physical separation between processes is negligible, the operating system at the receiving machine may defer the reading of the incoming clock value and thus introduce error. Ideally, we need to simulate a situation when a clock ticks in transit to account for these overheads. A single parameter ε (called the *reading error*) accounts for the inaccuracy.

Processing overhead: Every computation related to clock synchronization itself takes a finite amount of time that needs to be separately accounted for.

6.4.3 Algorithms for Internal Synchronization

Berkeley algorithm: A well-known algorithm for internal synchronization is the *Berkeley algorithm*, first used in Berkeley UNIX 4.3 BSD. The basic idea is as follows: the participating processes elect a leader that coordinates the clock synchronization. The leader process periodically reads the clocks from all the participant processes, computes the average of these values, and then reports back to the participants the *adjustment* that needs be made to their local clocks, so that clock skew never exceeds the permissible limit δ. The algorithm assumes that the condition holds in the initial state. A participant whose clock reading lies outside the maximum permissible skew δ is disregarded when computing the average. This prevents the overall system time from being unfavorably skewed due to one or more erroneous clocks. Figure 6.6 shows an example. The rationale behind sending the needed correction instead of the computed average value is that the absolute value is influenced by the propagation delay, whereas the needed correction hardly changes during the signal propagation time. One note of caution: the adjustments are to be applied to the clocks in such a way that the monotonicity property of clocks is not violated. Thus, negative corrections will be implemented via a slowdown of the clock, whereas positive corrections will be implemented via an appropriate speedup.

The algorithm handles the case where the notion of faults may be a relative one: for example, there may be two disjoint sets of clocks, and in each set, the clocks are synchronized with one another, but no clock in one set is synchronized with any other clock in the second set. Here, to every clock in one set, the clocks in the other set are faulty. The final outcome is determined by the choice of the leader process.

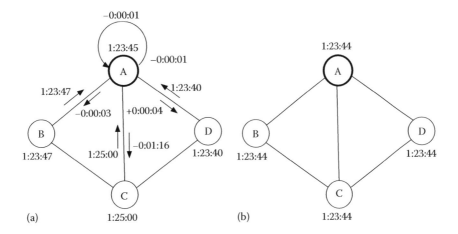

FIGURE 6.6   The readings of the clocks (a) before and (b) after one round of the Berkeley algorithm: A is the leader, and C is an outlier whose value lies outside the permissible limit of 0:00:06 chosen for this system.

Lamport and Melliar–Smith's algorithm: This algorithm is an adaptation of the Berkeley algorithm, with some extra features. It not only handles faulty clocks but also handles *two-faced* clocks, an extreme form of faulty behavior, in which two nonfaulty clocks obtain conflicting readings from the same faulty clock.* The algorithm assumes that the clocks are initially synchronized and that they are resynchronized often enough so that no two nonfaulty process clocks ever differ by more than δ. For the sake of simplicity, disregard the overhead due to propagation delay or computation overhead and consider clock reading to be an instantaneous action. Let $c_k[i]$ denote clock i's reading of clock k's value. Each clock i repeatedly executes the following three steps:

Step 1: Reads the value of every clock in the system.

Step 2: Discards outliers and substitutes them by the value of the local clock. Thus, if $|c_i[i] - c_j[i]| > \delta$, then $c_j[i] := c_i[i]$.

Step 3: Updates the clock reading using the *average* of these values.

The aforementioned algorithm guarantees that in a system of n processes, the clocks remain synchronized even if there are at most t two-faced clocks, when $n > 3t$. To verify this, consider two distinct nonfaulty clocks i and j reading a third clock k. Two cases are possible:

Case 1: If clock k is nonfaulty, then $c_k[i] = c_k[j]$.

Case 2: If clock k is faulty, then they can produce any reading. However, a faulty clock can make other clocks accept their readings as good values even if they transmit erroneous readings that are at most 3δ apart. Figure 6.7 illustrates such a scenario.

---

* Such a possibility is not a fantasy—it has been observed in real situations.

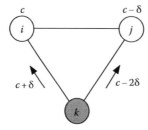

FIGURE 6.7   Two nonfaulty clocks i and j reading the value of a faulty clock k.

The following values constitute a feasible set of readings acceptable to every clock:

- $c_i[i] = c$

- $c_k[i] = c + \delta$

- $c_j[j] = c - \delta$

- $c_k[j] = c - 2\delta$

Now assume that at most t out of the n clocks are faulty, and the remaining clocks are nonfaulty. Then, the maximum difference between the averages computed by any two non-faulty processes is $3t\delta/n$. If $n > 3t$, then $(3t\delta/n) < \delta$, which means that the algorithm keeps $c[i]$ and $c[j]$ within the permissible skew δ and the nonfaulty clocks remain synchronized.

How often do we need to run the aforementioned algorithm? To maintain synchrony, the resynchronization interval R should be small enough so that the cumulative drift does not offset the convergence achieved after each round of synchronization. A sample calculation is shown in the following:

As a result of the synchronization, the maximum difference between two nonfaulty clocks is reduced from δ to $3t\delta/n$. Let $n = 3t + 1$. Then the amount of correction is

$$\delta - \frac{3t\delta}{3t+1} = \frac{\delta}{3t+1}$$

If ρ is the maximum rate at which two clocks drift apart, then it will take a time $\leq \delta/(\rho \cdot (3t + 1))$ before their difference grows to δ again and resynchronization becomes necessary. By definition, this is the upper bound of the resynchronization interval R. Therefore, $R \leq \delta/(\rho \cdot (3t + 1))$. If the resynchronization interval increases, then the system must be designed to tolerate a larger clock skew.

6.4.4 Algorithms for External Synchronization

Cristian's method: In this method, a client obtains the data from a special host (called the *time server*) that contains the reference time obtained from some precise external source. Cristian's algorithm compensates for the clock reading error. The client sends requests to

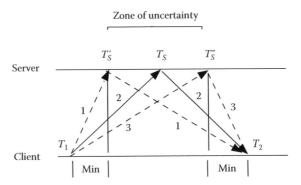

FIGURE 6.8   An illustration of Christian's algorithm for external synchronization: three different possibilities are shown.

the time server every R units of time where $R < \delta/2\rho$ (this follows from the fact that in an interval Δt, two perfectly synchronized clocks can be $2 \cdot \Delta t \cdot \rho$ apart and $2 \cdot \rho \cdot R \leq \delta$), and the server sends a response back to the client with the current time. For an accurate estimate of the current time, the client needs to estimate how long it has been since the time server replied. This is done by assuming that the client's clock is reasonably accurate over short intervals and that the latency of the link is approximately symmetric (the request takes as long to get to the server as the reply takes to get back). Given these assumptions, the client issues an RPC to measure the *round-trip time* ($RTT = T_2 - T_1$; see Figure 6.8) of the request using its local clock and then divide it by half (trajectory 2 of Figure 6.8) to estimate the propagation delay. As a result, if clock i receives the value T_S from the time server, then $c[i]$ corrects itself to $T_S + (RTT/2) = T_S + ((T_2 - T_1)/2)$.

However, this assumption about symmetric delays may not be realistic, and congestion in the network can increase these delays in one or both directions in unpredictable ways, leaving room for uncertainty in the calibration. To estimate this, let min be the minimum transit time in each direction. Two extreme possibilities are shown as cases 1 and 3 in Figure 6.8—for the same values of (T_1, T_2), the server may report any value in the interval $[T_S', T_S'']$. The length of this interval is ($T_2 - T_1 - 2$ min). Accordingly, the accuracy of the client clock is limited to $\pm(((T_2 - T_1)/2) - \text{min})$. The zone of uncertainty can be further reduced using repeated measurements of T_1 and T_2 and using the smallest of the measured intervals, which leads to a tighter estimate of the true server time. Synchronizing the clock from multiple time servers helps improve the accuracy and overcome server failures.

Network Time Protocol: NTP is an elaborate external synchronization mechanism designed to synchronize clocks on the Internet with the UTC. It is not practical to equip every computer with atomic clocks or GPS satellite receivers. Cost is a major factor. So, these computers use the NTP to synchronize the clocks. The time servers are located at different sites on the Internet. NTP architecture is a tiered structure of clocks, whose accuracy decreases as its level (defined by a *stratum number*) increases. In *stratum* 0, there are primary time servers that are devices of the highest precision, like cesium clocks or GPS-based clocks. The stratum 0 clocks are directly connected to the computers in the next level (i.e., *stratum* 1).

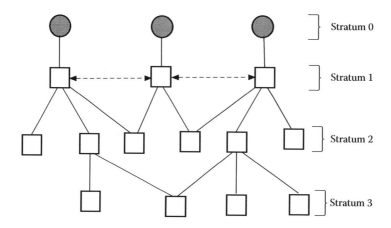

FIGURE 6.9 A network of time servers used in NTP. The top-level devices (stratum 0) have the highest precision.

The stratum 1 computers act as time servers for the computers belonging to the next level (i.e., *stratum* 2). In general, stratum *i* computers act as time servers for the stratum (i + 1) computers. NTP synchronizes clocks on the Internet despite occasional loss of connectivity, failure of some of the time servers, and malicious timing inputs from untrusted sources. Figure 6.9 shows the hierarchy with three stratum 0 nodes.

A computer will try to synchronize its clock with several servers and accept the best results to set its time. Accordingly, the synchronization subnet is dynamic. The quality of a time service depends on several factors, like the stratum of the server, the round-trip delay, and the consistency of the network transit times. In a sense, NTP is a refinement of Cristian's method. NTP provides time service using the following three mechanisms:

Multicasting: The time server periodically multicasts the current time to the client machines. All messages are delivered via UDP. This is the simplest method and perhaps the least accurate. The readings are not compensated for signal delays.

Procedure call: The client processes send requests to the time server, and the server responds by providing the current time. Using Cristian's method, each client can compensate for the propagation delay by using an estimate of the round-trip delay. The resulting accuracy is better than that obtained using multicasting.

P2P communication: NTP allows a time server to synchronize its clock with another time server operating at the same stratum. The P2P mode is used by master servers at lower strata, and it enables them to provide a more accurate time service to the client computers. This leads to the highest accuracy compared to the previous two methods.

As an example of this mode, consider the exchange of a pair of messages between two time servers P and Q as shown Figure 6.10. Server P sends a probe at time T_1, which is received by server Q at its local time T_2. Server Q responds to this query at its local time T_3, which is received by server P at its local time T_4. Define the *offset* between two servers P and Q as the difference between their clock values. Let T_{PQ} and T_{QP} be the message propagation

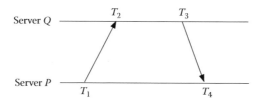

FIGURE 6.10   The exchange of messages between two time servers.

delays from P to Q and Q to P, respectively. Without loss of generality, assume that server Q's time is *ahead of* server P's time by an offset δ. Then

$$T_2 = T_1 + T_{PQ} + \delta \qquad (6.1)$$

$$T_4 = T_3 + T_{QP} - \delta \qquad (6.2)$$

Adding (6.1) and (6.2),

$$T_2 + T_4 = T_1 + T_3 + \left(T_{PQ} + T_{QP}\right)$$

So, the round-trip delay

$$T_{PQ} + T_{QP} = T_2 + T_4 - T_1 - T_3 \qquad (6.3)$$

Subtracting (6.2) from (6.1),

$$2\delta = \left(T_2 - T_4 - T_1 + T_3\right) - \left(T_{PQ} - T_{QP}\right)$$

Therefore, the offset

$$\delta = \frac{T_2 - T_4 - T_1 + T_3}{2} - \frac{T_{PQ} - T_{QP}}{2} \qquad (6.4)$$

Assume that $x = (T_2 - T_4 - T_1 + T_3)/2$ and the round-trip delay $y = T_{PQ} + T_{QP}$. Since $T_{PQ} > 0$ and $T_{QP} > 0$, the value of $(T_{PQ} - T_{QP})$ must lie between $+y$ and $-y$. Therefore, the actual offset δ must lie in the range $[x + (y/2), x - (y/2)]$. Note that T_{PQ} and T_{QP} are not individually measurable, but x and y can be calculated from the values of (T_1, T_2, T_3, T_4). Therefore, if each server bounces messages back and forth with another server and computes several pairs of (x, y), then a good approximation of the real offset δ can be obtained from that pair in which the round-trip delay y is the smallest, since that will minimize the dispersion in the window $[x + (y/2), x - (y/2)]$.

The *multicast mode* of communication is considered adequate for a large number of applications. When the accuracy from the multicast mode becomes inadequate, the *procedure call* mode is used. An example is a file server on a LAN that wants to keep track of when a file was created (by communicating with a time server at a higher level, i.e., a lower-level number). Finally, the P2P mode is used only with the higher-level time servers (stratum 1) for achieving the best possible accuracy. The synchronization subnet reconfigures itself when some servers fail or become unreachable. NTP can synchronize clocks within an accuracy of 1–50 ms.

6.5 CONCLUDING REMARKS

In asynchronous distributed systems, absolute physical time is not important, but the temporal order of events is significant for some applications. As an example, in replicated servers, each server is a state machine whose state is modified by the inputs from its clients. In order that all replicas *always* remain in the same state (so that one can seamlessly switch to a different server if one crashes), all replicas must receive the inputs from clients in the same order.

The performance of a clock synchronization algorithm is determined by how close two distinct clock times can be brought, the time of convergence, and the nature of failures tolerated by such algorithms. The adjustment of clock values may have interesting side effects. For example, if a clock is advanced from *171* to *174* during an adjustment, then the time instants *172* and *173* are lost. This will affect potential events scheduled at these times. On the other hand, if the clock is turned back from *171* to *169* during adjustment, then the time instants *169* through *171* appear twice. This causes the anomaly that an event at time *170 happens before* another event at time *169*! A simple fix for such problems is to appropriately speed up or slow down the clock for an appropriate number of ticks (until one catches up with the other) without violating the clock monotonicity property, instead of abruptly turning the clock forward or backward.

6.6 BIBLIOGRAPHIC NOTES

Lamport [L78] introduced logical clocks. In the year 2000, the distributed systems community adjudged this paper as the most influential paper in the field of distributed systems (later renamed as *Dijkstra Prize* after Edsger W. Dijkstra since 2003). In [M88], Mattern [M88] and Fidge [F88] independently proposed vector clocks. Gusella and Zatti [GZ89] developed the Berkeley algorithm for internal synchronization in 1989. The averaging algorithm for physical clock synchronization is the first of the three algorithms proposed by Lamport and Melliar-Smith in [LM85]. Cristian's algorithm is based on his work [C89]. David Mills designed the NTP, and a good introduction can be found in [M91].

EXERCISES

6.1 In a system of clocks, the maximum clock drift is 1 in 10^6. (a) What can be the maximum difference between the readings of the two clocks 24 h after they have been synchronized? (b) What should be the resynchronization interval, so that the skew does not exceed 20 ms?

6.2 Figure 6.11 shows the communication between a pair of processes P and Q. Calculate the *logical clock values* of the events $a - j$.

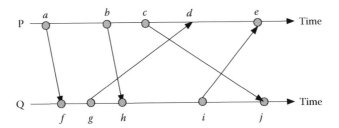

FIGURE 6.11   A sample communication between processes P and Q.

6.3 Calculate the *vector clock values* of the events $a - j$ in Figure 6.11. Use these vector clock values to prove that (d, h) are concurrent events, but f is causally ordered before e.

6.4 a, b, c are three events in a distributed system, and no two events belong to the same process. Using Lamport's definition of sequential and concurrent events, comment on the truth of the following statements:

a. $(a\|b)\wedge(b < c)\Rightarrow(a < c)$

b. $(a\|b)\wedge(b\|c)\Rightarrow(a\|c)$

(Here, $a\|b$ denotes that a, b are concurrent events.)

6.5 Vector clocks are convenient for identifying concurrent as well as causally ordered events. However, scalability is a problem, since the size of the clock grows linearly with the number of processes n. Is it possible to detect causality (or concurrency) using vector clocks of size smaller than n? Justify your answer.

6.6 Lamport and Melliar–Smith's algorithm for the internal synchronization of physical clocks safeguards against two-faced clocks. What kind of failures or problems can cause clocks to behave in such a strange manner? List the possible causes.

6.7 The averaging algorithm proposed by Lamport et al. works for a completely connected network of clocks. Will such an averaging algorithm for clock synchronization work on a *cycle* of n clocks, of which m can exhibit two-faced behavior and $n > 3m$? Assume that each link in the cycle allows bidirectional communication.

6.8 A limitation of time stamps is their unbounded size, since finite resources are inadequate to store or process them. The goal of this exercise is to explore if bounded-size time stamps can be used in specific solutions. Explore this possibility in the following scenario: Two processes (0, 1) compete with each other to acquire a shared resource that can be used by one process at a time. The life of the processes is as follows:

```
Program for process i ∈ {0,1}
do true →
       Request for a resource;
       Acquire and use the resource;
       Release the resource
od
```

To request the resource, a process sends a *time stamp request* to the other process, which grants the request only if (1) it is not interested in the resource at that moment or (2) its own time stamp for resource request is *larger* than the time stamp of the incoming request. In all other situations, the grant is deferred. After receiving the grant, a process acquires the resource. Once a process acquires a resource, it guarantees to release the resource within a finite amount of time. Thereafter, in a finite time, the resource is released.

Can you solve the problem using time stamps of bounded size? Explain your answer.

(Hint: First find out what is the maximum difference between the time stamps of the two processes if the time stamps are unbounded. The time stamp of bounded size must resolve the order of requests without any ambiguity.)

6.9 (Sequential time stamp assignment) There are n processes that are initially passive. At any time, a process may want to be active and execute an action, for which it has to acquire a time stamp that is *larger* than the time stamp of the remaining $n - 1$ processes. An allocator process will allocate a time stamp and can service one request for a new time stamp at any time.

The goal is to use time stamps of *bounded size*. The scenario can be viewed as a game between an adversary and the allocator: the adversary will identify a process that will request for a time stamp, and the allocator has to assign the time stamp to that process. For example, if $n = 2$, then a solution exists with time stamps 0, 1, 2, where $ts(i) = ts(j) + 1$ mod 3 implies that the time stamp (ts) of process i is *larger* than the time stamp of process j. Verify if this is true.

Can you find an algorithm for assigning bounded time stamps for arbitrary values of n?

6.10 An anonymous network is one in which processes do not have identifiers. Can you define a total order among events in an anonymous network of processes?

6.11 Five processes 0, 1, 2, 3, 4 in a completely connected network decide to maintain a *distributed bulletin board*. No central version of it physically exists, but every process maintains an image of it. To post a new bulletin, each process multicasts every message to the other four processes, and recipient processes willing to respond to an incoming message multicast their responses in a similar manner. To make any sense from a response, every process must *accept* every message and response in *causal order*, so a process receiving a message will postpone its *acceptance* unless it is confident that no other message causally ordered before this one will arrive in future.

To detect causality, the implementation uses *vector clocks*. Each message or response is tagged with an appropriate vector time stamp. Figure out (a) a rule for assigning these vector time stamps and (b) the corresponding algorithm using which a process will decide whether to accept a message immediately or postpone its acceptance.

(Hint: You may modify the scheme for assigning the vector time stamps.)

6.12 Describe an application in which the lack of synchronization among physical clocks can lead to a security breach.

6.13 In a network of n processes ($n > 2$), all channels are FIFO and of infinite capacity. Every process is required to accept messages from the other processes in *strictly increasing order* of time stamps. You can assume that (1) processes send messages infinitely often and (2) no message is lost in transit. Suggest an implementation to make it possible.

(Hint: Consider using *null* messages through a channel to signal the absence of a message from a sender.)

III

Important Paradigms

CHAPTER 7

Mutual Exclusion

7.1 INTRODUCTION

Mutual exclusion is a fundamental problem in concurrent programming and has been extensively studied under different contexts. Imagine that n users ($n > 1$) want to print data on a shared printer infinitely often. Since at most one user can print at any time, there should be a protocol for the fair sharing of that printer. As another example, consider a network of processes, where each process has a copy of a shared file F. To be consistent, all copies of F must be identical, regardless of how individual processes perform their read or write operations. Simultaneous updates of the local copies will violate the consistency of F. A simple way to achieve this is to give each process *exclusive write access* to its local copy of F during write operations and propagate all updates to the various local copies of F with the other processes, before any other process starts accessing its local copy. This shows the importance of studying the *mutual exclusion* problem. The problem can be generalized to the access of any shared resource on a network of processes. In multiprocessor cache coherence, at most one process has the right to update a shared variable. A well-known implementation of mutual exclusion is found in the CSMA/CD protocol used to resolve bus contention in Ethernets.

Most of the classical solutions to the mutual exclusion problem have been studied for shared-memory systems with read–write atomicity. In this chapter, we will examine both shared-memory and message-passing solutions. We begin with message-passing solutions.

7.2 SOLUTIONS ON MESSAGE-PASSING SYSTEMS

In the message-passing model of a distributed system, the mutual exclusion problem can be formulated as follows: Consider $n(n > 1)$ processes, numbered $0.. \ n - 1$ forming a distributed system. The topology is a completely connected graph, so that every process can directly communicate with every other process in the system. Each process periodically wants to enter a CS, executes the CS codes, and eventually exits the CS to do the rest of its work. The problem is to devise a protocol that satisfies the following three conditions:

ME1: [Mutual exclusion] At most, one process can remain in its CS at any time. This is a safety property.

ME2: [Freedom from deadlock] In every configuration, at least one process must be eligible to take an action and enter its CS. This is also a safety property.

ME3: [Progress] Every process trying to enter its CS must eventually succeed. This is a liveness property.

The violation of ME3 is known as *livelock* or *starvation*. In such a case, one or more processes may be prevented from entering their CSs for an indefinite period by other processes.

A measure of *fairness* is the criterion of *bounded waiting*. Let process i try to enter its CS. Then, the bounded waiting requirement specifies an upper bound on the number of times other contending processes may enter their CSs between two consecutive CS entries by process i. Most message-passing solutions implement *FIFO fairness*, where processes are admitted to their CS in the ascending order of their request time stamps. It is customary to assume that every process entering its CS eventually exits the CS—thus, process failure or deadlock within the CS is totally ruled out.

Many practical solutions to this problem rely on the existence of a *central coordinator* that acts as a manager of the CSs. This coordinator can be an extra process or one of the n processes in the system that has been assigned additional responsibilities. Any process trying to enter its CS sends a *request* to the coordinator and waits for the *ack* message from the coordinator, which is an approval for entering the CS. Similarly, any process willing to exit its CS sends out a *release* message. The coordinator monitors the status of the processes and decides when to send the ack to a certain process.

While such a solution is quite intuitive and criteria ME1, ME2, and ME3 can be easily satisfied, it is neither easy nor obvious how to implement FIFO fairness. To realize this, consider that process i sends a request x for entry into its CS and then sends a message m to process j. Process j, after receiving m, sends a request y for entry into its CS. Therefore $x \prec y$. However, even if the channels are FIFO, request x may not reach the coordinator before request y. Furthermore, if y reaches the coordinator first, then there is no way that the coordinator can anticipate the arrival of another request x with a lower time stamp.

In this chapter, we disregard centralized solutions using coordinators and present only decentralized algorithms, where every process has equal responsibility in the implementation of mutual exclusion.

7.2.1 Lamport's Solution

The first published solution to this problem is due to Lamport. It works on a completely connected network and assumes that interprocess communication channels are FIFO.

Each process maintains its own private request-queue Q. The algorithm is described by the following five rules:

LA1: To request entry into its CS, a process sends a time-stamped *request* to every other process in the system and also enters the request in its local Q.

LA2: When a process receives a *request*, it places it in its Q. If the process is not in its CS, then it sends a time-stamped *ack* to the sender. Otherwise, it defers the sending of the ack until its exit from the CS.

LA3: A process enters its CS, when (1) its request is ordered ahead of all other requests (i.e., the time stamp of its own request is *less than* the time stamps of all other requests) in its local Q and (2) it has received the *acks* from every other process in response to its current request.

LA4: To exit from the CS, a process (1) deletes the request from its local queue and (2) sends a time-stamped *release* message to all the other processes.

LA5: When a process receives a *release* message, it removes the corresponding request from its local queue.

Correctness proof: To prove correctness, we need to show that the program satisfies properties ME1–ME3.

Proof of ME1 (by contradiction): Let two different processes i and j enter their CSs at the same time. Since both processes received all the acks, both i and j must have received each other's requests and entered them in their local queues. If process i enters its CS, then $Q.i.ts < Q.j.ts$, and if j enters its CS, then $Q.i.ts > Q.j.ts$. Both of these cannot be true at the same time. Therefore, both i and j cannot enter their CSs at the same time. ■

Proofs of ME2 and ME3 (by induction): Since every request sent out by a process i is acknowledged and no message is lost, every process eventually receives $(n-1)$ ack signals. Our proof is based on the number of processes that are *ahead of* process i in its request queue.

Basis: When process i makes a request, there may be at most $(n-1)$ processes ahead of process i in its request queue.

Inductive step: Assume that there are K $(1 \leq K \leq n-1)$ processes ahead of process i in its request queue. In a finite time, process j with the lowest time stamp (1) enters CS, (2) exits CS, and (3) sends out the release message to every other process including i. Process i then deletes the entry for process j, and the number of processes ahead of process i is reduced from K to $K-1$. Subsequent requests from process j must be placed behind that of process i, since the time stamp of such requests is greater than that of i. It thus follows that in a bounded number of steps, the number of processes ahead of process i will be reduced to 0, and process i will enter its CS. ■

Proof of FIFO fairness (by contradiction): Let the time stamp of the request from process i be less than the time stamp of the request from process j. Assume that process j enters its CS before process i does so. This implies that when process j enters its CS, it has not received the request from process i. However, according to *LA3*, it must have received the ack of its own request from process i. By assumption, the channels are FIFO—so *ack of j's request from i* must be ahead of the *request from i*. However, no request can be acknowledged before it is received. So, *request from j* ≺ *ack of j's request from i*. It thus follows that *request from j* ≺ *request from i*, which contradicts our earlier assumption. Therefore, process i must enter its CS before process j. ■

Observe that when all requests are acknowledged, the request queue of every process is identical—so the decision to enter CS is based on local information that is globally consistent. Also, broadcasts are atomic—so all request messages bear the same time stamp and are transmitted without interruption. It leads to unnecessary complications if this view is overlooked, and each request message from one process to another is assumed to have different (i.e., progressively increasing) time stamps, or the broadcast is interrupted prematurely.

The message complexity is the number of messages required to complete one round trip (i.e., both entry and exit) to the CS. Each process sends $(n - 1)$ request messages and receives $(n - 1)$ acks to enter its CS. Furthermore, $(n - 1)$ release messages are sent as part of the exit protocol. Thus, the total number of messages required to complete one round trip to the CS is $3(n - 1)$.

7.2.2 Ricart–Agrawala's Solution

Ricart and Agrawala proposed an improvement over Lamport's solution to the distributed mutual exclusion problem. Unlike Lamport's algorithm, processes do not maintain local queues—instead, each process counts the number of acks that it receives from other processes, which determines whether it will enter its CS. Four rules form the basis of this algorithm:

RA1: Each process seeking entry into its CS sends a time-stamped request to every other process in the system.

RA2: A process receiving a request sends an ack back to the sender, only when (1) the process is not interested in entering its CS or (2) the process is trying to enter its CS, but its time stamp is larger than that of the sender. If the process is already in its CS or its timestamp is smaller than that of the sender, then it will buffer all requests until its exit from CS.

RA3: A process enters its CS, when it receives an ack from each of the remaining $(n - 1)$ processes.

RA4: Upon exit from its CS, a process must send ack to each of the pending requests before making a new request or executing other actions.

Intuitively, a process is allowed to enter its CS, only when it receives a go-ahead signal from every other process, since either these processes are not competing for entry into the CS or their requests bear a higher time stamp.

Proof of ME1: Two processes i and j can enter their CSs at the same time, only if both receive $(n − 1)$ acks. However, per RA2, both i and j cannot send acks to each other, so both cannot receive $(n − 1)$ acks and thus cannot be in their respective CS at the same time. ■

Proof of ME2 and ME3: Draw a directed graph G with the processes as nodes. Add an edge from node i to node j if j is already in its CS or j's request has a lower time stamp compared to i's request. Clearly, G is an acyclic graph. A process i trying to enter its CS is kept waiting, if there exists at least one other process j such that an edge $i → j$ exists and j will not send an ack to i until it completes its CS. Per RA3, the process represented by the node with out-degree 0 will receive all acks and enter its CS (unless it is already in its CS). Upon exit from its CS, it will send an ack to each of the waiting processes, which will decrement their out-degrees by 1 (RA4). Therefore, the waiting processes will enter their CSs in the order defined by their out-degrees. Deadlock is ruled out since G is acyclic, and every waiting process will enter its CS in a bounded number of steps. ■

Proof of FIFO fairness: It follows from the proof of ME2 and ME3. If the time stamp of the request from i < the time stamp of the request from j, then i has a smaller out-degree than j in the directed acyclic graph G. Therefore, i will enter its CS before j does. This proves progress as well as FIFO fairness. ■

Unlike Lamport's algorithm that explicitly creates consistent explicit local queues, Ricart and Agrawala's algorithm implicitly creates an acyclic *wait-for* chain of processes $i → j → k → \cdots$ where each process waits for the other processes ahead of it to send an ack. Note that this algorithm does not require the channels to be FIFO.

To compute the message complexity, note that each process sends $(n − 1)$ requests and receives $(n − 1)$ acks to complete one trip to its CS. Therefore, the total number of messages sent by a process to complete one trip into its CS is $2(n − 1)$. This is less than what was needed in Lamport's algorithm.

7.2.3 Maekawa's Solution

In 1985, Maekawa extended Ricart and Agrawala's algorithm and suggested the first solution to the n-process distributed mutual exclusion problem with a message complexity lower than $O(n)$ per process. The underlying principle is based on the theory of finite projective planes. As a clear deviation from the strategies adopted in the previous two algorithms, here, a process i is required to send request messages only to a subset S_i of the processes in the system, and the receipt of acks from each of the processes in this subset S_i is sufficient to allow that process to enter its CS.

Maekawa divides the processes into a number of subsets of identical size K. Each process i is associated with a unique subset S_i. The subsets satisfy the following three properties:

1. $\forall i, j : 0 \le i \le n − 1 :: S_i \cap S_j \ne \varnothing$
   Whenever a pair of processes i and j wants to enter their respective CSs, a process in $S_i \cap S_j$ takes up the role of an arbitrator and chooses only one of them by sending ack and defers the other one.

2. $i \in S_i$. It is only natural that a process gets an ack from itself for entering the CS. Note that this does not cost a message.

3. Every process i is present in the same number (D) of subsets. The fact that every node acts as an arbitrator for the same number of processes adds to the symmetry of the system.

$$
\begin{aligned}
S_0 &= \{0, 1, 2\} \\
S_1 &= \{1, 3, 5\} \\
S_2 &= \{2, 4, 5\} \\
S_3 &= \{0, 3, 4\} \\
S_4 &= \{1, 4, 6\} \\
S_5 &= \{0, 5, 6\} \\
S_6 &= \{2, 3, 6\}
\end{aligned}
$$

As an example, consider the table earlier showing the partition for seven processes numbered 0..6. Here, each set has a size $K = 3$, and each process is included in $D = 3$ subsets. The relationship $D = K$ is not essential, but we will soon find out that it helps in reducing the message complexity of the system. In the following, we present the first version of Maekawa's algorithm (call it *maekawa1*) that uses five basic rules:

MA1: To enter its CS, a process i first sends a time-stamped *request* message to every process in S_i.

MA2: A process receiving requests sends an *ack* to that process whose request has the lowest time stamp if it is outside its CS. It *locks* itself to that process and keeps all other requests waiting in a request queue. If the receiving process is inside its CS, then it defers this action until it leaves its CS.

MA3: A process i enters its CS when it receives acks from every member of S_i.

MA4: During exit from its CS, a process sends *release* messages to all members of S_i.

MA5: Upon receiving a release message from process i, a process *unlocks* itself, deletes the current request, and sends an ack to the process whose request has the lowest time stamp.

Proof of ME1 (by contradiction): Assume that the statement is false and two processes i and j ($i \neq j$) enter their CSs at the same time. For this to happen, every member of S_i must have received the request from i, and every member of S_j must have received the request from j. Since $S_i \cap S_j \neq \emptyset$, there is a process $k \in S_i \cap S_j$ that received requests from both i and j. Per MA2, process k will send ack to *only one* of them and refrain from sending the ack to the other, until the first process has sent a release signal. Since a process needs an ack from every member of its subset, both i and j cannot enter their CSs at the same time. ◼

The proposed algorithm *maekawa1*, however, does not satisfy the safety property ME2, since there is a potential for deadlock. The source of the problem is that no process is clair-voyant—so when a process receives a request, it does not know whether another request with a lower time stamp is on its way. Here is an example. Assume that processes 0, 1, and 2 have sent request to the members of S_0, S_1, and S_2, respectively. The following scenario is possible:

- From $S_0 = \{0, 1, 2\}$, processes 0 and 2 send *ack* to 0, but process 1 sends *ack* to 1.

- From $S_1 = \{1, 3, 5\}$, processes 1 and 3 send *ack* to 1, but process 5 sends *ack* to 2.

- Prom $S_2 = \{2, 4, 5\}$, processes 4 and 5 send *ack* to 2, but process 2 sends *ack* to 0.

Thus, 0 waits for* 1, 1 waits for 2, and 2 waits for 0. This circular waiting causes deadlock.

There are two possible ways of avoiding a deadlock. The first is to assume a system-wide order of message propagation, as explained in the following:

(Global FIFO) Every process receives incoming messages in strictly increasing order of time stamps.

We now demonstrate that if the global FIFO property holds, then algorithm *maekawa1* is deadlock-free and satisfies the FIFO fairness requirement.

Proof of ME2 and ME3 (by induction)
Basis: Consider a process i whose request bears the *lowest* time stamp, and assume that no process is currently in its CS. Because of the global FIFO assumption, every process in S_i must receive the request from process i before any other request with a higher time stamp. Therefore, process i will receive ack from every process in S_i and will enter its CS.

Induction step: Assume that a process m is already in its CS and the request from process j has the *lowest time stamp* among all the waiting processes. We show that process j eventually receives acks from every process in S_j and proceeds to its CS.

Because of the global FIFO assumption, request from process j is received by every process in S_j before any other request with a higher time stamp and ordered ahead of every other process seeking entry into the CS. Therefore, when process m exits the CS and sends release signals to every process in S_m, every process in $S_m \cap S_j$ sends ack to process j. In addition, every process in $(\neg S_m \cap S_j)$ has already sent acks to process j since the request from process j is at the head of the local queues of these processes. Therefore, process j eventually receives acks from every process in S_j.

This shows that after process m exits the CS, process j whose request bears the *next higher time stamp* eventually enters the CS. It therefore follows that every process sending a request eventually enters the CS.

The global FIFO property however neither holds in practice nor is easy to implement. Therefore, Maekawa presented a modified version of his first algorithm (call it *maekawa2*)

---

* Here "i waits for j" means that process i waits for process j to send a *release* message.

that does not deadlock. This version uses three additional signals: *failed, inquire,* and *relinquish.* The outline of the modified version is as follows:

MA1': To enter its CS, a process *i* sends a time-stamped request message to every process in S_i {same as MA1}.

MA2': A process receiving a request can take one of the following three steps when it is outside its CS:

- If lock has not yet been set, then the process sends an ack to the requesting process with the lowest time stamp and sets its own *lock* to the id of that process.

- If *lock* is already set and the time stamp of the incoming request is higher than the time stamp of the locked request, then the incoming request is enqueued and a *failed* message is sent to the sender.

- If *lock* is already set, but the time stamp of the incoming request is lower than that of the locked request, then the incoming request is queued, and an *inquire* message is sent to the sender of the locking request. This message is meant to check the current status of the process whose request set the lock.

MA3': When a requesting process *i* receives acks from *every member* of S_i, it enters its CS. However, if it receives an *inquire* message, it checks whether it can go ahead. If it has already received or subsequently receives at least one *failed* message, it knows that it cannot go ahead—so it sends a *relinquish* message back to the members of S_i, indicating that it wants to *give up*. Otherwise, it ignores the *inquire* message.

MA4': During exit from CS, process *i* sends *release* messages to each member of S_i (same as MA4).

MA5': Upon receiving a *release* message, a process deletes the currently locked request from its queue. If the queue is not empty, then it also sends an ack to the process with the lowest time stamp; otherwise, it resets the lock (same as MA5).

MA6': Upon receiving a *relinquish* message, the process sends ack to the waiting process with the lowest time stamp (but does not delete the current request from its queue) and sets its lock appropriately.

The same partial correctness proof is applicable to the modified algorithm also. Here is an informal argument about the absence of deadlock.

Assume that there is a deadlock involving k processes 0, 1, 2, ..., $k - 1$. Without loss of generality, we assume that process i waits for process $(i + 1)$ mod k. Let the request from process j bear the *lowest* time stamp. This means that process j cannot receive a *failed* message from any process in S_j. Since process j waits for process $(j + 1)$ mod k, there must exist a process $m \in (S_i \cap S_{j+1})$ that has set its lock to $(j + 1)$ mod k instead of j. When process m later receives the request from process j, it sends an *inquire* message to process $(j + 1)$ mod k, which eventually replies with either a *relinquish* or a *release* message. In either case, every

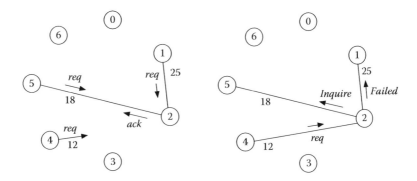

FIGURE 7.1 An example showing the first two phases of Maekawa's modified algorithm. Process 2 sends an *ack* to 5 and a *failed* message to 1. But when 2 later receives a request with a time stamp 12, it sends an *inquire* message to 5 to find out if it indeed entered its CS.

process belonging to $(S_j \cap S_{j+1})$ eventually locks itself to process j and sends ack to process j. As a result, the condition "process j waits for $(j + 1)$ mod k" ceases to hold, and the deadlock is avoided (Figure 7.1).* ■

Message complexity: Let K be the cardinality of each subset S_i. In the first algorithm *maekawa1*, each process (1) sends K request messages, (2) receives K acks, and (3) sends K release messages. When $D = K$, the relationship between n and K is $n = K(K - 1) + 1$. A good approximation is $K = \sqrt{n}$, which leads to the message complexity of $3 = \sqrt{n}$.

A more precise computation of worst-case complexity for the modified version of the algorithm can be found in Maekawa's original paper [M85]. The complexity is still $O = (\sqrt{n})$, but the constant of proportionality is larger.

7.3 TOKEN-PASSING ALGORITHMS

Another class of distributed mutual exclusion algorithms uses the concept of an explicit variable *token* that acts as a permit for entry into the CS and can be passed around the system from one requesting process to another. Whichever process wants to enter its CS must acquire the token. The first known algorithm belonging to this class is due to Suzuki and Kasami.

7.3.1 Suzuki–Kasami Algorithm

This algorithm is defined for a completely connected network of processes. It assumes that initially an arbitrary process has the token. A process i that does not have the token but wants to enter its CS broadcasts a request (i, num), where *num* is sequence number of that request. The algorithm guarantees that eventually process i receives the token.

Every process i maintains an array $req[0.. n - 1]$ of integers, where $req[j]$ designates the sequence number of the *latest* request received from process j. Note that although every process receives a request, only one process (which currently has the token) can grant the token.

---

* This informal argument echoes the reasoning in the original paper. Sanders [S87] claimed that the modified algorithm is still prone to deadlock.

As a result, some pending requests become stale or outdated. An important issue in this algorithm is to identify and discard these stale requests. To accomplish this, each process uses the following two additional data structures that are passed on with the token by its current holder:

- An array $last[0..n-1]$ of integers, where $last[k] = r$ implies that during its last visit to its CS, process k has completed its rth trip

- A queue Q containing the identifiers of processes with pending requests

When a process i receives a request with a sequence number num from process k, it updates $req[k]$ to $\max(req[k], num)$, so that $req[k]$ now represents the most recent request from process k. A process holding the token must guarantee (before passing it to another process) that its Q contains the most recent requests. To satisfy this requirement, when a process i receives a token from another process, it executes the following steps:

- It copies its num into $last[i]$.

- For each process k, process i retains process k's name in its local queue Q only if $1 + last[k] = req[k]$ (this establishes that the request from process k is a recent one).

- Process i completes the execution of its CS codes.

- If Q is nonempty, then it forwards the token to the process at the head of Q after deleting its entry.

To enter the CS, a process sends $(n-1)$ requests and receives one message containing the token. The total number of messages required to complete one visit to its CS is thus $(n-1) + 1 = n$. Readers are referred to [SK85] for a proof of this algorithm.

7.3.2 Raymond's Algorithm

Raymond suggested an improved version of a token-based mutual exclusion algorithm that works on a network with a *tree topology*. At any moment, one node holds the token, and it serves as a root of the tree. Every edge is assigned a direction, so that by following these directed edges, a request can be sent to that root. If there is a directed edge from i to j, then j is called the *holder* of i. As the token moves from one process to another, the root changes, and so do the directions of the edges (Figure 7.2).

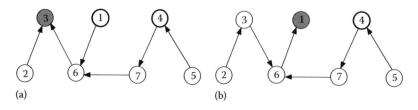

(a)   (b)

FIGURE 7.2   Two configurations in Raymond's algorithm: (a) process 3 holds the token and 1 and 4 make requests in that order; (b) the token is transferred to 1.

In addition to the variable *holder*, each node has a local queue Q to store the pending requests. Only the first request is forwarded to the holder. An outline of Raymond's algorithm is presented in the following:

R1: If a node j has the token, then it enters its CS. Otherwise, to enter its CS, a node enters the request in its local Q.

R2: When a node j (which is not holding the token) has a nonempty request Q, it sends a request to its holder, unless j has already done so and is waiting for the token.

R3: When the *root* receives a request, it sends the token to the neighbor at the head of its local Q after it has completed its own CS. Then, it sets its holder variable to that neighbor.

R4: Upon receiving a token, a node j forwards it to the neighbor at the head of its local Q, deletes the request from Q, and sets its holder variable to that neighbor. If there are pending requests in Q, then j sends another request to its holder.

Since there is a single token in the system, the proof of safety (ME1) is trivial. Deadlock is impossible (ME2) because the underlying directed graph is acyclic: a process i waits for another process j only if there is a directed path from i to j, which implies that j does not wait for i. Finally, fairness follows from the fact that the queues are serviced in the order of arrival of the requests, and a new request from a process that acquired the token in the recent past is enqueued behind the remaining pending requests in its local queue.

For a detailed proof of this algorithm, see [R89]. Since the average distance between pairs of nodes in a randomly generated tree is $O(\log n)$, the message complexity of Raymond's algorithm is claimed to be $O(\log n)$.

7.4 SOLUTIONS ON THE SHARED-MEMORY MODEL

Historically, the bulk of the work in the area of mutual exclusion has been done on shared-memory models. The Dutch mathematician Dekker was the first to propose a solution to the mutual exclusion problem using atomic read and write operations on a shared memory. The requirements of a correct solution on the shared-memory model are similar to those in the message-passing model, except that fairness is specified as *freedom from livelock* or *freedom from starvation*: no process can be indefinitely prevented from entering its CS by other processes in the system. This fits the definition of weak fairness: if a process remains interested in entering into its CS, then it must eventually be able to do so.

Of the many algorithms available for solving the mutual exclusion problem, we will only describe Peterson's algorithm.

7.4.1 Peterson's Algorithm

Gary Peterson's solution is considered to be the simplest of all the solutions to the mutual exclusion problem using atomic read and write operations. We first present his two-process solution here. There are two processes: 0 and 1. Each process *i* has a Boolean variable *flag*[*i*] that can be read by any process, but written by *i* only. To enter its CS, each process *i* sets *flag*[*i*] to true. To motivate the readers, we first discuss a naive approach in which each process, after setting its own *flag* to true, checks if the *flag* of the *other process* has also been set to true and jumps to a conclusion as summarized in the following:

```
program naïve;
define  flag[0], flag[1]: shared Boolean;
initially both are false;

{process 0 repeats the following forever}
flag[0]:=true;
do flag[1] → skip od;
critical section;
flag[0] := false

{process 1 repeats the following forever}
flag[1] := true
do flag[0] → skip od;
critical section;
flag[1] := false
```

The solution guarantees safety. However, it is not deadlock-free. If both processes complete their first steps in succession, then there will be a deadlock in their second steps.

To remedy this, Peterson's solution uses a shared integer variable *turn* that can be read and written by both processes. Since the writing operations on shared variables are atomic, in the case of a contention (indicated by *flag*[0] = *flag*[1] = *true*), both processes update *turn*, and the *last* write prevails. This delays the CS entry of the process that updated *turn* last. The program is as follows:

```
program    peterson;
define     flag[0], flag[1] shared Boolean;
           turn: shared integer
initially  flag[0] = false, flag[1] = false, turn = 0 or 1

       {program for process 0}
do     true→
1:     flag[0] := true;
2:     turn := 0;
3:     do (flag[1] ∧ turn = 0) → skip od
4:     critical section;
5:     flag[0] := false;
6:     non-critical section codes
od
```

```
     {program for process 1}
do   true →
7:   flag[1] := true;
8:   turn := 1;
9:   do (flag[0] ∧ turn = 1) → skip od;
10:  critical section;
11:  flag[1] := false;
12:  non-critical section codes
od
```

Note that the conditions in step 3 and step 9 need not be evaluated atomically. As a consequence, when process 0 has checked *flag*[1] to be *true* and is reading the value of *turn*, *flag*[1] may be changed to *false* by process 1 (step 11). It is also possible that process 0 has checked *flag*[1] to be *false* and is entering its CS, but by that time, process 1 has changed *flag*[1] to *true* (step 7). Despite this, the solution satisfies all the correctness criteria introduced earlier.

Proof of the absence of deadlock (ME2) (by contradiction): Process 0 can potentially wait in step 3, and process 1 can potentially wait in step 9. We need to show that they *both* cannot wait for each other. Suppose they both wait. Then the condition (*flag*[0] ∧ *turn* = 1) ∧ (*flag*[1] ∧ *turn* = 0) must be true. However, (*turn* = 1 ∧ *turn* = 0) = *false*. Therefore, deadlock is impossible. ■

Proof of safety (ME1): Without loss of generality, assume that process 0 is in its CS (step 4). This must have been possible because in step 3, either *flag*[1] was *false* or *turn* = 1 or both of these were true. The issue here is to demonstrate that process 1 cannot enter its CS.

To enter its CS, process 1 must read *flag*[0] as *false* or *turn* as 0. Since process 0 is already in its CS, *flag*[0] is *true*, so the value of *turn* has to be 0. Is this feasible?

Case 1: Process 0 reads *flag*[1] = *false* in step 3:

⇒ Process 1 has not executed step 7.

⇒ Process 1 eventually sets *turn* to 1 (step 8).

⇒ Process 1 checks *turn* (step 9) and finds *turn* =1.

⇒ Process 1 waits in step 9 and cannot enter its CS.

Case 2: Process 0 reads *turn* = 1 in step 3:

⇒ Process 1 executed step 8 after process 0 executed step 2.

⇒ In step 9, process 1 reads *flag*[0] = *true* and *turn* =1.

⇒ Process 1 waits in step 9 and cannot enter its CS. ■

Proof of progress (ME3): We need to show that once a process sets its *flag* to *true*, it eventually enters its CS. Without loss of generality, assume that process 0 has set *flag*[0] to true but is waiting in step 3 since it found the condition (*flag*[1] ∧ *turn* = 0) to be *true*. If process 1 is in its CS, then eventually in step 11, it sets *flag*[1] to *false* and gives an opportunity to process 1 to enter its CS. If process 0 notices this change, then it enters its CS. If process 0 does not utilize this opportunity, then subsequently process 1 will set *flag*[1] to true (step 6) again for its next attempt for entry to its CS. Eventually it sets *turn* = 1 (step 8). This stops process 1 from making any further progress and allows process 0 to enter its CS. ■

Peterson generalized his two-process algorithm to a system of n processes ($n > 1$) as follows: the program runs for $n - 1$ rounds—in each round, an instance of the two-process algorithm is used to prevent at least one process from advancing to the next round, and the winner after $n - 1$ rounds enters the CS. Like the two-process solution, the n-process solution satisfies all the required properties. The program is as follows:

```
program    Peterson n-process;
define     flag, turn: array [0.. n - 1] of shared integer;
initially  ∀k:flag[k] = 0, and turn = 0

        {program for process i}
do      true →
1:      j:=1;
2:      do j ≠ n - 1
3:          flag[i] := j;
4:          turn[j] := i;
5:          do (∃k ≠ i: flag[k] ≥ j ∧ turn[j] = i) → skip od;
6:          j := j + 1;
7:      od;
8:      critical section;
9:      flag[i] := false;
10:     non-critical section codes
od
```

7.5 MUTUAL EXCLUSION USING SPECIAL INSTRUCTIONS

Since it is not easy to implement shared-memory solutions to the mutual exclusion problem using read–write atomicity only, many processors include special instructions (with larger grains of atomicity) to facilitate such implementations. We will discuss two such implementations here:

7.5.1 Solution Using Test-and-Set

Let x be a shared variable and r be a local or private variable of a process. Then the instruction $TS(r, x)$ is an *atomic operation* defined as ($r := x; x := 1$). Instructions of this type are known as RMW instructions, as they package three operations: *read, modify,* and *write* as one indivisible unit. Using *TS*, the mutual exclusion problem can be solved for *n* processes ($n > 1$) as follows:

```
program    Test-and-set (for any process);
define     x: shared integer;
           r: integer (private);
initially  x = 0, r = 1;

do true →
       do r ≠ 0 → TS(r, x) od;
       critical section;
       x := 0
od
```

Due to the atomicity property, all TS operations are serialized, the first process that executes the TS instruction enters its CS. Note that the solution is deadlock-free and safe, but does not guarantee fairness since a process may be prevented from entering its CS for an indefinitely long period by other processes.

7.5.2 Solution Using Load-Linked and Store-Conditional

In many multiprocessors, atomic RMW instructions are difficult to implement. An alternative is to use *load-linked* (LL) and *store-conditional* (SC)—these are a pair of special instructions that achieve process synchronization by using some built-in features of the cache controllers for bus-based multiprocessors. LL and SC were first used by in the alpha processor, and similar instructions have since been used in MIPS, PowerPC, and ARM processors. Unlike TS, LL, and SC are not atomic RMW operations, but they simplify the implementation of atomic RMW operations and thus can be used to solve the mutual exclusion problem. If *x* is a shared integer variable in the memory and *r* is a private integer local to a process, then the semantics of *LL* and *SC* are as follows:

- $LL(r, x)$ is like a machine instruction *load* (i.e., $r := x$). In addition, the system automatically tracks changes to the *address x*.

- $SC(r, x)$ is like a machine instruction *store* (i.e., $x := r$). However, if the process executing SC is the *first process* to do so after the last LL executed by any process, then the store operation *succeeds*, and the success is reported by returning a value 1 into *r*. Otherwise, the store operation *fails*, the value of *x* remains unchanged, and a 0 is returned into *r*. The snooping cache controller responsible for maintaining cache coherence helps track changes to the address *x* during the $LL(r, x)$ and $SC(r, x)$ operations. In the following, we present a solution to the mutual exclusion problem using LL and SC:

```
program    mutex (for process i);
define     x: shared integer; r: integer (private);
initially  x = 0;
```

```
      do true →
1:    try:    do r ≠ 0 → LL(r, x) od;      {CS is busy}
2:      r = 1; SC(r, x);
3:      if r = 0 → goto try fi;             {SC did not succeed)
4:      critical section;
5:      x := 0;
6:      non-critical section codes;
7:      od
```

Observe that lines 1–3 essentially implement a TS instruction without blocking the bus for two consecutive bus cycles. Additionally, such instructions are easier to accommodate in the instruction set of reduced instruction set computers. Popular processors have adopted different variations of LL and SC instructions—for example, ARM version 6 and higher uses LDREX/STREX instructions that work similar to LL/SC.

7.6 GROUP MUTUAL EXCLUSION

The classical mutual exclusion problem has several variations, and group mutual exclusion is one of them. In this generalization, instead of trying to enter their individual CSs, processes opt to join one of distinct *forums*. Group mutual exclusion requires that at any time, at most, one forum should be *in session*, but any number of processes should be able to join the forum at a time. An example is that of a movie theater, where different people may want to schedule their favorite movies. Here, the forum is the set of viewers of a certain movie, and the forum is *in session* when the movie is screened in the movie theater. The problem was first proposed and solved by Joung in 1999. A more precise specification of the problem follows: let there be n processes $0.. n - 1$ each of which chooses to be in one of m distinct forums. Then the following four conditions must hold:

Mutual exclusion: At most, one forum must be in session at any time.

No deadlock: At any time, at least one process should have an eligible action.

Bounded waiting: Every forum that has been chosen by some process must be in session within a bounded time.

Concurrent entry: Once a forum is in session, concurrent entry into that session is guaranteed for all willing processes.

The group mutual exclusion problem is a combination of the *classical mutual exclusion* problem and the *readers and writers* problem (inasmuch as multiple readers can concurrently read a file), but the framework is more general. It reduces to the classical mutual exclusion problem if each process has its own forum not shared by any other process.

A centralized solution: To realize the challenges involved in finding a solution, we first attempt to solve it using a central coordinator. Let there be only two sessions F and F'. Each process has a *flag* $\in \{F, F', \bot\}$, which indicates its preference for the next session. Here, *flag* $= \bot$ implies that the process is not interested in joining any session. The coordinator will read the flags of the processes in ascending order from 0 to $n - 1$ and guarantee that the first active process always gets entry to its forum, followed by others requesting the same forum.

The simplistic solution will satisfy all requirements of group mutual exclusion except that of bounded waiting, since there is a possibility of starvation: when one forum is chosen to be in session, processes can collude to enter and leave that forum in such a manner that the other forum is never scheduled.

It is possible to resolve this problem by electing a *leader* for each forum that is scheduled. In fact, the very first process that enters a forum is the leader. When the leader leaves a forum, other processes are denied further entry into that forum. This prevents the processes joining a forum from monopolizing it.

Let us now study a decentralized solution to this problem.

A decentralized solution: The first version of the decentralized solution proposed by Joung follows the footsteps of the centralized solution. It works on a shared-memory model. Each process cycles through the following four states: (*request, in_cs, in_forum, passive*). Each process has a *flag* = (*state, op*), where *state* $\in$ {*request, in_cs, in_forum, passive*} and *op* $\in$ {$F, F', \bot$}. A process trying to join a forum F sets its flag to (*request, F*) and eventually moves to the *in_cs* state that gives it a temporary permit to attempt entry to the forum F. It enters forum F if no other process is in the *in_cs* state for the other forum F'. Finally, the process exits forum F and becomes passive. The solution is shown as follows:

```
      First attempt with two forums F and F'
      define   flag: array[1..n - 1] of (state, op), turn ∈ {F, F'}
               state ∈ {request, in_cs, in_forum, passive}
               op ∈ {F, F', ⊥}
      {Program for process i trying to attend forum F}
1     do    ∃j ≠ i: flag[j] = (in_cs, F') →
2              flag[i] := (request, F);                    {request phase}
3              do turn ≠ F ∧ ¬all_passive(F') → skip od;
4              flag[i] := (in_cs, F);                       {in_cs phase}
5     od;
6     attend forum F;                                       {in_forum phase}
7     turn := F';
8     flag[i] := (passive, ⊥)                               {passive phase}
```

The previous program uses the predicate:

$$all\_passive \; (F') \equiv \forall j \neq i: flag[j] = (state, \; op) \Rightarrow op \neq F'$$

The first version is fair with respect to forums. As in Peterson's two-process algorithm, contention is fairly resolved by using a variable *turn*. Note that reaching the *in_cs* phase gives the requesting process a temporary permit—it does not automatically qualify a process to attend its forum. For this, it also has to make sure that all processes in the other forum F' are out of the *in_cs* state.

The proposed however is not fair with respect to processes: If several processes request a forum F, then it is guaranteed that at least one of them will succeed, but we don't know who will. A process, while infrequently checking the predicate in line 3, may find that between two consecutive unsuccessful attempts, the forum has changed from F' to F and then back to F', giving it no chance to make progress. Thus, a requesting process may miss out the requested forum for an indefinite period of time. To make it fair with respect to processes, Joung's algorithm uses the idea from the centralized solution by introducing a *leader* for every session. Given a forum F, some process will lead others to F. For each process i, define a variable $successor[i] \in \{F, F', \perp\}$ to denote the forum that it is *captured* to attend by the leader. The very first process that enters a forum F is the leader of that forum, and the processes that are captured into F are the successors of the leader. Only a leader can capture successors. A process k for which $successor[k] = F$ gets direct entry into session F as long as the leader of F is in session. The permit is withdrawn as soon as the leader quits F. $successor[k] = \perp$ implies that process k is not currently captured. A description of the fair solution to the group mutual exclusion problem can be found in [J98].

7.7 CONCLUDING REMARKS

There are different metrics for evaluating the performance of mutual exclusion algorithms. In message-passing solutions, only the number of messages required to enter the CS is emphasized. For shared-memory systems, a metric is the maximum number of *steps* (atomic read or write operations) that a process takes to enter its CS in the absence of contention. Fairness is determined by the maximum number of other processes that may enter their CSs between two consecutive CS entries by a given process. To satisfy the progress property (ME3), this number has to be finite.

In message-passing solutions, using time stamps to determine the order of requests, the size of the logical clock (or the sequence number) can become unbounded. In [RA81], Ricart and Agrawala addressed the question of bounding the size of the logical clock. Assuming that the logical clock values are not incremented by any event outside the mutual exclusion algorithm, they argued that with n processes, the maximum difference between two different time stamps cannot exceed $(n - 1)$. Therefore, it is adequate to increment the logical clock in the mod M field, where $M = 2n - 1$, since it will resolve without ambiguity whether a time stamp is *ahead* or *behind* another time stamp.

In token-based algorithms, once a process acquires the token, it can enter the CS as many times as it wants unless it receives a request from another process. While reusing a permit is sometimes criticized as undemocratic, this reduces the message complexity when some processes access their CSs more often than others.

Shared-memory solutions to the mutual exclusion problem using read–write atomicity proved to be a fertile area of research and have been extensively studied. Joung's *group mutual exclusion* problem [J98] added a new twist to this problem by allowing a number of processes interested in the same *forum* to concurrently enter their CSs. This is relevant for computer-supported cooperative work.

Computer architects however decided to make the CS algorithms less painful by introducing special instructions in the instruction set. In addition to TS and (LL, SC) discussed in this chapter, there are other instructions like *compare-and-swap* (CAS) or *fetch-and-add* (FA) to facilitate process synchronization. A well-known software construct for process synchronization is *semaphore*. A semaphore s is a nonnegative shared integer variable that allows two atomic operations: $P(s) \equiv \langle s > 0 \rightarrow s := s - 1 \rangle$ and $V(s) \equiv \langle s := s + 1 \rangle$. Semaphores can be implemented using atomic read–write operations or more easily using special instructions like TS or CAS or LL/SC.

In both shared-memory and message-passing solutions to mutual exclusion, a desirable property is that no process that is outside its CS should influence the entry of another process into its CS. As a consequence, the following n-process solution (on the shared-memory model) is considered unacceptable:

```
program    round-robin (for process i}
define turn ∈ {0.. n - 1}

do true →
       do turn ≠ i → skip od;
       critical section;
       turn := turn + 1 mod n;
       non-critical section
od
```

This is because if process 0 wants to enter its CS more frequently than process 1 through $n - 1$, then process 0 may have to wait even if no process is in its CS, because each of the slower processes 1 through $n - 1$ has to take their turn to enter CS. Despite this, some practical networks use similar ideas to implement mutual exclusion. An example is the *token ring*. The acceptability hinges on the fact that a process not willing to use its turn promptly passes the token to the next process, minimizing latency.

7.8 BIBLIOGRAPHIC NOTES

Dijkstra presented the mutual exclusion problem in [D65] where he first described Dekker's two-process solution on the shared-memory model and then generalized it to n processes. His solution was not starvation-free. The first known solution that satisfied the progress property was due to Knuth [K66]. Peterson's algorithm [P81] is the simplest two-process algorithm on the shared-memory model. His technique for generalization to the n-process case is applicable to the generalization of other two-process algorithms too. The *bakery algorithm* was first presented by Lamport in [L74] and later improved in [L79]. It is the only known algorithm that solves the mutual exclusion problem without assuming read–write atomicity—that is, when a read overlaps with a write, the read is allowed to return *any* value, but still the algorithm works correctly. However, the unbounded nature of the shared variable poses a practical limitation. Ben-Ari's book [B82] contains a description of several well-known shared-memory algorithms for mutual exclusion.

Lamport's message-passing algorithm for mutual exclusion is described in [L78]; Ricart and Agrawala's algorithm can be found in [RA81]—a small correction was reported later. Carvalho and Roucairol [CR83] suggested an improvement of [RA81] that led to a message complexity between 0 and $2(n - 1)$. Maekawa's algorithm appears in [M85], and it is the first such algorithm with sublinear message complexity. Sanders [S87] presented a general framework for all message-based mutual exclusion algorithms. Suzuki and Kasami's algorithm [SK85] was developed in 1981, but due to editorial problems, its publication was delayed until 1985. Raymond's algorithm [R89] is the first algorithm with a message complexity of $O(\log n)$.

Joung [J98] introduced the group mutual exclusion problem. Hennessy and Patterson's book [HP11] contains a summary of various synchronization primitives used by historical and contemporary processors on shared-memory architectures.

EXERCISES

7.1 In Ricart–Agrawala's distributed mutual exclusion algorithm, show that processes enter their CSs in the order of their request time stamps even if the channels are not FIFO.

7.2 The *L-exclusion problem* is a generalized version of the mutual exclusion problem in which up to L processes ($L \geq 1$) are allowed to be in their CSs simultaneously. Precisely, if fewer than L processes are in the CS at any time and one more process wants to enter its CS, then it must be allowed to do so. Modify Ricart–Agrawala's algorithm to solve the L-exclusion problem.

7.3 Consider running Maekawa's algorithm on a system of 13 processes. Figure out the composition of the 13 subsets S_0–S_{12}, so that (1) each subset includes four processes, (2) there are exactly four subsets, and (3) process $i \in S_i$.

7.4 In the Suzuki–Kasami algorithm, prove the *liveness property* that any process requesting a token eventually receives the token. Also compute an upper bound on the number of messages exchanged in the system before the token is received.

7.5 In a network of processes, the *local mutual exclusion* problem guarantees that no two neighbors execute a critical action at the same time. Extend Ricart–Agrawala's mutual exclusion algorithm to solve the local mutual exclusion problem.

7.6 Here is a description of Dekker's solution, the first known solution to the mutual exclusion algorithm for two processes:

```
program    dekker (for two processes 0 and 1}
define     flag[0], flag[1]: shared Boolean; turn: shared integer
initially  flag[0] = flag[1] = false, turn = 0 or 1
```

```
       {program for process 0}
1    do   true →
2         flag[0] = true;
3         do flag[1] →
4             if turn = 1→
5                 flag[0] := false;
6                 do turn = 1 → skip od;
7                 flag[0] := true;
8             fi;
9         od;
10        critical section;
11        flag[0] := false; turn := 1;
12    non-critical section codes;
13    od
```

```
       {program for process 1}
14   do   true →
15        flag[1] := true;
16        do (flag[0] →
17            if turn = 0 →
18                flag[1] := false;
19                do turn = 0 → skip od;
20                flag[1] := true;
21            fi;
22        od;
23        critical section;
24        flag[1] := false; turn := 0;
25        non-critical section codes;
26   od
```

Check if the solution satisfies the necessary liveness and safety properties, and provide brief arguments.

7.7 Consider the following two-process mutual exclusion algorithm:

```
program    resolve{for process i ∈ {1, 2}}
define     x: integer, y: boolean
initially  y = false
```

```
do true →
start:   x := i;
         if y →
                do y → skip od;
                goto start;
         fi;
         y := true
         if x ≠ i→
                y := false;
                do x ≠ 0 → skip od;
                goto start;
         fi;
         critical section;
         y := false; x := 0
         non-critical section;
od
```

a. Does it satisfy the requirements of a correct solution?

b. If *n* processes 1, 2,.., *n* execute the previous algorithm, then what is the maximum number of processes that can be in their CSs concurrently?

7.8 In [L74], Lamport proposed the following solution to the mutual exclusion problem on the shared-memory model. It was named *bakery algorithm*, because to request access to the CS, each process has to take a ticket or a number as in a bakery.

```
program bakery {for process i}
define k: integer;
        choosing : shared array [1.. n] of boolean;
        number: shared array [1.. n] of integer;
initially k = 1;
do true →
        choosing[i] := true;
        number[i] := 1 + max{number[1], number[2],..., number[n]};
        choosing[i] := false
        do k ≠ n→
                do choosing[k] → skip od;
                do number[k] ≠ 0 ∧ (number[k], k) <lex (number[i], i)
                   → skip od;
                {Note: <lex means lexicographically smaller than}
                K := k + 1;
        od;
critical section;
number[i] := 0;
non-critical section;
od
```

a. Study the solution and explain why it works.

b. If read and write operations on variables are not atomic, then a read operation that is concurrent with a write on the same variable can return an arbitrary value. Show that bakery algorithm works even if the read and write operations are not atomic. (*Note*: Read the original paper for a proper understanding of the solution.)

7.9 In [L85], Lamport presented the following mutual exclusion algorithm for a shared-memory model:

```
program    fast mutex
define     x, y: integer
           z: array [1.. n – 1]of boolean
initially  y = false
do true→
start:     z[i] := true;
           x := i;
           if y ≠ 0→
                 z[i] := false;
                 do y ≠ 0 → skip od;
                 goto start;
           fi;
           y := i;
           if x ≠ i →
                 z[i] := false;
                 j := 1;
                 do j < n →
                       do z[j] → skip; j := j + 1 od
                 od;
                 if y ≠ 0 →
                       do y ≠ 0 → skip od;
                       goto start;
                 fi
           fi;
           critical section;
           y := 0; z[i] := false;
           non-critical section;
od
```

In previous algorithms, a process had to check with all other processes before entering its CS, so the time complexity was $O(n)$. This algorithm claims that in the absence of contention, a process can enter its CS in $O(1)$ time.

a. Show that in the absence of contention, a process performs at most *five* write and *two* read operations to enter its CS.

b. Is the algorithm starvation-free?

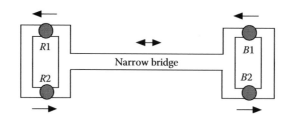

FIGURE 7.3   Cars crossing a narrow bridge on a river.

7.10 Some shared-memory multiprocessors have an *atomic instruction FA* defined as follows. Let x be a shared variable and v be a private variable local to a process. Then $FA(x, v) \equiv \langle return\ x;\ \ x := x + v \rangle$.

    a. Implement *FA* using LL and SC.

    b. Solve the mutual exclusion problem using FA.

7.11 Consider a bus-based multiprocessor, where processors use *TS* to implement CSs. Due to busy waiting, each processor wastes a significant fraction of the bus bandwidth. How can you minimize this using the private caches of the processors? Explain your answer.

7.12 A popular application of the mutual exclusion is on the Ethernet: when multiple processes try to enter their CSs at the same time, only one succeeds. Study how Ethernet works. Then write down the mutual exclusion algorithm used by the processes on the Ethernet to gain exclusive access to transmit data.

7.13 (*Programming project*) Figure 7.3 shows a section of a traffic route around the narrow bridge AB on a river. Two red cars ($R1$, $R2$) and two blue cars ($B1$, $B2$) move along the designated routes that involve indefinite number of trips across the bridge. The bridge is so narrow that at any time, multiple cars cannot pass in opposite directions.

    a. Using the message-passing model, design a decentralized protocol so that at most one car is on the bridge at any time and no car is indefinitely prevented from crossing the bridge. Treat each car to be a process and assume that their clocks are not synchronized.

    b. Modify the protocol so that multiple cars can be on the bridge as long as they are moving in the same direction, but no car is indefinitely prevented from crossing the bridge.

Design a graphical user interface to display the movement of the cars, so that the observer can visualize the cars, control their movements, and verify the protocol.

CHAPTER 8

Distributed Snapshot

8.1 INTRODUCTION

A computation is a sequence of atomic actions that transform a given *initial state* to the *final state*. While such actions are totally ordered in a sequential process, they are only partially ordered in a distributed system. It is customary to reason about the properties of a program in terms of *states* and *state transitions*.

In this context, the state (also known as *global state* or *configuration*) of a distributed system is the set of local states of all the component processes as well as the states of every channel through which messages flow. Since the local physical clocks are never perfectly synchronized, the components of a global state can never be recorded at the same time. In asynchronous distributed systems, actions are not related to time. So the important question is as follows: *When* or *how* do we record the states of the processes and the channels? Depending on *when* the states of the individual components are recorded, the value of the global state can vary widely.

The difficulty can be best explained using a simple example. Consider a system of three processes numbered 0, 1, and 2 connected by FIFO channels (Figure 8.1), and assume that an unknown number of indistinguishable *tokens* are circulating indefinitely through this network. We want the processes to cooperate with one another to count the exact number of tokens circulating in the system (without ever stopping the system). The task has to be initiated by an initiator process (say process 0) that will send *query messages* to the other processes to record the number of tokens sighted by them. Consider the case when there is *exactly one* token, and let n_i denote the number of tokens recorded by process i. In Figure 8.1, depending on *when* the individual processes count the tokens, each of the following situations is possible:

Possibility 1: Process 0 records $n_0 = 1$ when it receives the token. When process 1 records n_1, assume that the token is in channel $(1, 2)$, so $n_1 = 0$. When process 2 records n_2, the token is in channel $(2, 0)$, so $n_2 = 0$. Therefore, $n_0 + n_1 + n_2 = 1$.

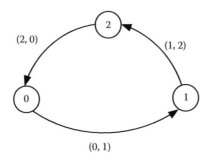

FIGURE 8.1   A token circulating through a system of three processes numbered 0, 1, and 2.

Possibility 2: Process 0 records $n_0 = 1$ when it receives the token. Process 1 records n_1 when the token has reached process 1, so $n_1 = 1$. Finally, process 2 records n_2 when the token has reached process 2, so $n_2 = 1$. Thus, $n_0 + n_1 + n_2 = 3$. Since tokens are indistinguishable, no process knows that the same token has been counted three times!

Possibility 3: Process 0 records $n_0 = 0$ since the token is in channel $(0, 1)$ at the time of recording. When processes 1 and 2 want to record the count, the tokens have already left them, so $n_1 = 0$ and $n_2 = 0$. Thus, $n_0 + n_1 + n_2 = 0$.

Clearly, possibilities 2 and 3 reflect incorrect views of the global state. How can we devise a scheme that always records a correct or a consistent view of the global state? In this chapter, we address this question.

The recording of the global state may look simple for some external observer who looks at the system from outside. The same problem is surprisingly challenging, when one takes a snapshot from *inside* the system. In addition to the intellectual challenge, there are many interesting applications of this problem. Some examples are as follows:

Deadlock detection: Any process that does not have an eligible action for a prolonged period would like to find out if the system has reached a deadlock configuration. This requires the recording of the global state of the system.

Termination detection: Many computations run in phases. In each phase, every process executes a set of actions. When every process completes all the actions belonging to phase i and there is no message in transit, phase i terminates and the next phase $(i + 1)$ begins. To begin a new phase, a process must therefore know whether the computation in the previous phase has terminated.

System reset or rollback: In case of a malfunction or a loss of coordination, a distributed system will need to *roll back* to a consistent global state and initiate a recovery. Previous snapshots may be used to define the point from which recovery should begin.

To understand the meaning of a consistent snapshot state, we will take a look at some of its important properties.

8.2 PROPERTIES OF CONSISTENT SNAPSHOTS

A critical examination of the three possibilities in Section 8.1 will lead to a better understanding of what is meant by a consistent snapshot state. Recall that in the absence of a global clock, concurrency implies the absence of a causal order (Section 3.6). A snapshot state *SSS* consists of a set of local states, where each local state is the outcome of a *recording event* that follows a send, or a receive, or an internal action. The important notion here is that of a *consistent cut*.

8.2.1 Cuts and Consistent Cuts

A *cut* is a set of events—it contains at least one event per process. Draw a timeline for every process in a distributed system, and represent events by points on the timeline as shown in Figure 8.2. Here, $\{c, d, f, g, h\}$ is a cut. A cut is called *consistent* if for each event that it contains, it also includes all events causally ordered before that event. Let a and b be two events in a distributed system. Then,

$$(a \in \text{consistent cut } C) \wedge (b \prec a) \Rightarrow b \in C$$

Thus, for a message m, if the state following *receive(m)* belongs to a consistent cut, then the state following *send(m)* also must belong to that cut. Of the two cuts in Figure 8.2, Cut 1 = $\{a, b, c, m, k\}$ is consistent, but Cut 2 = $\{a, b, c, d, g, m, e, k, i\}$ is not, since $(g \in \text{Cut 2}) \wedge (h \prec g)$ but $(h \notin \text{Cut 2})$. As processes make progress, new events update the consistent cut. The progress of a distributed computation can be visualized as the forward movement of the *frontier* (the latest events) of a consistent cut.

The set of local states following the most recent events (an event a is a *most recent* event if there is no other event b such that $a \prec b$) of a cut defines a *snapshot*. A consistent cut *induces* a *correct* or *consistent* distributed snapshot, that is, the set of local states following the recorded events of a consistent cut form a correct or consistent distributed snapshot.

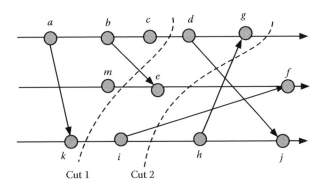

FIGURE 8.2 Two cuts of a distributed system. The broken lines represent cuts, and the solid-directed edges represent message transmission.

In a distributed system, many consistent snapshots can be recorded. A snapshot that is often of practical interest is the one that is *most recent*. Let C_1 and C_2 be two consistent cuts leading to two different snapshots S_1 and S_2, respectively, and let $C_1 \subset C_2$. Then, C_2 is *more recent* than C_1, and snapshot S_2 is *more recent* than snapshot S_1.

A computation (also called a *behavior* or a *run*) is a sequence of events in a distributed system. Let a and b be a pair of events in a computation. Then, the computation is *consistent* when $\forall a,b: a \prec b \Rightarrow a$ precedes b in it. Such a computation reflects one of the *feasible schedules* of a central scheduler. When the events are partially ordered, there may be multiple consistent runs. Given a consistent run X that contains a pair of concurrent events c and d, another consistent run Y can be generated from X by swapping their order and without violating any other causal order in the system.

Chandy and Lamport [CL85] addressed the issue of consistent snapshot and presented an algorithm for recording such a snapshot. The following section describes their algorithm.

8.3 CHANDY–LAMPORT ALGORITHM

Let the topology of a distributed system be represented by strongly connected graph. Each node represents a process and each directed edge represents a FIFO channel. The snapshot algorithm is superposed on the underlying application and is noninvasive inasmuch as it does not influence the underlying computation in any way. Note that temporarily freezing the entire system, taking a snapshot, and then restarting the system are not options, since freezing a nontrivial distributed system that provides a service is costly and impractical.

A process called the *initiator* initiates the distributed snapshot algorithm. Any process can be an initiator. The initiator process sends a special message, called a *marker* (*) that prompts other processes in the system to record their states. The markers are for instrumentation only—they neither force a causal order among events nor influence the semantics of the underlying computation in any way. The global state consists of the states of the processes as well as the channels. However, channels are passive entities—so the responsibility of recording the state of a channel lies with the process on which the channel is incident.

For the convenience of explanation, we will use the colors *white* and *red* with the different processes. Initially, every process is *white*. When a process receives the marker, it turns *red* if it has not already done so. Furthermore, every action executed by a process or every message sent by a process gets the color of that process, so both actions and messages can be red or white (Figure 8.3). The markers are used for instrumentation only and do not have any color. There are two actions in this algorithm as described in the following.

DS1: The initiator process, in one atomic action, does the following:

- Turns red

- Records its own state

- Sends a marker along all its outgoing channels

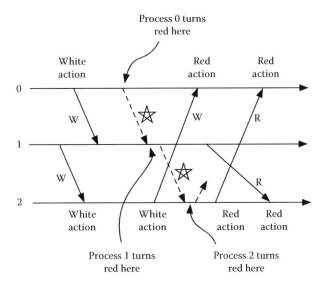

FIGURE 8.3 An example illustrating colors of action and messages: 0 is the initiator, W, white message; R, red message; and the star represents a marker.

DS2: Every process, upon receiving a marker for the *first time* and before doing anything else, does the following in one atomic action:

- Turns red

- Records its state

- Sends markers along all its outgoing channels

The state of a channel (p, q) is recorded as follows: let *sent*(p) denote the set of messages sent by process p along (p, q) and *receive*(q) denote the set of messages received by process q via (p, q). Then the state of the channel is recorded as *sent*(p)*receive*(q). Here, *sent*(p) and *received*(q) are locally recorded by the processes p and q, respectively. The snapshot algorithm terminates, when

- Every process has turned red

- Every process has received a marker through each of its incoming channels and sent a marker along each of its outgoing channels

The individual processes only record the *fragments* of a snapshot state *SSS*. It requires another phase of activity to collect these fragments and construct a composite view of *SSS*. Global state collection is not a part of the snapshot algorithm.

While the initiator turns red at the very beginning of the snapshot taking process, every other process changes its color from white to red is due to the arrival of the *first* marker.

Red processes do not change their colors to white during the lifetime of the algorithm. The following lemma is the cornerstone of the algorithm:

Lemma 8.1

No red message is received by a white action.

Proof: Before a process p receives a red message from a process q along a channel (p, q), process q must have turned red and sent a marker to p along channel (p, q) (see DS2). Since the channels are FIFO, the receiving process p must have received that marker and turned red in the mean time. Thus, the receive action by the process p cannot be white. ■

If all processes changed colors simultaneously, then the snapshot state SSS would have been determined by the states of all the processes just after this transition point. However, due to signal propagation delays, that is never the case. A reasonable alternative to this would be to record a sequence of white and red actions by all the processes, so that all white actions precede all red actions (Figure 8.4a). Here, SSS will correspond to the states of the processes after the last white actions and the first red actions by the processes. There is however no guarantee that an observer will be able to observe the actions in this sequence. In reality, an observer will record the *last white actions* and the *first red actions* of the processes in some order, as in Figure 8.4b. For each process i in an observed sequence, $w(i)$ must precede $r(i)$, since a causal order exists between them. Note that (a) markers do not change the state of a process, and (b) white messages received by white processes or red messages received by red processes are not interesting in the context of the present problem.

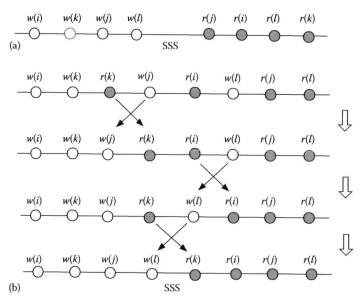

FIGURE 8.4   (a) The ideal view of the snapshot state: $w(i)$ and $r(i)$ denote, respectively, the last white action and the first red action by process i. (b) An observed sequence is being reduced to the ideal view via number of swap actions.

We will now show that the snapshot state recorded using *DS1* and *DS2* is indeed *equivalent to* the ideal view of Figure 8.4a. The equivalence hinges on the fact that a pair of actions (a, b) can be observed or scheduled in any order, as long as there is no causal order between them—so schedule (a, b) is equivalent to schedule (b, a). Now, the main theorem follows.

Theorem 8.1

The Chandy–Lamport algorithm records a consistent global state.

Proof: The snapshot algorithm records a state SSS' that consists of the states of every process when it turned red. Let $w(i)$ and $r(i)$ denote, respectively, the last white action and the first red action by process i. Without loss of generality, assume that an observer records the following partial schedule of actions (see Figure 8.4b) with processes $i, j, k, l, \ldots$:

$$w(i)\ w(k)\ r(k)\ w(j)\ r(i)\ w(l)\ r(j)\ r(l)\ldots$$

Call it a *break*, when a *red action* precedes a *white action* in a given schedule. The ideal view of the schedule (Figure 8.4a) has zero breaks, but in general, the number of breaks in the recorded sequence of actions can be a positive integer.

By Lemma 8.1, process j cannot receive a red message from process k in a white action, so there is no causal ordering between actions $r(k)$ and $w(j)$, and these actions can be swapped to produce an equivalent schedule. But this swap *reduces* the number of breaks by 1. It thus follows that in a finite number of swaps, the number of breaks in the equivalent schedule will be reduced to zero, and the resulting schedule will correspond to the ideal view of Figure 8.4. Therefore, the recorded state SSS' is equivalent to a consistent snapshot SSS. ▪

8.3.1 Two Examples

To study various features of the Chandy–Lamport algorithm, we present two examples of computing the global state.

8.3.1.1 Example 1: Counting of Tokens

Consider Figure 8.1 again. Assume that that only token is in transit in the channel $(0, 1)$ and process 0 initiates the snapshot algorithm. Then, the following is a valid sequence of events:

1. Process 0 turns red; records $n_0 = 0$, $|sent(0, 1)| = 1$, and $|received(2, 0)| = 0$; and sends the marker along $(0, 1)$.

2. Process 1 receives the token and forwards it along $(1, 2)$ before receiving the marker. Then, it receives the marker; turns red; records $n_1 = 0$, $|received(0, 1)| = 1$, and $|sent(1, 2)| = 1$; and sends the marker along $(1, 2)$.

3. Process 2 receives the token and forwards it along $(2, 0)$ before receiving the marker. Then, it receives the marker; turns red; records $n_2 = 0$, $|received(1, 2)| = 1$, and $|sent(2, 0)| = 1$; and forwards the marker to process 0.

4. Process 0 receives the token and then receives the marker along $(2, 0)$. The algorithm terminates here.

The total number of tokens recorded is as follows: $(n_0 + n_1 + n_2) + (|sent(0, 1)| - |received(0, 1)|) + (|sent(1, 2)| - |received(1, 2)|) + (|sent(2, 0)| - |received(2, 0)|) = 1$. This is consistent with the expected outcome.

8.3.1.2 Example 2: Communicating State Machines

This example is from [CL85]. Two state machines i and j communicate with each other by sending messages M and M' along channels $c1$ and $c2$ (Figure 8.5), respectively. Each state machine has two states: *up* and *down*. Let $s(i)$ and $s(j)$ denote the states of i and j, respectively, and S_0 represent the initial global state $s(i) = s(j) = down$ and $c1 = c2 = \emptyset$. A possible sequence of global states is shown in Figure 8.5. The global state returns to S_0 after machine j receives M, so that S_0, S_1, S_2, S_3, S_0 forms a cyclic sequence.

Now, use the Chandy–Lamport algorithm to record a snapshot SSS. Assume that in global state S_0, process i initiates the snapshot algorithm by sending a marker and then sending a message M along $c1$, which changes $s(i)$ from *down* to *up*. Before M reaches j, state machine j sends M' along $c2$, changing $s(j)$ from *down* to *up*. Then, j receives the marker and forwards it along $c2$. Thereafter, message M' and then the marker reach

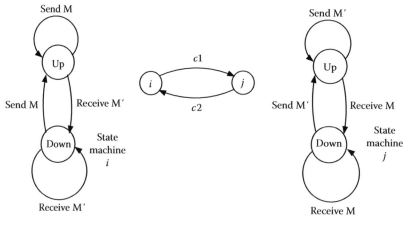

Global State	Process i	$c1$	Process j	$c2$
S_0	Down	$\emptyset$	Down	$\emptyset$
S_1	Up	M	Down	$\emptyset$
S_2	Up	M	Up	M'
S_3	Down	M	Up	$\emptyset$

FIGURE 8.5 A sequence of global states in a system of two communicating state machines.

process i in global state S_3. Finally, M reaches j, and the system returns to the state S_0. This leads to a following recording:

$$s(i) = down, \quad sent(c1) = \varnothing, \quad received(c2) = \varnothing \quad \{\text{Recorded by } i\}$$

$$s(j) = up, \quad sent(c2) = M', \quad received(c1) = \varnothing \quad \{\text{Recorded by } j\}$$

Accordingly, $SSS \equiv s(i) = down, c1 = \varnothing, s(j) = up, c2 = M'$.

Notice something unusual here: the snapshot state SSS does not belong to the cycle of states, so the system was never in that state! What good is a distributed snapshot if the system never reaches the resulting snapshot state?

To understand the significance of the recorded snapshot state, look at the partial history of the system (Figure 8.6). The behavior considered in our example is an infinite sequence of $S_0 S_1 S_2 S_3 S_0 \ldots$, but other behaviors are equally possible. In fact, the recorded snapshot state SSS corresponds to the state S_1' *that is reachable from the initial state* S_0, although this behavior was not observed in our example! Additionally, state S_3 that was a part of the observed behavior is reachable from the recorded snapshot state S_1'.

It follows from Figure 8.6 that if we could alter the course of the computation so that (j receives M′) precedes (i receives M), then the recorded state would be reachable from the initial state S_0. The unpredictability of the scheduling order of concurrent actions led to the anomaly. This is in tune with the *swapping argument* used in the proof of Theorem 8.1.

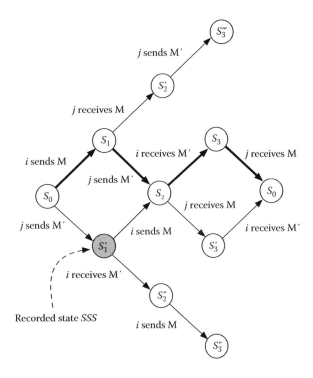

FIGURE 8.6   A partial history of the system in Figure 8.5.

The colors of the four actions in the state transitions $S_0 \to S_1 \to S_2 \to S_3 \to S_0$ are $r(i)$, $w(j)$, $r(i)$, $r(j)$, respectively. Using the notions of *equivalent computation*, this can be transformed into a sequence $w(j)$, $r(i)$, $r(i)$, $r(j)$ (by swapping the first two actions since by Lemma 8.1, no *red* message can be received in a *white* action). This alters the computation to $S_0 \to S_1' \to S_2 \to S_3 \to S_0$, and the recorded global state SSS will indeed be the same as S_1'. These observations can be summarized as follows:

CL1: Every snapshot state recorded by the Chandy–Lamport algorithm is reachable from the initial state through a feasible sequence of actions. However, there is no guarantee that this state will actually be attained during a particular computation.

CL2: Every final state that is reachable from the initial state is also reachable from the recorded snapshot state through a feasible sequence of actions.

Due to the second property, a malfunctioning distributed system can be reset to a consistent snapshot state *SSS* without altering the future course of the distributed system. Following such a reset action, the system has the potential to catch up with the expected behavior via a feasible sequence of actions.

Despite such anomalies, *SSS* indeed represents the actual global state, when the final state of the system corresponds to a *stable predicate*. A predicate P is called *stable* if once P becomes true, it remains true thereafter. This is different from nonstable predicates that may be true at a certain time, but change to false thereafter. Some examples of stable predicates are as follows: (1) the system is deadlocked and (2) the computation has terminated.

8.4 LAI–YANG ALGORITHM

Lai and Yang [LY87] proposed a modification of Chandy–Lamport's algorithm for distributed snapshot on a network of processes where the channels need not be FIFO. To understand Lai and Yang's modification, let us once again color the messages *white* and *red*: a message is *white* if it is sent by a process that has not recorded its state, and a message is *red* if the sender has already recorded its state. However, unlike Chandy–Lamport's algorithm, there are no markers—processes record their local states spontaneously and append this information as a Boolean flag (*red* or *white*) with any message they send. We will represent a message by (m, c) where m is the underlying message and c is the Boolean flag *red* or *white*.

The cornerstone of the algorithm is the condition Lemma 8.1: no red message is received in a white action. To make it possible, if a process that has not recorded its local state receives a message (m, red) from another process that has done so, then it first records its own state (this changes its color to *red* and the color of all its subsequent actions to *red*) and then accepts the message. No other action is required to make the global state consistent. The algorithm can be stated as follows:

LY1: The initiator records its own state. When it needs to send a message m to another process, it sends (m, red).

LY2: When a process receives a message (m, red), it first records its state if it has not already done so and then accepts the message m.

The approach is *lazy* inasmuch as processes do not send or use any control message for the sake of recording a consistent snapshot—the activity rides the wave of messages of the underlying algorithm. The good thing is that if a complete snapshot is taken, then it will be consistent. However, there is no guarantee that a complete snapshot will eventually be taken: if a process i wants to detect termination, then i will record its own state following its last action, but send no message, so other process may not record their states! To guarantee that eventually a complete and consistent snapshot is taken, dummy control messages may be used to propagate the information that some process has already recorded its state.

8.5 DISTRIBUTED DEBUGGING

A common problem of interest in the design and analysis of many distributed systems is to decide whether the global state will ever satisfy a predefined predicate ϕ. Such a question may have safety implications: for example, in an industrial control system, a state in which certain parameters attain predefined values can potentially lead to a hazard. In a traffic control network, certain combination of traffic signal values can lead to traffic congestion or a potential collision between vehicles. In a network protocol, certain global states may cause a safety breach. Observing and analyzing such potential threats are challenging due to two reasons: (1) the size of the state space of even a small-sized distributed system can be enormous and (2) the inherent nondeterminism prevents us from freely exploring the state space. Ideally, to identify a bug, any execution leading the system to the buggy state must be repeatable. Nondeterminism prevents us from exercising that kind of control in state exploration. Note that debugging encompasses both stable and unstable predicates.

The history of a distributed system can be represented by a lattice as shown in Figure 8.7. Assume that there is an *observer* (also called a *monitor*) who is observing the computations in the system. An observer is an *external process*[*] that receives notifications from every internal process (about the next state) whenever they execute an action. The observer has

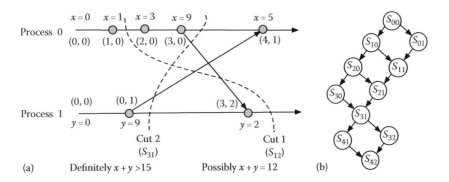

FIGURE 8.7 Illustration of distributed debugging. (a) Two communicating processes 0 and 1 updating variables x and y. (b) In the lattice of states, S_{ij} denotes a *consistent* global state after i actions by P and j actions by process Q. Inconsistent states are left out.

---

[*] So, it is a centralized instrumentation tool.

the following tasks: (1) determine which set of local states form a consistent snapshot, and construct the lattice of consistent states and (2) figure out if the predicate $\phi(S) = true$ for one or more snapshot states S.

8.5.1 Constructing the State Lattice

For this task, we use vector time stamps. Each event is labeled with its vector clock. Each process reports these events, along with their local sequence numbers and vector time stamps, to the monitor. The monitor receives them in some arbitrary order. In a set of n processes $V = \{0, 1, 2, \ldots, n-1\}$, let $VC(e_i)$ be the vector time stamp of event e_i in process i leading to its local state $s(i)$ and $VC(e_j)$ be the vector time stamp of event e_j in process j leading to its local state $s(j)$. Then, $s(i)$ and $s(j)$ belong to some consistent global state iff one of them is $(0, 0)$, or

$$\forall i, j \in V : VC_i(e_i) \sim VC_i(e_j)$$

Using this criterion, the monitor threads together a set of consistent global states as the nodes of a lattice. It also identifies the global state transitions by defining an edge from global state S to S' iff S' can be reached from S via a single action by some process. Figure 8.7 illustrates an example: part (a) shows a pair of communicating processes updating a pair of variables x and y and part (b) shows the corresponding state lattice. Cut 2 is consistent because the time stamps $(3, 0)$ (of event e_0 from process 0) and $(0, 1)$ (of event e_1 from process 1) satisfy the conditions $VC_0(e_0) \geq VC_0(e_1)$ and $VC_1(e_1) \geq VC_1(e_0)$. Cut 1 is inconsistent since $VC(e_0) = (1, 0)$ and $VC(e_1) = (3, 2)$, which means $VC_0(e_0) < VC_0(e_1)$. Inconsistent global states are excluded from the lattice.

Any traversal from the initial state S_{00} via the directed edges of the lattice defines a *feasible computation*. In the context of these computations, we evaluate the predicates of interest.

8.5.2 Evaluating Predicates

For debugging, let ϕ be a predicate of interest. Three kinds of queries about ϕ are often considered useful—*possibly ϕ*, *definitely ϕ*, and *never ϕ*:

Possibly ϕ: For a system, "*possibly ϕ*" holds if there exists at least one consistent global state S that is reachable from the initial global state, such that $\phi(S) = true$. This means that if there are many observers, then at least one of them may observe ϕ, although many others may not be able to observe it.

Definitely ϕ: This is different from "*definitely ϕ*" that signifies that all computations from the given initial state pass through some consistent global state S for which $\phi(S) = true$. This means every observer must be able to observe ϕ. Clearly, *definitely ϕ* $\Rightarrow$ *possibly ϕ*.

Never ϕ: For a system, the predicate "*never ϕ*" is true if no computation from the given initial state passes through a consistent global state S for which $\phi(S) = true$.

In Figure 8.7, let $\phi \equiv x + y = 12$. This predicate is true only at the global state S_{21} (i.e., after the second event of process 0 and the first event of process 1). Since there exists at least one computation (but not all) that passes through S_{21}, the condition *possibly* ϕ holds. However, for another predicate $\phi \equiv x + y > 15$, the stronger condition *definitely* ϕ holds, since ϕ is true in global state S_{32} and all computations pass through this S_{32}. Finally, for the predicate $\phi \equiv x = y = 5$, the condition *never* ϕ holds since none of the two global states S_{40} and S_{22} for which the predicate holds is consistent.

Distributed debugging using this approach uses a thorough state exploration that has a large time complexity—with n processes each having m actions, the time complexity of debugging is $O(m^n)$. Thus, scalability is a concern.

8.6 CONCLUDING REMARKS

The distributed snapshot algorithm clarifies what constitutes the global state of a distributed system when the clocks are not synchronized or processes do not have clocks. Several incorrect algorithms for distributed deadlock detection in the published literature have been attributed to the lack of understanding of the consistency of a global state.

An alternative suggestion is to freeze the entire system, record the states of the processes, and then resume the computation. However, a global freeze interferes with the ongoing computation and is not acceptable in most applications. Additionally, any algorithm for freezing the entire system requires the propagation of a *freeze* signal across the system and will be handled in a manner similar to the marker. The state recording algorithm should wait for all the computations to freeze and all the channels to be empty, which will rely upon a termination detection phase. Clearly, this alternative is inferior to the solution proposed in this chapter.

The snapshot algorithm captures the fragments of a consistent global state in the various component processes. It requires another broadcast algorithm (or state collection algorithm) to put these fragments together into a consistent global state and store it in the state space of the initiator.

8.7 BIBLIOGRAPHIC NOTES

Chandy–Lamport algorithm is described in [CL85]. In a separate note [D84], Dijkstra analyzed this algorithm and provided an alternative method of reasoning about the correctness. The proof presented here is based on Dijkstra's note. The algorithm due to Lai and Yang appears in [LY87]. Mattern's article [M89] summarizes several snapshot algorithms and some improvisations. Marzullo and Neiger [MN91] proposed the monitor-based approach for distributed debugging.

EXERCISES

8.1 Consider the Chandy–Lamport distributed snapshot algorithm running on a network whose topology is $G = (V, E)$, where V is the set of nodes representing processes and E is the set of directed edges (representing channels) connecting pairs of processes. Let $n = |V|$. Prove that in the snapshot the states of $(n - 1)$ channels will always be empty. Also, compute the message complexity when G is strongly connected.

8.2 Construct an example to show that Chandy–Lamport distributed snapshot algorithm does not work when the channels are not FIFO.

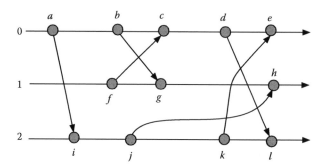

FIGURE 8.8   The events in a system of three processes numbered 0, 1, and 2.

8.3   In Figure 8.8, show all the consistent cuts that (a) include event d and (b) exclude event d but include event g.

8.4   Extend Chandy–Lamport algorithm so that it can be used on a network whose topology is a directed acyclic graph. Discuss any limitations of the extended version.

8.5   Assume that the physical clocks of all the processes are synchronized and channel propagation delays are known. Suggest an alternative algorithm for computing a distributed snapshot based on the physical clock and the channel delays.

8.6   (Programming exercise) Sunrise Bank wants to find out its cash reserve, so it initiates an audit in all of its branches. At any moment, a sum of money may be in transit from one branch of the bank to another. ATM transactions, customer transactions through bank tellers, and interbranch transactions are the only possible types of transactions.

   Define two teller processes $T1$ and $T2$ at two different branches and three ATM processes $A1$, $A2$, and $A3$, each with a predefined amount of cash. Assume that four customers have an initial balance of a certain amount in their checking accounts. Each user can deposit money into or withdraw money from their accounts through any ATM or a teller. In addition, any customer can transfer an amount of available cash from her account to any other customer, some of which may lead to interbranch transfers.

   Use Lai–Yang algorithm to conduct an audit of Sunrise Bank. Allow the users to carry out transactions at an arbitrary instant of time. The audit must reveal that at any moment, *total cash = initial balance + credit − debit* regardless of when and where users transact money.

8.7   While distributed snapshot is a mechanism for *reading* the global state of a system, there are occasions when the global state is to be *reset* to a new value. An example is a failure or a coordination loss in the system. As with the snapshot algorithms, we rule out the use of freezing the network.

   Explore if the Chandy–Lamport algorithm can be adapted for this purpose by replacing all *reads* by *writes*. Thus, when a process receives a marker, it will first reset its state and then send out the marker in one indivisible action.

8.8   You have been hired to lead to a project that measures how busy the Internet is. As a part of this project, you have to measure the maximum number of messages that in transit *at a time* in the Internet. Assuming that all machines on the Internet and their system administrators agree to cooperate with you, suggest a plan for measurement.

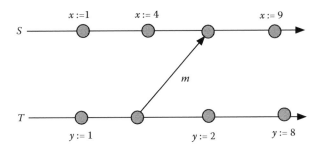

FIGURE 8.9   A pair of communicating processes S and T.

8.9 The Chandy–Lamport only helps record fragments of the global state. In the next phase, these fragments have to be collected to form a meaningful global state. Instead of making it a two-phase process, explore the possibility of computing the global state of a distributed system in a single phase using a mobile agent.

Assume that the topology is strongly connected and the edges represent FIFO channels. The initiator will send out a mobile agent with a briefcase that will store its data. The briefcase has a variable S—it is an array of the local states of the various processes. Before the agent is launched, $\forall i: S[i] = undefined$. At the end, the agent will return to the initiator, and S will represent a consistent global state (Figure 8.9).

8.10 In this system, consider the three predicates:

$$\phi1 \equiv x + y = 3$$

$$\phi2 \equiv x + y = 12$$

$$\phi3 \equiv x + y = 10$$

For each predicate, determine if it is *possibly true*, or *definitely true*, or *never true*. Briefly justify your answer.

CHAPTER 9

Global State Collection

9.1 INTRODUCTION

In a distributed system, each process executes actions on the basis of local information that consists of its own state and the states of its neighbors or messages through the incoming channels. Many applications need to find out the global state of the system by collecting the local states of the component processes. These include

- Computation of the network topology

- Counting the number of processes in a distributed system

- Detecting termination

- Detecting deadlock

- Detecting loss of coordination

The distributed snapshot algorithm (Chapter 8) clarifies the notion of a consistent global state and helps record the fragments of a consistent global state into the local state spaces of the individual processes but does not address the task of collecting these fragments. In this chapter, we address this issue and present several algorithms for global state collection.

The algorithms for global state collection have been classified into various types: *probe-echo*, *wave* or *PIF* algorithm, *heartbeat* algorithm, etc. The classification is usually based on the mechanism used to collect the states, and the classes are not mutually exclusive. For example, *wave algorithms* refer to a class of computations, in which an initiator starts a computation that triggers more actions in the adjacent noninitiator nodes, and in a bounded time, each node in the system reaches local termination when a predefined goal is reached. The resulting causal chain of events resembles the growth and decay of waves in a still pond—hence the name wave algorithm.

We will study a few algorithms that follow the various paradigms without paying much attention to the class to which they belong. We begin with the description and correctness proof of an elementary broadcasting algorithm, where the underlying message-passing model supports point-to-point communication only.

9.2 ELEMENTARY ALGORITHM FOR ALL-TO-ALL BROADCASTING

Consider a strongly connected network of n processes 0, 1, 2,..., $n-1$. Each process i has a stable value $s(i)$ associated with it. The goal is to devise an algorithm by which every process i can broadcast its value $s(i)$ to every other process in this system, so that at the end, each process i will have a set $V_i = \{s(k): 0 \leq k \leq n - 1\}$ of values. We will use a message-passing model.

Initially, $\forall i: V_i = s(i)$. To complete the broadcast, every process i will periodically (1) send its current V_i to each of its outgoing channels and (2) receive the values from its incoming channels to update V_i. A naive approach for broadcasting is *flooding*, where each process sends its value to all neighbors, the neighbors send that value to *their* neighbors, and so on. This approach is very inefficient in terms of message complexity, and controlling the termination is also a matter of concern. The following algorithm addresses both these issues.

To save unnecessary work, it makes little sense to send V_i, if it has not changed since the last send operation. Furthermore, even if V_i has changed since the last send operation, it suffices to send the incremental change only—this will keep the message size small.

To accomplish this, we associate two sets of values with each process i—the set V_i will denote the current set of values collected so far, and the set W_i will represent the last value of V_i sent along the outgoing channels so far. Let (i, j) represent the channel from process i to process j. The algorithm terminates when no process receives any new value and every channel is empty. The program for process i is given as follows:

```
program      broadcast (for process i}
define       Vᵢ, Wᵢ: set of values;
initially    Vᵢ = {s(i)}, Wᵢ = Ø {and every channel is empty}

1   do Vᵢ ≠ Wᵢ        →   send Vᵢ\Wᵢ to every outgoing channel;
                           Wᵢ:= Vᵢ
2   [] ¬empty (k,i)   →   receive X from channel (k,i);
                           Vᵢ:= Vᵢ ∪ X
od
```

Correctness proof

To prove the correctness, we first establish *partial correctness* and then prove *termination*. For partial correctness, we demonstrate that when all the guards are false, every process must have collected the value $s(i)$ from every other process i.

Lemma 9.1

$empty(i, k) \Rightarrow W_i \subseteq V_k$

Proof (by induction):

Basis: Initially, $W_i = \varnothing$ and $V_i = s(i)$, so $W_i \subseteq V_k$ holds.

Induction hypothesis: Assume $W_i \subseteq V_k$ holds after process i executes the first statement r times and channel (i, k) becomes empty. We will show that $W_i \subseteq V_k$ holds after process i executes the first statement $(r + 1)$ times and (i, k) becomes empty.

Induction step: Between two consecutive executions of statement 1, statement 2 must be executed at least once. Let V_i^r, W_i^r denote the value of V_i and W_i after process i executes statement 1 r times. After process i executes statement 1 for the $(r + 1)$th time, the set of messages sent down the channel $(i, k) = V_i^{r+1} \setminus W_i^r$ (Figure 9.1).

When every message in (i, k) is received by process k (statement 2), channel (i, k) becomes empty. This implies that $V_i^{r+1} \setminus W_i^r \subseteq V_k$. However, from the induction hypothesis, $W_i^r \subseteq V_k$. Therefore, $V_i^{r+1} \subseteq V_k$. Also, since $V_i^{r+1} = W_i^{r+1}$, $W_i^{r+1} \subseteq V_k$ holds. ■

Lemma 9.2

If algorithm *broadcast* terminates, then $\forall i: V_i = \{s(k): 0 \leq k \leq n - 1\}$.

Proof: When the first guard is false for each process, $\forall i: W_i = V_i$ holds. When the second guard is false for each process, all channels are empty, and from Lemma 9.1, $W_i \subseteq V_k$ where k is a neighbor of i. Therefore, $V_i \subseteq V_k$.

Now consider a directed cycle C that includes processes i and k. If for every pair of processes across a channel (i, k) the condition $V_i \subseteq V_k$ holds, then for $\forall i, k \in C: V_i = V_k$ must be true. Furthermore, in a strongly connected graph, every pair of processes (i, j) is contained in a directed cycle; therefore, the condition $V_i = V_j$ must hold for every pair of processes (i, j) in the system. Also, since $s(i) \in V_i$, and no element is removed from a set, finally, $\forall i: V_i = \{s(k): 0 \leq k \leq n - 1\}$. ■

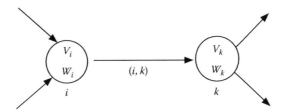

FIGURE 9.1 Two processes, i and k, connected by a channel.

Theorem 9.1

Algorithm *broadcast* terminates in a bounded number of steps.

Proof: Let $c_0, c_1, c_2, \ldots, c_{m-1}$ denote the m channels in the network. Consider the following variant function that is a tuple of sets:

$$Y = \left(V_0, V_1, V_2, \ldots, V_{n-1}, c_0, c_1, c_2, \ldots, c_{m-1} \right)$$

Initially, $Y = (s(0), s(1), s(2), \ldots, s(n-1), \varnothing, \varnothing, \varnothing, \ldots, \varnothing)$. After each execution of statement 1 by an eligible process, the content of some channel (in the second part of Y) grows in size. After the execution of statement 2 by some eligible process k, the set V_k (in the first part of Y) grows in size and all the input channels of process k become null. In either case, whenever an eligible process executes an action, the value of the Y grows in the lexicographic order. This growth is bounded since by Lemma 9.2, finally, $V_i = \{s(k): 0 \le k \le n - 1\}$ holds. Therefore, the algorithm terminates in a bounded number of steps. ■

The worst-case message complexity can be computed as follows: a process i sends something out only when V_i changes. Starting from the initial value $s(i)$, V_i can change at most $(n-1)$ times. Also, since each node can have at most $(n-1)$ neighbors, each execution of statement 1 sends at most $(n-1)$ messages. Thus, a process can send at most $(n-1)^2$ messages. So the message complexity is $O(n^2)$ per process or $O(n^3)$ for the entire system.

9.3 TERMINATION-DETECTION ALGORITHMS

Consider a computation running on a network of processes whose topology is $G = (V, E)$. One possible mechanism of distributing the computation to the various processes is as follows: The task is initially assigned to some node $i \in V$ that will be called an *initiator* node. The initiator delegates various parts of this task to its neighbors, which delegate parts of their work to *their* neighbors, and so on. As the computation makes progress, these nodes exchange messages among themselves. No one has knowledge about the entire topology of the network, but every node knows about its local neighborhood.

A node, when viewed in isolation, can remain in one of the two states: *active* and *passive*. A process is active when it has some enabled guards. A process that is not active is called passive. If a process is in a passive state at a certain moment, then it does not necessarily mean that the process will always remain passive—a message sent by an active neighbor may wake up the process and make it active. An active process, on the other hand, eventually switches to a passive state when it has executed all its local actions—these actions may involve the sending of zero or more messages.

In this setting, an important question for the initiator is to decide whether the present computation has terminated. Termination corresponds to the following three criteria: (a) every process is in a passive state, (b) all channels are empty, and (c) the global state of the system satisfies the desired postcondition. Note that the criteria for *termination* are similar to those for *deadlock*, with the exception that in deadlock, the desired postcondition

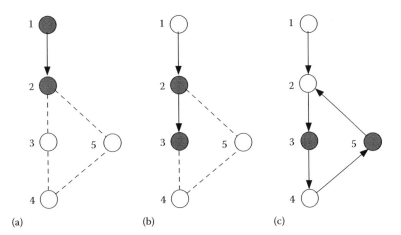

FIGURE 9.2 A computation graph with active (black) and passive (white) processes. (a) 1 engaged 2 and both are active, (b) 1 turned passive but 2,3 are active, (c) 1,4 became passive, but 5 is active and is trying to engage 2.

is not satisfied. Both termination and deadlock reflect quiescent conditions, and it is quiescence detection that we are interested in. The proposed detection method does not guarantee that the desired postcondition has been reached.

The network of processes participating in the computation forms a computation graph. An example is shown in Figure 9.2. The computation graph is a directed graph: if node i engages a neighboring node j by delegating a subtask, then we draw a directed edge from i to j. Messages propagate along the direction of the edges. Let us use colors to distinguish between active and passive processes: a process is *white* when it is passive, and *black* when it is active. In Figure 9.2a, only 1 and 2 are active. In Figure 9.2b, 1 has turned passive, but 2 and 3 are active. In Figure 9.2c, 2 turned passive, 4 turned passive after being active for a period, but 5 became active, and 5 is trying to engage 2. The picture constantly changes. Per our assumption, the computation is guaranteed to terminate, so eventually all nodes turn *white*, and all channels become empty. It is this configuration that the initiator node 1 wants to detect.

To see why termination detection is important, remember that many computations in distributed systems run in phases. Each phase of a computation runs over multiple processes in the system, and to launch phase $(i + 1)$, the initiator has to ascertain that phase i has terminated for every process.

9.3.1 Dijkstra–Scholten Algorithm

In [DS80], Dijkstra and Scholten presented a signaling mechanism that enables the initiator to determine whether the computation running on a network of processes has terminated. The computation initiated by a single initiator and spreading over to several other nodes in the network is called a *diffusing computation*, and its termination is reported to the initiator as a single event. The signaling mechanism is superposed on the underlying computation and is noninvasive in as much as it does not influence the underlying computation. We follow the original treatment in [DS80].

There are two kinds of messages in the network: *signals* propagate *along* the direction of the edges, and *acks* propagate in the *opposite* direction. The initiator is a special node (called the *environment* node) that has no edge directed toward it. Every other node is called an *internal* node and is reachable from the environment node via the edges of the underlying network.

For each directed edge (i, j), call node j a successor of node i and node s a *predecessor* of node j. The overall plan is as follows: The environment node initiates the computation by sending *signals* that engage its successors—this also initiates the termination-detection process. An internal node that receives a signal may send out signals to engage *its* successors. In this way, the computation spreads over a finite number of nodes in the network, and the computation graph grows. Eventually, each node sends *acks* to a designated predecessor to confirm the termination of the computation in the subgraph below it, and the computation subgraph shrinks. When the environment node receives acks from each of its successors, it detects the termination of the entire computation, and the computation subgraph becomes empty. The crucial issue here is to decide when and to whom to send the acks.

For an edge (i, j), the difference between the number of signals sent by i and the number of acks received from j will be called a *deficit*. A process keeps track of two different types of deficits:

C = total deficit along its incoming edges

D = total deficit along its outgoing edges

By definition, these deficits are nonnegative integers—no node sends out an ack before receiving a signal. This leads to the first invariant:

INV1. $(C \geq 0) \wedge (D \geq 0)$

Initially, for every node, $C = 0$ and $D = 0$. The environment node initiates the computation by spontaneously sending a message to each of its $k(k > 0)$ successors, so for that node, $C = 0$ and $D = k$. For every other node in the system, the proposed signaling scheme [DS80] preserves the following invariant:

INV2. $(C > 0) \vee (D = 0)$

INV2 is the cornerstone of the proposed algorithm. An internal node sends out signals only after it receives a signal from a predecessor node. This increases D for the sender, but it does not affect *INV2* as long as $C > 0$. The sending of an ack, however, reduces the sender's deficit C by 1—therefore, to preserve both *INV1* and *INV2*, an ack is sent when the following condition holds:

$$(C - 1 \geq 0) \wedge (C - 1 > 0 \vee D = 0) \quad \{\text{follows from } INV1 \text{ and } INV2\}$$

$$= (C > 1) \vee (C \geq 1 \wedge D = 0)$$

$$= (C > 1) \vee (C = 1 \wedge D = 0) \qquad (9.1)$$

This implies that an internal node returns an ack when its C exceeds 1 or when $C = 1$ and $D = 0$, that is, it has received acks from each of its successors (and, of course, the local computation at that node has terminated).

The proposed signaling scheme guarantees that the computation graph induced by the edges with positive deficits is a rooted spanning tree with the environment node as the root. The *parent* of an internal node is the first node from which it received a signal. The root does not have a parent. When a leaf node of this tree becomes passive, it sends an ack to its parent. This removes the node as well as the edge connecting it to its parent, and the tree shrinks. Again when an active node sends a message to a passive successor, the passive node becomes active, and the tree grows. By assumption, the underlying computation terminates. So during the life of the underlying computation, the computation graph may expand and contract a finite number of times and eventually becomes empty. This signals the termination of the underlying algorithm.

Initially for each internal node i, *parent*(i) is set to i. Equation 9.1 leads to the following program for each internal node:

```
program     detect {for an internal node i}
define      C, D : integer
            m: (signal, ack)  {represents the type of message received}
            state: (active, passive)
initially   C = 0, D = 0, parent(i) = i
do (m = signal) ∧ (C = 0)              →   C := 1; state := active;
                                           parent := sender
                                           {Read Note 1}
[] m = ack                             →   D := D - 1
[] (C = 1 ∧ D = 0) ∧ (state = passive) →   send ack to parent;
                                           C:= 0; parent(i) = i
                                           {Read Note 2}
[] (m = signal) ∧ (C = 1)              →   send ack to the sender
od                                         {Read Note 3}
```

Note 1: This node can send out messages to engage other nodes (which increases its D) or may turn passive. It depends on the computation.

Note 2: This node now returns to the initial state and disappears from the computation graph.

Note 3: This represents the action of sending an ack when $C > 1$ (see Equation 9.1).

By sending an ack to its parent, a node i provides the guarantee that all nodes that were engaged by i or its descendants are passive and no message is in transit through any channel leading to these descendants. The aforementioned condition will remain stable until node i receives another signal. A leaf node does not have any successor, so it sends an ack as soon as it turns passive. The environment node initiates the termination-detection algorithm by making $D = k$ ($k > 0$) and detects termination by the condition $D = 0$ after it receives k acks.

The last statement of the program needs some clarification. A node with $C = 1$ can receive a signal from a node, but it promptly sends an ack since the computation graph induced by the edges with positive deficits must always be a tree. Thus, in Figure 9.2c, if node 5 sends a signal to node 2, then node 2 becomes active but rejects the signal by sending an ack to 5 as a refusal to accept 5 as its parent, and the deficit on the edge (5, 2) remains 0. Thus, the computation graph remains a tree consisting of the nodes 1, 2, 3, 4, 5 and the edges (1, 2), (2, 3), (3, 4), (4, 5).

Dijkstra–Scholten algorithm is an example of a class of algorithms called *probe-echo* algorithm: The signals are the probes, and the acks are the echoes. One can use the basic messages from parents to their children as signals—the only control signals will be the acks. As the computation spreads over various nodes, the edges with positive deficits form a spanning tree that grows and shrinks. Due to the nondeterministic nature of the computation, different runs of the algorithm may produce different spanning trees. The number of acks will never exceed the number of messages exchanged by the underlying algorithm. This is because, for each message between nodes, an ack is generated. If the underlying computation fails to terminate (which violates our assumption), then the termination-detection algorithm will also not terminate. Note that the algorithm is not designed to report nontermination.

Complexity issues: Since the basic messages of the underlying computation are used as signals, and finally the deficits along all the edges must be zero, the number of control messages (i.e., acks) must equal the number of basic messages of the underlying computation. Chandrasekaran and Venkateshan [CV90] proved a lower bound that if the algorithm starts by sending out signals *after* the actual termination of the underlying computation, then it is possible to detect termination using only $O(|E|)$ messages—each edge will carry one signal and one ack. However, since the initiator does not correctly guess the time to reach termination, the lower bound will not be actually attained. Moreover, to avoid false detection, this scheme requires the edges to be FIFO.

The termination-detection algorithm can be easily modified to collect global states or computing other kinds of global predicates. For example, consider the problem of counting the number of processes in a network. If (1) the underlying computation is null, (2) the ack from each process k to its parent is tagged with an integer variable $size(k)$, the number of processes in the subtree under it, and (3) each parent i, after receiving the acks from all of its children, forwards the count

$$size(i) := 1 + \sum_{\forall j:\, i = parent(j)} size(j)$$

to its parent; then at the initiator node 0, $1 + size(0)$, will be the count of all the processes in the system.

9.3.2 Termination Detection on a Unidirectional Ring

Another class of termination-detection algorithms uses *token passing* (instead of probes and echoes) to detect termination. The main idea is as follows: Consider a distributed computation running on a strongly connected directed graph. An initiator node sends out a token to traverse the network and observe the states of the processes. Each node, after turning passive, forwards the token to the next process. When the token returns to the initiator, it contains useful information about whether the computation has terminated. We present here such an algorithm due to Dijkstra et al. [DFG83]. The algorithm is presented for a unidirectional ring that is embedded on the topology of the given network—the order of the processes in the ring is used to decide the order in which the token traverses the network. Assume that in a system of n processes 0, 1, 2,..., $n - 1$, the embedded ring is specified by $0 \rightarrow n-1 \rightarrow n-2 \rightarrow \cdots 2 \rightarrow 1 \rightarrow 0$. The ring topology has no connection with the sending and receiving of messages by the underlying algorithm—so a message can be sent by one process to another as long as a path exists, even if they are not neighbors in the embedded ring. All communication channels have zero delay, that is, message communication is assumed to be instantaneous. This also implies that messages are received in the order they are sent.

Without loss of generality, assume that process 0 is the initiator of termination detection (Figure 9.3). The initiator initiates termination detection by sending out a token—it traverses the network and eventually returns to the initiator. A process k accepting the token will not forward it to its ring successor $k - 1$ mod n until it becomes *passive*. When the initiator receives the token back, one may apparently believe that it detects termination.

However, this is too simplistic and is not foolproof. What if the token is currently with process k, but a process j ($n - 1 > j > k$) that was passive now becomes active by receiving a message from some process l ($k > l > 0$)? This could lead to a false detection, since the activation of process j will go unnoticed!

To prevent such a false detection, refine the scheme by assigning the colors *white* and *black* to processes and the token. Initially, all processes are white, and the initiator sends a *white token* to process ($n - 1$). Define the following two rules:

Rule 1: When a noninitiator process sends a message to a higher numbered process, it turns *black*.

Rule 2: When a black process sends a token, the token turns black. If a white process forwards a token, then the token remains white.

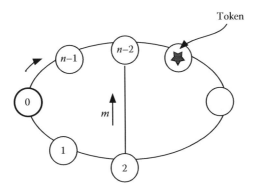

FIGURE 9.3 A ring of n processes: process 0 is the initiator. Process $n - 2$, after turning passive and releasing the token to its successor, received the message m from 2.

With these modifications, when the initiator receives a white token, termination is correctly detected. The scenario described earlier will now return a black token to the initiator (in Figure 9.3, process 2 turns black and transforms the white token into a black token), making the decision inconclusive. A fresh round of token circulation will be necessary.

If indeed all processes turn passive during the next traversal of the token, then a white token will return to process 0. However, a process like 2 needs to change its color to white before the next traversal begins. This leads to the last rule:

Rule 3: After sending a token to its ring successor, a black process turns white.

The final program is as follows:

```
program    term {for process i > 0}
define     color, token: (black, white) {colors of process and token}
           state : (active, passive)
do     (token = white) ∧ (state ≠ passive) → skip
[]     (token = white) ∧ (state = passive) →
           if color(i) = black → color(i) := white; send a black token
           [] color(i) = white → send a white token
           fi
[]     (token = black) →send a black token
[]     i sends a message to a higher numbered process → color(i) :=
           black
od

{for process 0}
send a white token;
do     (token ≠ white) → send a white token        od
{Termination is detected when process 0 receives a white token}
```

Theorem 9.2

Algorithm *term* is a correct termination-detection algorithm.

Proof: Assume that process 0 receives a white token. It means that

- Process 1 received a white token
- Process 1 is passive
- Process 1 did not send any message to process j, $(1 < j < n - 1)$

Since process 1 received a white token, it means that

- Process 2 received a white token
- Process 2 is passive
- Process 2 did not send any message to process j, $(2 < j < n - 1)$

Continuing these arguments, one can easily conclude that process 0 receives a white token when processes $1..n-1$ are all passive and none of them sent a message. This is a stable condition since processes $1..n-1$ cannot receive a message anymore. This corresponds to the termination of the computation. If some process i ($i \neq 0$) activates process 0 in the mean time, then process 0 will eventually turn passive and initiate a new round of token circulation until it receives a white token. ▪

Note 1: Since the underlying algorithm is guaranteed to terminate, the termination-detection algorithm will also terminate.

Note 2: The assumption about instantaneous message communication is a strong one but nevertheless necessary for correctness. To see why, assume that in Figure 9.3, the message m takes a very long time to reach its destination. In the mean time, the black token reaches the initiator, who initiates a fresh round of token circulation, while m is still in transit. Eventually, process 0 receives a white token back and declares termination. But clearly, this is false, since it will be negated when m will reach process $(n-2)$!

Complexity issues: Let t be the number of messages in the underlying computation. Once the initiator node starts the termination-detection algorithm by sending out a white token, each of the m messages can potentially turn the white token into a black one. Therefore, the maximum number of control messages is $n \cdot t$.

9.3.3 Credit-Recovery Algorithm for Termination Detection

A slightly different approach to termination detection has been taken in Mattern's *credit-recovery* algorithm. In this algorithm, there is an *initiator process* that is the custodian of a total *credit* share of 1 unit. The state of a process alternates between *active* and *passive*. While sending messages, credit shares get distributed across processes in the network according to some predefined rules. The activated processes eventually return their credit shares to its initiator. By definition,

- For each active process i, $credit(i) > 0$.

- For each passive process i, $credit(i) = 0$.

Termination is detected when the initiator has recovered the entire credit sum from the system of processes. The algorithm assumes that every process has a direct link with the initiator, so that at an appropriate time, any process can return its credit share directly to the initiator.

When one active process sends a message to another, it transfers a fraction of its credit to the activated process through the activation message. Thus, each activation message and each active process has a positive credit associated with it. When the initiator recovers the entire credit, it clearly implies that there is neither any active process nor any activation message in transit—this testifies the soundness of the approach. The rules for credit distribution and credit recovery are as follows:

Rule 1: An active process with credit X, while sending a message to another process, transfers a credit of $Y = X/2$ to it via the message. If the sender process sends the activation message immediately before becoming passive, it may also transfer its whole credit X with the activation message.

Rule 2: When a message with credit share Y arrives at a *passive* process, it becomes active and the credit share Y is transferred to the activated process.

Rule 3: When an active process receives a message with credit share Y, it remains active and Y is returned to the initiator. The algorithm assumes that each process has a direct communication channel with the initiator.

Since the credit shares are transferred from one process to another by piggybacking on the basic messages, the only control messages here are those that return the credit shares back to the initiator. It is easy to observe that the number of control messages equals the number of messages in the underlying computation. Thus, the message complexity is no better than of the Dijkstra–Scholten termination-detection algorithm. However, by modifying Rule 3, it is possible to reduce the message complexity. Consider the following modification:

Rule 3′: When an active process receives a message with credit share Y, it remains active and the credit share Y is added to the share of the recipient.

With this modification, the only time credits are returned to the initiator is when a process spontaneously turns passive from an active state. Accordingly, the message complexity of a typical computation is much lower. Another issue is the problem of bookkeeping with the fractional credits. While splitting a credit, it is not always possible to represent the fraction by a variable of limited precision. Mattern suggests a way to tackle this is by specifying the credit X as $c = -\log_2 X$. For further details, see [M89a].

9.4 WAVE ALGORITHMS

Various ideas for predicate detection or decision making based on state exploration have been proposed since Dijkstra–Scholten termination-detection algorithms were published—most of these are variations or generalizations of the basic ideas described in the previous section. One such generalization is *wave algorithms*.

An initiator process spontaneously starts a *wave algorithm* by executing a local action that triggers actions in the neighboring noninitiator nodes, which triggers further actions in *their* neighbors. The resulting computation is called a *wave*, and it satisfies the following criteria:

- Each computation is finite.
- Each computation contains at least one *decision* event.
- A decision event is causally preceded by some event at each process.

As an example, consider broadcasting a message M to every process in a unidirectional ring. The initiator sends out the message to its neighbor, who forwards it to its neighbor, and so on. When the initiator receives M, it stops forwarding M further, and the wave ends. This is a decision event, following which the initiator may initiate another wave to accomplish a different task. Here are a few examples of wave algorithms:

1. *Barrier synchronization*: Consider a distributed computation that runs in phases. An initiator process starts *phase* 0, and all other processes execute *phase* 0 thereafter. The constraint is as follows: no process starts *phase i* ($i > 0$) unless every process has completed its *phase* ($i - 1$). Here, the decision event is the end of a phase.

2. *Depth-first search* (DFS): Consider searching an object in a network of processes. An initiator initiates the search, and it ends with a decision event reporting to the initiator if the object was found or not.

3. *Propagation of information with feedback (PIF)* [S83]: PIF is a method for broadcasting a message to every node of a connected network and receiving a confirmation that *every process* received the broadcast. Such a mechanism can be used to wake up other nodes or start the execution of a new protocol at every node of a network. In the following, we present an example of a PIF from Segall's paper [S83].

9.4.1 Propagation of Information with Feedback

Consider a connected network $G = (V, E)$, where V is a set of nodes and E is a set of bidirectional edges. An initiator node $i \in V$ wants to broadcast a message M to every node of the network and receive a confirmation. The main idea is similar to that of Dijkstra–Scholten algorithm in Section 9.3.1; however, the feedback mechanism is slightly different—a process delays the sending of a copy of M (equivalent to an ack) to its parent until it receives a copy of M over *all links* incident on it. A PIF algorithm for broadcasting a value $\overset{.}{M}$ in a network is presented as follows:

```
program PIF {for the initiator node i}
define       count : integer
             N(i): set of neighbors of process i
send M to each neighbor; count := |N(i)|
do count ≠ 0 ∧ M received → count: = count - 1 od
{program for a non-initiator node j≠i}
if    message M received     → parent := sender
                               send M to each neighbor except parent;
                               count := |N(j)|;
[]    count > 0 ∧ M received → count: = count - 1
[]    count = 0              → send M to parent
fi
```

For every node, the condition *count* = 0 signals the end of PIF and is a confirmation that every node received M. Figure 9.4 illustrates the execution of the PIF algorithm on a network of five nodes.

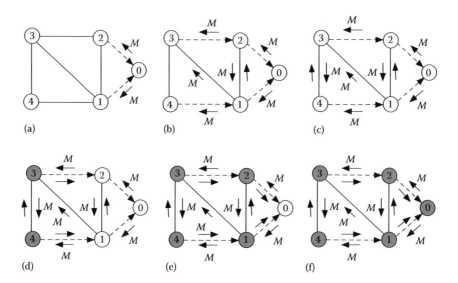

FIGURE 9.4 The execution of a PIF algorithm with node 0 as the initiator: the broken lines reflect the parent relationship, and the darkshade of the nodes indicates the reception of the message M through all the incident channels. (a)–(f) show the different phases of the algorithm.

9.5 DISTRIBUTED DEADLOCK DETECTION

In a distributed system, a set of processes is deadlocked, when the computation reaches some global state (different from the desired goal state) in which every process waits for some other process to execute an action—a condition that can never be satisfied. Since all well-behaved systems are expected to be free from deadlock, the motivation behind deadlock detection needs some justification. In resource-sharing distributed systems involving concurrent processes, there exist well-developed techniques for preventing the occurrence of deadlock. However, the overhead of deadlock prevention mechanisms is sometimes prohibitive, since it limits concurrency. Moreover, in real life, the occurrence of deadlock is somewhat rare. So a cheaper alternative is to avoid deadlock prevention techniques and let the computation take its own course with maximum possible concurrency—if deadlock occurs, then detect it and resolve it by preempting resources or preempting processes, as appropriate.

9.5.1 Resource Deadlock and Communication Deadlock

The detection of deadlock is simple when a single process acts as a central coordinator to oversee the usage of all the resources of the system. This is because the coordinator has a consistent picture of who is waiting for whom, which is represented using a *wait for graph* (WFG). Deadlock detection is more difficult in systems without a central coordinator, since fragments of the WFG are spread over various processes, and the computation of a consistent WFG is a nontrivial problem.

From an implementer's perspective, it is useful to study how the WFG is formed and maintained. Deadlock occurs when the following four conditions hold:

1. The access to each resource is mutually exclusive.

2. Requesting processes hold resources while requesting for more (hold and wait).

3. Resource scheduling is nonpreemptive.

4. Each process waits for another process to release a resource (circular waiting).

Consider a distributed database system where a set of processes executes a number of transactions. Each transaction needs to acquire one or more locks to gain exclusive access to certain types of objects—here, these locks are the resources. Let a process P execute a transaction T_1 for which it needs to acquire locks a, b, and c. Each process has a local resource controller: assume that the local controller of P manages lock a, but locks b and c are managed by the local controllers of processes Q and R, respectively. So after acquiring lock a through its local controller, P sends requests for locks b and c to the controllers of Q and R. This corresponds to the formation of two directed edges (P, Q) and (P, R) in the WFG. If Q concurrently executes another transaction s_2 that requires access to lock c, then the WFG will contain another directed edge (Q, R). Now, if the resource controller of R grants lock c to P, then Q will wait for P to release lock c. As a result, the edge (P, R) will disappear, and a new edge (Q, P) will be formed.

Before we search for a new algorithm for deadlock detection, let us ask: Couldn't we use Dijkstra–Scholten termination-detection algorithm for this purpose? This algorithm is certainly able to detect the condition when *every* process in the system is waiting. However, deadlock is also possible when a *subset of processes* is involved in a circular waiting condition. This is known as partial deadlock, and it cannot be readily detected using Dijkstra–Scholten's method. This motivates the search for other deadlock detection algorithms.

The choice of a proper algorithm also depends on the model of deadlock. Deadlocks arising out of the actions described in the transaction-processing scenario previously are characteristics of the *resource deadlock* model. In the *resource deadlock* model, a process waits until it has received *all* the resources that it has requested. The resource model is also called the AND model. In this model, a deadlock occurs if and only if there is a cycle of waiting processes, each dependent on the next process in the cycle to make progress. There is, however, another kind of deadlock model that has been considered in the present context—it is called the *communication deadlock* model. Consider the message-passing model of communication. The local states of the processes alternate between *active* and *passive*. A process P that is passive now may become active after receiving a message from *any one* of a set of processes—call it the *dependent set* *depend*(P) of P. In the corresponding WFG, a directed edge is drawn from each process in *depend*(P) to P. Since P can be activated by any of the processes in its dependent set, the corresponding model is called the OR model. We don't care about the activation mechanism but assume that everything is controlled by messages. In a subset S of the set of processes, a communication deadlock occurs when

1. Every process in S is passive

2. $\forall i \in S, depend(i) \subseteq S$

3. All communication channels between processes in S are empty

It is clear that since all incoming channels to the processes in S are empty, none of them can ever be active and the condition is stable. The corresponding subgraph is called a *knot*. This section discusses two distributed deadlock detection algorithms for these cases—these algorithms due to Chandy et al. [CMH83].

9.5.2 Detection of Resource Deadlock

In a system of n processes $0, 1, 2, \ldots, n - 1$, define $succ(i)$ to be the subset of processes that process i is waiting for. In the WFG, represent this by drawing a directed edge from process i to every process $j \in succ(i)$. An initiator node initiates deadlock detection by sending *probes* down the edges of the WFG—the receipt of a probe by the initiator of that probe indicates that the process is deadlocked—this is the main idea. These types of algorithms are also known as *edge-chasing* algorithms.

A *probe* $P(i, s, r)$ is a message with three components: i is the *initiator* process, s is the *sender* process, and r is the *receiver* process. The algorithm is initiated by a waiting process i, which sends $P(i, i, j)$ to every process j that it is waiting for. We use the following notation in the description of the algorithm:

Let $depend[j, i]$ be a Boolean that indicates that process i cannot progress unless process j releases a resource. Thus, $depend(j, i) \Rightarrow j \in succ^m(i)$ $(m > 0)$ in the WFG. Also, $depend[k, j] \wedge depend[j, i] \Rightarrow depend[k, i]$. The initiator process i is deadlocked when it discovers that $depend[i, i]$ is true.

Assuming that no process waits for an event that is internal to it, the program for a typical process k can be represented as follows:

```
program     resource deadlock {program for process k}
define      P() :        probe {has three fields initiator, sender,
                         receiver}
            depend[k]: array [0..n-1] of boolean
initially   ∀j: 0 ≤ j ≤ n-1, depend[k,j] = false
do      P(i,s,k) received ∧ k is waiting ∧ (k ≠ i) ∧ ¬depend(k, i)→
            ∀j ∈ succ(k): send P(i,k,j) to j;
            depend(k,i) := true
    [] P(i,s,k) received ∧ k is waiting ∧ (k = i) →
        process k is deadlocked
od
```

The aforementioned algorithm detects deadlock, only when the initiator process is contained in a cycle of the WFG. Thus, in Figure 9.5, if process 3 is the initiator, then it will

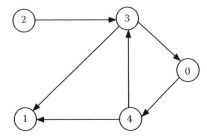

FIGURE 9.5 The WFG with five processes. Processes 0, 3, 4 are deadlocked.

eventually detect that it is deadlocked. However, process 2 is also unable to make progress, but using this algorithm, process 2 cannot detect it!* Another observation is this: The initiator will never know explicitly that it is *not* deadlocked—the absence of deadlock will only be signaled by the eventual availability of the resource that it requested. The proof of correctness follows.

Theorem 9.4

Deadlock is detected if and only if the initiator node belongs to a cycle of the WFG.

Proof: By definition, *depend*[*j, i*] implies that there is a directed path from process *i* to process *j* in the WFG. By the first action, every process forwards the probe to each of its successors in the WFG. Therefore, in a bounded number of steps, the initiator process *i* receives the probe and detects that it is deadlocked (second action). If the initiator does not belong to the cycle, then it will never receive its own probe, so deadlock will not be detected.

The first action guarantees that every probe is forwarded to its successors *exactly once*. Since the number of nodes is finite, the algorithm terminates in a bounded number of steps. ■

9.5.3 Detection of Communication Deadlock

The second algorithm proposed in [CMH83] detects the OR version of communication deadlock. As explained earlier, the OR model implies that when a process waits for messages from multiple processes belonging to its dependent set, it can go ahead when it receives *any one* of them. Consider the WFG in Figure 9.6. Here, process 3 will be able to move ahead when it receives a message from either 1 or 4 (in the AND model, both were necessary). Unfortunately, in this case, it will receive neither of them. In fact, no process will get the resource it needs. Notice the *knot* here: Each of the processes in the set {0, 1, 2, 3, 4} has a dependent set that belongs to S. This is an example of *communication deadlock*.

There are similarities between the termination-detection algorithm of Dijkstra and Scholten and the communication deadlock detection algorithm due to Chandy et al. A waiting process initiates the deadlock detection algorithm by first sending *probes*

---

* Of course this process is not included in any cycle in the WFG, so strictly speaking, this is not deadlocked, but it waits for some process that is deadlocked.

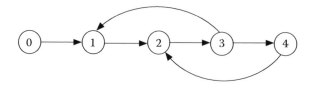

FIGURE 9.6   An example of a communication deadlock.

to all the processes in its dependent set. A process receiving the probe ignores it, if it is not waiting for any other process. However, if the recipient of a probe is waiting for another process, then it takes one of the following two steps:

- If this probe is received for the *first* time, then it marks the sender of the probe as its *parent* and forwards it to every process in its dependent set.

- If this is not the first probe, then it sends an ack to the sender.

When a waiting process receives acks from every process in its dependent set, it sends an ack to its parent. If the initiator receives ack from every process in its dependent set, then it declares itself deadlocked.

For any node, let *D* represent the *deficit* (i.e., *number of probes – number of acks*) along its outgoing edges. Starting from the initial condition in which an initiator node *i* has sent out probe *P(i, i, j)* to every node *j* in its dependent set *succ(i)*, the program is described as follows:

```
program   communication deadlock
define    P ()     : probe   {has three fields initiator, sender, receiver}
          parent : process
          ack      : message
          D        : integer
{program for the initiator node i}
initially node i send P(i,i,j) to each j ∈ succ(i), parent = null,
     D = |succ(i)|
do P(i,s,i)  →   send ack to s;
[] ack        →   D := D-1
[] D = 0      →   deadlock detected
od
{program for a non-initiator node k}
initially   D = 0, parent = k
do P(i,s,k) ∧ k is waiting ∧ (parent = k) →  parent := s;
                                             ∀j ∈ succ(k): send
                                                (i,k,j) to j;
                                             D := D + |succ(k)|

[] P(i,s,k) ∧ k is waiting ∧ (parent ≠ k)  →  send ack to s;
[] ack                                      →  D := D - 1
[] D = 0 ∧ k is waiting ∧ (parent ≠ k)      →  send ack to parent;
                                               parent := k

od
```

When the initiator sends out probes, the precondition ($D > 0$) holds for the initiator. Deadlock is detected when ($D = 0$) holds for the initiator and the algorithm terminates. The proof of correctness of the communication deadlock detection algorithm is similar to the proof of correctness of the termination-detection algorithm and is not separately discussed here.

Unlike the resource deadlock detection algorithm, the communication deadlock detection algorithm requires two different types of messages (*probe* and *ack*) but has wider applications. For example, using this algorithm, process 2 in Figure 9.5 can successfully detect that it cannot make progress, even if it is not part of a cycle in the WFG.

The message complexities of both the resource deadlock and the communication deadlock detection algorithms are $O(|E|)$, where E is the set of edges in the WFG.

9.6 CONCLUDING REMARKS

Most of these algorithms are superposed on an underlying basic computation that remains unaffected by the execution of the algorithms. This noninterference is much more pleasant than an alternative unrealistic approach in which the underlying computation is frozen, the desired global information is computed, and then the computation is restarted.

In deadlock detection, the presence of a cycle or a knot in the WFG is a stable property, since process abortion or resource preemption is ruled out. A frequently asked question is: What about possible modification of the proposed algorithms, so that not only the presence but also the absence of deadlocks is reported to the initiator? Remember that little is gained by detecting the absence of deadlock as it is an unstable property—if the initiator discovers that there is no deadlock, then there is no guarantee that the resource will be available, since much will depend on the pattern of requests after the absence of deadlock has been detected!

The edge-chasing algorithms for deadlock detection only record consistent global states—probes/acks report the finding *after* a cycle or a knot has been formed. Incorrect application of this rule or ad hoc detection techniques may lead to the detection of false deadlocks that plagued many old algorithms designed before the concept of consistent global states was properly understood.

9.7 BIBLIOGRAPHIC NOTES

The termination-detection algorithm presented in this chapter is described in [DS80]. The modified version with optimal message complexity appears in [CV90]. The token-based solution was proposed by Dijkstra et al. [DFG83]. Misra [Mi83] presented a marker-based algorithm for termination detection in completely connected graphs and suggested necessary extensions for arbitrary network topologies. Segall [Se83] introduced PIF algorithms. Chang [C82] explored probe-echo algorithms and demonstrated their applications for several graph problems. In [M87] and [M89a], Mattern described different variations of the termination-detection algorithm and introduced the credit-recovery algorithm. Tel [T00] formally defined the framework of wave algorithms. Both of the deadlock detection algorithms were proposed in [CMH83]. Knapp [K87] wrote a comprehensive survey of deadlock detection algorithms and its applications in distributed databases.

EXERCISES

9.1 Consider a unidirectional ring of n processes 0, 1, 2, ..., $n - 1$. Process 0 wants to detect termination, so after the local computation at 0 has terminated, it sends a token to process 1. Process 1 forwards that token to process 2 after process 1's computation has terminated, and the token is passed around the ring in this manner. When process 0 gets back the token, it concludes that the computation over the entire ring has terminated.

Is there a fallacy in the aforementioned argument? Explain.

9.2 Design a probe-echo algorithm to compute the topology of a network whose topology is an undirected connected graph. When the algorithm terminates, the initiator of the algorithm should have knowledge about all the nodes and the links in the network.

9.3 Design an algorithm to count the total number of processes in a unidirectional ring of unknown size. Note that any process in the ring can initiate this computation and more than one processes can concurrently run the algorithm. Feel free to use process ids.

9.4 Using well-founded sets, present a proof of termination of the Dijkstra–Scholten termination-detection algorithm on a tree topology.

9.5 In a distributed system, Figure 9.7 represents the WFG at a given time under the communication deadlock model. Use Chandy et al.'s algorithm to find out if node 1 will detect a communication deadlock. Briefly trace the steps.

9.6 In a resource-sharing system, requests for resources by a process are represented as ($R1$ and $R2$) or ($R3$ and $R4$) or....

a. How will you represent the WFG to capture the semantics of resource request?

b. Examine if Chandy–Misra–Haas algorithm for deadlock detection can be applied to detect deadlock in this system. If your answer is *yes*, then prove it. Otherwise propose an alternative algorithm for the detection of deadlock in this case.

9.7 Consider a network of processes: each process maintains a physical clock C and a logical clock LC. When a process becomes passive, it records the time C and sends a wave to other processes to inquire if all of them terminated by that time. In case the answer is not true, another process repeats this exercise. Devise an algorithm for termination detection using this approach. Note that there may be multiple waves in the system at any time. (This algorithm is due to Rana and is described in [R83].)

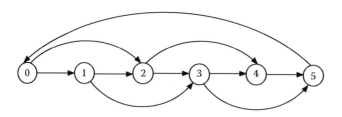

FIGURE 9.7   A WFG using the OR model.

CHAPTER 10

Graph Algorithms

10.1 INTRODUCTION

The topology of a distributed system is represented by a graph where the nodes represent processes and the links represent communication channels. Distributed algorithms for various graph theoretic problems have numerous applications in communication and networking. Here are some motivating examples.

The first example deals with routing in a communication network. When a message is sent from node i to a nonneighboring node j, the intermediate nodes route the message based on the information stored in the local routing table. This is called hop-by-hop or destination-based routing. An important problem is to compute these routing tables and maintain them, so that messages reach their destinations in the fewest number of hops, or with minimum delay. Finding the minimum hop route is equivalent to computing the *shortest path* between a pair of nodes using locally available information.

The second example focuses on the amount of space required by a node to store the routing table. Without any optimization, the space requirement is $O(n)$ where n is the number of nodes. But with the explosive growth of the Internet, n is increasing at a steep rate—therefore, the space requirement of the routing table as well as the overhead of maintaining the routing table are matters of concern. This leads to the following question: Can we reduce the size of the routing table? Given the value of n, what is the smallest amount of information that each individual node must store in their routing tables, so that every message eventually reaches its final destination?

The third example visits the problem of broadcasting in a network whose topology is represented by a connected graph. Uncontrolled transmission of messages leads to flooding, which wastes communication bandwidth. One way to save bandwidth is to transmit the messages along the edges of a *spanning tree* of the graph. How to compute the spanning tree of a graph? How to maintain a spanning tree when the topology changes?

The fourth example addresses the classical problem of computing of maximum flow between a pair of nodes in a connected network. Here the flow represents the movement of a certain commodity, like a bunch of packets, from one node to another. Each edge of the network has a certain capacity (read *bandwidth*) that defines the upper limit of the flow through that edge. This problem, known as the *maxflow problem*, is of fundamental importance in networking and operations research.

An important issue in distributed algorithms for graphs is that of static vs. dynamic topology. The topology is called *static* when it does not change. A topology that is not static is called *dynamic*—it is the result of spontaneous addition and deletion of nodes and edges, which reflects real-life situations. Clearly, algorithms for dynamic topologies are more robust than those for static topologies only. In this chapter, we will study distributed algorithms for solving a few graph theoretic problems.

10.2 ROUTING ALGORITHMS

Routing is a fundamental problem in networks. Each node maintains a *routing table* that determines how to route a packet to its destination. The routing table is updated when the topology changes. A route can have many attributes: These include the number of hops or the end-to-end delay. For efficient routing, a simple goal is to route a message using minimum number of hops. In a more refined model, a *cost* is associated with each link, and routing via the path of least cost may be required. For multimedia applications, routing delay is a major factor. Link congestion can influence the delay (so delay determines the cost). The path of minimum delay may change even if the topology remains unchanged. In this section, we will discuss a few algorithms related to routing.

10.2.1 Computation of Shortest Path

Let $G = (V, E)$ be an undirected graph where $V = \{0, 1, 2, ..., n - 1\}$ represents a set of processes and $E \subseteq V \times V$ represents a set of edges representing communication links. Define $N(i)$ to be the set of neighbors of node i. Each edge (i, j) has a weight $w(i, j)$ that represents the cost of communication through that edge. A simple path between a source and a destination node is called a *shortest path*, when the sum of all the weights in the path between them is the smallest of all such paths. The weight $w(i, j)$ of an edge is application dependent. For computing the path with minimum number of hops, we assume $w(i, j) = 1$. However, when $w(i, j)$ denotes the delay in message propagation through link (which depends on the degree of congestion), the shortest path computation can be regarded as the fastest path computation. To keep things simple, assume that $w(i, j) \geq 0$. Our goal is to present an asynchronous message-passing algorithm using which each node $i \in V$ can compute the shortest path to a designated node 0 (called the *source* or the *root* node) from itself. This is known as the *single-source shortest path* problem.

A well-known algorithm for computing single-source shortest path is the Bellman–Ford algorithm that was used to compute routes in the Advanced Research Projects Agency Network (ARPANET) during 1969–1979. In this algorithm, each process i maintains two variables:

- $D(i)$ is the *best knowledge* of node i about its shortest distance to node 0

- *parent(i)* is the neighbor that leads to node 0 via the shortest path

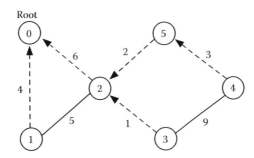

FIGURE 10.1   Shortest path computation in a weighted graph: For each node $i > 0$, the directed edge (represented by a broken line) points to its parent node.

Clearly $D(0) = 0$, and by definition, $parent(0) = null$. Initially $parent(i) = i$ and $\forall i > 0$: $D(i) = \infty$. As the computation progresses, $D(i)$ approaches its correct value (i.e., the shortest distance from node i to node 0). See Figure 10.1.

Denote a message from a sender by $(distance, sender\ id)$. The initiator node 0 initiates the algorithm by sending out $(D(0) + w(0, j), 0)$ to each node $j \in N(0)$. The program is described as follows:

```
program Bellman-Ford shortest path
{program for process 0}
send (D(0)+w(0,j),0) to each node in j ∈N(0)
{program for process j > 0, after receiving a message from process i}
do D(i)+ w(i,j) < D(j)→
        D(j):= D(i) + w(i,j);
        parent(j):= i;
        send the new D(j) to each node in N(j)\{i}
od
```

The action by a nonroot process is known as a *relaxation* step. When the computation terminates, for every process $i \in V$, $D(i)$ is the shortest distance from node i to node 0, and the path $(i, parent(i), parent(parent(i)) \ldots 0)$ defines the shortest path from node i to node 0. The algorithm works for both directed and undirected graphs. For directed graphs, the messages are sent along the outgoing edges and received via the incoming edges.

Lemma 10.1

When the algorithm terminates, let $k = parent(i)$. If $D(k)$ is the distance of the shortest path from k to 0, then $D(i) = D(k) + w(k, i)$ is the distance of the shortest path from i to 0 and the shortest path includes k.

Proof: Suppose this is false. Then the shortest path from i to 0 is via some neighbor j of i, where $j \neq k$. If $D(j) + w(j, i) < D(k) + w(k, i)$, then at some point, i must have received a message from j, and since $D(i) > D(j) + w(j, i)$, node i would have set $D(i)$ to $D(j) + w(j, i)$ and $parent(i)$ to j. Even if the message from k was received later, i would not have modified $D(i)$ any further, since $D(i) < D(k) + w(k, i)$ will hold. This contradicts the statement of the lemma. ▪

The shortest path from node i to node 0 must be acyclic. Since there are a finite number of acyclic paths from node i to node 0, and in each step $D(i)$ decreases, the algorithm terminates after a bounded number of steps.

10.2.1.1 Complexity Analysis

Under a *synchronous model* of computation, every eligible process executes a step in each round, and the messages reach their destinations before the next round begins. So, it takes at most $(n - 1)$ rounds for the algorithm to terminate. This is because, by Lemma 10.1, once $D(parent(i))$ attains its correct value, it takes one more round for $D(i)$ to become correct. Since (1) $D(0)$ is always correct, (2) the paths are acyclic, and (3) there are n nodes, it may take at most $(n - 1)$ rounds for every D to be correct and the algorithm to terminate. Since in every round every edge carries a message, the message complexity is $(n - 1) \cdot |E|$, i.e., $O(|V| \cdot |E|)$.

Under an *asynchronous model* of computation, the complexity of Bellman–Ford algorithm is higher. To visualize the worst case, consider the graph in Figure 10.2. Let n be odd. Assign the weights to the various edges as shown, where $k = (n - 3)/2$. The unlabeled edges have a weight 0. Now assume that the first message to reach node $(n - 1)$ from node 0 follows the path $0, 1, 2, 3, \ldots, (n - 2), (n - 1)$. This leads to

$$D(n-1) = \sum_0^k 2^k = 2^{k+1} - 1 = 2^{\frac{n-1}{2}} - 1$$

Let the next message reach node $(n - 1)$ following almost the same path, but only skipping node $(n - 2)$. The new value of $D(n - 1)$ will be $2^{k+1} - 2 = 2^{\frac{n-1}{2}} - 2$. Let the next message follow almost the same path, but in the last few steps, take the route $0, 1, 2, 3, \ldots, (n - 5)$, $(n - 3), (n - 2), (n - 1)$. This leads to $D(n-1) = 2^{\frac{n-1}{2}} - 3$. Continuing this style, at every subsequent step, create a path whose length is one less than the previous path. Since the shortest path has a length 0, it may take $2^{\frac{n-1}{2}} - 1$ steps for $D(n - 1)$ to settle down to the correct value. Therefore, the time complexity measured by the number of steps is exponential.

So far, we assumed that the edge weights are not negative. If this assumption is relaxed and cycles of negative weight are allowed to exist in a graph, then paths of arbitrarily small length can be created via repeated traversal of the cyclic path. Accordingly, the algorithm falls apart. Acyclic paths containing one or more edges with negative weights do not lead to this anomaly. Bellman–Ford algorithm has a mechanism to detect such cycles and aborting the computation.

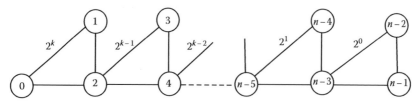

FIGURE 10.2 Analysis of the complexity of asynchronous Bellman–Ford algorithm.

10.2.1.2 Chandy–Misra Modification of the Shortest Path Algorithm

In [CM82], Chandy and Misra proposed a modification of this algorithm—the modified version not only detects the presence of cycles of negative weight but also detects the termination of the computation by leveraging Dijkstra–Scholten termination-detection algorithms discussed in Chapter 9. Corresponding to every message sent by a node, an *acknowledgment* signal (denoted by *ack*) is received. When the root node receives all acknowledgments, termination is detected, and the value of *D* at each node represents the distance of the shortest path between 0 and that node. To instrument this idea, define a variable *deficit*, representing the number of unacknowledged messages. Initially, for each node, *deficit* = 0. Node 0, after initiating the computation and sending a message to each of its neighbors, sets its own deficit to $|N(0)|$. The program is as follows:

```
program Chandy-Misra shortest path
{program for process 0}
send (D(0)+w(0,k),0) to each node in k ∈ N(0);
deficit:= |N(0)|;
do     D(i)+ w(i,0) ≥ D(0) → send ack to sender i
[]     deficit > 0 ∧ ack → deficit:=deficit - 1
od;    {deficit = 0 signals termination}
{program for process j > 0 after receiving a message from process i}
{initially ∀j:D(j) = ∞, deficit = 0}
do     D(i)+ w(i,j)< D(j)→
            if (deficit>0)∧(parent≠j)→ send ack to parent fi;
            D(j) := D(i)+ w(i,j);
            parent:=i; {the sender becomes the new parent}
            send the new D(j) to each node in N(j){i};
            deficit := deficit+|N(j)|-1
[]     D(i)+ w(i,j) ≥ D(j)→send ack to sender j
[]     deficit > 0 ∧ ack → deficit := deficit − 1
[]     (deficit=0) ∧ (parent≠j)→ send ack to parent; parent = j
od
```

To verify the termination-detection property, note that for a noninitiator node *j*, the predicate (*parent* ≠ *j*) indicates that node *j* has one unacknowledged message from its parent. When node *j* discovers another shorter path (to the initiator) through a different predecessor, it switches its parent. As in the Dijkstra–Scholten termination-detection algorithm, a node sends an acknowledgment to its *parent* only when its *deficit* = 0. If the initiator receives a message, then the condition *D*(*i*) + *w*(*i*, 0) ≥ *D*(0) must hold, and the initiator returns an acknowledgment to the sender. When the initiator receives a message for which the condition *D*(*i*) + *w*(*i*, 0) < *D*(0) holds, the existence of a cycle with negative weight is detected. This case is not discussed here.

When the weight of each edge is 1, the shortest path computation leads to the formation of the *breadth first search* (BFS) spanning tree with the initiator as the root. Every node with shortest hop distance D from the root has a parent whose shortest distance from the root is $D - 1$. The set of all nodes and the edges joining each node with its parent define the BFS spanning tree.

10.2.2 Distance-Vector Routing

Distance-vector routing uses the basic idea of shortest path routing but handles topology changes. The routing table of a node (i.e., a router) is an array of tuples (*destination, next hop, distance*). To send a packet to a given destination, it is forwarded to the node in the corresponding *next hop* field of the tuple. In a graph $G = (V, E)$ with $n = |V|$ nodes, the distance vector $D(i)$ of node i contains n elements $D(i, 0)$ through $D(i, n - 1)$, where $D(i, j)$ defines the shortest distance of node i from node j. The distance-vector elements are initialized as follows:

$$D(i, j) = 0 \quad \text{when } j = i,$$

$$= 1 \quad \text{when } j \in N(i), \quad \text{and}$$

$$= \infty \quad \text{when } j \notin N(i) \cup \{i\}$$

Each node j periodically advertises its distance vector to its immediate neighbors. Every neighbor i of j, after receiving the advertisements from *its* neighbors, updates its distance-vector elements (Figure 10.3a) as follows:

$$\forall k \neq i : D(i, k) = \min_j (w(i, j) + D(j, k))$$

When a node j or a link incident on j crashes, some neighbor k of it detects the event and sets the corresponding distance $D(j, k)$ to ∞. Similarly, when a new node j joins the network, or an existing node j is repaired, the neighbor k detecting it sets the corresponding distance to $D(j, k)$ to 1. Following this, the updated distance vector is advertised; the nodes receiving

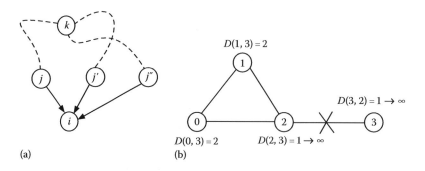

FIGURE 10.3 (a) Node i updates $D(i, k)$ from multiple advertisements received by it. (b) The distance vectors when link (2, 3) crashes.

the update appropriately modify their distance vectors and forward those values to their neighbors. In this way, eventually all nodes recompute their routing tables.

Unfortunately, depending on when a failure is detected, and when the advertisement is sent out, the routing table may not stabilize soon. Consider the network of Figure 10.3b. Initially, $D(1, 3) = 2$ and $D(2, 3) = 1$. As the link $(2, 3)$ fails, $D(2, 3)$ is set to ∞. But node 1 may still advertise $D(1, 3) = 2$ to nodes 0 and 2. As a result, node 2 will update $D(2, 3)$ to 3 and advertise it to nodes 1 and 0. Node 1, after receiving the advertisement from node 2, will subsequently update $D(1, 3)$ to 4. In this way, the values of $D(1, 3)$ and $D(2, 3)$ will slowly spiral upward until they become very large. This slow convergence is a major hurdle, and it is called the *count to infinity* problem. In the implementations of distance vector, ∞ is represented by a sufficiently large integer. The larger this integer, the slower the convergence.

A *partial* remedy to the slow convergence rate is provided by the *split horizon* method, where a node i is prevented from advertising $D(i, k)$ to a neighbor j if j is the first hop for destination k. For example, in Figure 10.3b, the split horizon method will prevent node 1 from advertising $D(1, 3) = 2$ to node 2. However, if node 1 detects a *direct connection* to 3, then node 2 is free to advertise $D(1, 3) = 1$ to node 1. Note that this only avoids delays caused by cycles of two nodes (i.e., 1 learning from 2 and 2 learning from 1), and not cycles involving more than two nodes. *Split horizon* is often used with *poison reverse,* where nodes adjacent to bad links (poisoned routes) advertise it back to other nodes in the network instructing that certain paths no longer exist and should be removed from their routing tables. In Figure 10.3b, node 2 will announce that the route $(2, 3)$ is poisoned (done by assigning a sufficiently large value *max* to $D(2, 3)$). After learning about a poisoned route, other nodes will remove all paths using this link from their tables. Normal operation resumes when node 2 notifies about the restoration of this link. This speeds up the convergence to some extent.

Distance-vector routing protocols mostly run on small networks (usually fewer than 100 nodes). Examples of distance-vector routing protocols used in practice include RIP (Routing Information Protocol) and IGRP (Interior Gateway Routing Protocol). RIP uses a limit on the hop count to determine how many nodes (routers) messages must go through to reach its destination. A node is considered unreachable if the hop count exceeds the limit. The scalability of these protocols is poor and the slow convergence following a topology change is an issue.

10.2.3 Link-State Routing

This is an alternative method of shortest path routing. In comparison with distance-vector routing, link-state routing protocol converges much faster and has better scalability. There are two phases in link-state routing. In *phase 1*, each node i periodically broadcasts a *link-state packet* (LSP) consisting of the weights of all edges (i, j) incident on it to every other node in the network. This is known as *reliable flooding*. In phase 2, each node collects the LSPs from other nodes, independently computes the topology of the network, and determines the shortest route between any pair of nodes using a known shortest path algorithm.

Reliable flooding must guarantee that eventually every node receives the LSPs from every other node and stores them in the array *local*. In this phase, there are two challenges: The *first challenge* is how to control the termination of flooding so that the LSPs do not roam around the network forever, and the second challenge is how to cope with node (i.e., router) failures and restarts.

The first challenge is handled by asking each node to forward every LSP to its neighbors *exactly once*—duplicate copies are discarded. Once each node receives every LSP, no packet is in circulation any more. An LSP $L(id, state, seq)$ is a message with three components: *id* is the identifier of the *initiator* process, *state* is the *link state* consisting of the weights of all edges incident on the initiator, and *seq* is the *sequence number* of the packet. These sequence numbers help detect duplicate copies of LSPs originating from a given node. The sequence numbers start from 0 and increase monotonically, so a new LSP with a larger sequence number reflects that it is more recent and replaces an old packet with a smaller sequence number. Only the most recent LSP is forwarded to the neighbors. When failures are not taken into consideration, the correctness follows trivially. The total number of LSPs circulating in the network due to the change of the neighborhood of a single node is $|E|$, since the packet traverses each edge exactly once. The protocol is as follows:

```
program  link state {for node i}
define   L() : link state packet LSP
         seq : integerz
         local : array [0..n−1] of LSP {local[k] is the LSP from node k}
         {initially, seq = 0, local[k] := (k, undefined for ∀k ≠ i, 0)}
do   neighborhood change detected →
                     compute link state S;
                     send L(i, S, seq) to k∈N(i);
                     local[0] := (i, S, seq)
                     seq := seq + 1
[]   L(j, S, seq) received →
         if (j = i) → discard L(j, S, seq)
         [] (j≠i) ∧ (L.seq > local[j].seq) →
                     enter L(j, S, seq) into the local database;
                     forward L(j, S, seq) to k∈N(i)\{sender}
         [] (j≠i) ∧ (L.seq ≤ local[j].seq) →
                     discard L(j, S, seq)
         fi
do
```

An interesting issue here is the issue of overflow of the sequence number field, which can cause the value of *seq* to abruptly change from its largest value to 0 and create confusion about whether the packet with *seq* = 0 is a new or an old packet. An apparent remedy is to use a large sequence number space—for example, with a 32-bit *seq*, even if a router creates a new LSP every 60 s, it will take more than 200 years for *seq* to overflow. This is more than the expected life of most routers. A 64-bit *seq* will be even better.

The second challenge, that is, the failure of a node (or the temporary unavailability of a router), can make the algorithm more complicated. The failure is detected by a neighbor, which marks the link to the faulty node as unavailable. Subsequently, the detecting nodes appropriately update their local states before the next broadcast. When a node i crashes, the LSPs stored in it are lost. Subsequently, when node i resumes operation, it has no clue about the previous value of its *seq*, so it reinitializes its *seq* to 0. As a consequence, other nodes will discard the newer packets from node i in favor of older packets transmitted by i in the past. This will continue until the value of its *seq* exceeds the *last value* of *seq* in the LSPs transmitted by i before the crash. To cope with such an anomalous behavior, each LSP also contains a *time-to-live* (TTL) field, which is an *estimate* of the time after which a packet should be considered stale and removed from the local databases.

Even with a large space for *seq*, sometimes a malfunction can push the value of *seq* to the edge of overflow. Before a node consciously causes an overflow, it should wait long enough for its past LSPs to age out. The actual version of the protocol uses several optimizations over the basic version described here.

Compared to distance-vector routing, link-state routing has better convergence rate. Link-state routing is used in Open Shortest Path First (OSPF).

10.2.4 Interval Routing

Consider a connected network of n nodes. The conventional routing table used to direct a message from one node to another is an array of $(n - 1)$ entries, one for each destination node. Each entry is of the type (*destination, port number*):*destination* = v and port number = k imply that to send a packet to its destination v, the node should forward it to port k. Since the size of the routing table grows linearly with the size of the network, scalability suffers. Can we do something to reduce the growth of the routing tables even if n is large? *Interval routing* is such a scheme.

Santoro and Khatib [SK85] first proposed interval routing for tree topologies only. To motivate the discussion on interval routing, consider the network shown in Figure 10.4. Each node has two ports: *port 0* is connected to a node of higher id, and *port 1* is connected with a node of lower id.

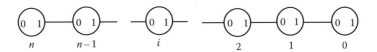

Condition	Port Number
Destination > id	0
Destination < id	1
Destination = id	Local delivery

FIGURE 10.4   An illustration of compact routing.

To take a routing decision, a process simply compares its own id with the id of the destination node in the incoming packet. If the destination id is larger than its own id, then the message is routed through port 0. If the destination id is smaller, then the message is forwarded through port 1. If the two ids are equal, then the message is meant for local delivery. Clearly, in this case, the number of entries in the routing table does not change with the size of the network. This is an example of a compact routing table.

Interval routing uses a similar concept. Each node has one or more ports, and each port is labeled with a nonnegative integer. In a network of n nodes numbered $0.. n - 1$, define the *interval* $[p, q)$ between ports p and q as follows:

$$\textbf{if } p < q \textbf{ then } [p,q) = p, p+1, p+2, \ldots, q-2, q-1$$

$$\textbf{if } p \geq q \textbf{ then } [p,q) = p, p+1, p+2, \ldots, n-2, n-1, 0, 1, 2, \ldots, q-2, q-1$$

As an example, if $n = 8$, then $[3, 7) = 3, 4, 5, 6$, $[5, 5) = 5, 6, 7, 0, 1, 2, 3, 4$ and $[6, 1) = 6, 7, 0$. Each node arranges the ports in the ascending order and computes the intervals between successive port numbers. Clearly, these intervals are nonoverlapping. Routing uses the following rule:

10.2.4.1 Interval Routing Rule
If the destination of a message belongs to the interval $[p, q)$, then send the message to port p.

The key problem in interval routing is the assignment of appropriate labels to the various nodes and their ports. Figure 10.5 shows a labeling scheme for a tree with $n = 11$ nodes 0, 1, 2, …, 9, 10 and illustrates how data will be forwarded.

In Figure 10.5, node 1 will send a message to node 5 through port 3, since the destination 5 is in the interval [3, 7). However, if node 1 wants to send the message to node 9,

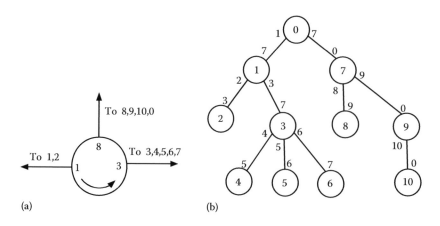

(a) (b)

FIGURE 10.5 An example of interval routing: (a) ports and message destinations. (b) A labeled tree of 11 nodes. Node labels appear inside the circles, and port numbers are assigned to each port connecting to a neighbor.

then it has to route it through port 7, since the destination 9 belongs to the interval [7, 2). Presented in the following is a labeling scheme for a rooted tree of *n* nodes:

1. Label the root as node 0, do a *preorder* traversal of the tree, and label the successive nodes in ascending order starting from 1.

2. For each node, label the port towards a child by the node number of the child. Then, label the port towards the parent by $L(i) + T(i) + 1$ mod n, where

   a. $L(i)$ is the label of the node i

   b. $T(i)$ is the number of nodes in the subtree under node i (excluding i)

As a consequence of the preorder traversal, the first child of node i has a label $L(i) + 1$, and the last child has a label $L(i) + T(i) + \mod n$. Thus, the interval $[L(i) + 1 \mod n, L(i) + T(i) + 1 \mod n)$ contains the labels of all the nodes in the subtree under i. The complementary interval $[L(i) + T(i) + 1 \mod n, L(i) + 1 \mod n)$ includes every destination node that does not belong to the subtree under node i.

For nontree topologies, a simple extension involves constructing a spanning tree of the graph and using interval routing on the spanning tree. However, this method does not utilize the nontree edges to reduce the routing distances. Van Leeuwen and Tan [LT87] proposed an improved labeling scheme for interval routing on nontree topologies—their method uses some nontree edges for efficient routing. Figure 10.6a illustrates an example of optimal labeling on a ring topology. Note that not all labeling leads to *optimal* routes towards the destination. For trees, this is a nonissue, since there is exactly one path between any pair of nodes.

Optimal paths can be more easily found in arbitrary networks if multiple labels are used for the various ports. This leads to *multiple intervals* for each destination: thus, if two successive ports p, q of a node bear the labels $L(p) = p1, p2$ and $L(q) = q1, q2$, then a packet can be

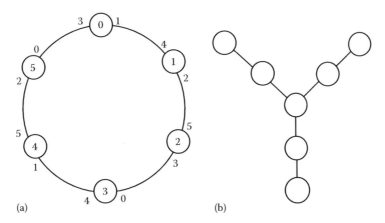

(a)   (b)

FIGURE 10.6   (a) An optimal labeling scheme on a ring of six nodes: Each node *i* has two ports with labels $(i + 1) \mod 6$ and $(i + 3) \mod 6$. (b) A network for which no *linear* interval-labeling scheme exists.

routed through port p when the destination id belongs to any one of the intervals $[p1, q1)$ or $[p2, q2)$. Optimal routing may need up to $\Theta(n)$ labels per node, but a slight compromise with optimality has the potential to drastically reduce the size of the routing table. A variant of the interval-labeling scheme uses only a linear interval for each port with no wraparound. This is called *linear interval routing*. In [BLT91], the authors show that although some well-known networks like hypercubes and grids have feasible linear interval routing schemes, there exist graphs in which no linear interval routing is possible. The example in Figure 10.4 demonstrates *linear* interval routing by assigning the interval $[i + 1, n - 1]$ to port 0 and $[0, i - 1]$ to port 1 of node i, but no linear interval routing exists for the network in Figure 10.6b.

10.2.5 Prefix Routing

While compactness of routing tables is the motivation behind interval routing, a major obstacle is its poor ability of adaptation to changes in the topology. Every time a new node or a link is added to a network, a very large fraction of the node and port labels have to be recomputed. This is awkward.

An alternative technique for routing using compact routing tables is *prefix routing*, which overcomes the poor adaptivity of classical interval routing to topology changes. Figure 10.7 illustrates the concept of prefix routing. The label of a node or a port is a string of characters from an alphabet $\sigma = \{a, b, c, d, ...\}$. The additional symbol λ designates the *empty string*, and $\forall x \in \sigma, \lambda \cdot x = x$. To assign labels, first construct a *spanning tree* of the given network, and designate a node as the *root*. Then, use the following rules:

1. Label the root by λ.

2. If a node has a label L, then label its child by $L \cdot x$ ($x \in \sigma$), that is, extend the label by an element of σ. The added label must be unique for each child. (Thus, the labels of two children of this node may be $L \cdot a$ and $L \cdot b$.)

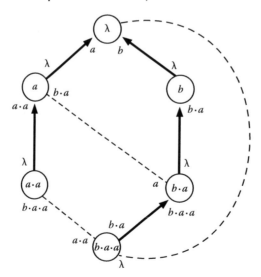

FIGURE 10.7 Prefix routing in a network. The broken lines denote the nontree edges and the directed edges point to the parent of a node.

3. Label every port from a parent to its child by the label of the child and every port from a child to its parent by the empty string λ.

4. If (u, v) is a nontree edge, then label the port of node u by the label of node v. If, however, node v is the root, then label the port from u to its parent p by the label of node p (instead of λ in Rule 4).

The first three rules are adequate for labeling a tree. Rule 4 is needed for dealing with non-tree topologies only.

Let us now examine the rules for routing a message to its destination. Let X be the label of the destination node. When a node with label Y receives this message, it makes the routing decision as follows:

```
program prefix routing
{Y = label of the current node, X = label of the destination}
if X = Y → deliver message locally
[] X ≠ Y → forward message to the port labeled with the longest
      prefix of X
fi
```

When a new node is added to a tree, new labels are assigned to that node and the edge connecting it, without modifying the labels of any of the existing nodes and ports. The size of the labels equals the depth of the spanning tree, and the size of the routing table at each node is $O(\Delta)$, where Δ is the degree of the node.

A variation of prefix routing is used in a class of structured overlay networks, commonly known as peer-to-peer networks. These are very large networks used for content sharing, so scalability is an important issue. The form of compact routing used in such networks guarantees that both the size of the routing table and the number of hops needed to reach one node from another are $O(\log n)$. We will address this in Chapter 21.

10.3 GRAPH TRAVERSAL

Given an undirected connected graph $G = (V, E)$, a traversal is the process of visiting all the nodes of the graph before returning to the initiator. A single initiator initiates each traversal. The visitor is a message (or a token or a query) that moves from one node to its neighbor in each hop. At any stage, there is a single message in transit. Since no node has global knowledge about the topology of G, the routing decision at each node is completely local. Traversal algorithms have numerous applications, starting from simple multicast and global state collection to web crawling, network routing, and solving game strategy–related problems.

Traversals on specific topologies like ring, tree, or clique are well covered in many textbooks. We will focus only on traversal of general graphs. The intellectually challenging task is the correctness of the traversal algorithm that will certify that all nodes will be visited and the visitor will eventually return to the initiator.

One approach is to construct a *spanning tree* of the graph and use a tree traversal algorithm. For a tree, the two important traversal orders are DFS and BFS. The shortest path algorithms generate a BFS tree when the edge weights are equal. In the following, we present a couple of algorithms for the construction of spanning trees.

10.3.1 Spanning Tree Construction

Let $G = (V, E)$ represent an undirected graph where V is the set of nodes and E is the set of edges. A spanning tree of G is a maximal connected subgraph $T = (V, E')$, $E' \subseteq E$ such that if one more edge from the set $(E \backslash E')$ is added to T, then the subgraph ceases to be a tree.

To construct a rooted spanning tree, a specific node is designated as the *root*. Several algorithms described in this chapter and in the previous chapter generate a spanning tree. For example, Dijkstra–Scholten termination-detection algorithm (Chapter 9) generates a spanning tree while detecting the termination of a diffusing computation. The Bellman–Ford shortest path algorithm also generates a BFS spanning tree when the weight of each edge is 1. In this section, we present another asynchronous message-passing algorithm for constructing a rooted spanning tree proposed by Chang [C82]. In this algorithm, the root node initiates the construction by sending out *empty messages* as *probes* to its neighbors. A process receiving a probe for the first time forwards it to all the neighbors except the one from which it received the probe; otherwise, it sends the probe back to the sender. Thereafter, each process counts the number of probes that they receive. When the number of probes received by the initiator node equals the number of its neighbors, the algorithm terminates. The steps are presented in the following:

```
program Changs's spanning tree
define probe, echo: messages, parent: process
initially ∀i>0, parent(i)=i, parent(0)=undefined
{program of the initiator node 0}
1  send probe to each neighbor j ∈ N(0)
2  do number of echoes ≠ number of probes →
3    echo received → echo:= echo + 1
4    probe received → send echo to the sender
5  od
6  {program for node j>0 , after receiving a probe}
7  first probe → parent:= sender; forward probe to non-parent
   neighbors;
8  do number of echoes ≠ number of probes →
9    echo received → echo:=echo+1
10   probe received → send echo to the sender
11 od
12 send echo to parent; parent(i):= i
```

The set of nodes and the edges connecting each node with its parent define the spanning tree. In the following, we argue about the termination of Chang's algorithm:

Define *deficit* = (*number of probes – number of echoes*) in the entire system. Let P denote the number of processes that have not received the probe m so far. We use $F = (P, deficit)$

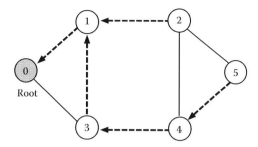

FIGURE 10.8 A spanning tree generated by Chang's algorithm. The directed edge from each nonroot node points to its parent.

as the variant function. Initially, $P = n - 1$ and *deficit* = 0. Observe that after every action, the value of F decreases lexicographically. Actions 3, 4, 9, 10, 12 decrease the value of *deficit* and thus reduce F. Actions 1, 7 increase the number of *probes* (and therefore the value of *deficit*) but, at the same time, reduce the value of P, so F decreases lexicographically. The smallest possible value of F is (0,0), which represents the terminal configuration. Therefore, the system reaches the terminal configuration in a bounded number of steps.

Figure 10.8 shows the result of such a construction with 0 as the root. The structure of spanning tree depends on the message propagation delays. Since these delays are arbitrary, different runs of the algorithm lead to different spanning trees.

The message complexity is $2 \cdot |E|$ since through each edge, a probe and an echo travel exactly once. If the root of the spanning tree is not designated, then to use the previous algorithm, a root has to be identified first. This requires a leader election phase. Leader election will be addressed in a subsequent chapter.

10.3.2 Tarry's Graph Traversal Algorithm

In 1895, Tarry proposed an algorithm [Ta1895] for graph traversal. It is the oldest known traversal algorithm and hence an interesting candidate for study. An initiator sends out a token to discover the traversal route. Define the *parent* of a node as one from which the token is received for the *first time*. All other neighboring nodes will be called *neighbors*. By definition, the initiator does not have a parent. The following two rules define the algorithm:

Rule 1: Send the token toward each neighbor exactly once.

Rule 2: If rule 1 cannot be used to send the token, then send the token to its parent.

When the token returns to the root, the entire graph has been traversed.

In the graph of Figure 10.9, a possible traversal route for the token is 0 1 2 5 3 1 4 6 2 6 4 1 3 5 2 1 0. Each edge is traversed twice, once in each direction, and the edges connecting each node with its parent form a spanning tree. Note that in different runs, Tarry's algorithm may generate different spanning trees, some of which are not DFS.

To prove that Tarry's algorithm is a traversal algorithm, we need to show that (1) at least one of the rules is applicable until the token returns to the root and (2) eventually every node is visited.

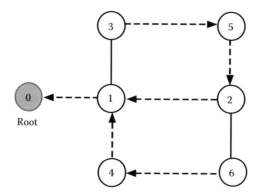

FIGURE 10.9   A possible traversal route 0 1 2 5 3 1 4 6 2 6 4 1 3 5 2 1 0. The directed edges show the parent relationship, and these edges induce a spanning tree.

Lemma 10.2

The token has a valid move until it returns to the root.

Proof: Initially when the token is at the root, rule 1 is applicable. Assume that the token reached a node $i \neq root$ from node $j = parent(i)$. It must have reached there. If rule 1 does not apply, then rule 2 must be applicable since the path from i to its parent node j remains to be traversed. It is not feasible for the token to stay at i if that path is already traversed. Thus, the token has a valid move.   ■

Lemma 10.3

Eventually, every node is visited by the token.

Proof (by contradiction): Consider a node j that has been visited, but a neighbor $k \in N(j)$ has not been visited and the token has returned to the root. Since the token finally leaves j via the edge toward its parent (rule 2), j must have forwarded the token to every neighbor (rule 1) prior to this. This includes k, and it leads to a contradiction.   ■

Since the token traverses each edge exactly twice (once in each direction), the message complexity of Tarry's algorithm is $2 \cdot |E|$.

10.3.3 Minimum Spanning Tree Construction

A given graph $G = (V, E)$, in general, can have many different spanning trees. To each edge of G, assign a weight to denote the cost of using that edge in an application. The weight of a spanning tree is the sum of the weights of all its edges. Of all the possible spanning trees of a graph, the spanning tree with the smallest weight is called the *minimum spanning tree* (MST). The MST has many applications. Consider building a subway system connecting a number of places of interest in a metro. There is a cost of digging underground tunnels and laying train lines, and this cost depends on the pair of endpoints chosen. An MST connecting the places of interest will keep the project cost to a minimum. In a communication network, if there is a predefined cost for sending a packet across the different edges, then the MST helps broadcast data packets to all nodes at minimum cost. Two well-known sequential algorithms for computing the MST are *Prim's algorithm* (also called Prim–Dijkstra algorithm) and *Kruskal's algorithm*.

Prim's algorithm of building the MST is a greedy algorithm and starts with a tree $T = (V', E')$ where $V' = \{i\}$ (the construction can start from any node $i \in V$) and $E' = \emptyset$. The MST construction augments T by adding an edge $(k, j) \in E$ such that (1) $k \in V'$ and (2) $j \notin V'$, and (k, j) has the smallest weight of all such edges. This augmentation clearly guarantees that no cycle is created. Recursive application of this step leads to the final MST, when $V' = V$. In case a choice has to be made between two or more edges with the same cost, any one of them can be chosen.

Kruskal's algorithm is also a greedy algorithm but works somewhat differently: It starts with a forest $G' = (V, E')$, where $E' = \emptyset$, and augments G' by adding the edge $(k, j) \in E$ such that (1) (k, j) has the smallest weight of all the edges not belonging to E' and (2) no cycle is created. When $(|V| - 1)$ edges have been added, the MST is formed. As in Prim's algorithm, when a choice has to be made between two or more edges with the same weight, anyone of them can be chosen.

Before we present a distributed algorithm for MST, consider the following lemma.

Lemma 10.4

If the weight of every edge is distinct, then the MST is unique.

Proof (by contradiction): Suppose the MST is not unique. Then, there must be at least two MSTs: MST1 and MST2 of the given graph. Let $e1$ be the edge of the *smallest* weight that belongs to MST1 but not MST2. Add $e1$ to MST2—this will form a cycle. Now, break the cycle by deleting an edge $e2$ that does not belong to MST1 (clearly, $e2$ must exist). This process yields a tree whose total weight is lower than that of MST1. So MST1 cannot be an MST. ■

In [GHS83], Gallager et al. proposed a distributed algorithm for constructing the MST of a connected graph in which the edge weights are unique. Their algorithm works on a message-passing model and can be viewed as a distributed implementation of Prim's algorithm. It uses a bottom-up approach (Figure 10.10). The main strategy is summarized later.

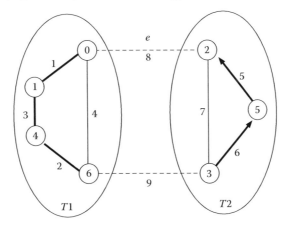

FIGURE 10.10 An example showing two fragments $T1$ and $T2$ being joined by a minimum cost edge e into a larger fragment.

10.3.3.1 Overall Strategy

Let $G = (V, E)$ be an undirected graph. Consider two MSTs $T1 = (V1, E1)$ and $T2 = (V2, E2)$ (called *fragments* in GHS83) involving the subsets of nodes $V1$ and $V2$, respectively, such that $V1 \cap V2 = \emptyset$. Let e be an edge with the *minimum weight* connecting a node in $T1$ with a node in $T2$. Then, the subgraph consisting of $T1$, $T2$, e is the MST covering the nodes $V1 \cup V2$.

Initially, each single node is a fragment. The repeated application of the previous strategy forms the MST of G. However, a distributed implementation of the merging algorithm involves the following challenges:

Challenge 1: How will the nodes in a given fragment identify the edge (of least weight) to be used for connecting with a different fragment?

The answer is that each fragment will designate a *coordinator* (also called the *root*) to initiate the search, and this coordinator will coordinate the task of choosing the least weight outgoing edge connecting to a different fragment.

Challenge 2: How will a node in fragment $T1$ determine if a given edge connects to a node of a different tree $T2$ or the same tree $T1$? In Figure 10.10, why will node 0 choose the edge e with weight 8, and not the edge with weight 4?

A solution is that all nodes in the same fragment must acquire the *same name* before the augmentation takes place. The augmenting edge must connect to a node belonging to a fragment with a *different* name.

In [GHS83], each fragment belongs to a *level*, which is a nonnegative integer. Initially, each individual node is a fragment at *level 0*. Fragments join with one another in the following two ways:

- (*Merge*) A fragment at level L connects to another fragment at the *same level*. The level of the resulting fragment becomes $(L + 1)$, and the resulting fragment is named after *the edge joining the two fragments* (which is unique since the edge weights are unique). In Figure 10.10, the combined fragment will be named 8, which is the weight of the edge e.

- (*Absorb*) A fragment at level L joins with a fragment at level $L' > L$. In this case, the level of the combined fragment becomes L'. The fragment at level L acquires the name of the fragment at level L'.

As a consequence of the aforementioned two operations, each fragment at level L has *at least* 2^L nodes in it. The grand plan is to generate the MST in *at most* $\log_2 n$ levels, where $n = |V|$. One can argue that instead of a larger fragment absorbing the smaller one, the smaller fragment could absorb the larger one. However, the number of messages needed for one fragment $T1$ to be absorbed by another fragment $T2$ depends on the size of the $T1$—so the proposed rule will lead to a lower message complexity.

Each fragment maintains a rooted spanning tree. Communication inside a fragment takes place via the edges of the spanning tree. Initially, every singleton node is a fragment, and this node serves as the root of its own fragment. Each node then looks for the least weight edge connecting to a neighbor. If both nodes pick each other, then a fragment of two nodes is formed at *level* 1. In this fragment, the node with *higher id* serves as the *new root*. In general, whenever two fragments merge, the node with higher id across the least weight outgoing edge serves as the new root. During an absorb operation, however, the root of the fragment with a larger level number continues to serve as the new root. This new *root* acts as the coordinator of that fragment for the next stage of expansion. The notification about the change of root within a fragment is sent out using the *changeroot* message. To facilitate communication within a fragment, every node keeps track of its *parent* and *children*—for multicast, the root uses the chain of child pointers, and for convergecast, nodes reach the root following the chain of parent pointers.

10.3.3.2 Detecting the Least Weight Outgoing Edge

When the root sends an *initiate* message, the nodes of that fragment search for the least weight outgoing edge (*lwoe*). Each node reports the finding through a *report* message to its parent. When the root receives the report from every process in its fragment, it determines the least weight outgoing edge for that fragment. The total number of messages required to detect the *lwoe* is $O(|V_i|)$, where V_i is the set of nodes in the given fragment.

To test if a given edge is *outgoing*, a node sends a *test* message through that edge. The node at the other end may respond with a *reject* message (when it belongs to the same fragment as the sender) or an *accept* message (when it is certain that it belongs to a different fragment). While rejection is straightforward, acceptance in some cases may be tricky. For example, it may be the case that the responding node belongs to a different fragment name when it receives the *test* message, but its fragment is in the process of merging with the fragment of the sending node. To deal with this dilemma, when node i sends a *test* message (containing its *name* and *level*) to node j, the responses from node j (containing its *name* and *level*) are as follows:

Case 1: **if** $name(i) = name(j)$, **then** send *reject*.

Case 2: **if** $(name(i) \neq name(j)) \wedge (level(i) \leq level(j))$, **then** send *accept*.

Case 3: **if** $(name(i) \neq name(j)) \wedge (level(i) > level(j))$, **then** delay sending a response until $level(j) \geq level(i)$ or $name(i) = name(j)$.

Note that the level numbers never decrease, and by allowing a node to send an accept message only when its level is at least as large as that of the sending node (and the fragment names are different), the dilemma is resolved.

To guarantee the absence of deadlock, we need to establish that the waiting period in case 3 is finite. Suppose this is not true. Then, there must exist a finite chain of fragments $T_0, T_1, T_2, \ldots, T_{k-1}$ of progressively decreasing levels, such that T_i $(0 \leq i \leq k-1)$ has sent a *test* message to T_{i+1}. But then the last fragment in the chain must also have sent a *test* message to another fragment of the same or higher level, and it is guaranteed to receive a response, enabling it to combine with another fragment and raise its level. Thus, the wait is only finite.

For the sake of bookkeeping, each edge is classified into one of the three categories: *basic*, *branch*, and *rejected*. Initially, every edge is a *basic* edge. When a *reject* message is sent through an edge, it is classified as *rejected*. Finally, when a basic edge becomes a tree edge, its status changes to *branch*. The following lemma is trivially true:

Lemma 10.5

The attributes *branch* and *rejected* are stable.

As a consequence of Lemma 10.5, while searching for the least weight output edge, *test* messages are sent through the basic edges only.

Once the *lwoe* has been found, the root node sends out a *changeroot* message to the nodes in its own fragment. This is the approval from the leader to go ahead. The node at the end of *lwoe* receiving the *changeroot* message sends out a *join* message to the node at the other end of the *lwoe*, indicating its willingness to join. The *join* message initiates a *merge* or an *absorb* operation. Some possible scenarios are summarized as follows:

Scenario 1: Merge: A node i in a fragment T at level L sends out a $(join, level = L, name = T)$ message to a node j in another fragment T' at the *same* level $L' = L$ and receives a $(join, level = L', name = T')$ message in return (Figure 10.11). Thereafter, the edge through which the *join* messages were exchanged becomes a tree edge and changes its status to *branch*. Between the two nodes (i, j) at the endpoints of the *lwoe*, the one with the larger id serves as the new root. The new name of the fragment corresponds to the weight of the *edge* (i, j) and its level changes to $(L + 1)$. The new root starts the next phase by broadcasting an $(initiate, L + 1, name)$ message to the nodes in the combined fragment.

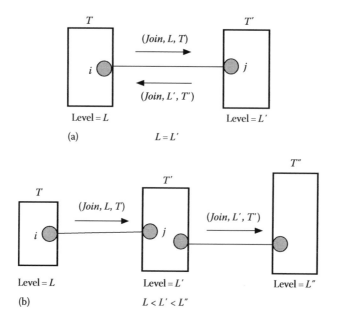

FIGURE 10.11　(a) *Merge* operation. (b) *Absorb* operation.

Scenario 2: Absorb: A node *i* in a fragment *T* at level *L* sends out a (*join, level = L, name = T*) message to a node *j* in another fragment *T'* at level *L' > L* across its *lwoe*. If the fragment *T'* has not completed its search for *lwoe*, then it absorbs *T* and includes it in the search by sending a (*join, level = L', name = T'*) message in return. The root of the fragment at level *L'* continues to serve as the root of the combined fragment. Fragment *T* at level *L* changes its level to *L'* and acquires the name *T'* of the other fragment. Then, they collectively search for the *lwoe*. The edge through which the *join* message is received becomes a tree edge and changes its status to *branch*.

If the fragment *L'* has already chosen its *lwoe*, then that must be distinct from the edge (*i, j*); otherwise, *T* would have already known about it. In this case, *T'* may be waiting to join with another fragment *T''*. Once *T'* joins with another fragment, it will initiate the next search of *lwoe* and will send an *initiate* message to the nodes in *T* to signal the absorption. The algorithm terminates and the MST is formed when no new outgoing edge is found in a fragment. A complete example of MST formation is illustrated in Figure 10.12.

What if every fragment sends a *join* message to a different fragment, but no fragment receives a reciprocating join message to complete the handshake? Can such a situation arise, affecting the progress property? The next lemma shows that this is impossible.

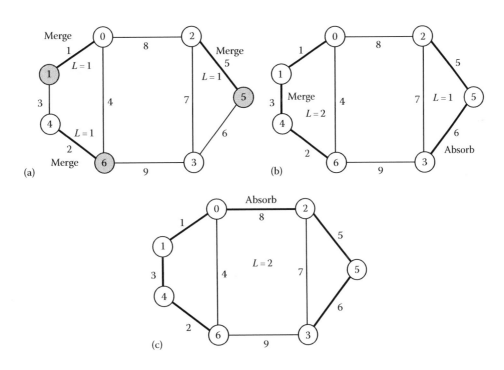

FIGURE 10.12 An example of MST formation using [GHS83]: The shaded nodes are the roots of the fragments and the thick lines denote the tree edges. In part (a), node 3 sends a *join* request to 5, but 5 does not respond until it has formed a fragment by joining with node 2 (part b). Part (c) shows the final absorb operation that leads to the formation of the MST.

Lemma 10.6

Until the MST is formed, there must always be a pair of fragments, such that the roots of these two fragments will send a *join* message to each other.

Proof: Consider the fragments across the edge of least weight. They must send *join* messages to each other. This will trigger either a merge or an absorb operation. ■

10.3.3.3 Message Complexity

Since a fragment at level k has at least 2^k nodes, the maximum level number cannot exceed $\log_2 n$. In each of these levels, each node receives (1) at most one *initiate* message from the root and (2) at most one *accept* message (response to a *test* message) and sends (3) one *report* message (response to *initiate*) toward the root, (4) at most one *test* message leading to an *accept* from the other end, and (5) one *changeroot* message (or a *join* message to a different fragment). Since there are n nodes, an upper bound of these messages is $5n \cdot \log_2 n$.

In addition to the above, count the test messages leading to a rejection. An edge is rejected only once in the entire algorithm. So the number of these *test* and *reject* messages will not exceed $2|E|$. Therefore, the overall message complexity of the MST algorithm will not exceed $5n \cdot \log_2 n + 2|E|$.

10.4 GRAPH COLORING

Graph coloring is a classic problem in graph theory and has been extensively investigated. The problem of *node coloring* in graphs can be stated as follows: Given a graph $G = (V, E)$, assign color to the nodes in V from a given set of colors, so that no two neighboring nodes have the same color. If we assume that each process has a unique id, and use it as node color, then it leads to a valid coloring. However, this is a trivial solution and not interesting at all. The design of coloring algorithms becomes particularly challenging when the color palette is small, and its size approaches the lower bound for a given class of graphs. The *chromatic number* of a graph is the size of the smallest set of colors that can be used to color the graph. In a distributed environment, knowledge is local—so no node knows anything about G beyond its immediate neighbors. This adds to the difficulty of designing graph-coloring algorithms in a distributed setting. To realize the difficulty, consider the graph in Figure 10.13.

Assume that nodes are anonymous and the process ids are being used for the purpose of identification only. It is easy to observe that the nodes of this graph can be colored using only two colors {0, 1}. Let $c(i)$ denote the color of node i and $N(i)$ denote the set of neighbors

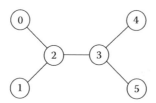

FIGURE 10.13   A graph that can be colored with two colors 0 and 1.

of node i. Assume that initially, $\forall i$, $c(i) = 0$. On the shared memory model of computation, let us try a naive algorithm:

```
program naive coloring
define c(i):color {of process i}
{program for process i}
do ∃j ∈ N(i):c(i)=c(j) → c(i):=1−c(i) od
```

Assume a central scheduler (so only one process executes a step at any time), and let each node examine only one neighbor at a time (i.e., fine-grain atomicity). Unfortunately, the naive algorithm does not terminate, since there exist infinite behaviors involving one or more nodes.* If instead we modify the algorithm by allowing a node to examine its entire neighborhood (i.e., coarse-grain atomicity) before choosing a new color that is distinct from the colors of its neighbors, then after node 0 executes an action, node 2 does not have any eligible action—so the computation hits a dead end.

10.4.1 $(D + 1)$-Coloring Algorithm

We now present a distributed algorithm for coloring the nodes of a graph with $(D + 1)$ colors, where D is the maximum degree of a node. We will designate the set of all colors by C. To make the problem a little more challenging, assume that the initial colors of the nodes are arbitrary.

The algorithm runs on a shared memory model under a central scheduler. No fairness is assumed. The atomicity is coarse-grained, so that a process can read the states of all its neighbors and execute an action in a single step. Define $nc(i) = \{c(j): j \in N(i)\}$. Then, the coloring algorithm is as follows:

```
program (D + 1) coloring
define c(i): color {of process i}, b: color
{program for process i}
do ∃j ∈ N(i):c(i)=c(j) → c(i):= b:b∈{C\nc(i)} od
```

Theorem 10.1

Program $(D + 1)$ *coloring* produces a correct coloring of the nodes.

Proof: Each action by a node correctly sets its color with respect to those of its neighbors. Once a node correctly sets its color, its guard is never enabled by the action of a neighboring node. So regardless of the initial colors of the nodes, each node executes its action *at most once*, and the algorithm requires at most $(n - 1)$ steps to terminate. ■

The size of the color palette used in the $(D + 1)$-coloring algorithm may be far from optimal. For example, consider a star graph where $(n - 1)$ nodes are connected to a single node that acts as the hub and $n = 100$. The $(D + 1)$-coloring algorithm will use 100 distinct

---

* Verify this before you proceed.

colors, whereas the graph can be colored using two colors only! Converting the graph into a *dag* (directed acyclic graph) helps reduce the size of the color palette. In the transformed dag, let $succ(i) = \{j: (i, j) \in E\}$ denote the *successors* of node i. Also, let $sc(i) = \{c(j): j \in succ(i)\}$ and the size of the color palette C exceed $|\max_i (succ(i))|$. Then, the following is an adaptation of the $(D + 1)$-coloring algorithm for a dag:

```
program dag coloring;
{program for node i}
initially ∀i :  c(i)=0;
do ∃j∈succ(i):c(i)=c(j) → c(i) := b:b ∈ {C\sc(i)} od
```

For the star graph, if all the edges are directed towards the hub, then each node at the periphery has only the hub as its successor, and the aforementioned algorithm can trivially produce coloring with two colors only.

Theorem 10.2

The dag-coloring algorithm produces a correct coloring.

Proof (by induction): Any dag has at least one node with no outgoing edges—call such a node a *leaf*. According to the program, the leaf nodes do not execute actions since they have no successors. So their colors are stable. This is the base case.

After every node $j \in succ(i)$ attains a stable color, it requires at most one more step for $c(i)$ to become stable, and such a color can always be found since the set $\{C\backslash sc(i)\}$ is nonempty. Thus, the nodes at distance one from a leaf node acquire a stable color in at most one step, those at distance *2* attain a stable color in at most $(1 + 2)$ steps, and so on. Eventually, all nodes are colored in at most $1 + 2 + 3 + \cdots + L = L(L + 1)/2$ steps where L is the length of the *longest directed path* in the dag. ■

Since $L \leq n - 1$, the dag-generation algorithm will terminate in $O(n^2)$ steps. To use this method for coloring *undirected* graphs, we need to devise a method for converting an undirected graph into a dag. A straightforward approach is to construct a BFS spanning tree and direct each edge toward a node of higher level, but the outdegree of some nodes (and consequently the size of the color palette) may still be large. In some cases, we can do much better. The next section addresses this issue with an example.

10.4.2 6-Coloring of Planar Graphs

In this section, we demonstrate a distributed algorithm for coloring the nodes of a *planar graph* with at most six colors (the color palette $C = \{0, 1, 2, 3, 4, 5\}$). The basic principle is to transform any given planar graph into a directed acyclic graph for which the degree of every node is <6 and execute the coloring algorithm on this dag. We begin with the assumption of *coarse-grain atomicity*—in a single step, each node examines the states of all its neighbors and, if necessary, executes an action. A central scheduler arbitrarily serializes the actions of the nodes.

For any planar graph $G = \{V, E\}$, if $e = |E|$ and $n = |V|$, then the following results can be found in most books on graph theory (e.g., see [Ha72]).

Theorem 10.3

(*Euler's polyhedron formula*) If $n \geq 3$, then $e \leq 3n - 6$.

Corollary 10.1

For any planar graph, there is at least one node with *degree* ≤ 5.

Call a node with *degree* ≤ 5 a *core* node. A distributed algorithm that assigns edge directions works as follows. Initially, all edges are undirected:

```
program undirected to dag;
initially all edges are undirected;
{program for each node i}
do  number of undirected edges incident on node i ≤ 5 →
        make all undirected edges outgoing
od
```

At the beginning, at least one core node of G will mark all undirected edges incident on them as outgoing. The *remainder graph* obtained by deleting the core nodes and the directed edges from G is also a planar graph, so the core nodes of the *remainder graph* now mark the undirected edges incident on them as outgoing. This continues until the remainder graph is empty and all edges are directed. Clearly, this will not take more than $n - 1 = |V| - 1$ steps. Figure 10.14 shows two steps of the dag-generation process.

The coloring algorithm will work on this dag. Since $\forall i \in V$: $sc(i) \leq 5$, the coloring algorithm will generate a valid node coloring using at most $(5 + 1) = 6$ colors. Interestingly, the coloring

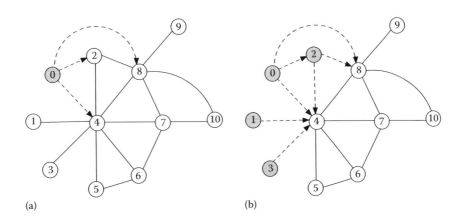

FIGURE 10.14 An example of generating a dag from a planar graph: The core nodes are shaded. (a) The core node 0 executes its action. (b) Core nodes 1, 3, 2 execute their actions. In fact, now all nodes are core nodes, and the execution of their actions in any order will lead to the final dag.

part of the algorithm need not wait for the dag-generation part of the algorithm to terminate—both of them can run concurrently. The composite algorithm will be as follows:

```
program planar graph coloring;
{program for node i}
do {Layer A: dag generation actions}
number of undirected edges incident on it ≤ 5 →
       make all undirected edges outgoing
{Layer B: coloring actions}
[] (outdgree(i)≤ 5)∧(∃j ∈ succ(i):c(i)=c(j))→ c(i) := b:b∈{C\sc(i)}
od
```

Theorem 10.4

The 6-coloring algorithm for planar graph terminates in $O(n^2)$ steps.

Proof: After *at most n* = $|V|$ steps of the dag-generation layer A, all edges will be directed. The actions of the coloring layer B cannot undo the effect of dag-generation layer A, or disable the actions of component A. Therefore, with a weakly fair scheduler, regardless of the progress of component B, in a bounded number of steps of the composite algorithm, all guards of component A will be disabled. Thereafter, in at most $O(n^2)$ steps, all nodes will be properly colored. Thus, the time complexity of the algorithm is $O(n^2)$. ■

The aforementioned proof uses the general idea of *convergence stairs* first proposed in [GM91]. The framework uses a finite sequence of predicates $H_0, H_1, H_2, ..., H_{k-1}$ for a layered construction involving a composite algorithm with k layers, where H_j ($0 \le j \le k - 1$) is the predicate that holds *after* layer j terminates. If H_j is closed under the actions of the layers ($j + 1$) through k, then the concurrent execution of the composite algorithm is guaranteed to reach a configuration that satisfies the postcondition H_k. Here, there are two layers:

1. $H_0 = \forall i \in V : outdegree(i) \le 5$ {postcondition for the dag-generation layer A}

2. $H_1 = \forall(i, j) : j \in succ(i) : c(i) \ne c(j)$ {postcondition for the coloring layer B}

These layers satisfy the composition rule of convergence stairs.

10.5 COLE–VISHKIN REDUCTION ALGORITHM FOR TREE COLORING

Consider a graph $G = (V, E)$, and assume that each node $v \in V$ has a unique *id*. Let $n = |V|$. It is obvious that by using the ids as node colors, one can always generate a legal node coloring for G. However, in general, coloring an n-node graph with n colors is hardly interesting. We therefore present an algorithm for *reducing* the size of the color palette without violating the constraint $\forall i, j \in V: (i, j) \in E, c(i) \ne c(j)$. It is a synchronous algorithm developed by Cole and Vishkin [CV86], and it demonstrates how any rooted tree can be colored using at most *three colors* in $\log^*(n)$ rounds. Let us first understand the $\log^*$ function: $\log^*(n)$ is the *smallest number* of log operations needed to bring n down to 2 or less. For example, consider n = one trillion. Now,

$$\log \text{(one trillion)} \cong 40,$$

$$\log (\log \text{(one trillion)}) \cong 5.322, \text{ and}$$

$$\log (\log (\log (\log \text{(one trillion)}))) < 2.$$

This means that $\log^*\text{(one trillion)} = 4$. This also illustrates that the $\log^*$ function grows very slowly with the value of the argument.

The reduction algorithm assumes that initially the color of each node is its id. Each nonroot node v is aware of its *parent $p(v)$*. Interpret each color c as a *little-endian* bit string $c_{k-1}c_{k-2}c_{k-3}...c_0$, and let $|c|$ denote the size of the bit string. In each round, every nonroot node v synchronously executes its actions. The algorithm terminates when $\forall v \in V: c(v) < 6$:

```
program reduce for a rooted tree: actions in each round;
{Program for the root node}
c(root):= 00 followed by bit 0 of c(root)
{Program for each non-root node v}
do c(v) ≥ 6 →
  {Let j = smallest index where the bit strings of c(v) and old
    c(p(v)) differ}
  c(v):= bit string for j followed by bit j of c(v)
od
```

Figure 10.15 shows an example. Consider the pair of nodes v, w. In Figure 10.15a, for node w, the minimum bit position in which its color label differs from that of its parent v is bit position 2. Accordingly, the first part (or the *head*) of the new label is the binary code for 2, that is, 10. This is followed by the second part (or the *tail*) of the color label, which is the value of bit 2 of node w. This leads to the new color label 100 for node w in the next

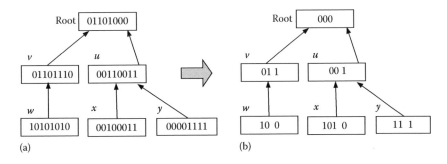

FIGURE 10.15   One step of the execution of algorithm reduce. (a) Initial colors. (b) After one step.

round. After each round, the algorithm leads to a valid color label whose size is roughly the base-2 logarithm of the size of the color label in the previous round. Therefore, in $\log^*(n)$ rounds, the algorithm terminates.

Theorem 10.5

Each round of algorithm *reduce* leads to a valid node coloring.

Proof: The new color label of any nonroot node v has two parts: (1) the *head* H, which is the *smallest index j* where the bit strings of the current $c(v)$ and $c(p(v))$ differ, and (2) the *tail* T, which is the value of bit j of $c(v)$. To prove that the new color labels form a valid coloring, we need to show that either the H- or the T-component of the new $c(v)$ and $c(p(v))$ is distinct.

Case 1: After one round, H-components of $c(v)$ and $c(p(v))$ are not identical. This automatically leads to a valid coloring.

Case 2: After one round, H-components of $c(v)$ and $c(p(v))$ are identical. In this case, by definition, the T-components of $c(v)$ and $c(p(v))$ must be different. ■

Why does the algorithm allow up to six colors for the nodes? Observe that as long as the size of the current color label $|c(v)|$ is ≥ 6, the new color label will shrink in size, and the program execution will continue to the next round. Once $c(v) < 6$ holds, the continuation of the program execution will be useless, since it will not reduce the values of c. At this point, there will be three choices for the H-component (0, 1, 2) and two choices for the T-component (0, 1). This leads to a total of six possible colors.

An additional procedure is needed for further reducing the size of the color palette from six to three. It begins with a *push down* mechanism that preserves the legality of the colors but leads to a configuration where all the children of the same parent have the same color. Thereafter, a simple manipulation of the color labels leads to a valid 3-coloring of the tree. The shift-down operation is as follows:

```
The shift-down operation {concurrently executed}
The root picks a color < 6 and different from its current color.
Each nonroot node v concurrently executes c(v):= c(p(v)).
```

It is trivial to show that shift-down leads to a valid coloring of the tree. Consider a node v and its parent $w = p(v)$. The current colors are legal, so $c(v) \neq c(w)$. After the shift-down operation, $c(v)$ equals the value of old $c(w)$, but node w gets the color of its parent $p(w)$. This color must be different from the old $c(w)$ and hence must be distinct from the new $c(v)$. In case w is the root, it chooses a new color, so $c(v) \neq c(w)$ will hold. Thus, the shift-down operation guarantees that for every nonroot node v, $c(v) \neq c(p(v))$ is true. So, the new color labels form a valid coloring.

Since the operation $c(v) := c(p(v))$ makes the colors of all the children of a given node identical, it opens up the possibility of further reducing the size of the color palette from six to three. Consider using the colors {0, 1, 2} instead of {0, 1, 2, 3, 4, 5}. Each node has to find a *free color* that is different from (a) the color of its parent and (b) the color of its children. Since there are three colors, such a free color must be available. This forms the basis of the final reduction scheme:

```
Reduction of the color palette size from six to three
z:= 3;
do z ≤ 5 →
        c(v) = z → pick a color from {0, 1, 2} not used by the
          neighbors of v;
        z := z + 1
od
```

This will run for at most three rounds. The shift-down and the palette reduction steps are shown in Figure 10.16.

It is now possible to put the pieces together as a single algorithm for 3-coloring the nodes of a tree in $O(\log^* n)$ rounds:

```
program 3-coloring of a tree in O(log*n) rounds
{Program for node v}
Execute Algorithm Reduce for log* (n) rounds;
Execute shift-down;
Reduce the color palette size from six to three
```

The primary merit of the aforementioned algorithm is its near-constant time complexity, while keeping the size of the color palette close to optimal.

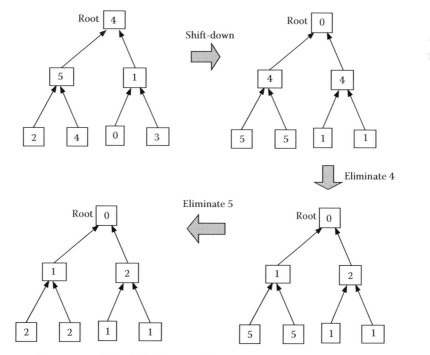

FIGURE 10.16   Illustration of the shift-down and the palette reduction steps.

10.6 MAXIMAL INDEPENDENT SET: LUBY'S ALGORITHM

Given a graph $G = (V, E)$, an *independent set* is a subset of nodes $W \subseteq V$ such that no two nodes in W are adjacent to one another. An independent set is *maximal* if no node can be added to W without violating its independence. Figure 10.17 shows an example.

There are many applications of independent and maximal independent sets (MISs). Consider the problem of scheduling an examination. In Figure 10.17, assume that each node represents a course and an edge between a pair of nodes (v, w) denotes that at least one student has registered for both v and w. Then an independent set denotes which examinations can be concurrently scheduled. A set of examinations will form an MIS if no additional examinations can be scheduled at that time.

Note that there is a difference between MIS and *maximum* independent set. Maximum independent set refers to the independent set with largest cardinality, and its computation is NP-hard. We focus only on the MIS.

A simple distributed algorithm for computing the MIS of an undirected connected graph $G = (V, E)$ is as follows. An initiator node uses a traversal algorithm to send a token to visit the various nodes of the graph. The token, while visiting a node, determines if the node can be included in the independent set or not. Nodes that are included in the MIS are assigned a binary tag: $f(v) = 1$ means the node v is included in the MIS, and $f(v) = 0$ means that the node v is excluded from the MIS. The algorithm is outlined here:

```
A simple algorithm for computing MIS
{Program for each node v upon receiving a token}
Initially ∀v ∈ V: f(v) = 0
∀w ∈ N(v): f(w) = 0 → f(v) := 1;
Continue with the next node until the traversal is complete;
{The MIS is the set {v ∈ V: f(v) = 1}}
```

The algorithm is clearly sequential with a time complexity $O(|E|)$ and is not very interesting. Subsequently, faster algorithms for solving the MIS problem have been devised using the synchronous round-based model of computation. One such algorithm is due to Luby [Luby86].

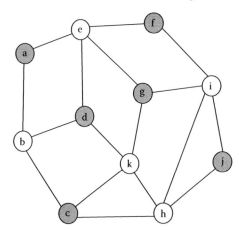

FIGURE 10.17 Examples of independent and MISs. *Note:* {a, d, h} is an independent set; {a, c, d, f, g, j} is a maximal independent set; {a, d, h, f} is also a maximal independent set.

Luby's algorithm is randomized and it operates in synchronous rounds, grouped into phases. In each phase, every node tries to join the MIS by marking itself with a specific probability that is related to its degree. A marked node joins the MIS if none of its higher-degree neighbors is marked. Otherwise, it unmarks itself. Each successful node removes itself and all its neighboring nodes before the next phase begins. The algorithm terminates when the remainder set of nodes becomes empty:

```
Luby's MIS Algorithm
{Program for node v: all nodes synchronously execute their
actions}
Define V': set of nodes
       d(v): degree of v
V':=V;
do V' ≠ ∅ → {A single phase}
1  Node v marks itself with probability  1/(2d(v)) ;
2  if (∀w ∈ N(v): d(w)>d(v)  w is not marked) ∧ (v is marked) →
      node v joins the MIS
3  [] (∃ w ∈ N(v): d(w)>d(v)  w is marked) ∧ (v is marked) →
      node v unmarks itself
4  fi {If d(v)=d(w) then break the tie using the node identifiers}
5  if v joins the MIS → remove v and its neighbors from V' fi
od {End of phase}
```

Note: In any phase, a node v with $d(v) = 0$ automatically joins the MIS. No marking is necessary.

We first argue that the algorithm indeed produces an MIS. Steps 2, 3, and 4 affirm that no two neighboring nodes will join the MIS. Step 5 ensures that if v joins the MIS, then none of its neighbors will. Since each of the remaining nodes gets a chance to mark itself and join the MIS, the algorithm terminates when no more eligible nodes are left.

The important question here is, how fast does the algorithm terminate? Let $L(v)$ be the set of neighbors of v whose degree is larger than $d(v)$. Also let M denote the set of marked nodes.

Lemma 10.7

The probability that a node v joins the MIS $\geq 1/4d(v)$.

Proof: $P[v \notin MIS | v \in M]$

$$= P\big[\exists w \in L(v): w \in M \mid v \in M\big]$$

$$\leq P\big[\exists w \in L(v): w \in M\big] = \sum_{w \in L(v)} P[w \in M] = \sum_{w \in L(v)} \frac{1}{2d(w)} \cdots \tag{1}$$

$$\leq \sum_{w \in L(v)} \frac{1}{2d(v)} \big\{\text{Since } d(v) < d(w)\big\} \leq d(v)\frac{1}{2d(v)} \big\{\text{Since } |L(v)| \leq d(v)\big\}$$

that is, $P\left[v \notin MIS \mid v \in M\right] \leq \dfrac{1}{2}$

Therefore, $P\left[v \in MIS \mid v \in M\right] \geq \dfrac{1}{2}$

Per (1), $P[v \in M] = \dfrac{1}{2d(v)}$, which implies $P[v \in MIS] \geq \dfrac{1}{2} \cdot \dfrac{1}{2d(v)}$, that is, $\geq \dfrac{1}{4d(v)}$ ■

Definition

A *good* node v is one for which $\displaystyle\sum_{w \in N(v)} \dfrac{1}{2d(w)} \geq \dfrac{1}{6}$.

A *good* node has a chance of being removed due to one of its neighbors the *MIS* joining. A node that is not good will be called a *bad* node.

Lemma 10.8

A good node will be removed in step 5 with probability $\geq 1/12$.

Proof: A node v is removed if at least one of its neighbors joins the MIS. The probability that at least one neighbor w of a *good node* v joins the MIS in step 2 is

$$P\left[\exists w \in N(v): w \in MIS\right] \geq \sum_{w \in N(v)} \dfrac{1}{4d(w)} \quad \{\text{Lemma 10.7 and sum rule}\}$$

$$= \dfrac{1}{2} \cdot \sum_{w \in N(v)} \dfrac{1}{2d(w)}$$

$$\geq \dfrac{1}{2} \cdot \dfrac{1}{6} = \dfrac{1}{12} \quad \{\text{Definition of good node}\}$$

The inclusion of w into the MIS triggers the removal of node v in step 5. ■

How many good nodes are there? If we could show that a constant fraction of the nodes is removed in each phase, then we could claim that the algorithm terminates in $O(\log n)$ rounds. However, this is not necessarily true. In a star graph with $n > 4$, only the hub is a good node.

To work around this, we prove that a constant fraction of the edges is removed in each phase. Define a *bad edge* as one that connects two bad nodes. An edge that is not bad is a *good* edge. We now prove the following lemma:

Lemma 10.9

At least half of the edges are good.

Proof (by contradiction): Construct a directed graph from G by directing each edge from a node towards the higher-degree neighbors (break any tie using node ids). We first show that for a bad node v: *outdegree*$(v) \geq 2 \cdot indegree(v)$.

To prove it by contradiction, assume that this claim is not true, which means that of all edges incident on v, $> \frac{1}{3}d(v)$ are incoming and $< \frac{2}{3}d(v)$ are outgoing edges. Now, define a set $T = \{w: (w, v) \text{ is an incoming edge}\}$. Note that

$$\sum_{w \in N(v)} \frac{1}{2d(w)} \geq \sum_{w \in T} \frac{1}{2d(w)} \quad \{\text{Since } T \subseteq N(v)\}$$

$$\geq \frac{1}{3}d(v) \cdot \frac{1}{2d(v)} \quad \left\{\text{Since } d(w) < d(v) \quad \text{and} \quad |T| > \frac{1}{3}d(v)\right\}$$

Therefore, $\displaystyle\sum_{w \in N(v)} \frac{1}{2d(w)} \geq \frac{1}{6}$.

However, this implies that node v is a good node and leads to a contradiction. So, $indegree(v) \leq \frac{1}{3}d(v)$, $outdegree(v) \geq \frac{2}{3}d(v)$, and $outdegree(v) \geq 2 \cdot indegree(v)$.

Since v is a bad node, $indegree(v) \leq \frac{1}{3}d(v) < \frac{1}{2}d(v)$. So, in the directed version of G, *at most half* of the edges are directed towards bad nodes (Figure 10.18). This means, *at least half* of the edges are directed towards nodes that are not bad, that is, good nodes. By definition, these are good edges. ■

Theorem 10.6

Luby's algorithm terminates in an expected number of $O(\log n)$ rounds.

Proof: Per Lemma 10.8, in each phase, a good node is removed with a constant probability $>1/12$. By definition, any edge incident on a good node is a good edge, and when a node is removed, all of its neighbors (i.e., all edges incident on it) are also removed (step 5).

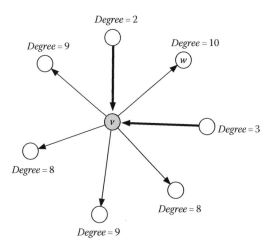

FIGURE 10.18   At least 1/3 of the edges incident on a bad node are bad.

It follows from Lemma 10.9 that at least half of the edges are removed in each phase. Upon termination, all the $|E|$ edges will be removed, which will take $O(\log|E|)$ phases. Since $|E| \leq n^2$ and each phase has a constant number of rounds, the algorithm terminates in an expected number of $O(\log n)$ rounds. ■

Note: This proof is due to Roger Wattenhofer [Lecture notes: Chapter 12: Summer 2003], who attributes it to a technique taken from Israeli and Itai developed for solving matching problems.

10.7 CONCLUDING REMARKS

Many applications in distributed computing center around a few common graph problems—this chapter addresses a few of basic algorithms. An algorithm is considered robust, when it works on dynamic graphs, that is, it handles (or survives) changes in topology. Mobile ad hoc networks add a new dimension to the fragility of the network topology, because their topologies continuously change due to the limited transmission range of each node. In recent times, embedded systems have witnessed significant growth—such systems use sensors that monitor environmental parameters and relay the values to a base station. The nodes of sensor networks run on limited battery power, so power consumption is a major issue—a low consumption of power adds to the life of the system. Therefore, in addition to space, time, and message complexities, a useful performance metric for sensor networks is the amount of power used by the sensor nodes during the execution of an algorithm.

The graphs represent not only physical networks but also logical networks (like overlay networks), where the neighborhood relationships are user defined and change over time. One classic example is a social network. The scale of these networks is constantly increasing—for example, as of 2012, Facebook has nearly 600 million users. As a result, scalability of common tasks (like multicasting) is a major issue. For an algorithm to be of practical use at that scale, space and time complexities of at most $O(\log^k n)$ $(k \geq 1)$ are considered to be acceptable.

The GHS algorithm for MST construction has been extensively studied in the published literature. This algorithm is an interesting case study of techniques that are valuable in the design of many distributed algorithms.

Distance-vector and link-state routings (and their variations) have been the two main contenders in network routing. Compared to distance-vector algorithm, the link-state algorithm has the merit that it does not suffer from the *counting-to-infinity* problem when there is a change in topology. The main disadvantage of a link-state routing protocol is that it does not scale well as more routers are added to the routing domain. Increasing the number of routers increases the size and frequency of the topology updates and also the length of time it takes to calculate end-to-end routes. This lack of scalability means that a link-state routing protocol is unsuitable for routing across the Internet at large, which is the reason why the Internet, for the purpose of routing, is divided into *autonomous systems*. Internet Gateway Protocols like OSPF is a link-state protocol that only route traffic within a single AS. An Exterior Gateway Protocol like BGP routes traffic between autonomous systems. These are primarily vector routing protocols and are more scalable.

Interval routing is a topic that has generated some interest among theoreticians. It addresses the scalability problem of routing tables. However, as of now, its limited ability to adapt to changes in topology restricts its applicability. So far, it has been used only for communication in some transputer-based distributed systems.* Some attempts of using it in sensor networks have recently been reported. Compared to interval routing, prefix routing is much more popular and used in many structured peer-to-peer networks.

Cole and Vishkin's algorithm for tree coloring introduces a technique for solving a problem in $O(\log^* n)$ rounds. This technique has been leveraged to solve a few related problems in $O(\log^* n)$ rounds. For example, it is possible to show that a graph with degree D can be colored with $(D + 1)$ colors in $O(\log^* n)$ rounds. The proof of Luby's algorithm for MIS construction is due to Wattenhofer.

10.8 BIBLIOGRAPHIC NOTES

Chandy–Misra's shortest path algorithm [CM82] is an adaptation of the Bellman–Ford algorithm used with ARPANET during 1969–1979. The adaptation included a mechanism for termination detection and a mechanism to deal with negative edge weights. The link-state routing algorithm also originated from ARPANET and was first proposed by McQuillan et al. and is described in [MRC80]. After several modifications, it was adopted by the ISO as an OSPF protocol. Santoro and Khatib [SK85] introduced interval routing. This chapter demonstrated the feasibility for tree topologies only. Van Leeuwen and Tan [LT87] extended the idea to nontree topologies. The probe algorithm for computing the spanning tree is originally due to Chang [Ch82]—Segall [Se83] presented a slightly different version of it. In [GHS83], Gallager et al. presented their MST algorithm—it has been extensively studied in the field of distributed algorithms, and many different correctness proofs have been proposed. Tarry's traversal algorithm [Ta1895], proposed for exploring an unknown graph, is one of the oldest known distributed algorithms. The distributed algorithm for coloring planar graphs is due to Ghosh and Karaata [GK93]—the original paper proposed a self-stabilizing algorithm for the problem. The version presented in this chapter is a simplification of that algorithm. Luby's algorithm for constructing an MIS can be found in [Luby86]; the complexity analysis presented here is due to Wattenhofer [Lecture notes: Chapter 12: Summer 2003], who attributes it to a technique taken from Israeli and Itai developed for solving matching problems.

EXERCISES

10.1   Let $G = (V, E)$ be a directed graph. A maximal strongly connected component of G is a subgraph $G' = (V', E')$ such that (1) for every pair of vertices $u, v \in V'$, there is a directed path from u to v and a directed path from v to u and (2) no other subgraph of G has G' as its subgraph. Propose a distributed algorithm to compute the maximal strongly connected component of a graph.

---

* Transputers were introduced by INMOS as building blocks of distributed systems.

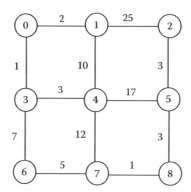

FIGURE 10.19   Compute a spanning tree of this graph using Chang's algorithm.

10.2   Consider the graph in Figure 10.19, where each edge is labeled with the signal propagation delay on that edge. With node 0 as the initiator, execute Chang's algorithm and compute the spanning tree of this graph. Assume that nodes forward the probes as soon as they receive it, and local computation takes zero time. Also determine how much time it will take to generate the spanning tree.

10.3   Let $G = (V, E)$ be an undirected graph and let $V' \subset V$ represent the membership of a group. The members of V' want to communicate with one another via a *multicast tree*, which is a *minimal subgraph* of G containing all members of V'—between any two members of the group, there is exactly one simple path, and if you remove a single node or edge from the multicast tree, that at least one member of the group becomes unreachable.

   a.   For the graph in Figure 10.19, show a multicast tree for $V' = \{0, 5, 6, 7\}$.

   b.   Given a graph G and a subset V' of k nodes, suggest a distributed algorithm for constructing a multicast tree. Briefly argue why your solution will work.

10.4   In a spanning tree of a graph, there is exactly one path between any pair of nodes. If a spanning tree is used for broadcasting a message and a node crashes, then some nodes will not be able to receive the broadcast. Our goal is to improve the connectivity of the subgraph used for broadcast, so that it can tolerate the crash of a single node. Such a subgraph is a *biconnected subgraph* of the given graph.

   Given the graph in Figure 10.19, give an example of such a minimal graph. Then suggest a distributed algorithm for constructing such a subgraph. Argue why your algorithm will work.

10.5   In most of the algorithms for rooted spanning tree generation, only one designated process can be the initiator. For the sake of speedup, consider a modification where there are more than one initiator nodes. Explain your strategy and illustrate a construction with two initiators. Does it lead to speedup? Justify your answer.

10.6   Devise an interval-labeling scheme for *optimal routing* on a (1) 4 × 4 grid of 16 processes and (2) a 3-cube of 8 processes. For each topology, show three test cases to show that the routes are optimal.

10.7 Propose an algorithm for locally repairing a spanning tree by restoring connectivity when a single node crashes. Your algorithm should complete the repair by adding the fewest number of edges. Compute the time complexity of your algorithm.

10.8 a. Produce two different labeling of the nodes and the ports of following graph for the purpose of interval routing. In the first, begin by labeling node *u* as 0, and in the second, start with labeling node *v* as 0. Check the correctness of these labels by verifying the routes between different pairs of nodes (Figure 10.20).

b. Propose a *prefix routing scheme* for the following two networks. In each case, you have to label the nodes (not the ports) of the graph in such a way that a message can be routed from node X to node Y by forwarding it to the neighbor that has the largest prefix match with the destination node (Figure 10.21).

10.9 Decide if a *linear* interval-labeling scheme exists for the tree of Figure 10.22. Explain your decision.

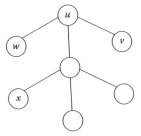

FIGURE 10.20   A graph for interval routing in problem 10.8.

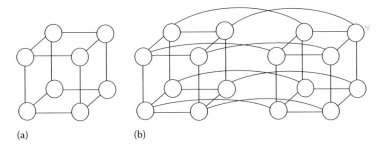

(a)   (b)

FIGURE 10.21   (a) 3-cube. (b) 4-cube.

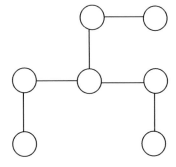

FIGURE 10.22   A tree for linear interval routing in problem 10.9.

10.10 Figure 10.2 illustrates the *counting-to-infinity* problem with distance-vector routing. Suggest a method to fix this problem when the routing graph contains k-cycles ($k > 2$).

10.11 Most of the classical algorithms for generating a spanning tree require $O(|E|)$ messages and are completed in $O(|E|)$ time. Devise an algorithm that generates the spanning tree in $O(n)$ time, where $n = |V|$.

(Hint: When the empty message m visits a node i for the first time, it lets every non-parent neighbor j know that it has been visited. The token is not forwarded until it has received an acknowledgment from node j. Since j knows that i has been visited, it will not forward the token to i.) Show your analysis of the time and the message complexities.

10.12 Given an undirected graph $G = (V, E)$, a matching M is a subset of E, such that no two edges are incident on the same vertex. A matching M is called *maximal* if there is no other matching $M' \supset M$. Suggest a distributed algorithm for computing a *maximal matching*. When the algorithm terminates, each node must know its matching neighbor, if such a match exists.

10.13 Devise a distributed algorithm for computing a spanning tree of a graph in which no root is designated. You can assume that the nodes have unique names.

10.14 Using the ideas of the $O(\log^* n)$-round coloring algorithm for trees, devise an algorithm for coloring the nodes of a ring of size n using six or fewer colors in $O(\log^* n)$ rounds.

10.15 Use the Cole–Vishkin $O(\log^* n)$ coloring algorithm to devise an $O(\log^* n)$-round algorithm for computing the MIS of a tree.

10.16 The *eccentricity* of a vertex v in a graph G is the maximum distance from v to any other vertex. Vertices of minimum eccentricity form the *center*. A tree can have one or two centers. Design a distributed algorithm to find the center(s) of a tree. Present arguments about why your algorithm works.

CHAPTER 11

Coordination Algorithms

11.1 INTRODUCTION

Distributed applications rely on specific forms of coordination among processes to accomplish their goals. Some tasks of coordination can be viewed as a form of preprocessing. Examples include *clock synchronization, spanning tree construction*, and *leader election*. In this section, we single out a couple of specific coordination algorithms and explain their construction. Our first example addresses *leader election*, where one among a designated set of processes is chosen as leader and assigned special responsibilities. The second example addresses a problem of model transformation: Recall that asynchrony is hard to deal with in real-life applications due to the lack of temporal guarantees—it is simpler to write algorithms on the synchronous process model (where processes execute actions in lockstep synchrony) and easier to prove their correctness. This motivates the design of *synchronizers* that transform an asynchronous model into a synchronous one and help run synchronous algorithms on asynchronous systems. In this section, we will discuss several algorithms for leader election and synchronizer construction.

11.2 LEADER ELECTION

A wide variety of distributed applications rely on the existence of a *leader among* its constituent process. The leader is invariably the focus of control and entrusted with the responsibility of system-wide management. Consider the client–server model of resource management: Here, the server can be viewed as a leader. Client processes send requests for resources to the server, and based on the information maintained by the server, a request may be either granted or deferred or denied. As another example, consider a centralized database manager—the shared data can be accessed or updated by client processes in the system. This manager maintains a queue of pending read and writes and processes these requests in an appropriate order. Such a centralization of control is not essential, but it offers a simple solution with manageable complexity.

When the leader (i.e., coordinator) fails or becomes unreachable, a new leader is elected from among nonfaulty processes. Failures affect the topology of the system—even the partitioning of the system into a number of disjoint subsystems is not ruled out. For a partitioned system, some applications elect a leader from each connected component. When the partitions merge, a single leader remains, and others drop out.

It is tempting to compare leader election with the problem of mutual exclusion and use a mutual exclusion algorithm for electing a leader. They are not exactly equivalent: The similarity is that whichever process enters the critical section becomes the leader. However, there are three major differences between the two paradigms:

1. Failure is not an inherent part of mutual exclusion algorithms. In fact, failure within the critical section is typically ruled out.

2. Starvation is an irrelevant issue in leader election. Processes need not take turns to be the leader. The system can happily function for an indefinite period with its original leader, as long as there is no failure.

3. If leader election is viewed from the perspective of mutual exclusion, then exit from the critical section is unnecessary. On the other hand, the leader needs to inform every active process about its identity, which is a nonissue in mutual exclusion.

A formal specification of leader election follows: Let $G = (V, E)$ represent the system topology, and each process $i \in V$ has a unique identifier. Each process i has a variable $L(i)$ that represents the identifier of *its* leader. Also, let $ok(i)$ denote the predicate that process i is nonfaulty. Then, the following three conditions must hold:

1. $\forall i, j \in V: ok(i) \wedge ok(j) \Rightarrow L(i) = L(j)$

2. $L(i) \in V$

3. $ok(L(i)) = true$

11.2.1 Bully Algorithm

The bully algorithm is due to Garcia–Molina [G82] and works on a *completely connected* network of processes. It assumes that (1) communication links are fault-free, (2) processes can only fail by stopping, and (3) failures can be correctly detected using some mechanism like time-out. Once a failure of the current leader is detected, the bully algorithm allows the nonfaulty process with largest id to eventually elect itself as the leader.

The algorithm uses three different types of messages: *election, reply,* and *leader*. A process initiates the election by sending an *election* message to every other process with a *higher* id. By sending this message, a process effectively asks, "Can I be the new leader?" A *reply* message is a response to the election message. To a receiving process, a reply implies, "No, you cannot be the leader." Finally, a process sends a *leader* message when it believes that it is the leader. The algorithm can be outlined as follows:

Step 1: Any process, after detecting the failure of the leader, bids for being the new leader by sending an *election* message to every process with a higher identifier.

Step 2: If any process with a higher id responds with a *reply* message, then the requesting process gives up its bid for becoming the leader. Subsequently, it waits to receive a *leader* message (*I am the leader*) from some process with a higher identifier.

Step 3: If no higher-numbered process responds to the *election* message sent by node i within a time-out period, then node i elects itself as the leader and sends a *leader* message to *every process* in the system.

Step 4: If no *leader* message is received by process i within a time-out period after receiving a reply to its *election* message, then process i suspects that the winner of the election failed in the mean time and reinitiates the election.

```
program    bully {program for process i}
define     failed : Boolean {set if the failure of the leader is detected}
           L:       process {identifies the leader}
           m:       message {election | leader | reply}
           state : idle | wait for reply | wait for leader
initially  ∀i ∈ V state = idle, failed = false

do failure of L(i) detected ∧ failed:= false → failed:= true
[] failed → ∀j > i:send election to j; state := wait for reply;
   failed:= false
[] (state = idle) ∧ (m = election) → send reply to sender; failed := true
[] (state = wait for reply) ∧ (m = reply) → state := wait for leader
[] (state = wait for reply) ∧ timeout → L(i):=i; state := idle;
   send leader to all
[] (m = leader) → L(i):=sender; state := idle
[] (state = wait for leader) ∧ timeout → failed:= true; state := idle
od
```

Theorem 11.1

Every nonfaulty process eventually elects a unique leader.

Proof: A process i sending out an *election* message may or may not receive a reply.

Case 1: If i receives a reply, then i will not send the *leader* message. In that case, $\forall j \in V, L(j) \neq i$ will eventually hold.

Case 2: If i does not receive a reply, then i must be unique in as much $\forall j \in V: ok(j), j \leq i$ holds. Note that there will always be a process like this. In this case, i elects itself as the leader, and after the *leader* message is sent out by i and received by all nonfaulty processes, the condition $\forall j \in V: ok(j), L(j) = i$ holds.

If case 1 holds, but no *leader* message is subsequently received before time-out, then the would-be leader itself must have failed in the mean time. This sets *failed* to true for every process that was waiting for the *leader* message. As a result, a new election is initiated and one of the earlier two cases will eventually hold, unless every process has failed, and the problem becomes vacuous. ■

Message complexity: Each process i (except the process with the largest id) can send an *election* message to $(n - i)$ other processes. Also, each process j receiving an election message can potentially send the reply message to $(j - 1)$ processes. If the would-be leader does not fail in the mean time, then it sends out $(n - 1)$ leader messages. However, in the worst case, the would-be leader can also fail before sending the leader messages. In this case, a time-out occurs, and every process repeats the first two steps. Since out of the n processes at most $(n - 1)$ can fail, the first two steps can be repeated $O(n)$ times. Thus, the worst-case message complexity of the bully algorithm is $O(n^3)$.

11.2.2 Maxima Finding on a Ring

Once we disregard failures and fault detection,* the task of leader election reduces to finding a node with a unique (like maximum or minimum) id. While the bully algorithm works on a completely connected graph, there are several algorithms for maxima finding that are designed to work on ring topologies. These solutions are conceptually simple, but differ from one another in message complexity. We will discuss three algorithms in this section.

11.2.2.1 Chang–Roberts Algorithm

In [CR79], Chang and Roberts presented a leader election algorithm for a *unidirectional* ring—it is an improvement over the first such algorithm proposed by LeLann [Le77]. Assume that a process can have one of two colors: *red* or *black*. Initially, every process is *red*, which implies that every process is a potential candidate for being the leader. A red process initiates the election by sending a token, which means, *I want to be the leader.* Any number of red processes can initiate the election. If, however, a process receives a token *before* initiating the algorithm, then it knows that there are other processes running for leadership—so it quits and turns *black.* A black process never turns red and acts as a router. At the end, only one process remains red, and it is the leader.

```
program    Chang-Roberts
define     token: process id, color ∈ {red,black}
initially  all processes are red and i sends a token ⟨i⟩ to its neighbor;
(for a red process i)
do token ⟨j⟩ received ∧ (j < i)→ skip {j's token removed, so j quits}
[] token ⟨j⟩ received ∧ (j > i)→ send ⟨j⟩; color := black {i quits}
[] token ⟨j⟩ received ∧ (j = i)−L(i):=i {i becomes the leader}
od
{for a black process}
do token ⟨j⟩ received → send ⟨j⟩ od
```

Let us examine why the algorithm works. A token initiated by a *red* process j will be removed when it is received by a process $i > j$. So ultimately, the token from a process with the largest id will prevail and will return to the initiator. Thus, the process with the largest id elects itself as the leader. It will require another round of *leader* messages to inform the identity of the leader to every other process.

To analyze the complexity of the algorithm, consider Figure 11.1. Assume that every process is an initiator, and their tokens are sent in the anticlockwise direction around the ring.

---

* We will deal with failures in Chapters 12 through 17.

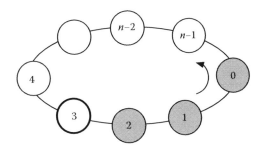

FIGURE 11.1   Illustration of the execution of Chang–Robert's election algorithm: the token from process 3 reached process $(n - 1)$ and processes 2 and 1 turned black.

Before the token from process $(n - 1)$ reaches the next process $(n - 2)$, the tokens from *every other* process reach node $(n - 1)$ in the following order: *Token*$\langle 0 \rangle$ reaches $(n - 1)$, *token*$\langle 1 \rangle$ reaches $(n - 1)$, *token*$\langle 2 \rangle$ reaches $(n - 1)$, and, finally, *token*$\langle n-2 \rangle$ reaches $(n - 1)$, and all these tokens get removed. The worst-case message complexity is therefore $1 + 2 + 3 + \cdots +(n - 1) = n(n - 1)/2$.

The algorithm can be naturally extended to an arbitrary graph topology for which a Hamiltonian cycle exists.

11.2.2.2 Franklin's Algorithm

Franklin's election algorithm works on a ring that allows *bidirectional* communication. Compared to Chang–Robert's algorithm, it has a lower message complexity. Processes with unique identifiers are arranged in an arbitrary order in the ring. There are two possible colors for each process: *red* or *black*. Initially, each process is *red* which implies that every process is a potential candidate for leadership.

The algorithm is synchronous and works in rounds. In each round, to bid for leadership, each red process sends a *token* containing its unique id to both neighbors and then examines the tokens received from other processes. As process i receives a token from process j and $j > i$, it quits the race and turns black. A black process remains passive and acts as a router only.

Since tokens are sent in both directions, whenever two adjacent red processes exchange tokens, one of them must turn black. In each round, a fraction of the existing red processes turn black. The algorithm terminates when there is only one red process in the entire system. This is the leader. The program for a red process i is as follows:

```
program Franklin
define token: process id, color ∈ {red,black}, r: integer {round number}
initially all processes are red, r = 0 and i sends a token ⟨i⟩ to its neighbor;
{for a red process i in round r ≥ 0}
send token⟨i⟩ to both neighbors;
receive token⟨j⟩ from both neighbors;
if ∃ token⟨j⟩:j > i  →  color := black
[]∀ token⟨j⟩:j < i  →  r: = r + 1 {move to the next round}
[]∀ token⟨j⟩:j = i  →  L(i):= i {algorithm terminates}
fi
```

Theorem 11.2

Franklin's algorithm elects a unique leader in at most $(1 + \log_2 n)$ rounds.

Proof: For a red process i, in each of the two directions, define a *red neighbor* to be a red process that is closest to i in that direction. Thus, in Figure 11.2, after round 0, processes 7 and 9 are the two red neighbors of the process 2.

After each round, every red process i that has at least one red neighbor $j > i$ turns black. Therefore, in a ring with $k(k > 1)$ red processes, at least $\lfloor k/2 \rfloor$ turn black. Initially, $k = n$. Therefore, after at most $\log_2 n$ rounds, the number of red processes is reduced to one. In the next round, it becomes the leader. ■

The algorithm terminates in $O(\log_2 n)$ rounds, and in each round, every process sends (or forwards) a message in both directions. Therefore, the worst-case message complexity of Franklin's algorithm is $O(n \cdot \log n)$.

11.2.2.3 Peterson's Algorithm

Like Franklin's algorithm, Peterson's algorithm works on a ring topology and operates in synchronous rounds. Interestingly, it elects a leader using only $O(n \cdot \log n)$ messages even though it runs on a *unidirectional* ring. Compared to Franklin's algorithm, there are two distinct differences:

1. A process communicates using an *alias* that changes from one round to another during the progress of the computation.

2. A unique leader is eventually elected, but that is not necessarily the process with the largest identifier in the system.

As before, we assume that processes can have two colors: *red* or *black*. Initially, every process is red. A red process turns black when it quits the race for becoming a leader. A black process is passive—it only acts as a router and forwards incoming messages to its neighbor.

Assume that the ring is oriented in the clockwise direction. Any process will designate its anticlockwise neighbor as the *predecessor* and its clockwise neighbor as the *successor*.

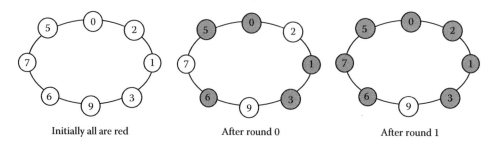

| Initially all are red | After round 0 | After round 1 |

FIGURE 11.2 Execution of Franklin's algorithm: The shaded processes are black. After two rounds, process 9 is identified as the leader (maxima).

Designate the *red predecessor* (the closest red process in the anticlockwise direction) of *i* by $N(i)$ and the *red predecessor of N(i)* by $NN(i)$. Until a leader is elected, in each round, every red process *i* receives *two* messages: one from $N(i)$ and the other from $NN(i)$—these messages contain *aliases* of the senders. The channels are FIFO. Depending on the relative values of the aliases, a red process either decides to continue with a new alias to the next round or quits the race by turning black.

Denote the alias of a process *i* by *alias(i)*. Initially, *alias(i) = i*. The idea is comparable to that in Franklin's algorithm, but unlike Franklin's algorithm, a process cannot receive a message from *both* neighbors. So, each process determines the local maxima by comparing *alias(N)* with its own alias and *alias(NN)*. If *alias(N)* happens to be larger than the other two, then the process continues its run for leadership by assuming *alias(N)* as its new alias. Otherwise, it turns black and quits. In each round, every red process executes the following program:

```
program     Peterson{for a process in each round}
define      alias: process id, color: black or red
initially   ∀i:color(i)=red, alias(i)=i
send alias; receive alias(N);
if alias = alias(N)→ I am the leader
[]alias ≠ alias(N)→ send alias(N); receive alias(NN);
               if alias(N) > max (alias,alias(NN))→alias:=alias(N)
               {In the above case, i remains red and moves to the
                next round}
               []alias(N) < max (alias,alias(NN)) → color:=
               black {i quits}
               fi
fi
```

Figure 11.3 illustrates the execution of one round of Peterson's algorithm.

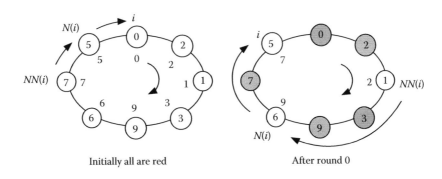

Initially all are red   After round 0

FIGURE 11.3 One round of execution of Peterson's algorithm: For every process, its id appears inside the circle, and its alias appears outside the circle. Shaded circles represent black processes. After one more round, only 5 remains red.

Proof of correctness

Let i be a red process before round $r (r \geq 0)$ begins. Then, after round r, process i remains red if and only if the following condition holds:

$$LocalMax \equiv alias(N(i)) > alias(i) \wedge alias(N(i)) > alias(NN(i))$$

We show that in each round, at least one of the two red processes i and $N(i)$ will turn black.

Lemma 11.2

Let i be a red process, and $j = N(i)$ before a round begins. Then at the end of the round, either i or j must turn black.

Proof (by contradiction): Assume that the statement is false. Then both i and j will remain red after that round. From *LocalMax*, it follows that if i remains red after that round, then $alias(j) > alias(i) \wedge alias(j) > alias(N(j))$ must hold. Again, if j remains red after that round, then $alias(N(j)) > alias(j) \wedge alias(N(j)) > alias(NN(j))$ must also hold. Both of these cannot hold at the same time, so the statement is true. ■

It is not impossible for two or more neighboring red processes to turn black in the same round (see Figure 11.3). In fact, this helps our case and accelerates the convergence.

Theorem 11.3

Peterson's algorithm elects a unique leader at most $(1 + \log_2 n)$ rounds.

Proof: It follows from Lemma 11.2 that after every round, *at least* half of the existing red processes turn black. So, after at most $\log_2 n$ rounds, only one red process i remains, $i = N(i)$, and the condition $alias(i) = alias(N(i))$ holds. In the next round, process i elects itself as the leader. ■

Message complexity: Since there are *at most* $(1 + \log_2 n)$ rounds, and in each round every process sends (or forwards) two messages, the number of messages required to elect a leader is bounded from above by $O(n \cdot \log_2 n)$. Despite the fact that the communication is unidirectional, the message complexity is not inferior to that found in Franklin's algorithm for a bidirectional ring.

11.2.3 Election in Arbitrary Networks

For general networks, if a ring is embedded on the given topology, then a ring algorithm can be used for leader election. The orientation of the embedded ring helps messages propagate in a predefined manner. As an alternative, one can use flooding to construct a leader election.

Assume that the algorithm runs in rounds. Initially, $\forall i \in V: L(i) = i$. In each round, each node i sends its $L(i)$ to every node $j \in N(i)$. After a node i received the messages from its neighbors, it picks the largest id from the set $\{L(i) \cup L(j): j \in N(i)\}$, assigns it to $L(i)$, and sends the updated $L(i)$ to its neighbors. The algorithm will terminate after D rounds, where D is the diameter of the graph. Here is an outline:

```
program election in an arbitrary network {for each process i}
define r: integer {round number},L: process id {identifies the leader}
initially r = 0, L(i)=i
do r < D→
      send L(i) to each j ∈N(i);
      receive L(j) from each j ∈N(i);
      L(i):= max{L(i)∪L(j:j ∈N(i)}
      r:=r + 1
od
```

The algorithm requires processes to know the diameter (or at least the total number of processes) of the network. The message complexity is $O(\Delta \cdot D)$ where Δ is the maximum degree of a node.

11.2.4 Election in Anonymous Networks

In *anonymous* networks, process identifiers are not used. Therefore, the task of leader election becomes an exercise in symmetry breaking. Randomization is a widely used tool for breaking symmetry. Here, we present a randomized leader election algorithm that uses a single bit $b(i)$ for each process. Algorithm works in synchronous rounds and works on *completely connected* networks only.

A process can be in one of two states: *active* or *passive*. Initially, every process is active and is a contender for leadership. In each round, each active process i randomly chooses a value for $b(i)$ from the set $\{0,1\}$. Define the set $S = \{i: b(i) = 1\}$. If $|S| = 1$, then process $i \in S$ becomes the leader, and the algorithm terminates. Otherwise, the processes in S move to the next round, and those not belonging to S become passive and quit the race. This reduces the competition. There is one exception: When $\forall i,j: b(i) = b(j)$, no progress is made, and all processes must repeat the random selection. However, this event cannot continue forever, and in a bounded number of rounds, the condition $\exists i,j: b(i) \neq b(j)$ will hold. The active processes moving to the next round continue the algorithm, until one process is elected as the leader. The description of the algorithm uses ids for the purpose of identification only.

Note that after each round, half of the processes are expected to advance to the next round. So in a system of n processes, a leader is elected after an expected number of $\log n$ rounds. Since in each round, every active process communicates with every other active process, the message complexity is $O(n^2 \log n)$.

```
program election_anonymous {for each active process i in every round}
define    b: bit, state ∈{active,passive}
          S={i:b(i)=1}
initially every process is active, ∀i:b(i) = 0, S = ∅
b(i):=random{0,1}
send b(i) to every active process;
receive b(i) from every active process
if (|S|=1)∧(i ∈ S)→i is the leader
[] (1<|S|<n)∧(i ∈ S)→i moves to the next round
[] (1<|S|<n)∧(i ∉ S)→state:=passive{i quits}
fi
```

11.3 SYNCHRONIZERS

Compared to synchronous algorithms, asynchronous algorithms are often more difficult to design and analyze and have a higher complexity. Since most distributed computing platforms are naturally asynchronous, it is tempting to devise a mechanism by which synchronous algorithm could be run on asynchronous platforms. A *synchronizer* is a protocol that transforms an asynchronous model into a synchronous process model (i.e., processes operate in lockstep synchrony) and enables synchronous algorithms to run on it. A synchronous algorithm runs in discrete steps known as rounds or clock *ticks* (a.k.a. clock pulse). In each tick, a process can

- Perform a local computation

- Send out messages to its neighbors and receive messages from its neighbors

Messages sent out in clock tick *i* are assumed to reach their destinations in the same clock tick—based on these, processes update their states in the next clock tick ($i + 1$). A *synchronizer* performs the model transformation by simulating the synchronous rounds on an asynchronous network. This results in a two-layered design. Synchronizers provide an alternative technique for designing algorithms on asynchronous distributed systems. Assume that each node has a clock pulse generator and these clocks tick in unison. Actions are scheduled with these clock ticks. The implementation of a synchronizer must guarantee the condition that a new pulse is generated at a node only after it receives all the messages of the synchronous algorithm, sent by its neighbors at the previous pulse. However, the difficulty in providing this guarantee is that no node knows which messages were sent to it by its neighbors, and the message propagation delays can be arbitrarily large.

Contrary to apprehensions about the complexity of the two-layered algorithm, the complexity figures of algorithms using synchronizers are quite encouraging. In this chapter, we will present the design of a few basic types of synchronizers.

11.3.1 ABD Synchronizer

An asynchronous bounded delay (ABD) synchronizer [CCGZ90] [TKZ94] can be implemented on a network where every process has a physical clock, and the message propagation

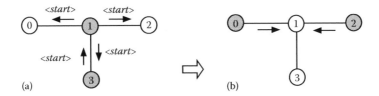

FIGURE 11.4 Simulation of a clock tick by an ABD synchronizer: (a) 1 and 3 spontaneously initiate the synchronizer operation, initialize themselves, and send start signals to the noninitiators 0 and 2; (b) 0 and 2 wake up and complete the initialization. This completes the action of tick 0.

delays have a known upper bound δ. In real life, all physical clocks tend to drift. However, to keep our discussion simple, we assume that once initialized, the difference between a pair of physical clocks does not change during the life of the computation.

Let C denote the physical clock of a process. One or more processes spontaneously initiate the synchronizer actions by assigning $C := 0$, executing the actions for *tick* (0), and sending a *<start>* signal to its neighbors (Figure 11.4). By assumption, actions take zero time. Each noninitiating neighbor j wakes up when it receives the *<start>* signal from a neighbor, initializes its clock C to 0, and executes the actions for *tick* 0. This completes the initialization.

Before a process p simulates the actions of *tick*$(i + 1)$, p along with its neighbors must send and receive all messages corresponding to *tick i*. If p sends the *<start>* message to q, q wakes up and sends a message that is a part of initialization actions of *tick* 0, then p will receive it at time $\leq 2\delta$. Therefore, process p will start the simulation of the next pulse (*tick* 1) at time 2δ. Eventually, process p will simulate *tick k* of the synchronous algorithm at local clock time $2k\delta$. The permission to start the simulation of a tick thus entirely depends on the local clock value.

11.3.2 Awerbuch's Synchronizers

When neither the physical clocks are synchronized nor the upper bound of the message propagation delays is known, the ABD synchronizer does not work. In [Aw85], Awerbuch addressed the design of synchronizers for such weaker models and proposed three different kinds of synchronizers with varying message and time complexities.

The key idea behind Awerbuch's synchronizers is the determination of when a process is *safe* to schedule the actions of the next clock tick. By definition, a process is *safe* for a given clock tick when (1) it has received an *ack* for every message that it sent during the current tick and (2) its neighbors have sent out their messages for the current tick. A safe process announces this by sending a *< safe>* message to its neighbors. A violation of this policy may prompt a process to start scheduling the actions of *tick* $(j + 1)$ before receiving a message for *tick j* from a neighbor. Here, we describe three different types of synchronizers. Each has a different strategy for using the topological information and detecting safe configurations.

11.3.2.1 α-Synchronizer

Before incrementing *tick*, each node needs to ensure that it is safe to do so. The α-synchronizer implements this by asking each process to send a *<safe>* message after it has sent and received *all messages* for the current clock tick. Each process executes the following three steps for the simulation *tick i*:

1. Send and receive messages <*m,i*> for the current tick *i*.

2. Send <*ack,i*> for each incoming message, and receive <*ack,i*> for each outgoing message for the current tick *i*.

3. Send <*safe,i*> to each neighbor.

When a process receives <*safe,i*> messages from every neighbor for tick *i*, it increments its *tick* to $(i + 1)$ and starts the simulation of tick $(i + 1)$. Figure 11.5 shows a partial trace of the execution of the α-synchronizer.

Complexity issues: For the α-synchronizer, the *message complexity* $M(\alpha)$ is the number of messages passed around the entire network for the simulation of *each* tick. It is easy to observe that $M(\alpha) = O(|E|)$—in addition to the original messages of the synchronous algorithm, two other messages <*ack*> and <*safe*> are sent through every edge of the network. Similarly, the *time complexity* $T(\alpha)$ is the maximum number of asynchronous rounds required by the synchronizer to simulate each tick across the entire network. Since each process exchanges three messages <*m*>,<*ack*>,<*safe*> for every round of the synchronous algorithm, $T(\alpha) = 3$.

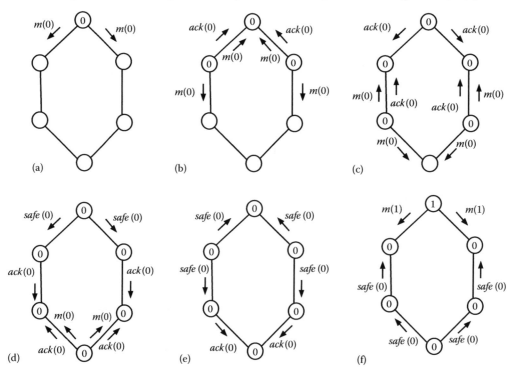

FIGURE 11.5 Partial trace of the execution of an α-synchronizer: the numbers inside the circles indicate the tick number that they are simulating. The process at the top starts the computation. The message *m*(0) sent by it wakes up its neighbors. (a)–(f) show six different phases.

Let M_S and T_S be, respectively, the message and the time complexities of a synchronous distributed algorithm. Then using a synchronizer, the same algorithm can be implemented on an asynchronous model with a message complexity M_A and a time complexity T_A where

$$M_A = M_S + T_S \cdot M(\alpha)$$

$$T_A = T_S \cdot T(\alpha)$$

It is possible that in the original synchronous algorithm, one or more processes do not send out a message to its neighbors in one or more rounds. However, in these rounds, the α-synchronizer must send out a blank message on behalf of these processes.

11.3.2.2 β-Synchronizer

The β-synchronizer needs an initialization phase before the simulation of the clock ticks begins. The initialization phase involves constructing a spanning tree of the network. A designated initiator is the root of the spanning tree. The initiator starts the simulation by sending out to its children a *next* message that prompts them to start the simulation for tick 0. Thereafter, the operations are similar to those in the α-synchronizer, except that the control messages (*next, safe,* and *ack*) are sent along the *tree edges* only. A safe process sends a *<safe,i>* message to its parent to indicate that the entire subtree under it is safe for tick *i*. If the root receives a *<safe,i>* message from each child, then it knows that every node in the spanning tree is safe for tick *i*—so it sends a *next* message to start the simulation of the next tick $(i + 1)$.

The message complexity $M(\beta)$ can be estimated as follows. Each process exchanges the following messages:

1. Sends and receives messages *<m,i>* for tick *i* to and from its neighbors.

2. Sends and receives *<ack,i>* to and from its neighbors. Thereafter it sends a *<safe,i>* message to its parent.

3. Receives a *<safe,i>* message from each child via the tree edges. The *safe* signals are convergecast via tree edges. If the process itself is safe, and it is not the root, then it sends a *safe* message to its parent.

4. When the root receives the *safe* messages from every child, it knows that the entire tree is safe. Then it sends out a *<next,(i+1)>* message via the tree edges to the nodes of the network. After receiving the *<next,(i+1)>* message, a node begins the simulation of tick $(i + 1)$.

In a spanning tree of a graph with *n* nodes, there are $(n - 1)$ edges. The three control messages (*ack, safe, next*) flow through each of the $(n - 1)$ tree edges. So, the extra message complexity is $M(\beta) = O(n)$. The time complexity $T(\beta)$ is proportional to the height of the tree, which is at most $(n - 1)$, but often much smaller when the tree is balanced.

The method of computing the complexity of an asynchronous algorithm using a β-synchronizer is similar to that using the α-synchronizer, except that there is an overhead for the construction of the spanning tree. This is a one-time overhead, and its complexity depends on the algorithm chosen for it.

11.3.2.3 γ-Synchronizer

Comparing the α-synchronizer with the β-synchronizer, we observe that the β-synchronizer has a lower message complexity, but the α-synchronizer has a lower time complexity. The γ-synchronizer utilizes the best features of both β- and α-synchronizers.

In a γ-synchronizer, there is an initialization phase, during which the network is divided into *clusters of processes*. Each cluster is a subgraph containing a subset of processes. The β-synchronizer protocol is used to synchronize the processes *within* each cluster, whereas the α-synchronizer is used to synchronize processes *between* clusters. To implement this idea, each cluster identifies a leader—it acts as the *root* of a spanning tree for that cluster. Neighboring clusters communicate with one another through a designated edge known as an *intercluster edge* (see Figure 11.6).

Using the β-synchronizer protocol, when the leader of a cluster finds that each process in the cluster is safe, it broadcasts to all processes in the cluster that the cluster is safe. Processes incident on the intercluster edges forward this information to the neighboring clusters using a *cluster.safe* message. The nodes receiving the *cluster.safe* message from the neighboring clusters convey to their leaders that the neighboring clusters are safe. The sending of a *cluster.safe* message from the leader of one cluster to the leader of a neighboring cluster simulates the sending of a *safe* message from one node to its neighbor in the α-synchronizer.

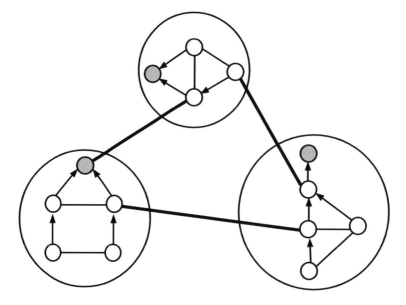

FIGURE 11.6   Clusters in a γ-synchronizer: The shaded nodes are the leaders in the clusters, and the thick lines are the intercluster edges between clusters.

We now compute the message and time complexities of a γ-synchronizer. Partition the graph $G = (V, E)$ into a number of clusters; adjacent clusters are connected by a set of preferred edges called *intercluster* edges. Within a cluster, the coordination progresses in the usual way. The intercluster coordination is carried out using the following control messages for clock tick i:

1. A *<safe,i>* message is sent to the parent, after a *<safe,i>* message is received from all children in the cluster. When the leader in a cluster receives the *<safe,i>* messages from every child, it knows that the entire cluster is safe. To advertise this, the leader broadcasts a *<cluster.safe,i>* message to every node in the cluster via the spanning tree.

2. When a node incident on the intercluster edge receives the *<cluster.safe,i>* message from its parent, it forwards that message to the neighboring cluster(s) through the intercluster edges to indicate that this cluster is safe.

3. The nodes receiving the *<cluster.safe,i>* message from the neighboring clusters convergecast it toward their leaders via the tree edges. The leader of a cluster eventually learns about the receipt of the *<cluster.safe,i>* messages from *every* neighboring cluster. At this time, the leader broadcasts the *next* message to the nodes in its cluster, and the simulation of the next tick ($i + 1$) begins.

Table 11.1 shows the chronology of the various control signals. Here, let T be the time when the simulation of the current clock starts and H_p be the maximum height of the spanning tree in each cluster for the given partition, and let each hop take 1 unit of time.

It follows from Table 11.1 that the *time complexity* of γ-synchronizer is $4H_p + 3$. To compute the *message complexity*, let E_p represent the set of all tree edges and intercluster edges in a given partition P of the network. Through each tree edge within a cluster, the messages *ack*, *safe*, *cluster.safe*, and *next* are sent out exactly once. In addition, through each intercluster edge, the message *cluster.safe* is sent twice, once in each direction. Therefore, the message complexity of the γ-synchronizer is $O(|E_p|)$. Note that if there is a single cluster, then the γ-synchronizer reduces to a β-synchronizer. On the other hand, if each node is treated as a cluster, then the γ-synchronizer reduces to an α-synchronizer. To reduce both time and space complexities of a γ-synchronizer, it is important to look for a partition so that both E_p and H_p are small. In [Aw85], Awerbuch presented a partition algorithm that leads to $E_p \leq k \cdot n$ and $H_p \leq \log n/\log k$, where $n = |V|$ and $2 \leq k < n$. To compute the overall complexity, one should

TABLE 11.1   Timetable of the Control Signals in a γ-Synchronizer

Time	Control Action
$T + 2$	All processes receive *<safe>* messages.
$T + H_p + 2$	Leaders receive *<safe>* messages and find that their clusters are safe.
$T + 2H_p + 2$	Processes in a cluster receive *<cluster.safe>* message from the leader.
$T + 2H_p + 3$	*<cluster.safe>* is received from (and sent to) neighboring clusters.
$T + 3H_p + 3$	The leaders receive the *<cluster.safe>* message.
$T + 4H_p + 3$	All processes receive the *<next>* message and simulate the next tick.

also take into account the initial *one-time* overhead of building these clusters, designating the intercluster edges, and computing the spanning trees within each cluster.

11.3.2.4 Performance of Synchronizer-Based Algorithms

The use of a synchronizer for running synchronous algorithms on asynchronous systems does not necessarily incur a significant performance penalty. To demonstrate this, consider a synchronous BFS algorithm, and transform it to an asynchronous version using a synchronizer. The synchronous algorithm works as follows:

1. A designated root starts the algorithm by sending a probe to each neighbor in *tick 0*.

2. All nodes at distance $d(d > 0)$ receive the *first* probe in *tick $(d - 1)$* and forward that probe to all neighbors (other than the sender) in *tick d*.

3. The algorithm terminates when every node has received a probe, and the BFS tree consists of all edges through which nodes received their first probes.

The BFS tree is computed in D rounds (where D is the diameter of the graph), and it requires $M_S = O(|E|)$ messages.

Now, consider running this algorithm on an asynchronous system using an α-synchronizer—the synchronizer will simulate the clock ticks, and the action of the synchronous algorithm will be scheduled at the appropriate ticks. Since $T(\alpha) = 3$, the time complexity of the overall algorithm is $T_A = T_S \cdot T(\alpha) = 3D$ asynchronous rounds. Also, $M(\alpha) = O(|E|)$, so the message complexity of the composite algorithm will be

$$M_A = M_S + T_S \cdot M(\alpha)$$

$$= O(|E|) + O(|E|) \cdot D$$

$$= O(n \cdot |E|) \quad \{\text{since } D = O(n) \text{ for sparse graphs}\}$$

If instead of an α-synchronizer a β-synchronizer is used, then the time complexity of the composite algorithm will be $T_A = T_S \cdot T(\beta) = D \cdot O(height) = O(n^2)$, and the message complexity of the composite algorithm will be

$$M_A = M_S + T_S \cdot M(\beta)$$

$$= O(|E|) + O(n) \cdot D$$

$$= O(n^2)$$

Compare these figures with the complexities of a solution to the same problem *without* using the synchronizer. A straightforward algorithm for the asynchronous model starts in the same way as the synchronous version, but the additional complexity is caused by the arbitrary propagation delays of the channels. It is possible that a node receives a probe from some node with distance d and assigns to itself a distance $(d + 1)$ but later receives another probe from a node with a distance less than d, thus revoking the earlier decision. It is possible to devise an asynchronous algorithm than constructs a BFS tree in $O(n^2)$ time using $O(n^2)$ messages. As another example, consider the asynchronous Bellman–Ford algorithm whose time complexity is $2^{O(n)}$ steps. The synchronous version of the Bellman–Ford algorithm has a time complexity of $T_S = O(n) \cdot O(|E|)$ rounds—so by running a synchronous Bellman–Ford algorithm on an α-synchronizer, it is possible to reduce the time complexity to $O(n^3)$ asynchronous rounds.

11.4 CONCLUDING REMARKS

Numerous types of coordination problems find use in the design of distributed applications. In this section, we singled out two such problems with different flavors.

Extrema (i.e., maxima of minima) finding is a simple abstraction of the problem of leader election, since it disregards failures. The problem becomes more difficult, when failures or mobility affects the network topology. For example, if a mobile ad hoc network failure gets partitioned into two disjoint components, then separate leaders need to be defined for each connected component to sustain a reduced level of performance. Upon merger, one of the two leaders will relinquish leadership. Algorithms addressing such issues differ in message complexities and depend on topological assumptions.

Synchronizer is an example of simulation. Such simulations preserve the constraints of the original system on a *per node* (i.e., local) basis, but may not provide any global guarantee. An example of a local constraint is that no node in a given tick will accept a message for a different tick from another node. A global constraint for a truly synchronous behavior is that when one node executes the actions of *tick k*, then every other node will execute the actions of the same tick. One can observe that the α-synchronizer does not satisfy this requirement, since a node at distance $d > 1$ may simulate an action of an earlier (or a later) clock tick. This hardly has an impact on the correctness of the simulation, since no node communicates with nonneighbors.

11.5 BIBLIOGRAPHIC NOTES

LeLann [Le77] presented the first solution to the maxima finding problem: Every initiator generates a token and sends to its neighbors, and a neighbor forwards a token when its own id is lower than that of the initiator; otherwise, it forwards a token with its own id. This leads to a message complexity of $O(n^2)$. Chang and Roberts algorithm [CR79] is an improvement over LeLann's algorithm—its worst-case message complexity is $O(n^2)$, but its average-case complexity is $O(n \cdot \log n)$. Garcia–Molina [G82] proposed the bully algorithm. Hirschberg and Sinclair [HS80] described the first algorithm for leader election on a bidirectional ring. Their algorithm had a message complexity of $O(n \cdot \log n)$, the constant factor being approximately 8. Franklin's algorithm [F82] had an identical complexity measure, but the constant was reduced to 2. Peterson's algorithm [P82] was the first algorithm on a *unidirectional* ring with a message complexity $O(n \cdot \log n)$ with a constant factor of 2. Subsequent modifications lowered this constant factor. The randomized algorithm for leader election is a simplification of the stabilizing version presented by Dolev et al. in [DIM91].

Chou et al. [CCGZ90] and subsequently Tel et al. [TKZ94] studied the design of synchronization on the ABD model of networks. The α-, β-, γ-synchronizers were proposed and investigated by Awerbuch [Aw85].

EXERCISES

11.1 In a network of 100 processes, specify an initial configuration of Franklin's algorithm so that the leader is elected in the *second* round.

11.2 Consider Peterson's algorithm for leader election on a unidirectional ring of 16 processes 0 through 15. Describe an initial configuration of the ring so that a leader is elected in the *fourth round*.

11.3 Show that Chang–Roberts algorithm has an average complexity of $O(n \cdot \log n)$.

11.4 Election is an exercise in symmetry breaking: Initially, all processes are equal, but at the end, one process stands out as the leader. Assume that instead of a single leader, we want to elect k leaders ($k \geq 1$) on a unidirectional ring. Modify *Chang–Roberts* algorithm to elect k leaders. (Do not consider the obvious solution in which first a single leader gets elected and this leader picks ($k - 1$) other processes as leaders. The goal is to explore if there is a solution to the k leader election problem that needs *fewer messages* than the single leader algorithm.)

11.5 In a hypercube of n nodes, suggest an algorithm for leader election with a message complexity of $O(n \cdot \log n)$.

11.6 Design an election algorithm for a tree of anonymous processes. (Of course, the tree is not a rooted tree; otherwise, the problem would have been trivial.) Think of orienting the edges of the tree so that (1) eventually there is exactly one process (which is the leader) with all incident edges directed toward it and (2) every leaf process has outgoing edges only.

11.7 The problem of leader election has some similarities with the mutual exclusion problem. Chapter 7 describes Maekawa's distributed mutual exclusion algorithm with $O(\sqrt{n})$ message complexity. Can we use similar ideas to design a leader election with sublinear message complexity? Explore this possibility.

11.8 In the following is the proposed design of a simple synchronizer:

  a. Each process has a variable *tick* initialized to 0.

  b. A process with *tick* = j exchanges messages for that tick with its neighbors.

  c. When a process has sent and received all messages for the current tick, it increments tick (the detection of safe configuration is missing from this proposal).

  Clearly, a process with *tick* = j can receive messages for both *tick* = j and *tick* = $j + 1$ from its neighbors. To satisfy the requirements of a synchronizer, the process will buffer the messages for *tick* = $j + 1$ for later processing, until the process has exchanged all messages for *tick* = j and incremented its tick.

  Comment on the correctness of the simple synchronizer, and calculate its time and message complexities.

11.9 In the ideal ABD synchronizer, the physical clocks do not drift. As a result, after the initial synchronization, no messages are required to maintain the synchronization.

Assume now that the physical clocks drift, so that the difference between a pair of clocks grows up to 1 in R time units, and each process simulates a clock tick every 2δ time unit, where δ is the upper bound of the message propagation delay along any link. Calculate the maximum time interval, after which the ABD synchronizer will start malfunctioning.

11.10 Consider an array of processes $0, 1, 2, \ldots, 2n - 1$ that has a different type of synchrony requirement: We will call it *interleaved synchrony*. Interleaved synchrony is specified as follows: (1) neighboring processes should *not* execute actions simultaneously, and (2) between two consecutive actions by any process, all its neighbors must execute an action. When $n = 4$, a sample schedule is as follows:

0,	2,	4,	6,	:	tick	0
1,	3,	5,	7,	:	tick	0
0,	2,	4,	6,	:	tick	1
1,	3,	5,	7,	:	tick	1

The computation is an infinite execution of such a schedule. Design a synchronizer to implement interleaved synchrony. Consider two different cases: (1) an ABD system and (2) a system without physical clocks.

11.11 In the following network, a synchronous distributed algorithm takes 10 rounds. The edges are full duplex. The *first round* begins with each process sending a message to its neighbors. Thereafter, in each round, every process executes the actions ⟨*receive message, update a local variable, send message*⟩, until at the end of the last round, every process receives the messages from its neighbors, updates its local state, and then the computation terminates (Figure 11.7).

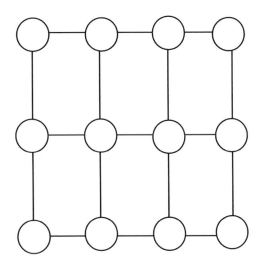

FIGURE 11.7 A network of processes.

Assume that you are using an α-*synchronizer* to run the same algorithm on an asynchronous version of the system where message delays and processor speeds are arbitrary but finite and messages are never lost:

a. How many *asynchronous rounds* will the algorithm need to complete?

b. How many messages will be used in the synchronous version of the algorithm?

c. How many messages will be used in the asynchronous version of the algorithm?

Briefly justify your calculations.

IV

Faults and Fault-Tolerant Systems

CHAPTER 12

Fault-Tolerant Systems

12.1 INTRODUCTION

A fault is the manifestation of an unexpected behavior, and fault tolerance is a mechanism that masks or restores the expected behavior of a system following the occurrence of faults. Attention to *fault tolerance* or *dependability* has drastically increased over the recent years due to our increased dependence on computers to perform critical as well as noncritical tasks. Also, the increase in the scale of such systems indirectly contributes to the rising number of faults. Advances in hardware engineering can make the individual components more dependable, but it cannot eliminate them altogether. Bad system designs and behavioral patterns like mobility can also contribute to failures.

Historically, models of failures have been linked with the level of abstraction in the specification of a system. A VLSI designer may focus on *stuck-at-0* and *stuck-at-1* faults where the outputs of certain gates are permanently stuck to either a 0 or a 1 regardless of input variations. A system-level hardware designer, on the other hand, may be ready to view a failure as any arbitrary or erroneous behavior of a module as a whole. A dip in the power supply voltage or radio interferences due to lightning or a cosmic shower can cause transient failures by perturbing the system state without causing any permanent damage to the hardware system. Messages propagating from one process to another may be lost in transit. Finally, even if hardware does not fail, software may fail due to code corruption, system intrusions, improper or unexpected changes in the specifications of the system, environmental changes, or human error.

Failures are a part of any system—the real issue is the frequency of the failures and their consequences. The computer system ENIAC had a *mean time between failures** (MTBF) of 5 min. The real interest on dependable computing started from the time of space exploration, where the cost of a failure was unacceptably high. The widespread use of computers in the financial world as well as in critical systems like nuclear reactors, air-traffic

---

* MTBF is a well-known metric for system reliability. The associated term is mean time to repair (MTTR). In any viable system, MTTR has to be much less than MTBF.

control, avionics, or hospital patient monitoring systems where human lives are directly affected has renewed interest in the study of fault tolerance. Despite all textbook studies, the sources of failures are sometimes obscure or at best unforeseen, although the consequences can be devastating. This calls for sound system design methodology and sound engineering practices.

Before we discuss how to tolerate faults, we present a characterization of the various kinds of faults that can occur in a system.

12.2 CLASSIFICATION OF FAULTS

Our view of a distributed system is a process-level view, so we begin with the description of certain types of failures that are visible at the process level. Note that each type of failure at any level may be caused by a failure at some lower level of design. Thus, a process may cease to produce an output when a wire in the circuit breaks or a dust speck bridges two wires on a printed circuit board. A complete characterization of the relationship between faults at different levels is beyond the scope of our discussion. The major classes of failures are as follows:

Crash failure: A process undergoes crash failure when it permanently ceases to execute its actions. This is an *irreversible* change. There are several variations in this class. In one variation, crash failures are treated as reversible, that is, a process may play dead for a finite period of time, and then resume operation, or it may be repaired. Such failures are called *napping failure*.

In an asynchronous model, crash failures cannot be detected with total certainty, since there is no lower bound of the speed at which a process can execute its actions. The design of systems tolerating crash failures would have been greatly simplified if processes could correctly detect whether another process has crashed. In a synchronous system where processor speeds and channel delays are bounded, crash failure can be detected using *timeout*. One such implementation requires processes to periodically broadcast a heartbeat signal that signifies *I am alive*. When other correct processes fail to receive the heartbeat signal within a predefined timeout period, they conclude that the process has crashed.

In general, internal failures within a processor may not lead to a *nice* version of a faulty behavior—it can sometimes be quite messy. Since most fault-tolerant algorithms are designed to handle crash failures only, it would have been nice if any arbitrary internal fault within a processor could be transformed to a crash failure (by incorporating extra hardware and/or software in the processor box). This is the motivation behind the more benign model of *fail-stop* processors. A fail-stop processor has two properties: (1) it halts program execution when a failure occurs, and (2) the internal state of the volatile storage is irretrievably lost. Schlichting and Schneider [SS83] described an implementation of a *k-fail-stop* processor—it satisfies the fail-stop properties with high probability when k or fewer faults occur, and the system can detect when another fail-stop processor halts. Fail-stop is a simple abstraction meant for simplifying the design of fault-tolerant algorithms. If a system cannot tolerate fail-stop failures, then there is no way that it can tolerate crash failures.

Omission failure: Consider a transmitter process sending a sequence of messages to a receiver process. If the receiver does not receive one or more of the messages sent by the transmitter, then an omission failure occurs. In real life, this can be either caused by transmitter malfunction or due to the properties of the medium. For example, limited buffer capacity in the routers can cause some communication systems to drop packets. In wireless communication, messages are lost when collisions occur in the MAC layer or the receiving node moves out of range. Techniques to deal with omission failures form a core area of research in networking.

Transient failure: A transient failure can perturb the global state in an arbitrary way. The agent inducing this failure may be momentarily active (like a power surge, or a mechanical shock, or lightning), but it can make a lasting effect on the global state. In fact, omission failures are special cases of transient failures, when the channel states are perturbed.

Empirical evidence suggests that transient faults occur frequently. Transient faults capture the effects of environmental hazards such as gamma rays, whose duration in time is limited. Transient failures are also caused by an overloaded power supply or due to weak batteries. Hardware faults are not the only source of transient faults. Transient failures also capture state corruption that occurs when software components fail. Gray [G85a] called them *Heisenbugs*, a class of temporary internal faults that are intermittent in nature. They are essentially permanent faults whose conditions of activation occur rarely or are not easily reproducible and so difficult to detect in the testing phase. Gray and Reuter [GR93] estimated that over 99% of bugs in IBM's DB2 production code are supposedly nondeterministic in nature and are thus transient.

Byzantine failure: Byzantine failures represent the weakest of all the failure models that allow *every* conceivable form of erroneous behavior. As a specific example, assume that process i multicasts the value x of a local variable to each of its neighbors. Then the following are examples of inconsistent behaviors:

- Two distinct neighbors j and k receive values x and y, where $x \neq y$.

- Every neighbor receives a value z, where $z \neq x$.

- One or more neighbors do not receive any data from process i.

Some possible causes of the aforementioned kind of byzantine failures are as follows: (1) Total or partial breakdown of a link connecting i with its neighbors. (2) Software problems in process i. (3) Hardware synchronization problems—assume that every neighbor is connected to the same bus, and reading the same copy of a variable x sent out by i, but since the clocks are not perfectly synchronized, they may not read the value of x exactly at the same time. If the value of x varies with time and the local clocks are not perfectly synchronized, then different neighbors of i may receive different values of x from process i. (4) Malicious actions by process i.

Software failure: There are several reasons that lead to software failure:

1. *Coding errors or human errors*: There are documented horror stories of losing a spacecraft because the program failed to use the appropriate units of physical parameters. As an example, on September 23, 1999, NASA lost the $125 million Mars orbiter spacecraft because one engineering team used metric units while another used English units leading to a navigation fiasco, causing it to burn in the atmosphere.

2. *Software design errors*: Mars pathfinder mission landed flawlessly on the Martial surface on July 4, 1997. However, later its communication failed due to a design flaw in the real-time embedded software kernel VxWorks. The diagnosis was that the problem occurred due to priority inversion, when a medium priority task could preempt a high priority one.

3. *Memory leaks*: The execution of programs suffers from the degeneration of the runtime system due to memory leaks, leading to a system crash. Memory leak is a phenomenon by which processes fail to free up the entire physical memory that has been allocated to them. This effectively reduces the size of the available physical memory over time. When the available memory falls below the minimum required by the system or an application, a crash becomes inevitable.

4. *Problem with the inadequacy of specification*: Assume that a system running program S is producing the intended results. If the system suddenly fails to do so even if there is no hardware failure or memory leak, then there may be a problem with specifications. If $\{P\}\ S\ \{Q\}$ is a theorem in programming logic, and the precondition P is inadvertently weakened or altered, then there is no guarantee that the postcondition Q will always hold!

A classic example of software failure is the so-called *Y2K* bug that rocked the world and kept millions of service providers in an uneasy suspense for several months as the year 2000 dawned. The problem was one of inadequate specifications: When programs were written in the twentieth century for financial institutions, power plants, or process control systems, the year 19xy was most often coded as xy so 1999 will appear as 99. At the beginning of the year 2000, this variable will change from 99 to 00 and the system will mistake it for the year 1900. This could potentially upset schedules, stall operations, and trigger unknown side effects. At least one airline grounded all of its flight on January 1, 2000, for the fear of the unknown. Prior to January 1, 2000, millions of programmers were hired to fix the potential problem in programs, even if in many cases the documentation was missing. As a result, the damages were miniscule compared to the hype.

Note that many of the failures like crash, omission, transient, or byzantine can be caused by software bugs. For example, a badly designed loop that does not terminate can mimic a crash failure in the sender process. An inadequate policy in the router software can cause packets to drop and trigger omission failure. And we have already observed that *Heisenbugs* (permanent faults whose conditions of activation occur rarely or are not easily reproducible) cause transient failures.

Temporal failure: Real-time systems require actions to be completed within a specific amount of time. When this deadline is not met, a *temporal failure* occurs. Like software failures, temporal failures also can lead to other types of faulty behaviors.

Security failure: Virus and other kinds of malicious software creeping into a computer system can lead to unexpected behaviors that conform to the definition of a fault. The effects range from allowing an intruder to eavesdrop or steal passwords to granting a complete takeover of the computer system. Such compromised systems can exhibit arbitrary behavior.

Some failures are repeatable, whereas others are not. Failures caused by incorrect software are often repeatable, whereas those due to transient hardware malfunctions or due to race conditions may not be so and therefore not detected during debugging. In the domain of software failures, Heisenbugs are difficult to detect. Finally, human errors play a big role in system failure. In November 1988, much of the long-distance service along the East Coast was disrupted when a construction crew accidentally severed a major fiber optic cable in New Jersey; as a result, 3,500,000 call attempts were blocked. On September 17, 1991, AT&T technicians in New York attending a seminar on warning systems failed to respond to an activated alarm for 6 h. The resulting power failure blocked nearly 5 million domestic and international calls and paralyzed air travel throughout the Northeast, causing nearly 1170 flights to be canceled or delayed.

12.3 SPECIFICATION OF FAULTS

For the purpose of modeling fault tolerance, a system specification consists of (1) a set of system actions S representing the fault-free (also called failure-intolerant) system and (2) a set of fault actions F that mimic the faulty behavior. The faulty system consists of the union of all the actions of both S and F. Readers are cautioned that this is only a *simulation* of the faulty behavior and has no connection with how the fault actually occurs in the physical system. Such simulations use *auxiliary variables* that are not a part of the real system. Also, such specifications are not unique—many different specifications are possible for the same faulty system. We present a few examples here:

Example 12.1

Assume that a system, in the absence of any fault, sends out the message "*a*" infinitely often (i.e., the output is an infinite sequence *a a a a*...). However, a failure *occasionally* causes a message to change from "*a*" to "*b*". Such a faulty system can be specified as follows:

```
      program      faulty system 1;
      define       x : boolean; a b: messages
      initially    x = true;
S:    do     x → send a
F:    []     true → send b
      od
```

With a scheduler that is at least *weakly fair*, the difference between faulty and the fault-free systems becomes perceptible to the outside world. For the same system, one can model a *crash failure* by specifying the *F*-action as *true* → *x*: = *false*. After *F* is executed, the system ceases to produce an output—a condition that cannot be reversed using the actions of *S* or *F*.

Example 12.2

Under the broad class of *byzantine failures*, specific faulty behaviors can be modeled using appropriate specifications. In the present example, the fault-free system executes a nonterminating program by sending out an infinite sequence of integer values 0, 1, 2, 0, 1, 2, ... The system failure changes the 2s to 9s.

```
   program faulty system 2
   define k: integer {value contained in a message}
          x: boolean {initially x = true}
   initially x = true, k = 0
S: do        k<2              →    send k; k: = k + 1
   []        x ∧ (k = 2)      →    k: = 0
F: []        x                →    x: = false
   []        ¬x ∧ (k = 2)     →    send 9; k: = 0
   od
```

Example 12.3

Temporal failures are detected using a special predicate *timeout*, which becomes true when an event does not take place within a predefined deadline. Consider a process *i* broadcasting a message every 60 s to all of its neighbors. Assume that the message propagation delay is negligibly small. If process *j* does not receive any message from process *i* within, say, 61 s (i.e., it keeps a small allowance before passing a verdict), a temporal failure occurs. The specification follows:

```
   program faulty system 3
   {for process i};
   define T(i): timer (value in seconds}
S: do    T(i) = 0 → send message to j; T(i): = 60
         {Every second T(i): = T(i)-1}
F  []    true→T(i): = 65
   od
   {for process j};
   define T(j): timer (value in seconds}
S: do    (T(j)>0)∧message received from process i → T(j): = 61
         {Every second T(j): = T(j)-1}
F: []    true →T(j): = 55
   od
```

Define *timeout(i, j)* ≡ (T = 0)∧ *no message received from process i.* The predicate *timeout(i, j)* will be true whenever *i* or *j* executes the *F*-action once, and then the normal countdown continues. In actual practice, a temporal failure can be caused by an increase in the propagation delay or by a loss of synchronization between the sender and the receiver clocks.

12.4 FAULT-TOLERANT SYSTEMS

A system that does not tolerate failures is a *fault-intolerant* system. In such systems, the occurrence of a fault violates some liveness or safety property. Let *P* be the set of configurations of the fault-intolerant system reachable via the specified system actions *S*. Given a set of fault actions *F*, the *fault span Q* corresponds to the largest set of configurations that the system can get into. It is a measure of how *bad* the system can become. The following two conditions are trivially true:

1. $P \subseteq Q$.

2. Q is closed under the actions of both S and F.

It is not possible to design a system that can tolerate all kinds of failures. A system is called *F-tolerant* (i.e., tolerates the fault action *F*) when the system maintains or returns to its original configuration (i.e., *P* holds) after all *F*-actions stop executing. There are four major types of fault tolerance:

1. Masking tolerance

2. Nonmasking tolerance

3. Fail-safe tolerance

4. Graceful degradation

12.4.1 Masking Tolerance

A fault *f* is masked if its occurrence has no impact on the application, that is, $P = Q$. Masking tolerance is important in many safety-critical applications where the failure can endanger human life or cause massive loss of property. An aircraft must be able to fly even if one of its engines malfunctions. A patient monitoring system in a hospital must not record patient data incorrectly even if some of the sensors or instruments malfunction, since this can potentially cause an improper dose of medicine to be administered to the patient and endanger his or her life. Masking tolerance preserves both liveness and safety properties of the original system.

12.4.2 Nonmasking Tolerance

In nonmasking fault tolerance, faults may temporarily affect the application and violate the safety property, that is, $P \subset Q$. However, liveness is not compromised, and eventually,

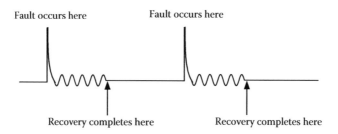

FIGURE 12.1 An illustration of fault recovery.

normal behavior is restored (Figure 12.1). Consider that while you are watching a movie, the server crashed, but the system automatically restored the service by switching to a standby server. The temporary glitch may be acceptable, since the failure did not have any catastrophic consequence. As another example, consider the routing of packets from a source to a destination node. Let a failure create a cycle in the route and trap a few packets. However, the algorithms for routing table computation broke the cycle and the packets eventually reached the destination. The net impact of the failure was an increase in the message propagation delay.

There are different types of nonmasking tolerance. In *backward error recovery*, snapshots of the system are periodically recorded on an incorruptible form of storage, called stable storage. When a failure is detected, the system *rolls back* to the last configuration saved on the stable storage (to undo the effect of the failure), and the computation progresses from that point. The important issue here is to make the history of the computation right. In *forward error recovery*, when a fault occurs, the system does not look back or try a rerun, since minor glitches are considered inconsequential, as long as the normal operation is eventually restored. Forward recovery systems that guarantee recovery from an *arbitrary initial configuration* are known as *self-stabilizing* systems.

12.4.3 Fail-Safe Tolerance

Certain faulty configurations do not affect the application in an adverse way and are therefore considered harmless. A *fail-safe system* relaxes the tolerance requirement by only avoiding those faulty configurations that may have catastrophic consequences, even when failures occur. For example, if at a four-way traffic crossing, the lights are green in both directions, then a collision is possible. However, if the lights are red in both directions, then at best traffic will stall, but will not have any catastrophic side effect.

Given a safety specification *P*, a fail-safe system preserves *P* despite the occurrence of failures. However, there is no guarantee that liveness will be preserved. Sometimes, halting progress and leaving the system in a safe state may be the best possible way to cope with failures. The ground control system of the *Ariane 5* launcher was designed to mask all single faults, but when two successive component failures occur, it would postpone the launch (this is safer than a mishap in the space).

12.4.4 Graceful Degradation

There are systems that neither mask nor fully recover from the effect of failures. Such systems exhibit a degraded behavior that falls short of the normal behavior, but are still considered acceptable. The notion of acceptability is highly subjective and is entirely dependent on the user running the application. Some examples of degraded behavior are as follows:

1. Consider a taxi booth where customers call to order a taxi. Under normal conditions, (i) each customer ordering a taxi must eventually get it, and (ii) these requests must be serviced in the order in which they are received at the booth. In case of a failure, a degraded behavior that satisfies only condition (i) but not (ii) may be acceptable.

2. While routing a message between two points in a network, a program computes the shortest path. In the presence of a failure, if this program returns another path that is not the shortest path but one that is marginally longer than the shortest one, then this may be considered acceptable.

3. A pop machine returns a can of soda when a customer inserts 75 cents in quarters, dimes, or nickels. After a failure, if the machine refuses to accept dimes and nickels, but returns a can of soda only if the customer deposits three quarters, then it may be considered acceptable.

4. An operating system switches to a *safe mode* when users cannot create or modify files, but can only read the files that already exist.

12.4.5 Detection of Failures in Synchronous Systems

The implementation of fault tolerance becomes easier if there exists some mechanism for detecting failures. This in turn depends on specific assumptions about the *degree of synchronization*: like the existence of synchronized clocks, or the lower bound on the processor speed, or upper bound on message propagation delays, as described in Chapter 4. Consider, for example, a *crash failure*. Without any assumption about the lower bound of process execution speeds and the upper bound of message propagation delays, it is impossible to detect a crash failure, because it is not feasible to distinguish between a crashed process and a nonfaulty process that is executing actions very slowly. When these bounds are known, timeout can be used to detect crash failures.

As another example, consider how *omission failures* can be detected. The problem is as hard as the detection of crash failures, unless the channels are FIFO or an upper bound of the message propagation delay δ is known. With a FIFO channel that is initially empty, a sender process i tags every message m with a sequence number seq as described in the following:

```
do true → send ⟨m[seq],seq⟩; seq: = seq + 1 od
```

If a receiver process receives two consecutive messages whose sequence numbers are i and j, and $j \neq i + 1$, then an omission failure is detected. With non-FIFO channels, if a message with sequence number i is received, but the previous message does not arrive within δ time units, then the receiver detects an omission failure.

Compared to crash failures, byzantine failures are much harder to deal with. In synchronous distributed systems where an upper bound on the message propagation delays or lower bound on processor speeds exists, some forms of byzantine failures can be detected. These are based on the various consensus protocols available in the published literature for tolerating byzantine failures. For asynchronous distributed systems, *failure detectors* and their use in solving consensus problems have received substantial attention. We postpone further discussions on failure detectors to Chapter 13.

The design of a fault-tolerant system requires knowledge about the application and its expected behavior, the fault scenario, and the type of tolerance desired. In the following sections, we will discuss the implementation of fault tolerance for crash and omission failures, and examine their complexities. Byzantine agreement and self-stabilizing systems will be presented in Chapters 13 and 17, respectively.

12.5 TOLERATING CRASH FAILURES

A simple and age-old technique of tolerating crash failures is to replicate the process or functional modules, with the idea that the functioning modules will make up for the nonfunctional ones. *Double modular redundancy* (DMR) can mask a single crash, and TMR can mask any single fault. The ideas are explained in the following text:

12.5.1 Double and Triple Modular Redundancy

Consider a process B that receives an input value x, computes the function $y = f(x)$, and sends it to process C. If B crashes, then C does not receive the value of y. To mask the effect of B's crash, replace process B by a pair of replicas $B0$ and $B1$ (Figure 12.2a)—such a system can tolerate a single crash. With n such replicas ($n > 1$) of B, it is possible to tolerate the crash of m replicas, where $m \leq n - 1$.

Extending this idea, if there are three replicas of B (Figure 12.2b) and any one of them malfunctions (this includes not only crash but also arbitrary failures), then process C will

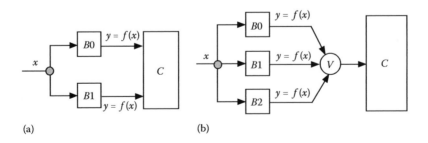

(a) (b)

FIGURE 12.2   (a) DMR that tolerates a single crash. (b) TMR that tolerates a single failure of arbitrary type: V is the voting unit.

receive the correct value of $y = f(x)$ since the majority voting unit V will vote out the bad value. This is the basic idea of TMR. A generalization of this approach is *n-modular redundancy* that masks m arbitrary failures, where $n \geq 2m + 1$.

Can the unit that computes the majority vote also fail? Yes. However, the voting unit is considered to be a logical part of C. If C fails, then the result of the vote does not matter.

12.6 TOLERATING OMISSION FAILURES

Assume that a sender process S sends out a sequence of messages $m[0]$, $m[1]$, $m[2]$, ... to a receiver process R (Figure 12.3). If the upper bound of the message propagation delay is not known and the channels are not FIFO, then the receiver must send acknowledgments to the sender to notify the arrival of a message. To detect the nonarrival of a message, process S can at best make a good guess of the round-trip propagation delay and set a timer accordingly. If there is a timeout, the sender retransmits the message. Due to the transient nature of the failure, it is assumed that if a message is sent a bounded number of times, then it will eventually reach the receiver and will be delivered to the application at the receiving end. In case the sender's guess about the round-trip delay is incorrect and the timeout is too early, retransmissions may lead to duplicate messages on the channel. In this context, a *reliable channel* from S to R should meet the following three requirements:

1. *No loss*: The receiver eventually delivers each message sent out by the sender.

2. *No duplication*: The receiver delivers each message sent by the sender exactly once.

3. *No reordering*: The receiver always delivers the message $m[i]$ before $m[i + 1]$.

Thus, if S sends out the sequence $a\ b\ c\ d\ e$, then the first condition rules out R accepting this sequence as $a\ c\ d\ e$, the second condition rules out the possibility of R accepting it as $a\ b\ b\ c\ c\ c\ d\ e$, and the third condition prevents R from accepting it as $a\ c\ e\ b\ d$.

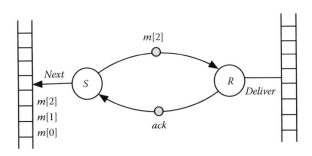

FIGURE 12.3   Implementation of a reliable channel from S to R.

12.6.1 Stenning's Protocol

Stenning's protocol provides an implementation of a reliable channel. The receiving process R, after receiving each message $m[i]$, sends an acknowledgment back to S and decides whether to deliver it to the upper layer. However, not only messages but also acknowledgments can disappear, making the game a bit tricky. Here is a description of Stenning's protocol:

```
program Stenning's protocol;
{program for the sender S}
define ok : boolean, next, j : integer
     m: array[0..∞] of messages, ack: acknowledgment
initially next = 0, ok = true, and both channels (S,R) and (R,S) are empty
do    ok →    send (m[next],next); ok: = false
[]    (ack, j) received → if (j = next)→ ok: = true; next: = next + 1
                            [] (j≠next)→ skip
                         fi
[]    timeout(R,S)  →  send (m[next], next)
od

{program for the receiver R}
define s,r : integer {message sequence numbers}
initially r = 0;
do    (m[],s) received ∧ (s = r) → deliver m[]; send (ack,r); r: = r+1
[]    (m[],s) received ∧ (s ≠ r)→ send (ack,r - 1)
od
```

Theorem 12.1

Stenning's protocol implements a reliable channel.

Proof: (*No loss*) If R does not receive $m[i]$, then it does not send (ack, i), which triggers a retransmission of $m[i]$. After a bounded number of retransmissions, one copy of $m[i]$ must be received and delivered by R.

(*No duplication and no reordering*) Once R delivers $m[i]$, it increments r to $(i + 1)$. Note that R does not deliver the received message unless its sequence number matches r, so duplicate messages are rejected. For the same reason, R always delivers the message $m[i]$ before $m[i + 1]$. Therefore, the *no reordering* criterion holds.

Finally, we demonstrate that the progress property is satisfied. Assume that S sent $m[i]$, it was received by R, and R sent (ack, i), but the ack was lost in transit. So S resent $m[i]$. But in the meantime, R increased the value of r to $(i + 1)$ and was expecting to receive $m[i + 1]$, so it would reject the duplicate copy of $m[i]$. By sending a replica of (ack, i), R guarantees that eventually S receives it and then sends $m[i + 1]$. This guarantees progress. ■

Stenning's protocol is also commonly known as the *stop-and-wait* protocol. It is inefficient, since even in the best case, the sender has to wait for one round-trip delay to send each message, leading to a low throughput. In real life, different variations of this basic idea are used. One such variation is the sliding window protocol.

12.6.2 Sliding Window Protocol

The *sliding window protocol* is a widely used transport layer protocol for implementing a reliable channel between a pair of processes. It is a generalization of Stenning's protocol and uses the same mechanism to deal with the loss or reordering of messages. In addition, the sliding window protocol has provisions to improve the transmission rate and restore the message order at the receiving end without overflowing its buffer space. The primary feature is as follows: The sender continues the send action without receiving the acknowledgments of *at most w* ($w > 0$) outstanding messages, where w is called the *window size*. If no acknowledgment to the previous w messages is received within the timeout period, then the sender resends those w messages (Figure 12.4).

```
program sliding window;
{program for process S}
define next, last, j, w : integer; {w is the constant window size}
        m: array[0..∞] of messages, ack: acknowledgment
initially next = 0, last = -1, w > 0, channels (S,R) and (R,S) are empty;
do last + 1 ≤ next ≤ last + w → send (m[next],next); next: = next + 1
[] (ack,j) received → if j>last → last: = j
                         []j ≤ last → skip
                      fi
[] timeout(R,S) → next: = last + 1   {Retransmission begins}
od

{program for process R}
define s,r : integer {message sequence numbers}
initially r = 0;
do      (m[],s) received ∧ (s = r) → deliver m[]; send (ack,r); r: = r + 1
[]      (m[],s) received ∧ (s ≠ r)→ send (ack,r - 1)
od
```

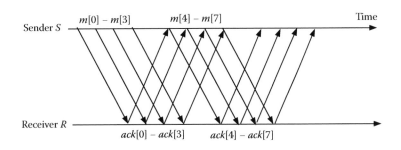

FIGURE 12.4   The trace of a sliding window protocol with window size 4.

Theorem 12.2

Program *sliding window* implements a reliable channel.

Proof (by induction): *Basis*: R eventually receives and delivers $m[0]$. This is because, if R does not receive $m[0]$, then it does not send any acknowledgment, which triggers a time-out at S followed by a retransmission of $m[0]$. In a bounded number of transmissions, R receives and delivers $m[0]$. Using the arguments in the proof of Stenning's protocol, it is easy to establish that $m[0]$ is delivered exactly once.

Induction hypothesis: Assume that R has delivered every message $m[0]$ through $m[k]$ ($k > 0$) exactly once and in the ascending order. S now sends $m[k + 1]$ since the condition $last < k + 1 \leq last + w$ holds. We need to show that eventually R accepts $m[k + 1]$.

Induction step: If $m[k + 1]$ is lost in transit, then no guard of R is not enabled. When the remaining messages in the current window are sent by S, the acknowledgment (ack, k) returned by R does not cause S to increment the value of $last$ beyond k. Eventually, *timeout* is enabled for process S, and it retransmits messages $m[last + 1]$ through $m[last + w]$—this includes $m[k + 1]$. In a finite number of rounds of retransmission, $m[k + 1]$ is received and delivered by R. Since the condition $r = k + 1$ is asserted exactly once, the message $m[k + 1]$ will be accepted exactly once. ▪

A concern here is that the sliding window protocol uses *unbounded* sequence numbers. This raises the question: Is it possible to implement a window protocol using *bounded sequence numbers*, such that it withstands the loss, duplication, and reordering of messages? The answer is *no*. To understand why, consider the informal argument: To withstand loss (which cannot be distinguished from indefinitely large propagation delay), messages have to be retransmitted, so for every message $(m[], k)$ sent by the sender, there may be a duplicate (call it $(m'[], k)$ in the channel. Also, if the sequence numbers are bounded, then they will be recycled, so at some point in the future, there will be another new message $(m''[], k)$ sent *after* $(m[], k)$ and it bears the same sequence number k.

Suppose that m, m', and m'' are in the channel. Since the channel is not FIFO, these messages can arrive in any order. If the receiver receives and delivers m, then it should not deliver m' to avoid duplicate reception of the same message, but then it will also not deliver the new message m'' since m' and m'' have the same sequence numbers, and they are indistinguishable! Using bounded sequence numbers, there is no way to institute a mechanism by which m'' is delivered but m' is rejected. This leads to the following theorem:

Theorem 12.3

If the communication channels are non-FIFO, and the message propagation delays are arbitrarily large, then using bounded sequence numbers, it is impossible to design a window protocol that can withstand the loss, duplication, and reordering of messages.

Several variations of the sliding window protocol are used in practice. One common variation if for the receiver to maintain a window that defines the *range of message sequence numbers* (instead of a single number) that the receiver is ready to accept. The messages received within this window are buffered until they can be delivered. In the *selective repeat* version, when the sender notices the absence of an acknowledgment for one or more messages during a timeout, it selectively transmits only the missing messages. This conserves bandwidth at the expense of buffer space.

12.6.3 Alternating Bit Protocol

The alternating bit protocol is a special version of the sliding window protocol, for which $w = 1$. It works only when the *links are FIFO*, which rules out message reordering. It is a suitable candidate for use in the data-link layer.

The unbounded sequence number was a major hurdle in implementing the sliding window protocols on non-FIFO channels. With FIFO channels, the alternating bit protocol overcomes this problem by appending only a 1-bit sequence number to the body of the message. The protocol is described as follows:

```
program ABP;
{program for the sender S}
define sent, b, j ∈ {0, 1}, next : integer;
initially next = 0, sent = 1, b = 0, and channels (S,R) and (R,S) are empty;
do    sent ≠ b → send (m[next], b); next: = next+1; sent: = b
[]    (ack,j) received ∧ (j = b) → b: = 1 - b
[]    timeout(R,S) → send (m[next - 1], b)
od

{program for the receiver R}
define s,r ∈ {0, 1}
initially r = 0;
do    (m[],s) received ∧ (s = r) → deliver m[]; send (ack,r); r: = 1 - r
[]    (m[],s) received ∧ (s ≠ r)→ send (ack, 1 - r)
od
```

Without going through a formal proof, we demonstrate why the FIFO property of the channel is considered essential for the alternating bit protocol.

Consider the global state of Figure 12.5 reached in the following way: $m[0]$ was transmitted by S and accepted by R, but its acknowledgment $(ack, 0)$ was delayed—so S transmitted $m[0]$ once more. When $(ack, 0)$ finally reached S, it sent out $m[1]$. In general, the state of channel (S, R) is a concatenation of zero or more $(m[i], b)$, followed by zero or more $(m[i + 1], (1 - b))$.

If the channels are *not* FIFO, then $(m[1], 1)$ may reach R *before* the duplicate copy of $(m[0], 0)$ reaches R. R will accept it and send it back $(ack, 1)$ to S. On receipt of this $(ack, 1)$, S will send out $(m[2], 0)$. When the duplicate copy of $(m[0], 0)$ reaches R, and R accepts it, as it mistakes it for $m[2]$ since both $m[0]$ and $m[2]$ have the same sequence number 0! Clearly,

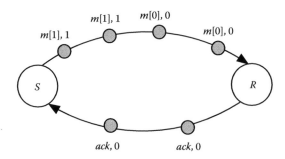

FIGURE 12.5   A global state of the alternating bit protocol.

this possibility is ruled out when the channels are FIFO—the receiver R is certain that the received packet contains $m[2]$ and not a duplicate copy of $m[0]$.

12.6.4 How TCP Works

TCP is the most widely used connection management protocol used on the Internet. It supports the end-to-end logical connection between two processes running on any two computers on the same network or on different networks. It uses the main idea of sliding windows and handles a wide variety of channel failures or perturbations that include the loss and reordering of packets. The link is created by a connection establishment phase and terminated by a connection-closing phase.

A key issue is the ability to generate unique sequence numbers for packets. A sequence number is *safe* to use if it is not identical to one of the sequence numbers that is currently in use. There are various approaches for generating unique sequence numbers. Some of these do not provide absolute guarantees, but sequence numbers are unique with a high probability. For example, if the sequence number is a randomly chosen 64-bit pattern, then it is highly unlikely that the same sequence number will be used again during the lifetime of the application. The guarantee can be further consolidated if the system knows the upper bound (δ) of the message propagation delay across the channel. Every sequence number is automatically flushed out of the system after an interval 2δ from the time it was generated. It is even more unlikely for two identical random 32-bit or 64-bit sequence numbers to be generated within a time 2δ.

TCP uses a *connection establishment phase* outlined in Figure 12.6. The sender S sends a synchronization message (*SYN, seq = x*) to request a connection, x being the sequence number proposed by S. The receiver returns its acceptance by proposing a new sequence number *seq = y* (the sender can verify its uniqueness by checking it against a pool of used ids in the past 2δ time period; otherwise, a *cleanup* is initiated) and appending it to x, so that the sender can recognize that it was a response to its recent request. The (*ack = x + 1*) message is a routine response reflecting the acceptance of the connection request with a sequence number x. If the receiver responds with (*seq = y, ack = z*) and $z \neq x + 1$, then the sender recognizes it as a bad message, ignores it, and waits for the correct message. To complete the connection establishment, the sender sends an (*ack, y + 1*) back to the

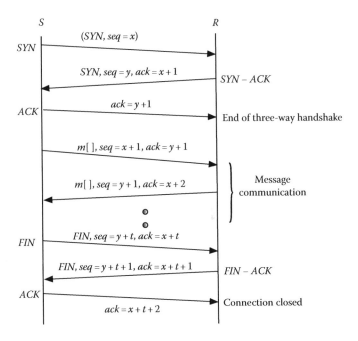

FIGURE 12.6   The exchange of messages in TCP.

receiver, indicating that it accepted the proposed sequence number y. This is called a *three-way handshake*. The sender starts sending data using the starting sequence number $x + 1$. To close a TCP connection, the initiator sends a *FIN* packet. The connection closes only when an *ack* has been sent by R and received by the S.

The initial 32-bit sequence numbers x and y are randomly chosen integers. When a machine crashes and reboots, a randomly chosen 32-bit sequence number is most likely to be different from any sequence number used in the recent past. In case a *request* packet or an *accept* packet is lost, the packet is retransmitted after a timeout period.

A knowledge of δ (obtained by monitoring the round-trip delay) helps choose an appropriate value of timeout. If the timeout period is too small, then unnecessary retransmissions will drastically increase the congestion, whereas a large timeout period will unduly slow down the throughput. The choice is made using *adaptive retransmission*. TCP also allows flow control by permitting the receiver to throttle the sender and control the window size depending on the amount of buffer space it has to store the unprocessed data. For details on flow control, read a standard textbook on networking like [PD96].

12.7 CONCLUDING REMARKS

The specification of faulty behavior using auxiliary variables and fault actions only mimic the faulty behavior and has no connection with the physical cause of the failure. Adding additional actions to overcome the effect of a failure mostly works at the model level—only in limited cases, they can be translated into the design of a fault-tolerant system.

There are several different views regarding the taxonomy of failures. For example, omission failures may not always result in the loss of a message, but may include the skipping of one or more steps by a process, which leads to an arbitrary pattern of behavior that fits the byzantine failure class.

No fault-tolerant system can be designed without some form of redundancy. The redundancy can be in hardware (like spare or standby units), or in the amount of space used per process, or in the time taken by the application. The type of fault tolerance to be used largely depends on the application and its users. It is quite possible that different types of tolerances are desirable against different kinds of faults. As an example, one may expect a system to exhibit masking tolerance against all single faults, but accept a stabilizing behavior when multiple faults occur, since the cost of implementing masking tolerance for multiple faults may be too high. This is known as *multitolerance*.

Stabilization and checkpointing represent two opposing kinds of scenario in nonmasking tolerance. Checkpointing relies on history and recovery leads to an auditable history of the computation. Stabilization, on the other hand, is history-insensitive and does not worry about behavioral errors as long as eventual recovery is guaranteed.

Finally, security breach can lead to different kinds of failures. While many failures are caused due to dust, humidity, cobwebs on the printed circuit boards, or spilled coffee on the keyboard, a security loophole that allows a virus can force the system to shut down services or make the system behave in a bizarre way. However, so far, the fault-tolerance community and the security community have maintained separate identities with minimal overlap between them.

12.8 BIBLIOGRAPHIC NOTES

Ezhilchelvan and Srivastava [ES86] presented a taxonomy of failures that closely resembles the taxonomy presented in this chapter. Rushby [R94] wrote an exhaustive and thoughtful article on faults and related terminologies. Gray's article [G85a] examined many intricate aspects of why computers stop working, and discussed about Heisenbugs and Bohrbugs (unlike Heisenbugs, these are solid bugs that can be consistently reproduced during the testing phase). The general method of fault specification has been adapted from Arora and Gouda's work [AG93]. TMR and *n*-modular redundancy have been in use from the days of World War II to current applications in aircraft control and website mirroring—the original idea is due to von Neumann [VN56]. Lampson [LPS81] introduced stable storage for implementing atomic transactions. Sliding window protocols and their use in the design of TCP and other end-to-end protocols can be found in every textbook on networking— see, for example, Peterson and Davie's book [PD96]. The alternating bit protocol was first introduced as a mechanism of supporting full-duplex communication on half-duplex channels—the earliest work is due to Lynch [Ly68].

Herlihy and Wing [HW91] presented a formal specification of graceful degradation. The work on multitolerance is due to Arora and Kulkarni [AK98]. Dega [De96] reported how fault tolerance was built into the design of Ariane 5 rocket launcher. Unfortunately, despite such a design, the project failed. On June 4, 1996, only 40 s after the launch, the launcher veered off its path and exploded—which was later attributed to software failure. Gärtner [G99] wrote a survey on the design of fault-tolerant systems.

EXERCISES

12.1 A sender P sends a sequence of messages to a receiver Q. In the absence of any failure, the communication from P to Q is FIFO—however, due to failure, the messages can reach Q out of order. Give a formal specification of this failure using normal and fault actions.

12.2 Buffer overflows (also known as *pointer* or *heap* or *stack smashing*) are a well-known source of program failure. What kind of fault models would capture their effects on the rest of the system?

12.3 Consider the following real-life failure scenarios and examine if these can be mapped to any of the known fault classes introduced in this chapter:

a. On January 15, 1990, 114 switching nodes of the long-distance system of AT&T went down. A bug in the failure recovery code of the switches was responsible for this. Ordinarily, when a node crashed, it sent *out-of-service* message to the neighboring nodes, prompting the neighbors to reroute traffic around it. However, the bug (a misplaced *break* statement in C code) caused the neighboring nodes to crash themselves upon receiving the *out-of-service* message. This further propagated the fault by sending an *out-of-service* message to nodes further out in the network. The crash affected the service of an estimated 60,000 people for 9 h, causing AT&T to lose $60 million revenue.

b. A program module of the Arianne space shuttle received a numerical value that it was not equipped to handle. The resulting variable overflow caused the shuttle's onboard computer to fail. The rocket went out of control and subsequently crashed.

12.4 No fault-tolerant system can be implemented without some form of redundancy. The redundancy could be in spare hardware or extra space or extra time or extra messages used in the implementation. In this context, revisit the sliding window protocol and identify the redundancies.

12.5 Consider the following specification of a faulty process:

```
program what-fault
define : x: integer {initially x = 1}
do      x = 0→x: = 1; send "hello"   {S-actions}
[]      x = 1→x: = 2                  {S-action}
[]      x = 2→x: = 0                  {S-action}
[]      x = 1→x: = 3                  {F-action}
od
```

What kind of fault does the F-action induce?

12.6 The fault-tolerance community sometimes uses a term: *repair* fault. How will you justify repair as a failure so that it can trigger unexpected behavior? Present an example.

12.7 Failures are not always independent—the effect of one type of failure can sometimes be manifested as another kind of failure. Create two different scenarios to illustrate how temporal failures (i.e., a missed deadline) can lead to (1) crash failures and (2) omission failures.

12.8 Explain how software errors can lead to (1) omission failure and (2) byzantine failure in a process. Provide examples to explain the complete scenario in each case.

12.9 A sender P sends a sequence of messages to a receiver Q. Each message m is tagged with a unique sequence number seq that increases monotonically, so the program for P can be specified as follows:

```
program sender
define    seq : integer {message sequence number}
initially seq = 0
do true→send(m[seq],seq); seq: = seq + 1 od
```

The messages may reach Q out of order, but they are never lost. To accept messages in strictly increasing order of sequence numbers, Q uses a *sequencer* that resequences the message delivery:

a. Describe the program for the sequencer. Calculate its buffer requirement.

b. Now, assume that the sequencer has the ability to hold *at most two messages* in its buffer. Rewrite the programs of P and the sequencer, so that Q receives the messages in strictly increasing order of sequence numbers.

12.10 Communication links are generally reliable, and omission failures are generally rare. To conserve bandwidth, sometimes *acks* are replaced by *nack* (negative acknowledgment), where the receiver only sends a *nack* to inform the sender that it did not receive an expected message. If no nack is received, the sender is happy, assuming that the receiver has received its messages. Rewrite Stenning's protocol using *nack*.

12.11 Consider a soda machine that dispenses a can of soda to a customer who deposits 50 cents using some combination of quarters, dimes, and nickels. The program is as follows:

```
program   soda machine
define    credit : integer {initially credit = 0)
do        quarter deposited  → credit: = credit+25
[]        dime deposited      → credit: = credit+10
[]        nickel deposited    → credit: = credit+5
[]        credit≥50           → dispense a can of soda; credit: = 0
od
```

a. The following fault action $F \equiv credit = 45 \rightarrow credit: = 0$ triggered a malfunction of the soda machine. Describe the nature of the malfunction. What notice will you post on the machine to alert the customers until the problem is fixed?

b. Modify the program for the soda machine so that it tolerates the specific failure and customers are not affected at all.

12.12 The alternating bit protocol described earlier is designed for a window size 1. This leads to a poor throughput. To improve the throughput, generalize the protocol to a window size $w > 1$ and calculate the lower bound of the size of the sequence number.

12.13 A system of n processes can mask up to m arbitrary failures using majority voting when $n \geq 2m + 1$. Consider downgrading the tolerance from *masking* to *fail-safe* and assume that *no output* is safe. Can the system now tolerate the crash of a larger number of processes? Explain your answer.

CHAPTER 13

Distributed Consensus

13.1 INTRODUCTION

Consensus problems have widespread applications in distributed computing. Before introducing the formal definitions or discussing possible solutions, we first present a few motivating examples of consensus, some of which have been visited in Chapters 7 and 13:

Example 13.1

A number of processes in a network decide to elect a leader. Each process begins with a bid for leadership. At the end, one of these processes is elected the leader, and it reflects the final decision of every process.

Example 13.2

Alice wants to transfer a sum of $1000 from her savings account in Las Vegas to a checking account in Iowa City. There are two components of this transaction: *debit $1000* and *credit $1000*. Two distinct servers handle the two components. Regardless of failures, a transaction must be atomic—it will be a disaster if the debit operation is completed but the credit operation fails, or vice versa. Accordingly, two different servers have to reach an agreement about whether to *commit* or to *abort* the transaction.

Example 13.3

Five independent sensors measure the temperature T of a furnace. Each sensor checks if T is greater than 1000°C. Some sensors may be faulty, but it is not known which are faulty. The nonfaulty sensors have to agree about the truth of the predicate $T > 1000$°C (so that the next course of action can be decided).

Example 13.4

Consider the problem of synchronizing a set of phase clocks that are ticking at the same rate in different geographical locations of a network. Viewed as a consensus problem, the initial values (of phases) are the set of local readings (which can be arbitrary), but the final phases must be identical.

Consensus problems are far less interesting in the absence of failures. This chapter studies distributed consensus in the presence of failures. The problem can be formulated as follows: a distributed system contains n processes $\{0, 1, 2, \ldots, n-1\}$. Every process has an *initial value* in a mutually agreed domain. The problem is to devise an algorithm such that despite the occurrence of failures, processes eventually agree upon an irrevocable *final decision value* that satisfies the following three conditions:

Termination: Every nonfaulty process must eventually decide.

Agreement: The final decision of every nonfaulty process must be identical.

Validity: If every nonfaulty process begins with the same initial value v, then their final decision must be v.

The validity criterion adds a dose of sanity check—it is silly to reach agreement when the agreed value reflects nobody's initial choice. Also, the irrevocability of the final decision is important: If a stock broker gets confirmation from his or her system that Alice bought 50 shares of Wind Energy Inc. online, then even if the broker's computer is destroyed by a tsunami, Alice should be able to go to any other broker and sell those shares.

All of the previous examples illustrate *exact* agreement. In some cases, *approximate agreement* is considered adequate. For example, physical clock synchronization always allows the difference between two clock readings to be less than a small skew—exact agreement is not achievable. The agreement and validity criteria do not specify whether the final decision value has to be chosen by a majority vote—the lower threshold of acceptance is defined by the validity rule. Individual applications may fine-tune their agreement goals within the broad scopes of the three specifications. In Example 13.2, if the server in Iowa City decides to *commit* the action but the server in Las Vegas decides to *abort* since Alice's account balance is less than $1000, then the consensus should be an *abort* instead of *commit*—this satisfies both termination and validity. Validity does not follow from agreement. Leaving validity out may lead to the awkward possibility that even if both servers prefer to *abort* the transaction, the final decision may be a *commit*.

In this chapter, we will address various problems related to reaching an agreement about a value or an action in a distributed system. We will separately discuss consensus in asynchronous and synchronous systems.

13.2 CONSENSUS IN ASYNCHRONOUS SYSTEMS

Seven members of a busy household decided to hire a cook, since they do not have time to prepare their own food. Each member of the household separately interviewed every applicant for the cook's position. Depending on how it went, each member formed his or her independent opinion *yes* (means hire) or *no* (means don't hire). These members will now have to communicate with one another to reach a uniform final decision about whether the applicant will be hired. The process will be repeated with each applicant, until someone is hired.

Consider an instance of this hiring process that deals with a single candidate. The members may communicate their decisions in arbitrary ways. Since they do not meet each other very often, they may decide to communicate by phone, or through letters, or by posting a note on a common bulletin board in the house. The communication is completely asynchronous, so actions by the individual members may take an arbitrary but finite amount of time to complete. Also, no specific assumption is made about how the final decision is reached, except that the final decision is irrevocable (so you cannot decide to hire the candidate and later say *sorry*). Is there a guarantee that the members will eventually reach a consensus about whether to hire the candidate?

If there is no failure, then reaching an agreement is trivial. Each member sends his or her initial opinion to every other member. Define V to be the bag of initial opinions by all the members of the household. Due to the absence of failure, every member is guaranteed to receive an identical bag V of opinions from all the members. To reach a common final decision, every member will apply the same choice function f on this bag of opinions.

Reaching consensus, however, becomes surprisingly difficult, when one or more members fail to execute actions. Assume that at most, k members ($k > 0$) can undergo *crash failure*. An important result due to Fischer et al. [FLP85] states that in a *fully asynchronous* system, it is impossible to reach consensus even if $k = 1$.* The consensus requires agreement among nonfaulty members only—we do not care about the faulty members.

For an abstract formulation of this impossibility result, treat every member as a process. The network of processes is completely connected. Assume that the initial opinion or the final decision by a process is an element of the set $\{0, 1\}$ where $0 \equiv$ *don't hire*, $1 \equiv$ *hire*. We demonstrate the impossibility result for the shared-memory model only, where processes communicate with one another using *read* and *write* actions. The results hold for the message-passing model too, but we will skip the proof for that model.

13.2.1 Bivalent and Univalent States

The progress of a consensus algorithm can be abstracted using a binary *decision state*. The two possible decision states are *bivalent* and *univalent*. A decision state is *bivalent* if starting from that state, there exist at least two distinct executions leading to two distinct

---

* Remember that crash failures cannot be reliably detected in asynchronous systems. It is not possible to distinguish between a processor that has crashed and a processor that is executing its actions very slowly.

decision values 0 or 1. What it means is that from a bivalent state, there is a *potential* for reaching any one of the two possible decision values. A state from which only one decision value can be reached is called a *univalent* state. Univalent states can be either *0-valent* or *1-valent*. In a 0-valent state, the system is committed to the irrevocable final decision 0. Similarly, in a 1-valent state, the system is committed to the irrevocable final decision 1.

As an illustration, consider a best-of-five-sets tennis match between two players A and B. If the score is 6-3, 6-4 in favor of A, then the decision state is bivalent, since any-one could win. If, however, the score becomes 6-3, 6-4, 6-1 in favor of A, then the state becomes univalent, since even if the game continues till the fifth set, only A can win. At this point, the outcome of the remaining two sets becomes irrelevant. This trivially leads to the following lemma.

Lemma 13.1

No execution leads from a 0-valent to a 1-valent state or vice versa.

Lemma 13.2

Every consensus protocol must have a bivalent initial state.

Proof (by contradiction): An initial state consists of a vector of n binary values—each value is the initial choice of a process. Assume that no initial state is bivalent. Now, consider an *array* $s[0 .. n - 1]$ of n carefully chosen initial states as shown in Table 13.1. Note that the vectors in the successive rows differ by one bit only. Per the validity criterion of consensus, $s[0]$ is 0-valent, and $s[n - 1]$ is 1-valent. In the above array, there must exist states $s[n - i - 1]$ and $s[n - i]$ such that (1) $s[n - i - 1]$ is 0-valent, (2) $s[n - i]$ is 1-valent, and (3) the vectors differ in the value chosen by some process i. Now consider a computation e (not involving process i—thus, e mimics the crash of i) that starts from $s[n - i - 1]$ and leads to the final decision 0. The same computation e must also be a valid computation starting from $s[n - i]$, which is 1-valent. This contradicts Lemma 13.1. Therefore, no consensus protocol should have a univalent initial state. ◾

TABLE 13.1 Array of Possible Initial States for a System of n Processes

Process →	0	1	2		i	$i + 1$		$n - 3$	$n - 2$	$n - 1$	Type
$s[0]$	0	0	0	⋯	0	0	⋯	0	0	0	0-valent
$s[1]$	0	0	0	⋯	0	0	⋯	0	0	1	0-valent
$s[2]$	0	0	0	⋯	0	0	⋯	0	1	1	0-valent
$s[3]$	0	0	0	⋯	0	0	⋯	1	1	1	0-valent
⋯	⋯	⋯	⋯	⋯	⋯	⋯	⋯	⋯	⋯	⋯	0-valent
$s[n - i - 1]$	0	0	0		0	1	1	1	1	1	0-valent
$s[n - i]$	0	0	0	⋯	1	1	⋯	1	1	1	1-valent
⋯	⋯	⋯	⋯	⋯	⋯	⋯	⋯	⋯	⋯	⋯	1-valent
$s[n - 1]$	1	1	1	1	1	1	1	1	1	1	1-valent

Lemma 13.3

In a consensus protocol, starting from any initial bivalent state I, there must exist a reachable bivalent state T, from where every action taken by some process p leads to either a 0-valent or a 1-valent state.

Proof: If every state reachable from I is bivalent, then no decision is reached—this violates the termination criteria of the consensus protocol. In order that decisions are reached, every execution must eventually lead to some univalent state. ■

Assume that there is a bivalent state T reachable from I (Figure 13.1)—from T *action* 0 by process p leads to the 0-valent state $T0$, and *action* 1 by process $q \neq p$ leads to the 1-valent state $T1$. This implies that the guards of both p and q are enabled in state T. We will first argue that p and q cannot be distinct processes. Assume a shared-memory model* and consider three cases:

Case 1: At least one of the actions (action 0 or action 1) is a read action.
Without loss of generality, let *action* 0 by p be a *read* operation and *action* 1 by q be a *write* operation. The read action does not have any impact on any other process in the system, so the write action by process q in state T remains a feasible action in state $T0$ also. Consider now a computation $e1$ from state T (Figure 13.2a) where q executes action 1 followed by a suffix that *excludes* any step by process p (which mimics the crash of p). Such a computation must exist if consensus is possible when a process crashes. Let $e1$ lead to decision 1. Now, consider another computation $e0$ that starts with action 0 by p. To every process $r \neq p$, the states $T0$ and T are indistinguishable. So one can use $e1$ as the *suffix* of the computation $e0$ from state $T0$ onwards. The final decision now is 0. To every process other than p, the two computations $e0$ and $e1$ are indistinguishable, but the final decisions are different. This is not possible. Therefore, $p = q$.

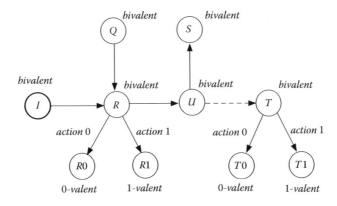

FIGURE 13.1   The transition from bivalent to univalent states.

---

* Similar arguments can be made for the message-passing model also. See [FLP85] for details.

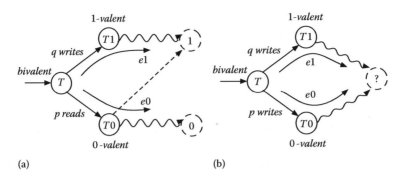

FIGURE 13.2 The different scenarios of Lemma 13.3. (a) p reads and q writes. (b) Both p and q write, but on different variables.

Case 2: Both action 0 and action 1 are write operations, but on different variables.
In this case, the two writes are noninterfering, that is, no write by one process negates the effect of write by the other process. Now, consider two computations $e0$ and $e1$ from state T: In $e0$, p executes action 0, q executes action 1, and then the remaining actions lead to the final decision 0. In $e1$, the first two actions are swapped, that is, the first action is by q and the second action is by p, and the remaining actions lead to the final decision 1 (see Figure 13.2b). However, in both cases, after the first two steps, *the same global state is reached*, so the final outcomes cannot be different! Therefore, $p = q$.

Case 3: Both action 0 and action 1 are write operations, but on the same variable.
If p, q execute actions 0 and 1 one after another, then the second action will overwrite the result of the first action. Consider two computations $e0$ and $e1$ from state T: In $e1$, q executes action 1 followed by a suffix that excludes any step by process p (which mimics the crash of p), and it leads to decision 1. In $e0$, use $e1$ as the suffix of the computation after action 0 (i.e., p writes). Observe that in $e0$, after q writes following p's write, the state is the same as $T1$. To every process other than p, the two computations $e0$ and $e1$ are indistinguishable, but $e0$ leads to a final decision 0, whereas $e1$ leads to decision 1. This is impossible. Therefore, $p = q$.

Process p is called a *decider* process for the bivalent state T.

Theorem 13.1

In asynchronous distributed systems, it is impossible to design a consensus protocol that will tolerate the crash failure of a single process.

Proof: Let T be a bivalent decision state reachable from the initial state and p be a decider process, such that action **0** by process p leads to the decision **0**, and action **1** by process p leads to the decision **1**. What if p crashes in state T, that is, beginning from state T, process p does not execute any further action? Due to actions taken by the remaining processes, the system can possibly move to another bivalent state U. Since the computation must terminate, that is, the system will eventually reach consensus, bivalent states cannot form a cycle. This implies there must exist a reachable bivalent state R, in which (1) no action leads to another bivalent state, and (2) each enabled process r is a decider. If r crashes, then no consensus is reached. ■

This explains why the family members may never reach a consensus about hiring a cook as long as one of them postpones his or her decision indefinitely or does not respond (equivalent to a crash). For a more rigorous proof of this result, see [Ly96]. Note that the impossibility result holds only for deterministic solutions—there are probabilistic solutions for reaching consensus in the presence of crash failures [B83].

13.3 CONSENSUS IN SYNCHRONOUS SYSTEMS: BYZANTINE GENERALS PROBLEM

The *byzantine generals problem* deals with reaching consensus in the presence of byzantine failures. Byzantine failure model (Chapter 12) captures any arbitrary form of erratic behavior: a faulty process may send arbitrary messages, or undergo arbitrary state transitions not consistent with its specification, or decide not to execute any action. Even malicious behavior is not ruled out. A solution to the consensus problem in the presence of byzantine failures should be applicable to systems where no guarantees about process behaviors are available. Purely asynchronous models are so weak that it is impossible to solve the consensus problem even for simple failures like crash. In Lamport et al. [LSP82], a solution for *synchronous distributed systems* using the message-passing model was proposed. This section presents the solution in [LSP82]. The story line is as follows:

In a particular phase of a war, *n* generals 0, 1, 2, ..., *n* – 1 try to reach an agreement about whether to *attack* or to *retreat*. Each general may have his own preference about whether to attack or to retreat. However, a sensible goal in warfare is that all generals agree upon the same plan of action. The generals will do so by exchanging their individual preferences. However, some of these generals may be traitors—their goal is to prevent the loyal generals from reaching an agreement. For this, a traitor may indefinitely postpone their participation or send conflicting messages. No one knows who the traitors are. The problem is to devise a strategy, by which every loyal general eventually agrees upon the same plan, regardless of the action of the traitors.

We assume that the system is synchronous: computation speeds have a known lower bound and channel delays have a known upper bound so that the absence of messages can be detected using timeout.

13.3.1 Solution with No Traitor

In the absence of any traitor, every general sends his input (*attack* or *retreat*) to every other general. We will refer to these inputs as *orders*. These inputs need not be the same. The principle of reaching agreement is that every loyal general eventually collects the *identical bag S* of orders and applies the same choice function on S to decide the final strategy.

13.3.2 Solution with Traitors: Interactive Consistency Criteria

The task of collecting S can be divided into a number of subtasks. Each subtask consists of a particular general *i* broadcasting his order to every other general. Each general plays the role of the *commander* when he broadcasts and the role of a *lieutenant* when he receives an order from another general (playing the role of the commander).

Some generals can be traitors. When a loyal commander broadcasts his order to the lieutenants, every loyal lieutenant receives the *same* order. This may not be true when the commander is a traitor, since he can say *attack* to some lieutenants and *retreat* to the others. The precise requirements of the communication between a commander and his lieutenants are defined by the following two *interactive consistency* criteria:

IC1: Every loyal lieutenant receives the same order from the commander.

IC2: If the commander is loyal, then every loyal lieutenant receives the order that the commander sends.

Regardless of whether the commander is a traitor, IC1 can be satisfied as long as every loyal lieutenant receives the same order. However, it does not preclude the awkward case in which the commander sends the order *attack*, but all lieutenants receive the order *retreat*, which satisfies IC1! This justifies the inclusion of IC2. When the commander is loyal, IC2 holds, and IC1 follows from IC2.

Satisfying the interactive consistency criteria while collecting the order from a commander who is a potential traitor can be complex. This is illustrated in the subsequent sections.

13.3.3 Consensus with Oral Messages

The solution to the byzantine generals problem depends on the model of message communication. The *oral message* model satisfies the following three conditions:

1. Messages are not corrupted in transit.

2. Messages can be lost, but the absence of message can be detected.

3. When a message is received (or its absence is detected), the receiver knows the identity of the sender (or the defaulter).

The corruption of a message can be attributed to the sender sending a different order—a behavior that is well captured under the byzantine failure model, so allowing the messages to be corrupted in transit does not add any more value to the fault model. Let m represent the number of traitors. An algorithm that satisfies IC1 and IC2 using the oral message model of communication in the presence of at most m traitors will be called an $OM(m)$ algorithm. The $OM(0)$ algorithm is based on direct communication only: The commander sends his order to every lieutenant, and the lieutenants receive these orders that are sent.

However when $m > 0$, direct communication is not adequate. To check for possible inconsistencies, a loyal (and wise) lieutenant would also like to know from every other lieutenant, "What did the commander tell you?" The answers to these questions constitute the indirect messages. Note that not only the commander but also some of the fellow lieutenants may be traitors. The actual *order* accepted by a lieutenant from the commander will depend on both direct and indirect messages.

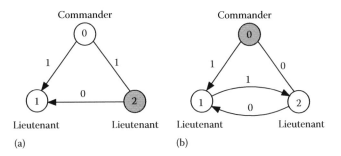

Lieutenant   Lieutenant   Lieutenant   Lieutenant

(a)   (b)

FIGURE 13.3   (a) Commander is loyal. (b) Commander is a traitor.

13.3.3.1 Impossibility Result

An important impossibility result is that, using oral messages, no solution to the byzantine generals problem exists with three or fewer generals and one traitor.

The two-general case is clearly not interesting—if there is a traitor, then the loyal general does not have to agree with anyone, and the problem becomes vacuous. To motivate the readers, consider an example with *three* generals and *one* traitor. Figure 13.3a shows the case when the commander is loyal, but lieutenant 2 is a traitor. We use the values 0 and 1 in place of *attack* and *retreat*, respectively. For lieutenant 1, the direct message from the commander is 1, and the indirect message received through lieutenant 2 (who altered the message) is 0. To satisfy IC2, lieutenant 1 must choose the first element from the ordered set (direct message, indirect message) as the order from the commander.

Now consider the scenario of Figure 13.3b, where the commander is a traitor. If the lieutenants use the same choice function as in the previous case, then IC1 is violated. Since no loyal lieutenant has a prior knowledge about who is faulty, there will always exist a case in which either IC1 or IC2 will be violated. These form the seeds of the impossibility result. A more formal proof of this impossibility result (adapted from [FLM86]) is presented in the following.

Theorem 13.2

Using oral messages, there is no solution to the byzantine generals problem when there are three generals and one traitor.

Proof (by contradiction): Assume that there exists a solution to the byzantine generals problem for a system S consisting of the generals P, Q, R (Figure 13.4a), of which one is a traitor. In Figure 13.4b, define another system SS that consists of two copies of each general—P',Q',R' are clones of P, Q, R, respectively. Note that based on the local knowledge about his own neighborhood, no general can distinguish whether it belongs to S or SS. We call them look-alike systems.

Assume that in SS, each process is nonfaulty. Also, each of the generals P,Q,R' has an initial value x, and each of the other generals P',Q',R has a different initial value y (i.e., $x \neq y$). System SS can mimic at least three separate instances of a three-general system in which one

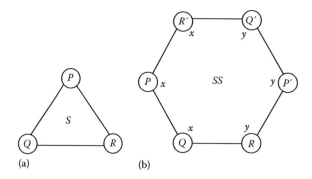

FIGURE 13.4   (a) and (b) Two look-alike systems.

can be a potential traitor: for example, R and R' present to P, Q with two different versions of their values (orders), Q and Q' do the same to R, P', and so on.

To P and Q in SS, it will appear that they are a part of a three-general system like S in which R is faulty. Since both P and Q have initial values x, their final decision must also be x. To P' and R in SS, it will appear that they are a part of a three-general system like S in which Q is faulty. Since both P' and R have initial values y, their final orders must also be y.

Now, to Q and R in SS, it will appear that they are a part of a three-general system like S in which P is faulty. But they cannot reach consensus, since Q has already chosen x, and R has already chosen y as their final decision. Therefore, a three-general system with one traitor cannot reach consensus.   ■

Corollary 13.1

Using oral messages, no solution to the byzantine generals problem exists with $3m$ or fewer generals and m traitors ($m > 0$).

Proof (contradiction): Assume that the statement is false, that is, a solution to the byzantine generals problem exists with $3m$ or fewer generals and m traitors. Then we can use it to construct a solution to the case of three generals and one traitor in the following way: Divide the $3m$ generals into three groups of m generals each, such that all the traitors are put in one group. Let a general simulate the members of each of these groups. The general that simulates the group of m traitors mimics the actions of a single traitor, whereas the other two mimic the loyal generals.

Since we know that it is impossible to construct a solution with three generals and one traitor, we have a contradiction.   ■

13.3.3.2 OM(m) Algorithm
It follows from Corollary 13.1 that in order to reach a consensus in the presence of m traitors, $n \geq 3m+1$ generals are needed. Recall that $OM(m)$ refers to an algorithm that satisfies the interactive consistency criteria IC1 and IC2 in the presence of at most m traitors. In Lamport et al. [LSP82], proposed the following version of $OM(m)$. The algorithm is recursive: $OM(m)$ invokes $OM(m - 1)$, which invokes $OM(m - 2)$ and so on.

Algorithm *OM*(0)

1. Commander *i* sends out a value $v \in \{0, 1\}$ to every lieutenant $j \neq i$.
2. Each lieutenant *j* accepts the value from *i* as the *order* from commander *i*.

Algorithm *OM*(*m*)

1. Commander *i* sends out a value $v \in \{0, 1\}$ to every lieutenant $j \neq i$.
2. If $m > 0$, then each lieutenant *j*, after receiving a value from the commander, starts a new phase by broadcasting it to the *remaining* lieutenants using $OM(m - 1)$. In this phase, *j* acts as the commander. Each lieutenant thus receives $(n - 1)$ values: (a) a value *directly* received from the commander *i* of $OM(m)$ and (b) $(n - 2)$ values *indirectly* received from the $(n - 2)$ lieutenants resulting from their broadcast $OM(m - 1)$. If a value is not received, then it is substituted by a *default* value.
3. Each lieutenant chooses the *majority* of the $(n - 1)$ values received by it as the *order* from the commander *i*.

Figure 13.5 illustrates the *OM*(1) algorithm with four generals and one traitor. The commander's id is 0. In part (a), lieutenant 3 is a traitor, so he broadcasts a 0 even if he has received a 1 from the commander. Per the algorithm, each of the loyal lieutenants 1 and 2 chooses the majority of {1, 1, 0}, that is, 1. In part (b), commander 0 is a traitor and broadcasts conflicting messages. However, all three loyal lieutenants eventually decide on the same final value, which is the majority of {1, 0, 1}, that is, 1.

The algorithm progresses in rounds. Algorithm *OM*(0) requires one round, and *OM*(*m*) requires $m + 1$ rounds to complete. As the recursion unfolds, *OM*(*m*) invokes $n - 1$ separate executions of $OM(m - 1)$, each $OM(m - 1)$ invokes $n - 2$ separate executions of $OM(m - 2)$, and so on. This continues until *OM*(0) is reached. The total number of messages required in the *OM*(*m*) algorithm is therefore $(n-1)(n-2)(n-3)\cdots(n-m+1)$, that is, $O(n^m)$.

Figure 13.6 illustrates the execution of the *OM*(*m*) algorithm with $m = 2$ and $n = 7$. Each level of recursion is explicitly identified. Commander 0 initiates *OM*(2) by sending a 1 to

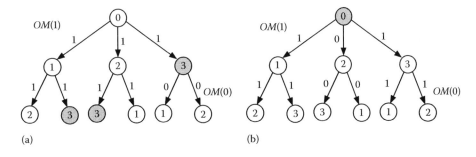

(a) (b)

FIGURE 13.5 An illustration of *OM*(1) with four generals and one traitor: the messages at the upper level reflect the opening messages of *OM*(1), and those at the lower level reflect the *OM*(0) messages that are triggered by the upper level messages. (a) Lieutenant 3 is the traitor. (b) Commander 0 is the traitor.

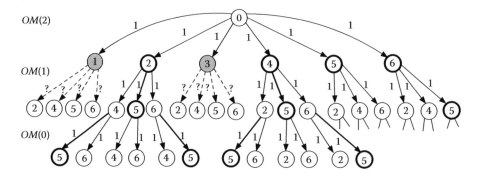

FIGURE 13.6   A partial trace of the *OM*(2) algorithm with seven generals and two traitors: The messages received by the traitors at the lowest level are not shown, since they do not matter.

every lieutenant. The lieutenants 1 and 3 are traitors. The diagram does not show values of the messages from the traitors, but you can assume any arbitrary value for them. Loyal commanders 2, 4, 5, 6 receive the direct message 1 and initiate *OM*(1). Each *OM*(1) triggers an instance of *OM*(0), and the unfolding of the recursion ends.

At the *OM*(1) level, let us focus on 5. What message did 5 receive from 2 via *OM*(1)? Lieutenant 5 is eligible to receive five messages at this level: a direct message 1 from 2 and four other messages from 4, 6, 1, and 3. Nothing can be said about what 5 would receive from 1 or 3, since they are traitors. But as Figure 13.6 shows, 5 received a 1 from each of 4 and 6. So 5 received three 1s out of the five values that it can receive, and the majority is 1. Use similar arguments and observe that 5 received a 1 from 4 and 6 too via *OM*(1). At the *OM*(2) level, each lieutenant collects at least four 1s out of the maximum six values that it can receive, and the majority is 1. This reflects the *order* from commander 0. Any misinformation sent by the traitors is clearly voted out, and the final results satisfy the interactive consistency criteria IC1 and IC2.

In case the decision value is not binary, the function *majority* may be replaced by the function *median* to reach agreement.

Proof of the oral message algorithm

Lemma 13.4

Let the commander be loyal, and $n > 2m + k$, where m = maximum number of traitors. Then *OM*(*k*) satisfies IC2.

Proof (by induction):

Basis: When $k = 0$, the theorem is trivially true.

Induction hypothesis: Assume that the theorem holds for $k = r$ ($r > 0$), that is, *OM*(*r*) satisfies IC2. We need to show that it holds for $k = r + 1$.

Inductive step: Consider an *OM*(*r* + 1) algorithm initiated by a loyal commander. Here, $n > 2m + r + 1$ (from the statement of the theorem), so $n - 1 > 2m + r$ (Figure 13.7).

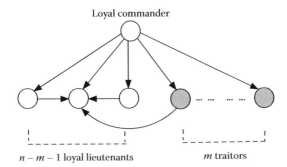

FIGURE 13.7   The $OM(r + 1)$ algorithm with n generals and m traitors.

After the commander sends out a value, each of the remaining $(n - 1)$ lieutenants initiates $OM(r)$. According to the induction hypothesis, $OM(r)$ satisfies IC2—so every loyal lieutenant receives the same order via $OM(r)$ from every other loyal lieutenant. Now $n - 1 > 2m + r$, which implies $n - m - 1 > m + r > m$. This shows that a majority of the lieutenants are loyal. So, regardless of the values sent by the m traitors, the majority of the $(n - 1)$ values collected by each lieutenant, including the direct message from the loyal commander, must be v. So, IC2 is satisfied.   ■

Theorem 13.3

If $>3m$ where m is the maximum number of traitors, then $OM(m)$ satisfies both IC1 and IC2.

Proof (by induction):

Basis: When $m = 0$, the theorem trivially holds.
Induction hypothesis: Assume that the theorem holds for $m = r$. We need to show it holds
   for $OM(r + 1)$ too.
Inductive step: Substitute $k = m$ in Lemma 13.4 and consider the following two cases:

   Case 1: The commander is loyal. Then $OM(m)$ satisfies IC2 and hence IC1.
   Case 2: The commander is a traitor. Then there are more than $3(r + 1)$ generals and
      at most $(r + 1)$ traitors. This implies that there are more than $(3r + 2)$ lieutenants
      remaining, of which at most r are traitors, and more than $(2r + 2)$ are loyal. Since
      $(3r + 2) > 3r$, by the induction hypothesis, $OM(r)$ satisfies IC1 and IC2.

In $OM(r + 1)$, a loyal lieutenant chooses the majority from (1) the values from more than $(2r + 1)$ loyal lieutenants obtained via $OM(r)$, (2) the values received from at most r traitors via $OM(r)$, and (3) the value sent by the commander, who is a traitor. Since $OM(r)$ satisfies IC1 and IC2, the values collected in part (1) are the same for all loyal lieutenants—it is the same value that the lieutenants received from the commander. Also, by the induction hypothesis, the loyal lieutenants will receive the same value via $OM(r)$ from each of the r traitors, regardless of what they send. So every loyal lieutenant will collect the same set of values, apply the same choice function, and reach the same final decision.   ■

13.3.4 Consensus Using Signed Messages

The communication model of *signed message* satisfies all the conditions of the model of oral message. In addition, it satisfies two extra conditions:

1. A loyal general's signature cannot be forged, that is, any forgery can be detected.

2. Anyone can verify the authenticity of a signature.

In real life, signature is an encryption technique. Authentication methods to detect possible forgery are discussed in Chapter 19. The existence of a safeguard against the forgery of signed messages can be leveraged to find out simpler solutions to the consensus problem. So if the commander sends a 0 to every lieutenant $\{0, 1, 2, ..., n - 1\}$, but lieutenant 2 tells lieutenant 1: "the commander sent me a 1," then lieutenant 1 can immediately detect it and discard the message. The signed message algorithm also exhibits better resilience to faulty behaviors.

We begin with an example to illustrate interactive consistency with three generals and one traitor using signed messages. The notation $v\{S\}$ will represent a signed message, where v is a value initiated by general i and S is a signature list consisting of the sequence of signatures by the generals $i, j, k, ...$ Each signature is essentially an encryption mechanism using the sender's private key. Loyal generals try to decrypt or decipher it using a public key. If that is not possible, then the message must be forged and is discarded. The number of entries in S will be called the *length* of S.

In Figure 13.8a, lieutenant 1 will detect that lieutenant 2 forged the message from the commander, and reject the message. In Figure 13.8b, the commander is a traitor, but no message is forged. Both 1 and 2 will discover that the commander sent out inconsistent values. To reach an agreement, both of them will apply some mutually agreed choice function f on the collected bag of messages and make the same final decision. Note that this satisfies both IC1 and IC2.

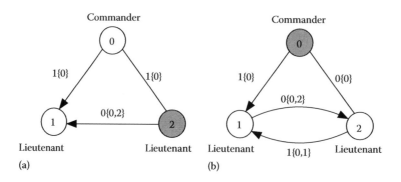

FIGURE 13.8 Interactive consistency using signed messages. (a) Lieutenant 1 detects that the message from 2 is forged, (b) both 1 and 2 accept all messages since no lieutenant forged any message.

In contrast with the *OM(m)* algorithm, the signed message version *SM(m)* satisfies both IC1 and IC2 whenever $n \geq m + 2$, so it has a much better resilience against faulty behavior. The algorithm *SM(m)* is described in the following:

Algorithm *SM(m)*

1. Commander *i* sends out a signed message $v\{i\}$ to every lieutenant $j \neq i$.
2. Lieutenant *j*, after receiving a message $v\{S\}$, adds it to a set $V \cdot j$, only if
   (i) It is not forged
   (ii) It has not been received before
3. If the *length* of S is less than $m + 1$, then lieutenant *j*
   (i) Appends his own signature to S
   (ii) Sends $V \cdot j$ it to every other lieutenant whose signature does not appear in S
4. When lieutenant *j* receives no more messages, he applies a choice function on $V \cdot j$ to generate the final decision.

The only requirements on the choice function are (1) when $V \cdot j$ contains a single element *v*, $choice(V \cdot i) = v$, (2) when $V \cdot i = \emptyset$, $choice(V \cdot i) = 0$ and (3) when $|V \cdot i| > 1$, the median element is returned, assuming that the elements are ordered.

Theorem 13.4

If $n \geq m + 2$, where *m* is the maximum number of traitors, then *SM(m)* satisfies both IC1 and IC2.

Proof: Consider the following two cases:

Case 1: Assume that the commander is loyal. Then every loyal lieutenant *i* must receive the value *v* from the initial broadcast. Since forged messages are discarded, any other value indirectly received and accepted by a loyal lieutenant must be *v*. Since the only message in the set $V \cdot i$ is *v*, $choice(V \cdot i) = v$. This satisfies both IC1 and IC2.

Case 2: Assume that the commander is a traitor. We argue that for a pair of loyal lieutenants *i* and *j*, $V \cdot i = V \cdot j$. If $length(S) < m + 1$, then *i* sends to *j* every message that he accepts, and vice versa. If $length(S) = m + 1$, then at least one loyal lieutenant must have signed the message and sent it to *i* and *j*, so it appears in both $V \cdot i$ and $V \cdot i$. Therefore, $V \cdot i = V \cdot j$. Application of the choice function on the sets $V \cdot i$ and $V \cdot j$ will lead to the same final decision by both *i* and *j*. This satisfies IC1. ■

SM(m) begins by sending up to $(n - 1)$ messages, each recipient of this message sends up to $(n - 2)$ messages and so on. So, to reach an agreement, it will cost $(n - 1)(n - 2) \cdots (n - m + 1)$ messages. The message complexity is thus similar to that of *OM(m)*. However, for the same value of *n*, the signed message algorithm has much better resilience, that is, it tolerates a much larger number of traitors.

13.4 PAXOS ALGORITHM

Paxos is an algorithm for implementing fault-tolerant consensus. It runs on a completely connected network of n processes and tolerates up to m failures, where $n \geq 2m + 1$. Processes can crash and messages may be lost, but byzantine failures are ruled out, at least in the current version. The algorithm solves the consensus problem in the presence of these faults on an asynchronous system of processes. Although the requirements for consensus are *agreement*, *validity*, and *termination*, Paxos primarily guarantees agreement and validity, and not termination—it allows for the possibility of termination only if there is a sufficiently long interval during which no process restarts the protocol.

A process may play three different roles: *proposer, acceptor,* and *learner.* Each role has a different responsibility: Proposers submit proposed values on behalf of clients, acceptors decide the candidate values for the final decision, and learners collect these information from the acceptors and report the final decision back to the clients. A proposal sent by a proposer is a pair (v, n) where v is a value and n is a sequence number. If there is only one acceptor that decides which value will be chosen as the consensus value, then it will be too simplistic. What if the acceptor crashes? To deal with this, there are multiple acceptors. A proposal must be endorsed by at least one acceptor before it becomes a candidate for the final decision. The sequence number is used to distinguish between successive attempts to invoke the protocol. Upon receiving a proposal with a larger sequence number from a given process, acceptors discard the proposals with lower sequence numbers. Eventually, an acceptor accepts the *majority's choice.* The sequence of actions is as follows:

Phase 1: The preparatory phase

> *Step 1.1*: Each proposer sends a proposal (v, n) to each acceptor (Figure 13.9).

> *Step 1.2*: If n is the largest sequence number of a proposal received by an acceptor, then it sends an $ack(n, \perp, \perp)$ to its proposer, which is a *promise* that it will ignore all proposals numbered lowered than n. However, in case an acceptor has accepted a proposal with a sequence number $n' < n$ and a proposed value v, it responds with $ack(n, v, n')$. This implies that the proposer has no point in trying to submit another proposal with a larger sequence number.

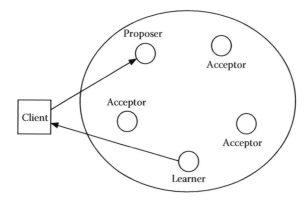

FIGURE 13.9 The setup in Paxos: each circle is a process that can act as any one of the three agents: proposer, acceptor, or learner.

An alternative is to respond to the proposer with a negative acknowledgment (*nack*) that discourages the proposer to force an agreement with new input values. Note that this is a performance issue and does not affect the correctness of the protocol. When there are two proposals with identical sequence numbers, the tie is broken using the process ids.

Phase 2: Request for acceptance of an input value

Step 2.1: If a proposer receives $ack(n, \perp, \perp)$ from a majority of acceptors, then it sends $accept(v, n)$ to all acceptors, asking them to *accept* this value. If however, an acceptor returned an $ack(n, v, n')$ to the proposer in phase 1 (which means that it already accepted proposal with value v) then the proposer must include the value v with the highest sequence number in its request to the acceptors.

Step 2.2: An acceptor *accepts* a proposal (v, n) unless it has already promised to consider proposals with a sequence number greater than n.

Phase 3: The final decision

When a majority of the acceptors accept a proposed value, it becomes the final decision value. The acceptors multicast the accepted value to the *learners*, which enables them to determine if a proposal has been accepted by a majority of acceptors. The learners convey it to the client processes invoking the consensus.

The following two observations highlight the properties of the algorithm:

Observation 1: An acceptor accepts a proposal with a sequence number n if it has not sent a promise to any proposal with a sequence number $n' > n$.

Observation 2: If a proposer sends an $accept(v, n)$ message in phase 2, then either no acceptor in a majority has accepted a proposal with a sequence number $n' < n$, or v is the value in the highest numbered proposal among all accepted proposals with sequence numbers $n' < n$ accepted by at least one acceptor in a majority.

13.4.1 Safety Properties

The two important safety properties are as follows:

Validity: Only a proposed value can be chosen as the final decision.

Agreement: Two different processes cannot make different decisions.

The validity part is obvious, so let us argue that the agreement property will hold. By definition, a decision value is endorsed by a majority of acceptors. Since the intersection of two majorities is nonempty, and any acceptor will accept only one value in one instance of the protocol, the final decision is unique.

Note that no process learns that a value has been decided unless it actually has been decided. For this, a learner has to find out that a proposal has been accepted by a majority of acceptors. Acceptors can communicate directly to the learners or communicate through some distinguished learner—the latter approach will reduce the message complexity, but can fall apart when the distinguished learner fails. Since messages can be lost, a value can be chosen without any learner finding out about it. The failure of one or more acceptors can make it impossible for a learner to find out whether a value received the endorsement of the majority, even if it did so. After a while, the proposer will restart the protocol with a higher sequence number, until a decision is reached.

13.4.2 Liveness Properties

It is possible for multiple proposers to submit proposals with increasing values of sequence numbers in such a manner that none of those are ever accepted. Consider the following sequence of events:

- (Phase 1) Proposer 1 sends out *prepare*($n1$).

- (Phase 1) Proposer 2 sends out *prepare*($n2$) where $n2 > n1$.

- (Phase 2) Proposer 1's *accept*($n1$) is declined, since the acceptor has already promised to proposer 2 that it will not accept any proposal numbered lower than $n2$ so proposer 1 restarts phase 1 with a higher number $n3 > n2$.

- (Phase 2) Proposer 2's accept request is now declined on a similar ground.

The race can go on forever, stalling progress. One way out is to elect a single proposer and entrust it with the responsibility of sending proposals on behalf of clients. However, leader election itself is a consensus problem that may not be solvable in an asynchronous setting using deterministic algorithms. To elect a leader, either a randomized algorithm or a timeout-based solution should be adopted. Multiple proposers can however be present when one crashes at the middle of a negotiation, leading another proposer to take over. A sample trace is illustrated in the following:

- (Phase 1) Proposer sends *prepare*($n1$) to acceptors 1, 2, 3.

- (Phase 1) Acceptors respond with the promise of not accepting a proposal with a sequence number less than $n1$.

- (Phase 2) Proposer receives the promise, sends *accept*($n1$), and then becomes nonresponsive.

At this time, a new proposer will be elected who will take over the unfinished task from proposer 1. But the old proposer may not have crashed; it was perhaps executing its actions slowly. One method of terminating a race among multiple proposers is to use the idea of exponential backoff as in the Ethernet—a proposer receiving a *nack* will randomly choose a large delay before starting the next round. With probability 1, this will create a large-enough time gap between the two proposers caught in a race, and one will soon get done without interference.

13.5 FAILURE DETECTORS

The design of fault-tolerant systems will be simple if faulty processes can be reliably detected. Here are three scenarios highlighting the importance of failure detection:

Scenario 1: In a sensor network, a base station delegates the task of monitoring an environment to a set of geographically dispersed sensor nodes. These sensors send the monitored values back to the base station. If a sensor node crashes, and the crash is reliably detected, then its task can be assigned to another sensor.

Scenario 2: In group-oriented activities, sometimes a task is divided among the members of a group. If a member crashes, or a member voluntarily leaves the group, then the other members can take over the task of the failed member only if they can detect the failure.

Scenario 3: Distributed consensus, which is at the heart of numerous coordination problems, has trivially simple solution if there is a reliable failure detection service. In the byzantine generals problem, if the traitors could be reliably identified, then consensus could be reached by ignoring the inputs from the traitors.

We focus on the detection of crash failures only. In synchronous distributed systems where message delays have upper bounds and processor speeds have lower bounds, timeouts are used to detect faulty processes. In purely asynchronous systems, we cannot distinguish between a process that has crashed and one that is running very slowly. Consequently, the detection of crash failures in asynchronous systems poses an intriguing challenge.

A *failure detector* is a service that generates a list of processes that are *suspected* to have failed. Each process has a local detector that coordinates with its counterparts in other processes to provide the failure detection service. For asynchronous systems, the individual detection mechanisms are unreliable and error prone. Processes are *suspected* based on local observations or indirect information, and different processes may have different list of suspects. One way to compile a suspect list is to send a probe to all other processes with a request to respond. If a process does not respond within a specified period, then it is suspected to have crashed. The list of suspects is influenced by how long a process waits for the response, and the failure scenario. If the waiting time is too short, then every process might appear to be a suspect. Again, if a process i receives two probes from j and k, responds to j, and then crashes—then j will treat i to be a correct process, but k will suspect i as a crashed process. Even the detector itself might be a suspect. The study of failure detectors examines how such unreliable failure detectors may coordinate with one another in the design of fault-tolerant systems.

To relate the list of suspects with processes that have actually crashed, we begin with two basic properties of a failure detector:

Completeness: Every crashed process is suspected.

Accuracy: No correct process is suspected.

Completeness alone is of little use, since a detector that suspects every process is trivially complete. Similarly, accuracy alone is of little use, since a detector that does not suspect any process is trivially accurate. It is a combination of both properties that makes a failure detector meaningful, even if it is unreliable.

Note that completeness is a liveness property. Correct processes may not be able to suspect the crashed process immediately after the crash occurs—it is ok if it is recognized after a finite period of time. Accuracy, on the other hand, is a safety property. Two extreme forms of completeness are as follows:

> *Strong completeness*: Every crashed process is eventually suspected by *every* correct process and remains a suspect thereafter.

> *Weak completeness*: Every crashed process is eventually suspected by *at least one* correct process and remains a suspect thereafter.

The above definitions are completely independent of the actual mechanism for constructing the suspect list, although readers may find it helpful to visualize heartbeat messages and timeout as tools. Based on the communication history over a period of time, each process draws such a list. Later, if a process receives a message from a member of its suspect list, or learns from another process that the suspect has not crashed, then it will remove that member from the suspect list. There is a straightforward implementation of strong completeness using weak completeness:

```
program strong completeness (program for process i};
define D: set of process ids (representing the set of suspects);
initially D is generated by the weakly complete detector;
do true →
        send D(i)) to every process j ≠ i;
        receive D(i) from every process j ≠ i;
        D(i) : = D(i) ∪ D(j);
        if j ∈ D(i) →D(i):=D(i)\{j} fi {sender of a message is not a
          suspect}
od
```

The rationale behind the implementation is that, since every crashed process is suspected by at least one correct process, at the end, every $D(i)$ of every correct process will contain all the suspects. Henceforth, we will only consider the strong completeness property.

Like completeness, one can define two forms of accuracy:

> *Strong accuracy*: No correct process is ever suspected.

> *Weak accuracy*: There is at least one correct process that is never suspected.

Both versions of accuracy can be further weakened using the attribute *eventually*. A failure detector is *eventually strongly accurate* if there exists a time T after which no correct process is suspected. Before that time, a correct process may be added to and removed from the list of suspects any number of times. Similarly, a failure detector is *eventually weakly accurate* if there exists a time T after which at least one correct process is no more suspected, although before that time, every correct process could be a suspect. We will use the symbol ◊ (borrowed from temporal logic) to represent the attribute *eventually*.

By combining the strong completeness property with the four types of accuracy, we can define the following four classes of fault detectors:

Perfect P: (Strongly) Complete and strongly accurate

Strong S: (Strongly) Complete and weakly accurate

Eventually perfect ◊P: (Strongly) Complete and eventually strongly accurate

Eventually strong ◊S: (Strongly) Complete and eventually weakly accurate

Admittedly, other classes of failure detectors can also be defined. One class of failure detector that combines the *weak completeness* and *weak accuracy* properties has received substantial attention. It is known as the *weak (W) failure detector*. The *weakest detector* in this hierarchy is the *eventually weak failure detector ◊W*.

Chandra and Toueg introduced failure detectors to tackle the impossibility of *consensus* in *asynchronous distributed systems* [FLP85]. Accordingly, the issues to examine are as follows:

1. Given a failure detector of a certain type, how can we solve the consensus problem?

2. How can we implement these classes of failure detectors in asynchronous distributed systems?

13.5.1 Solving Consensus Using Failure Detectors

Any implementation of a failure detector, even an unreliable one, uses timeout in a direct or indirect way. Synchronous systems with bounded message delays and processor speeds can reliably implement a perfect failure detector using timeout. Given a *perfect* failure detector (P), we can easily solve the consensus problem in both synchronous and asynchronous distributed systems. However, on an asynchronous model, we cannot implement P—its properties are too strong. This motivates us to look for weaker classes of failure detectors that can solve asynchronous consensus. For such weak classes, we hope to find an *approximate* implementation—and the weaker the class, the narrower is the gap between the constraints they satisfy, and what a real implementation can offer (although such approximations have not been quantified). The interesting observation is that we know the end result—no failure detector that can be implemented will solve asynchronous consensus, because that would violate the FLP impossibility result. The intellectually stimulating question is the quest for the *weakest class* of failure detectors that can solve consensus. A related issue is to explore the implementation of some of the stronger classes of failure detectors using weaker classes.

In this section, we present two different solutions to the consensus problem, each using a different type of failure detector.

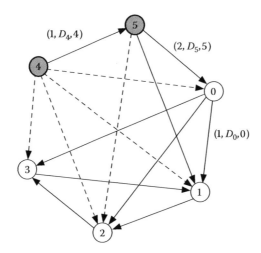

FIGURE 13.10   A scenario with $n = 6$ and $t = 2$ showing a fraction of the communications: Process 4 sends data to 5 in round 1 and then crashes. Process 5 sends data in round 2 to 0 and 1 and then crashes. In spite of the crashes, processes 0–3 receive the input values of 4 and 5.

13.5.1.1 Consensus Using P

A perfect failure detector lets each correct process receive inputs from every other correct process. Assume that up to t out of the n processes can crash. The underlying mechanism behind reaching consensus is *reliable multicast*. For every process p, let V_p be a vector of size n, the ith component of which represents the input from process i. If each process multicasts its input value to others, then everyone will receive an identical set of values, apply a choice function, and reach a consensus. But each multicast is a sequence of unicasts, and the sender may crash at the middle, which complicates matters. Reliable multicast overcomes this, and implements atomicity of the multicast by continuing the multicast for $(t + 1)$ asynchronous rounds (Figure 13.10), after which the V_p of each correct process p becomes identical, and contains the inputs from all processes whose multicasts reached some correct process at least once.

The algorithm runs in two phases. The computation progresses in rounds. The rounds are asynchronous, and each process p keeps track of its own round number r_p. In each round of phase 1, every process p exchanges its V_p with other processes. If p receives no message from $q \neq p$, and no other process reports to p about receiving a message from q, then p sets $V_p[q]$ to $\perp$ (undefined input). The algorithm guarantees that at the end of phase 1, $\forall p, q: V_p = V_q$. Phase 2 generates the final decision value for each process. The steps are outlined as follows:

```
program      Consensus with P {program for process p}
define       Vₚ, Dₚ: array [0..n - 1] of input values
initially    Vₚ= (⊥,⊥,⊥,…,⊥)
Vₚ[p] :=input of p; Dₚ:=Vₚ; rₚ:=1
(Phase 1)
        do   rₚ<t+1→
             send (rₚ,Dₚ,p) to each q ≠ p;
             wait for (rₚ,Dₚ,q) from all each q ≠ p;{or q becomes a
               suspect}
```

```
        for k = 0 to n - 1
        if Vp[k]=⊥ ∧ ∃ (rq,Dq,q):Dq[k]≠⊥ →
                Vp[k]:=Dq[k];  Dp[k]:=Dq[k]
        fi
        rp:=rp + 1
    od
```
{Phase 2}   Final decision value is the *first* element of $V_p[]$:
             $V_p[] \neq \bot$

13.5.1.2 Consensus Using S

Chandra and Toueg proposed the following algorithm for asynchronous consensus using a strong failure detector S. This algorithm is structurally similar to the previous algorithm, but runs in three phases. Phase 1 is similar to the first phase of the previous algorithm. However, recall that the *weak accuracy* property of S allows some correct processes to be suspected by other correct processes—somehow, a few correct processes may not receive the messages from a few other correct processes. As a result, the values of V_p for each process p may not be identical after phase 1. Phase 2 takes additional steps that enable each nonfaulty process p to refine its V_p, so that they become identical for all nonfaulty process. Phase 3 generates the final decision value using a choice function.

```
program     Consensus with S {program for process p}
define      Vp, Dp: array[0..n−1] of input values
initially   Vp=(⊥,⊥,⊥,...,⊥)
Vp[p]:= input of p; Dp:=Vp; rp:=1
(Phase 1)
Same as phase 1 of consensus with P but runs for n (instead of t + 1)
   rounds
(Phase 2)
        send (rp,Dp,p) to each q ≠ p;
        wait for (rq,Dq,q) from all each q ≠ p; {or q becomes a suspect}
        for k = 0 to n - 1
            if ∃ (rq,Dq,q):Dq[k] = ⊥ ∧ Vp[k] ≠ ⊥ →
                    Vp[k]: = ⊥; Dp[k]: = ⊥
            fi
(Phase 3) Decide on the first element Vp [j]: Vp [j] ≠ ⊥
```

13.5.1.3 Rationale

At the end of phase 1, V_p may not be identical for every correct process p. The action $\exists (r_q, D_q, q): D_q[k] = \bot \wedge V_p[k] \neq \bot \rightarrow V_p[k]:= \bot; D_p[k]:= \bot$ (in phase 2) helps p realize that q did not receive the input from process k and prompts p to delete the controversial input from V_p. As a result, when phase 2 ends, $\forall p,q : V_p = V_q$ holds. Since there is at least one correct process that is not suspected by any correct process, at least of one of the elements in V_p contains a valid input value. Therefore, phase 3 produces a common value for every process.

13.5.1.4 Implementing a Failure Detector

There is no magic in the implementation of a failure detector. Such implementations invariably depend on timeout and some secondary information about suspects observed and propagated by other processes. By extending the timeout period, one can only get a better estimate of the list of suspects. Since a weak failure detector has fewer requirements, the implementations better approximate the weaker version than the stronger versions.

13.6 CONCLUDING REMARKS

While distributed consensus in asynchronous distributed systems has been a rich area of theoretical research, most practical applications rely on synchronous or partially synchronous models. For consensus in the presence of byzantine failure, the signed message algorithm has much better resiliency compared to the oral message version. Dolev and Strong [DS82] demonstrated a solution to byzantine consensus in polynomial time. Dolev [D82] also studied the byzantine consensus problem in topologies that are not completely connected and demonstrated that to tolerate m failures, the graph should be at least $(2m + 1)$ connected.

The Paxos algorithm has several variations. Google used Paxos in their Chubby fault-tolerant system [B06] that implements a distributed locking mechanism for data centers. Chubby uses replication for fault tolerance.

Failure detector is an important component of group communication service. The emphasis of the work is on the classification of detectors, on their computing powers, and on the various methods of simulating stronger classes of detectors from the weaker ones. In [CT96], Chandra and Toueg showed that the eventually weak failure detector $\Diamond W$, which provides very little information about which processes have crashed, is sufficient to solve consensus in asynchronous systems with a majority of correct processes. Also, failure detectors that are perpetually accurate (like P, S) can solve consensus with any number of failures, whereas failure detectors with eventual accuracy solve consensus if and only if a majority of the processes are correct.

All results in this chapter use the deterministic model of computation. Many impossibility results can be circumvented when a probabilistic model of computation is used (see Ben-Or's paper [B83]). These solutions have not been addressed here.

13.7 BIBLIOGRAPHIC NOTES

Fischer et al. [FLP85] presented the impossibility of distributed consensus in asynchronous distributed systems. During the ACM PODC 2001 Conference, the distributed systems (PODC) community voted this paper as the most influential paper in the field (later renamed as *Dijkstra Prize* after Edsger W. Dijkstra since 2003).

Byzantine failures were identified by Wensley et al. in the SIFT project [WLGt78] for aircraft control. The two algorithms for solving byzantine agreement are due to Lamport et al. [LSP82]. This paper started a flurry of research activities during the second half of the 1980s. The proof of the impossibility of byzantine agreement with three generals and one traitor is adapted from Fischer et al. [FLM86], where a general framework of such proofs has been suggested. Lamport [L83] also studied a weaker version of byzantine agreement that required the validity criterion to hold only when there are no faulty processes.

The polynomial time algorithm for byzantine consensus can be found in [DS82]. Dolev [D82] also studied byzantine agreement in networks that are not completely connected. Coan et al. [CDD+85] studied a different version of agreement called the distributed firing squad problem, where the goal is to enable processes to start an action at the same time: If any correct process receives a message to start DFS synchronization, then eventually all correct processes will execute the fire action at the same step.

A simple description of the basic Paxos algorithm is available in Lamport's article [L01]. Burrows described the Chubby distributed locking service in [B06].

Chandra and Toueg [CT96] introduced failure detectors. Their aim was to cope with the well-known impossibility result by Fischer et al. [FLP85] for asynchronous distributed systems. The result on the weakest failure detector for solving consensus appears in [CHT96]. In 2010, Chandra and Toueg received the Dijkstra prize for their contributions in failure detectors.

EXERCISES

13.1 Seven members of a family interviewed a candidate for the open position of a cook. If the communication is purely asynchronous and message-based, and decisions are based on *majority votes*, then describe a scenario to show how the family can remain undecided, when one member disappears after the interview.

13.2 Present an example to show that in a synchronous system, byzantine agreement *cannot* be reached using the oral message algorithm when there are *six* generals and *two* traitors.

13.3 Two loyal generals are planning to coordinate their actions for conquering a strategic town. To conquer the town, they need to *attack at the same time*; otherwise, if only one of them attacks and the other does not attack at the same time, then the generals are likely to be defeated. To plan the attack, they send messages back and forth via trusted messengers. The communication is asynchronous. However, the messengers can be killed or captured—so the communication is unreliable.

Argue why it is impossible for the two generals to coordinate their actions of attacking at the same time.

(Hint: Unlike the byzantine generals problem, here all generals are loyal, but the communication is unreliable. If general A sends a message *attack at 2 a.m.* to general B, then he will want an acknowledgment from B; otherwise, he won't attack in the fear of moving alone. But B also will ask for an acknowledgment of the acknowledgment. Now you see the rest of the story.)

13.4 Prove that in a connected network of generals, it is impossible to reach byzantine agreement with n generals and m traitors, the connectivity of the graph $\leq 2m + 1$. (The connectivity of a graph is the minimum number of nodes whose removal results in a partition.) (See [D82]).

13.5 A synchronous distributed system consists of 2^n processes, and its topology is an n-dimensional hypercube. What is the maximum number of byzantine failures that can be tolerated when the processes want to reach a consensus? [The result should follow from Q4.]

13.6 Using oral messages, the byzantine generals algorithm helps reach a consensus when less than one-third of the generals are traitors. However, it does not suggest how to identify the traitors. Examine if the traitors can be identified without any ambiguity.

13.7 In the context of failure detectors in asynchronous distributed systems, eventual accuracy is a weaker version of the accuracy property of failure detectors; however, there is no mention of the *eventual completeness* property. Why is it so?

13.8 If one can design a perfect failure detector, then consensus problems can be easily solved. What are the hurdles in the practical implementation of a perfect failure detector?

13.9 Implement a perfect failure detector P using an eventually perfect failure detector $\Diamond P$.

13.10 In an asynchronous distributed system whose topology is a completely connected network, each process has a perfect failure detector. How can you elect a leader using the perfect failure detector?

CHAPTER 14

Distributed Transactions

14.1 INTRODUCTION

Amy wakes up in the morning and decides to transfer a sum of $200 from her savings account in Colorado to her checking account in Iowa City where the balance is so low that she cannot write a check to pay her apartment rent. She logs into her home computer and executes the transfer that is translated into a sequence of the following two operations:

- Withdraw $200 from Colorado State bank account 4311182.

- Deposit the said amount into Iowa State bank account 6761125.

The next day, she writes a check to pay her rent. Unfortunately, her check bounced. Furthermore, it was found that $200 was debited from her Colorado account, but due to a server failure in Iowa City, no money was deposited to her Iowa account.

Amy's transactions had an undesirable end result. A *transaction* is a sequence of server operations that must be carried out *atomically* (which means that either all operations must be executed or none of them will be executed at all). Amy's bank operations violated the *atomicity property* of transactions and caused problems for her. Certainly this is not desirable. It also violated a *consistency property* that corresponds to the fact that the total money in Amy's accounts will remain unchanged during a transfer operation. A transaction *commits* when all of its operations complete successfully and the states are appropriately updated. Otherwise, the transaction *aborts*, which implies no change will take place and the old state will be retained.

In real life, all transactions must satisfy the following four properties regardless of server crashes or omission failures:

Atomicity: Either all operations are completed or none of them is executed.

Consistency: Regardless of what happens, the database must remain consistent. In this case, Amy's total balance must remain unchanged.

Isolation: If multiple transactions run concurrently, then it must appear as if they were executed in some arbitrary sequential order. The updates of one transaction must not be visible to another transaction until it commits.

Durability: Once a transaction commits, its effect will be permanent.

These four properties are collectively known as ACID properties, a term coined by Härder and Reuter [HR83]. The implementation of these goals is the main task in the implementation of a transaction. Readers may wonder whether these properties could be enforced via mutual exclusion algorithms, by treating the transaction as a critical section. In principle, this would have been possible, if there were no failures. However, mutual exclusion algorithms discussed so far rule out failures, so newer methods are required for the implementation of transactions.

14.2 CLASSIFICATION OF TRANSACTIONS

Transactions can be classified into several different categories. A few of these are listed in the following:

14.2.1 Flat Transactions

A *flat transaction* consists of a set of operations on objects. For example, Amy has to travel to Stuttgart from Cedar Rapids, so she books flights in the following three sectors: Cedar Rapids to Chicago, Chicago to Frankfurt, and Frankfurt to Stuttgart. These three bookings collectively constitute a flat transaction (Figure 14.1a). No operation on the objects of a flat transaction is itself a transaction.

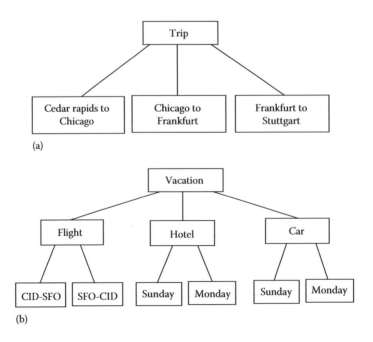

FIGURE 14.1 (a) A flat transaction and (b) a nested transaction.

14.2.2 Nested Transactions

A *nested transaction* is an extension of the transaction model. It allows a transaction to contain other transactions. For example, in Figure 14.1a, the trip from Frankfurt to Stuttgart can itself be a transaction, if this trip is composed of a number of shorter train rides. Clearly, a nested transaction has a multilevel tree structure. The nonroot transactions are called *subtransactions*. As another example, note that many airlines offer the vacationers hotel accommodation and automobile reservation along with flight reservation. Here, the top-level transaction *vacation* is a nested transaction that consists of three subtransactions (Figure 14.1b):

- A flight from Cedar Rapids to San Francisco and back
- A hotel room in San Francisco for 2 days
- A car in San Francisco for 2 days

Such extensions of the transaction model are useful in the following way: It may be the case that the flights and the hotel room are available, but not the car—so the last subtransaction will abort. In such cases, instead of aborting the entire transaction, the main transaction may (1) look for alternative modes of sightseeing (like contracting a tour company) or (2) observe that the car is not available, but still commit the top-level transaction with the information about possible alternative choices and hope that the availability of the transport can be sorted out later.

Subtransactions can commit and abort independently, and these decisions can be taken concurrently. In case two subtransactions access a shared object, the executions will be serialized.

14.2.3 Distributed Transactions

Some transactions involve a single server, whereas others involve objects managed by multiple servers. When the objects of a transaction are distributed over a set of distinct servers, the transaction is called a *distributed transaction*. A distributed transaction can be either flat or nested. Amy's money transfer in the introductory section illustrates a flat distributed transaction.

14.3 IMPLEMENTING TRANSACTIONS

Transaction processing may be centralized or distributed. In centralized transaction processing, a single *transaction manager* (TM) manages all operations. We outline two different methods of implementing such transactions. One implementation uses a *private workspace* for each transaction. The method consists of the following steps:

1. At the beginning, allocate the transaction a *private workspace,* and copy all files or objects that it needs.

2. Read/write data from the files or carry out appropriate operation on the objects. Note that changes will take place in the private workspace only.

3. If all operations are successfully completed, then *commit* the transaction by writing the updates into the permanent record. Otherwise, the transaction will abort, which is implemented by not writing the updates. The original state will remain unchanged.

Another method of implementing transactions uses a *write-ahead log*. A write-ahead log records the current state of the transaction *before* it updates the objects. The content of the log is

(Transaction id, disk block number, old value, new value)

By assumption, these logs are not susceptible to failures. If the transaction commits, then the log is discarded. On the other hand, if the transaction aborts, then the log is used to undo the changes and restore the database to a consistent state. The log can also be used to rerun the transaction after a failure.

The implementation of distributed transactions is much more tricky. Here, each computer has a local *TM*, and these TMs interact with the *data managers* that are geographically distributed (Figure 14.2). The problem of atomic commitment becomes challenging when one TM fails but the others continue to work as usual. Thus, in a transaction $(x := x - 100;$ $y := y + 100)$ where x and y belong to different databases, if the TM crashes after completing the operation $x := x - 100$, then the database becomes inconsistent.

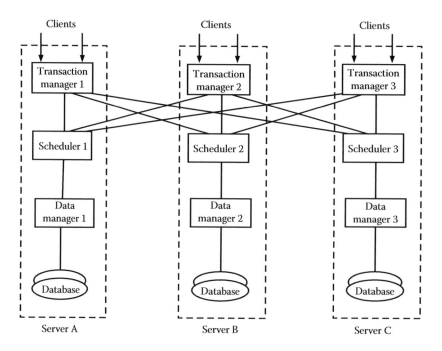

FIGURE 14.2   The handling of distributed transactions.

14.4 CONCURRENCY CONTROL AND SERIALIZABILITY

The goal of concurrency control is to guarantee that when multiple transactions are concurrently executed, the net effect is equivalent to executing them in some serial order. This is the essence of the *serializability* property.

Concurrent transactions dealing with disjoint objects are trivially serializable. However, when objects are shared among concurrent transactions, an arbitrary interleaving of the operations may not satisfy the serializability property. The consistency properties can be violated in several ways, two of which are illustrated here.

Lost update problem: Amy and Bob have a joint account in a bank. Each of them independently deposits a sum of \$250 to their joint account. Let the initial balance in the account be $B = \$1000$. After the two updates, the final balance should be \$1500. Each deposit is a transaction that consists of a sequence of three operations as shown in the following:

No.	Amy's Transaction	No.	Bob's Transaction
1	$local \leftarrow B$	4	$local \leftarrow B$
2	$local \leftarrow local + \$250$	5	$local \leftarrow local + \$250$
3	$B \leftarrow local$	6	$B \leftarrow local$

The final value of B will very much depend on how these operations are interleaved. When the interleaving order is (1 2 3 4 5 6) or (4 5 6 1 2 3), the final value of B becomes \$1500. However, if the interleaving order is (1 4 2 5 3 6) or (1 2 4 5 6 3), then the final value will be only \$1250! What happened to the other deposit? This is the essence of the *lost update* problem.

Serializability can be satisfied by properly scheduling the *conflicting operations*. The operations on a shared object are either read or write. Two operations on a shared object are said to be *conflicting*, when at least one of them is a *write* operation. In the example of Amy or Bob's transactions, B is the shared object and (3, 4), (1, 6), or (3, 6) are conflicting operations. The improper scheduling of these operations can be avoided via *locking* (as in two-phase locking [2PL]) or by *time stamp ordering*.

Dirty read problem: Even when serializability property is satisfied, the abortion of a transaction can lead to sticky situations. For example, suppose the operations in Amy and Bob's transactions are scheduled as (1 2 3 4 5 6), which is a correct schedule satisfying the serializability criterion. Consider the scenario in which first Amy's transaction aborts and then Bob's transaction commits. The balance B will still be set to \$1500, although it should have been set to only \$1250. The cause of the anomaly is that Bob's transaction read a value of B (updated by step 3) before Amy's transaction decided to abort. This is called the *dirty read* problem.

Premature write problem: As another example of a problem situation caused by the abortion of a transaction, consider a shared variable B whose initial value is 0, and there are two concurrent transactions $T_1 \equiv B \leftarrow 500$ and $T_2 \equiv B \leftarrow 1000$. Depending on the order in which the two transactions complete, we will expect that B will be set to either 500 or 1000. Assume that they are executed in the order $(T_1; T_2)$. Let T_2 commit first (raising the balance B to 1000), but T_1 abort thereafter (changing the value of B to 0). Clearly, this is not acceptable. This problem is called the *premature write* problem. The problem could be avoided, if T_2 delayed the writing of B until T_1's decision to commit or abort was taken. If the earlier transaction aborts, then the later transaction must also abort. To avoid dirty reads and premature writes, a transaction must delay its read or write operations until the transactions scheduled earlier either commit or abort.

14.4.1 Testing for Serializability

Concurrent transactions commonly use locks with conflicting operations on shared objects. One method of testing whether a schedule of concurrent transactions satisfies the serializability property is to create a *serialization graph*. The serialization graph is a directed graph $G = (V, E)$, where V is the set of transactions and E is the set of directed edges between transactions: a directed edge $(T_j \rightarrow T_k)$ implies that transaction T_k acquired a lock only after transaction T_j released that lock. The following theorem tests the serializability property:

Theorem 14.1

For a schedule of concurrent transactions, the serializability property holds if and only if the corresponding serialization graph is acyclic.

For a proof of this theorem, see [BHG87].

14.4.2 Two-Phase Locking

Prior to performing an operation on an object, a transaction will request for a lock for that object. A transaction will be granted an *exclusive lock* for all the objects that it will use. Other transactions using one or more of these objects will wait until the previous transaction releases the lock. In the case of Amy or Bob, the locks will be acquired and released as follows:

```
lock B
     local ← B
     local ← local + $250
     B ← local
unlock B
```

Accordingly, the only two feasible schedules are (1 2 3 4 5 6) and (4 5 6 1 2 3), which satisfy the serializability property.

The previous example is in a sense trivial, since only one lock is involved and atomicity of the transaction implies serializability. To implement serializability with multiple objects and multiple locks, a widely used method is *2PL*. Two-phase locking guarantees that all pairs of conflicting operations on shared objects by concurrent transactions are always executed *in the same order*. The two phases are as follows:

Phase 1: The scheduler acquires all locks one after another without releasing any lock. This is called the *growing* phase or *acquisition* phase.

Phase 2: The scheduler releases all locks acquired so far without acquiring any new lock. This is called the *shrinking* phase or the *release* phase.

Two-phase locking prevents a transaction from acquiring any new lock after it has released a lock, until all locks have been released. Locking can be fine grained. For example, there can be separate *read locks* or *write locks* for an object. A read operation does not conflict with another read operation, so the data manager can grant multiple read locks for the same shared data. Only in the case of conflicts, the requesting transaction has to wait for the lock.

The actions of the two phases alone do not rule out the occurrence of deadlocks. Here is a simple example of how deadlock can occur. Suppose there are two transactions T_1 and T_2, each needs to acquire the locks on objects x and y. Let T_1 acquire the lock on x and T_2 acquire the lock on y. Since each will now try to acquire the other lock and no transaction will release a lock until its operation is completed, both will wait forever resulting in a deadlock. However, a sufficient condition for avoiding deadlocks is to require all transactions to acquire the required locks in the *same global order*—this avoids circular waiting. This completes the final requirement of 2PL.

Theorem 14.2

Two-phase locking guarantees serializability.

Proof (by contradiction): Assume that the statement is not true. From Theorem 14.1, it follows that the serialization graph must contain a cycle

$$\cdots T_i \to T_j \to T_k \to \cdots \to T_m \to T_i \cdots$$

This implies that T_i must have released a lock (that was later acquired by T_j) and then acquired a lock (released by T_m). However, this violates the condition of 2PL that rules out acquiring any lock after a lock has been released. ■

14.4.3 Concurrency Control via Time Stamp Ordering

A different approach to concurrency control avoids any form of locking and relies solely on time stamps. Each transaction T_i is assigned a time stamp $TS(i)$ when it starts, and the same time stamp $TS(i)$ is assigned to all read and write accesses of every object x by T_i. Time stamp ordering makes an a priori selection of the serialization order and forces the transaction execution to obey that order. In particular, *conflicting operations* on a shared variable x will be granted read or write access only if it comes in time stamp order—otherwise, the access will be denied, and the transaction will abort and restart at a later time with a new time stamp. There are several different versions of scheduling reads and writes based on time stamps—here, we only discuss the basic version of time stamp ordering.

Time stamp ordering assigns two time stamps RTS and WTS to each object x—their values are determined by the time stamps of the transactions performing read or write on x. Let $RTS(x)$ be the *largest* time stamp of any transaction that reads x and $WTS(x)$ be the *largest* time stamp of any transaction that updates x. Now, if a transaction T_j requests access to x, then the time stamp ordering protocol will handle the request as follows:

```
{Transaction T_j wants to read x}
if  TS(j) ≥ WTS(x)  →  allow read; RTS(x) := max(RTS(x),TS(j))
[] TS(j) < WTS(x)   →  deny read {transaction aborts}
fi
{Transaction T_j wants to write x}
if  TS(j) ≥ max(WTS(x),RTS(x))  →  allow write; WTS(x) := max
    (WTS(x), TS(j))
[] TS(j) < max(WTS(x),RTS(x))   →  deny write {transaction aborts}
fi    {After a transaction aborts, it restarts with a new larger
    time stamp.}
```

Consider the examples in Figure 14.3, where there are three concurrent transactions T_1, T_2, and T_3. In Figure 14.3a, all transactions will commit without any problem. In Figure 14.3b, both T_1 and T_2 will abort—the schedulers will not schedule their $r[x]$ operations since the largest WTS is 50, which is greater than the time stamps of the transactions T_1 and T_2. If these reads were scheduled, then they would have returned a value different from what T_1 or T_2 wrote and thus violate atomicity. In Figure 14.3c, once again, both T_1 and T_2 will abort, but for a different reason, the schedulers will not schedule their $w[x]$ operations since the largest WTS is 50. These writes, if scheduled, would have affected the outcome of the $r[x]$ by T_3.

14.5 ATOMIC COMMIT PROTOCOLS

A distributed transaction deals with data from multiple servers. Each server completes a subset of the operations belonging to the transactions. At the end, for each transaction, all servers must *agree* to a final *irrevocable decision* regardless of crash or omission

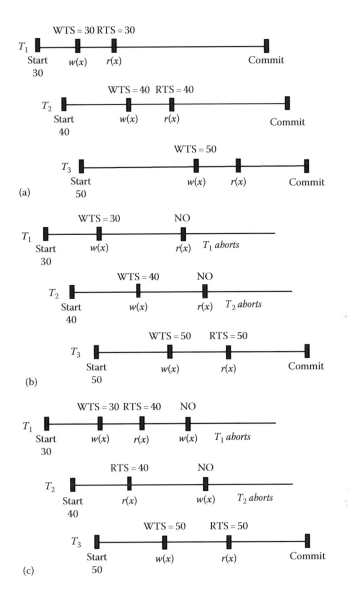

FIGURE 14.3   Examples of concurrency control using time stamp ordering. Here, $r[x]$ and $w[x]$ refer to read and write operations on the shared variable x. (a)–(c) illustrate three different scenarios.

failures—the final decision must be either *commit* or *abort*. An *atomic commit protocol* solves a variation of the consensus problem (Chapter 13) with the following specifications:

Termination: All nonfaulty servers must eventually reach an irrevocable decision.

Agreement: If any server decides to *commit*, then every server must have voted to *commit*.

Validity: If all servers vote *commit* and there is no failure, then all servers must *commit*.

The agreement property implies that all servers reach the same decision. In the classic consensus problem, if some servers vote *commit* but others vote *abort*, then the final irrevocable decision could have been *commit*—however, here, the agreement clause of the atomic commit problem rules out this possibility. We now look into the implementation of atomic commit protocols.

14.5.1 One-Phase Commit

In implementing atomic commit protocols, one of the servers is chosen as the *coordinator*. A client sends a request to *open a transaction* to an accessible server. This server becomes the coordinator for that transaction. Other servers involved in this transaction are called *participants*. The coordinator will assign relevant components of this transaction to its participants and ask them to commit when they are done. This is the essence of *one-phase commit* protocol. It is expected that the participants will be able to complete the tasks assigned to them. If due to local problems (like failure or concurrency control conflict) some participants are unable to commit, then neither the coordinator nor other participants will know about this (since everything has to be completed in a single phase), and atomicity will be violated. So each participant has to assume the responsibility for local failure recovery. This is a somewhat naive approach for implementing atomic transactions. A much more pragmatic approach is the *two-phase commit* (2PC) protocol.

14.5.2 Two-Phase Commit

2PC protocols (Figure 14.4a) overcome the limitations of one-phase commit protocol. The two phases of the protocol are as follows:

> *Phase 1*: The coordinator sends out a *PREPARE* message to each participant, asking them to respond whether they are ready to *commit* (by voting *yes*) or prefer to *abort* (by voting *no*). When the participants respond, the coordinator collects their votes.

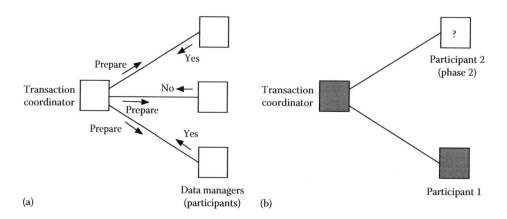

FIGURE 14.4   (a) Implementation of 2PC and (b) participant 2 is blocked.

Phase 2: If all servers vote *yes* at the end of phase 1, then in phase 2, the coordinator will send a *COMMIT* message to all participants. Otherwise, if at least one server votes *no*, then the coordinator will send out an *ABORT* message to all participants. Each participant sending a *yes* vote waits to receive an *ABORT* or a *COMMIT* message from the coordinator and takes appropriate actions thereafter. The steps are as follows:

```
program 2PC
{program for the coordinator}
{Phase 1: prepare}
    Send PREPARE to all participants;
    Wait to receive the vote (yes or no) from each participant
      {Phase 2: commit}
if  ∀participant j: vote(j) = yes → multicast COMMIT to all
  participants
[]  ∃ participant j: vote (j) = no → multicast ABORT to all
  participants
fi
{program for each participant}
if  {Phase 1} message from coordinator = PREPARE → send yes or no
[]  {Phase 2} message from coordinator = COMMIT → commit local actions
[]  {Phase 2} message from coordinator = ABORT → abort local actions
fi
```

In phase 2, if the coordinator sends a COMMIT message, then each participant updates its local log, releases all locks, and sends a DONE message to the coordinator. After the coordinator receives DONE messages from all participants, it writes the complete record into its log and deletes the record from its volatile store.

Failure handling: The basic two-phase protocol may run into difficult situations when failures occur. A server may fail by crashing and thus may fail to respond. Also, messages may be lost. This makes the implementation of the 2PC protocol nontrivial. In a purely asynchronous environment, when the coordinator or a participant is blocked while waiting for a message, it cannot figure out if the sender crashed or the message is lost or delayed. Although time-out is used to detect message losses and process failures, it is not foolproof. Therefore, in case of any doubt, the system should ideally revert to a fail-safe configuration by aborting all transactions. Some specific failure scenarios and their remedies are summarized as follows:

Case 1: (*Phase 1*) The coordinator times out waiting for a *vote* message from a participant. To deal with possible message loss, the coordinator may send the PREPARE message a bounded number of times. If there is no response, then the coordinator decides to abort the transaction and sends ABORT to all.

Case 2: (*Phase 2*) A participant times out waiting for a COMMIT or an ABORT message from the coordinator. The participant has no knowledge of the coordinator's decision, so it cannot unilaterally decide since that may violate the atomicity. The participant *remains blocked* while sending a *get status* message to the coordinator asking for its decision. When the coordinator receives such a message, it checks if a record exists in its volatile memory and replies accordingly. Otherwise, it sends an ABORT to the participant. However, if the coordinator crashes, then the participant remains undecided until the coordinator recovers from the failure. It has no clue if other participants received a COMMIT or an ABORT message before the coordinator crashed. So, for an indefinite period, resources may remain locked, affecting other transactions. One possible solution for the blocked participant is to query a nonfaulty participant: *What message (COMMIT or ABORT) did you get from the coordinator?* If the answer is COMMIT, then the participant will commit its local transaction; otherwise, it will abort.

But now consider the more difficult case of multiple server failures as described in the following:

Case 3: (*Phase 2*) Consider the case when both the coordinator and a participant have crashed (Figure 14.4b), but a second participant has voted *yes* in phase 1 and is waiting for a *commit* or an *abort* message in phase 2 from the coordinator. If one of the crashed participants decided to abort, then the second participant must abort. On the other hand, if the crashed participant decided to commit before the crash, then the second participant must commit. With these two conflicting possibilities, the second participant has little choice but to wait, until the crashed servers recover.

The previous situation may be rare, but it reflects that 2PC is a blocking protocol: the operational servers may sometimes have to wait indefinitely on the recovery of a failed server. Locks must be held on the database while the transaction is blocked. To improve failure handling, Skeen and Stonebraker [SkS83] proposed *three-phase commit* (3PC) protocol. It is a *nonblocking protocol*. An atomic commitment protocol is called *nonblocking*, if in spite of crashes, *every nonfaulty server eventually decides*—this implies that the operational servers will not block until the recovery of the failed servers.

14.5.3 Three-Phase Commit

The 3PC protocol removes the uncertainty in the two-phase protocol caused by the failures of the coordinator and a participant. Phase 1 is similar to that in 2PC: A time-out in this phase will prompt the coordinator to *abort* of the transaction. Phase 2, the *precommit* phase, is the new addition. This phase is meant to remove the uncertainty period for participants that have committed and are waiting for the final commit or abort message from the coordinator. If a precommit message is not received, then the participant will abort and release any blocked resources. When receiving a precommit message, participants know that all others have voted to commit. After this, even if the coordinator crashes or times out, the participant goes forward with the commit.

```
program 3PC
{program for the coordinator}
{Phase 1: prepare} Send PREPARE to all participants;
                   Wait to receive the vote from each participant
{Phase 2: prepare to commit}
if ∀participant j: vote(j) = commit → multicast PRECOMMIT to all
                                                participants;
                                        wait to receive ack from
                                            the participants
[] ∃participant j:vote (j)=abort → multicast ABORT to all participants
fi
{Phase 3: commit}
if majority of participants sends ack → multicast COMMIT to all
  participants;
[] majority does not send ack → multicast ABORT to all participants
fi
{program for each participant}
if {Phase 1} message from coordinator = PREPARE → send commit or abort
[] {Phase 2} message from coordinator = PRECOMMIT → send ack
[] {Phase 3} message from coordinator = COMMIT → commit local actions
[] {Phases 2 and 3} message from coordinator = ABORT → abort
  local actions
fi
```

14.6 RECOVERY FROM FAILURES

Atomic commit protocols enable transactions to tolerate server crashes or omission failures, so that the system remains in a consistent state. However, to make transactions durable, we would expect servers to be equipped with incorruptible memory. A random access memory (RAM) loses its content when there is a power failure. An ordinary disk may crash, and thus, its contents may be lost forever—so it is unsuitable for archival storage. A form of archival memory that can survive all single failures is known as a *stable storage*, introduced by Butler Lampson.

14.6.1 Stable Storage

A *stable storage* can be implemented using a pair of ordinary disks. Let A be a composite object with components $A_0, A_1, A_2, ..., A_{n-1}$. Consider a transaction that assigns a new value x_i to each component A_i. We will represent this by $A := x$. This transaction is *atomic*, when either *all* assignments $A_i := x_i$ are completed or *no* assignment takes effect. Ordinarily, a crash failure of the updating process will potentially allow a fraction of these assignments to be completed and violate the atomicity property.

The *stable storage* maintains two copies* of the object A (i.e., two copies of each component A_i) and allows two operations *update* and *inspect* on A. Designate these two

---

* The technique is called *mirroring* and is used in RAID1 disk systems.

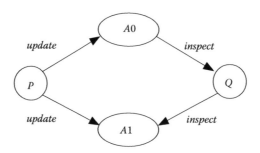

FIGURE 14.5 The model of a stable storage: P performs the *update* operation and Q performs the *inspect* operation.

copies by A0 and A1 (Figure 14.5). When a process P performs the *update* operation, it updates the two copies alternately and stamps these updates with (1) the time stamp T and (2) a unique signature S called *checksum*, which is a function of x and T.

A fail-stop failure can halt the update operation after any step. Process Q, which performs the inspect operation, checks both copies and, based on the times of updates as well as the values of the checksums, chooses the correct version of A.

```
{procedure update}
1       A0 := x;                  {copy 0 updated}
2       T0 := timestamp;          {timestamp assigned to copy 0}
3       S0 := checksum(x,T0);     {signature assigned to copy 0}
4       A1 := x;                  {copy 1 updated}
5       T1 := timestamp;          {timestamp assigned to copy 1}
6       S1 := checksum(x,T1)      {signature assigned to copy 1}

{procedure inspect}
[a] if  S0 = checksum(A0,T0) ∧ S1 = checksum(A1,T1) ∧ T0 > T1 → accept A0
[b] []  S0 = checksum(A0,T0) ∧ S1 = checksum(A1,T1) ∧ T0 < T1 → accept any
[c] []  S0 = checksum(A0,T0) ∧ S1 ≠ checksum(A1,T1) → accept A0
[d] []  S0 ≠ checksum(A0,T0) ∧ S1 = checksum(A1,T1) → accept A1
fi
```

Case [a] corresponds to a possible failure between steps 3 and 4 of an update. Case [b] represents no failure—so any one of the copies is acceptable. Case [c] indicates a failure between steps 4 and 6, and case [d] indicates a failure between steps 1 and 3. As long as the updating process fails by stopping at any point during steps 1–6 of the update operation, one of the guards A–D becomes true for process Q. The application must be able to continue regardless of whether the old or in the new state is returned.

The two copies A0 and A1 are stored on two disks mounted on separate drives. Data can also be recovered when instead of a process crash one of the two disks crashes. Additional robustness can be added to the previous design by using extra disks.

14.6.2 Checkpointing and Rollback Recovery

Consider a distributed transaction and assume that each process has access to a stable storage to record its local state. Failures may hit one or more processes, making the global state inconsistent. *Checkpointing* is a mechanism that enables transactions to recover from such inconsistent configurations using backward error recovery. When a transaction is in progress, the states of the participating servers are periodically recorded on the local stable storages. These fragments collectively define a *checkpoint*. Following the detection of a failure, the system state *rolls back* to the most recent checkpoint, which completes the recovery. The technique is not limited to transactions only, but is applicable in general to all message-passing distributed systems.

In a simple form of checkpointing, each process has the autonomy to decide when to record checkpoints. This is called *independent* or *uncoordinated* checkpointing. For example, a process may decide to record a checkpoint when the information to be saved is small, since this will conserve storage. However, a collection of unsynchronized snapshots may not represent a meaningful or consistent global state (see Chapter 8). To restore the system to a consistent global state, intelligent rollback is necessary. If the nearest checkpoint does not reflect a consistent global state, then rollback must continue until a consistent checkpoint is found. In some cases, it may trigger a domino effect. We illustrate the problem through an example.

In Figure 14.6, three processes P, Q, and R communicate with one another via message passing. Let each process spontaneously record three checkpoints for possible rollback—these are marked with bold circles. Now, assume that process R failed after sending its *last* message $m7$. So its state will roll back to $r2$. Since the sending of the message by R is nullified, its reception of $m7$ by Q also must be nullified, and the state of Q will roll back to $q2$. This will nullify the sending of the last message $m5$ by Q to P, and therefore, the state of P has to roll back to $p2$.

But the rollbacks do not end here. Observe that $(p2, q2, r2)$ does not represent a consistent global state. Process R has to undo the sending of its previous message $m6$, and its local state now has to revert to $r1$. This will cause Q to revert its state to $q1$. Eventually, the processes have to roll back all the way to their initial checkpoints $(p0, q0, r0)$, which by

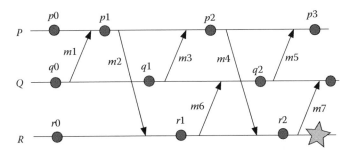

FIGURE 14.6   An example of domino effect in uncoordinated checkpointing: The dark circles represent the local states of processes saved on stable storage. If R crashes after sending its last message $m7$ to Q, then the global state of the system will eventually roll back to $(p0, q0, r0)$.

definition is a valid checkpoint (since no messages were exchanged before it). So the entire computation is lost despite all the space used to record the nine checkpoints. If process P recorded its last checkpoint *after* sending the last message, then the computation could have rolled back only to $(p2, q2, r2)$. This would have been possible with *coordinated checkpointing*. Coordinated checkpoints can be recorded using Chandy–Misra's distributed snapshot algorithm (when the channels are FIFO) or Lai–Yang's algorithm (when the FIFO guarantee of channels does not exist) (see Chapter 8).

There are some computations that communicate with the outside world and cannot be reversed during a rollback. A printer that has printed a document cannot be asked to reverse its action for the sake of failure recovery, nor can an ATM machine that has dispensed some cash to a customer be expected to retrieve the cash from a customer. In such cases, *logging* is used to replay the computation. All messages received from the environment are logged before use, and all messages sent to the environment are logged before they are sent out. During a replay, these logged messages are used to substitute the interaction with the outside world.

14.6.3 Message Logging

Message logging is a general technique for improving the efficiency of checkpoint-based recovery following crash failures. Checkpointing involves writing a complete set of states on the stable storage, and frequent checkpointing slows down the normal progress of the computation. For coordinated checkpointing, add to that the overhead of synchronizing the recording instants. On the other hand, infrequent checkpointing may cause significant amount of rollback when failures occur, which adds to the cost of recovery. Message logging helps us strike a middle ground—even if the checkpointing is infrequent, starting from a consistent checkpoint P, a more recent recovery point Q can be reconstructed by replaying the logged messages in the same order in which they were delivered before the crash.

Either the sender or the receiver can log messages into a stable storage. Logging takes time, so whether a process should wait for the completion of the logging of messages before delivering it to the application is an important design decision. All message-logging protocols have a common goal: Once a crashed process recovers, its state is consistent with the states of the other processes. Surviving processes whose states are inconsistent with those of other process are known as *orphan processes*. Inadequate message logging can lead to the creation of *orphan processes*. Message-logging protocols must guarantee that upon recovery, no process remains an orphan.

Figure 14.7 illustrates three messages exchanged among the processes 0, 1, 2, and 3. Assume that $m1$ and $m3$ were logged (into the stable storage) by the receiving processes, but $m2$ was not. If the receiver of the message $m2$ (i.e., process 0) crashes and later recovers, the state of the system can be reconstructed up to the point when $m1$ was received. Neither $m2$ nor $m3$ (which is causally ordered after $m2$) can be correctly replayed. However, process 3 already received message $m3$. Since the sending of $m3$ cannot be replayed using the log, process 3 becomes an *orphan process*.

Note that, if process 1 or 0 logged $m2$ before the crash, then that could prevent 3 from being an orphan. In general, for any given message, let *depend*(m) be the set of *processes* that

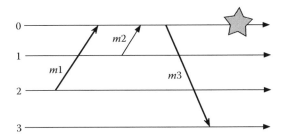

FIGURE 14.7 Messages $m1$ and $m3$ have been logged by their receiving processes, but not $m2$. Process 0 crashes and then recovers. From the message log, $m1$ will be replayed, but not $m2$. This means the sending of $m3$, which is causally dependent on $m2$, may not be accounted for, and process 3 becomes an orphan.

(1) delivered m or (2) delivered a message m' such that $m \prec m'$. The reconstruction of the computation of every process in $depend(m)$ will depend on the ability to replay m. Define $copy(m)$ to be the set of processes that sent or received m, but did not log it. In Figure 14.5, $depend(m2) = \{0, 1, 3\}$ and $copy(m2) = \{0, 1\}$. If every process in $copy(m)$ crashes, then the transmission of m cannot be replayed. As a result, every process $j \in depend(m)\backslash copy(m)$ becomes an orphan.

There are two versions of logging that deal with this. In the *pessimistic protocol*, each process delivers a message only after every message delivered before it has been logged. *Optimistic protocols* cut down on the logging overhead by taking some risk that improves the speed during failure-free runs. When a failure occurs, the protocol determines if there are orphans, and the system rolls back to make the state consistent.

Many computations in a distributed system are nondeterministic. Two different runs of the system may lead to two different global states both of which are consistent. The order of arrival of the messages may vary from one run to another. In the recovery of nondeterministic computations using message logging, the replayed messages simply reflect the last run. It does not lead to any new nondeterminism other than whatever was present in the original system prior to the crash.

14.7 CONCLUDING REMARKS

Time stamp ordering and 2PL are two different techniques for concurrency control, and these work differently. In 2PL, transactions wait for locks from time to time, but do not abort and restart unless a deadlock occurs. In time stamp ordering, deadlock is not possible. An *optimistic version* of time stamp ordering thrives on the observation that conflicts are rare, so all transactions run without restrictions. Each transaction keeps track of its own reads and writes. Prior to committing, each transaction checks if those values have been changed (by some other transaction). If not, then the transaction commits; otherwise, it aborts.

For distributed commit, the 2PC protocol has been extremely popular, although the basic version is prone to blocking when the coordinator crashes. The 3PC, originally proposed by Skeen [S83], resolves the blocking problem, but its implementation involves additional states and extra messages. Another solution is the nonblocking 2PC by Babaõglu and Toueg [BT93], which is technically sound but the additional message complexity is unavoidable. None of these two improvisations seem to be very much in use, since blocking due to the crash of the coordinator is rare. Distributed systems with *communication failures* do not have nonblocking solutions to the atomic commit problem.

Checkpointing is widely used for transaction recovery. The frequent writing of global states on the stable storage tends to slow down the computation speed. A combination of infrequent checkpointing and message logging prevents massive rollbacks and potentially improves performance.

14.8 BIBLIOGRAPHIC NOTES

Härder and Reuter [HR83] coined the phrase ACID properties. Papadimitriou [P79] formally introduced the *serializability property* to characterize the logical isolation of concurrent transactions. Bernstein et al.'s book [BGH87] describes many important methods of concurrency control. Eswaran et al. [EGT76] proved that 2PL satisfies serializability. Gray and Reuter [GR93] discussed the time stamp ordering protocol for serializable transactions. Kung and Robinson [KR81] described several optimistic versions of concurrency control.

The 2PC protocol is due to Gray [G78]. The blocking properties of this protocol are also discussed in [BGH87]. Skeen [S83] introduced the 3PC protocol.

Lampson et al. [LPS81] introduced the idea of *stable storage* and suggested an implementation of it. Randell [R75] wrote an early paper on designing fault-tolerant system by checkpointing. Strom and Yemini [SY85] studied optimistic recovery methods. Elnozahy et al.'s paper [EJZ92] contains a good survey of checkpointing and logging methods. Alvisi and Marzullo [AM98] described various message-logging methods and orphan elimination techniques.

EXERCISES

14.1 Figure 14.8 shows three transactions $T1$, $T2$, and $T3$. Consider concurrency control by time stamp ordering. Which of these three concurrent transactions will commit?

14.2 Three variables x, y, and z belong to three different servers. Now, consider transactions $T1$ and $T2$ defined as follows:

$$T1: \quad R(x); R(y); W(x := 10); W(y := 20)$$

$$T2: \quad R(z); W(x := 30); R(x); W(z := 40)$$

Also, consider the following interleavings of the different operations:

a. $R(x)_{T1}; R(y); R(z); W(x := 30); W(x := 10); W(y := 20); R(x)_{T2}; W(z := 40)$

b. $R(z); W(x := 30); R(x)_{T1}; R(y); W(x := 10); W(y := 20); R(x)_{T2}; W(z := 40)$

Are these traces serializable? Justify your answer.

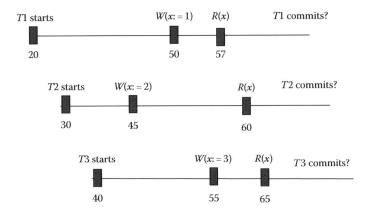

FIGURE 14.8   Three concurrent transactions: $T1$, $T2$, and $T3$.

14.3   In concurrency control using time stamp ordering, whenever there is a conflict, some transaction aborts and later restarts with a new time stamp. Consider three transactions, $T1$, $T2$, and $T3$, in a concurrent run. Is it possible to have a scenario where every transaction periodically aborts and then restarts, and this behavior continues forever? In other words, how can you prove time stamp ordering guarantees termination?

14.4   How will you extend 2PL to nested transactions? Explain your answer with respect to the example of the two-level transaction shown in Figure 14.9:

14.5   Two-phase locking works for distributed transactions. As an alternative, consider this: There are n ($n > 1$) servers; each server manages m objects ($m > 1$). Each server will allow transaction sequential access to the locks on the objects managed by it. Is this sufficient for serializability? Is this sufficient to avoid deadlocks? Explain.

14.6   Specify two execution histories $H1$ and $H2$ over a set of transactions, so that (a) $H1$ is permissible under 2PL, but not under basic time stamp ordering, and (b) $H2$ is permissible under basic time stamp ordering, but not under 2PL.

14.7   Consider the *uncoordinated checkpointing* in Figure 14.10 where processes P, Q, and R spontaneously record three checkpoints each as shown by the bold dots:
    If process P crashes after recording its last state $p2$ as shown in Figure 14.10, then to which *consistent checkpoint* will the system state roll back to?

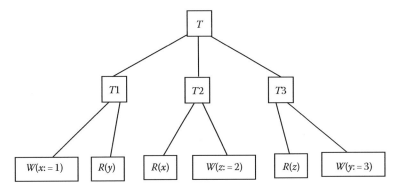

FIGURE 14.9   An example of a nested transaction.

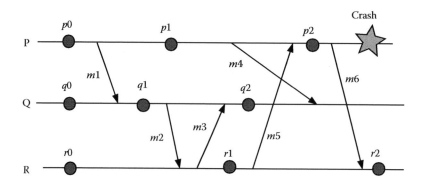

FIGURE 14.10   A set of checkpoints in three concurrent processes: P, Q, and R.

14.8   Some servers are *stateful,* and some others are *stateless.* Is checkpointing relevant to stateless servers too?

14.9   (Programming exercise) There are four travel agencies in a city. Each offers three kinds of services: airline reservation, hotel reservation, and car rental. Create a small database for these services. A query for a service will either return a booking or respond with a failure. When a failure occurs, the entire transaction is aborted.

Simulate this transaction using the 2PC protocol. Include a 5% possibility of server crashes. To simulate the crash, each server will randomly pick a number between 1 and 20. If the random value <20, then the server provides the normal response. If the random value = 20, then the server stops responding, which mimics a crash.

Run your simulation with different possible queries, and verify that your simulation preserves the ACID property in spite of failures.

CHAPTER **15**

Group Communication

15.1 INTRODUCTION

A group is a collection of users or objects sharing a common interest or working toward a common goal. With the rapid growth of the WWW and electronic commerce, group-oriented activities have substantially increased in recent years. Examples of groups are as follows: (1) the batch of students who graduated from a high school in a given year, (2) the members of a particular travel club, (3) the students of a long-distance education course, (4) the members participating in a videoconference, (5) a set of replicated servers forming a highly available service, etc.

Groups may be classified into various types. One classification is based on whether the composition of members is fixed or variable. A group is called *closed*, when its membership remains unchanged. An example is the batch of students who graduated from the Sleepy Hollow High School in the year 2014. The other type of group, in which the membership changes from time to time, is called *open*. The members of the travel club Himalayan Hikers form an open group since the existing members can leave the group, and new members can join at any time. Another classification is based on the nature of communication: A group is called a *peer-to-peer* group when all members have equal rights and privileges in the group and members communicate with their peers to sustain the group activities. An example of an activity is the maintenance of an electronic *bulletin board* by the graduating students of the class of 2014 from a high school—any member could post items of common interest to be viewed by every other member of the group. In contrast, a group is called a *hierarchical* group, where one member is distinguished from the rest, and typically communication takes place between this distinguished member and the rest of the group. A stockbroker communicating with his clients forms a hierarchical group. Note that a hierarchical group can have multiple levels—the president of a company may want to communicate with the managers only, and each manager may communicate with the employees belonging to his or her team.

The members of a group communicate with one another using some form of multicast[*] that is restricted to the members of that group. A message sent by any member of a group must be received by every member of that group or by none at all. This atomicity property is at the heart of group communication—but there are other issues too. These include communication complexity, the ability to tolerate various kinds of failures, and network partition tolerance. Furthermore, when group membership changes in a planned or an unplanned way, certain guarantees must hold to preserve the integrity of group services. We will elaborate these issues later.

15.2 ATOMIC MULTICAST

A multicast in a group is called *atomic*, when the message is received either by every nonfaulty (i.e., functioning) member or by no member at all. *Atomic multicast* is a basic requirement of all group-oriented activities. Cases where some nonfaulty members receive a particular message but others do not lead to inconsistent update of the states of the members and are not acceptable. For example, consider a group of people forming a travel club. If a multicast about a special travel opportunity for the coming Christmas season is sent out, then every member of the travel club should receive it. Another example is a group of replicated servers. If the primary server fails, then a backup server is expected to take over. For this to happen, the states of all servers must be identical to that of the primary server before the failure occurred (via the multicasts from the primary server). However, this will not be possible if one or more backup servers fail to receive some of the updates from the primary server.

Clearly, failures play a prominent role in the implementation of atomic multicast. Accordingly, we will consider two broad classes of atomic multicasts: *basic* and *reliable*. Basic multicast rules out process crashes (or does not provide any guarantee when processes crash), whereas reliable multicasts take process crash into account and provide guarantees. If failures are ruled out, then every basic multicast is trivially atomic. Reliable atomic multicasts should satisfy the following three properties:

Validity: If a correct process multicasts a message m, then it eventually delivers m.

Agreement: If a correct process delivers m, then all correct processes eventually deliver m.

Integrity: Every correct process delivers a message m at most once, only if some process in the group multicasts that message. The reception of spurious messages is ruled out.

The *delivery* of a message affects the state of the recipient and the application supported by it. Note that agreement does not follow from validity or vice versa—a correct process may start a multicast operation as a sequence of point-to-point communications and then crash.

We first address *basic multicasts*. At the data link layer, shared media like Ethernet LAN provide a natural support for multicast. In wireless networks, if each member is within the broadcast signal range from every other member, then every process receives the messages that are sent out. Hosts deliver only those messages that are addressed to them.

---

[*] Defined as one-to-many communication.

15.3 IP MULTICAST

IP *multicast* is a popular technique for multicasting at the network layer. It is a bandwidth-conserving technology that reduces traffic by simultaneously delivering a single stream of information to multiple clients. This is particularly suitable for large-scale applications—examples include distance learning, videoconferencing, and the distribution of software, stock quotes, and news. The source sends only one copy, which is replicated by the routers.

An arbitrary set of clients forms a group before receiving the multicast. The *Internet Assigned Numbers Authority (IANA)* has assigned class D IP addresses for IP multicast. This means that all IP multicast group addresses belong to the range of 224.0.0.0 to 239.255.255.255. The multicast group address serves as a virtual channel. Group members select the channel by selecting the appropriate address, and the network configures itself to deliver the multicast traffic to the group members. The data are distributed via a *distribution tree*. Members of groups can join or leave at any time, so the distribution trees must be dynamically updated. When all active receivers connected to a particular edge of the distribution tree leave the multicast group, the routers prune that edge from the distribution tree (i.e., stop forwarding traffic down that edge). If some receiver connected to that edge becomes active again and resume their participation, then the router modifies the distribution tree and starts forwarding traffic again.

Two widely used forms of distribution trees are *source trees* and *shared trees*. A *source tree* is a *shortest path tree* rooted at the source. Figure 15.1a shows a source tree where a host connected to router *B* is the *source*. For a different source, the tree will be different. The source sends one copy to each neighboring router across the shortest path links. These routers replicate it and forward a copy to each of *their* neighbors. The shortest path property optimizes network latency and works well for streaming data. This optimization does come with a price, though: The routers must maintain path information for each source. In a network that has thousands of sources and thousands of groups, this quickly becomes a resource issue for the routers.

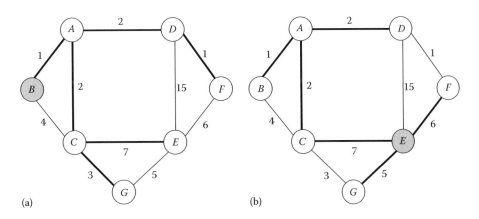

(a)   (b)

FIGURE 15.1   Two distribution trees for multicast in a network of seven nodes: the thick lines are the tree edges (a) a source tree rooted at *B* and (b) a shared tree with *E* as RP.

An alternative is to use *shared trees*. In a shared tree, a specific router is chosen as the *rendezvous point* (*RP*), which becomes the root of all spanning trees used for multicasting (Figure 15.1b). All routers must forward the group communication traffic from their local hosts toward the (*RP*), which forwards them to the appropriate destinations via a common shortest path tree rooted at the RP. The overall memory requirement for the routers of a network that allows only shared trees is lower. The disadvantages of shared trees are primarily twofold: (1) the load on the RP is large and (2) the paths between the source and the destination nodes may not be optimal, introducing additional latency (notice the path from *B* to *F* in Figure 15.1b where *E* is the RP).

15.3.1 Reverse Path Forwarding

In point-to-point communication, traffic from a source is routed through the network along a unique path to the destination. Here, routers do not care about the source address—they only care about the destination address. Each router looks up its routing table and forwards a single copy of the packet toward a router in the direction of the destination.

In multicast routing, the source sends a packet to an arbitrary group of hosts identified by a multicast group address. For each source, the multicast router must be able to distinguish between upstream (toward the source) and downstream (away from the source) links. If there are multiple downstream links, the router replicates the packet and forwards the traffic down the appropriate downstream links—which is not necessarily all links. This concept of forwarding multicast packets away from the source is called *reverse path forwarding* (RPF).

The RPF algorithm that builds a shortest path tree is constructed for each source. Whenever a router *R* receives a multicast packet from the source *S* on a link *L*, it checks to see if the link *L* belongs to the shortest path from *R* to *S* (which is the *reverse path* back to the source). If this is the case, then the packet is forwarded on all links except *L*. Otherwise, the packet is discarded as a *forged* or a *spurious* packet or a redundant one (Figure 15.2). This strategy helps avoid the formation of circular paths in the routes.

Multicast backbone (*Mbone*) is a popular multicast protocol to distribute audio and video to group members over the Internet. Mbone first creates an overlay network implemented on top of some portions of the Internet that includes the group members. An overlay network

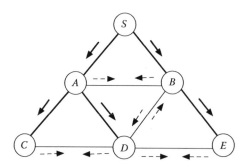

FIGURE 15.2   An example of RPF with the source node *S*: The recipients discard the incoming packets represented by the broken arrows.

is a logical network defined on top of the IP network—the nodes are islands of multicast-capable subnetworks, and the logical links connecting the nodes are *tunnels* through which multicast messages are forwarded via non-multicast-capable portions of the Internet. The multicast routers have the ability to interpret the multicast command, but the unicast routers are unable to do so—therefore, multicast packets are encapsulated inside unicast packets and dispatched along the tunnels. The unicast routers simply forward these packets, but the multicast routers interpret them and locally distribute them to the members of the group. Each router performs an RPF check on each received packet to select only packets on the interface that is the most efficient path back to the source. All other packets are discarded.

15.4 APPLICATION LAYER MULTICAST

While IP multicast is bandwidth efficient and many routers are equipped with the facility of IP multicast, one significant impediment is the growing size of the state space that each router has to maintain per group. The rapid growth of group-oriented activities is making the problem worse. Application layer multicast is an alternative approach that overcomes this problem.

Basic application layer multicast works on an *overlay network* connecting the members of the group. Data or messages are replicated and managed at the end hosts instead of the routers, and the routers are no more required to maintain group-specific states. Application-level multicasts are implemented as a series of point-to-point messages. The network only needs to provide the basic stateless, unicast, best-effort delivery. The downside is an increase in the consumption of network resources like bandwidth, since the same message may be routed multiple times across a link (the replication factor is known as *stress*). Figure 15.3 shows an example. Here, host 0 multicasts a message to hosts 1, 2, and 3. In Figure 15.3a, the same message is sent to router A three times (so the stress on the link is 3). Figure 15.3b uses a different routing strategy (using the paths 0 A C 2 and 0 A B 1 B D 3), which reduces the load on the link from 0 to A and A to B.

The basic scheme is sometimes called the P2P (*peer-to-peer*) scheme where only the peers maintain group-specific information and manage the routing process. An alternative

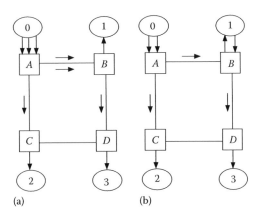

(a)   (b)

FIGURE 15.3   Two examples of application layer multicast with host 0 as the source. (a) The *stress* on the link from host 0 to router A is 3. (b) The *stress* on no link exceeds 2.

is the *overlay multicast* scheme [AMM+04], where a number of strategically deployed proxy nodes are used to back up the hosts. The content is distributed to these proxy servers that are connected by an overlay network. Akamai [NSS10] uses this approach in their *content distribution network* for video streaming, where surrogate servers maintain copies of the content and thus maintain a better quality of service. Various other improvements are feasible in the basic application layer multicast scheme. For example, using cross-layer communication, hosts may collect information from the routers to improve the efficiency of IP multicast by reducing packet duplication.

15.5 ORDERED MULTICASTS

In multicast groups, there are two orthogonal issues: *reliability* and *order*. Reliable multicast addresses only the reliability issue by guaranteeing that each member receives every message sent out by the other members in spite of process failures. So far, it is silent about the order in which these messages are delivered. However, even in basic multicast, many applications require a guarantee stronger than atomicity—here, the *order* of message delivery becomes important. Even if the underlying communication is reliable, guaranteeing the order of message delivery can be far from trivial. One such version requires all messages to be delivered to every group member in the *same total order*. For example, a group of replicated files cannot be in the same state unless all replicas apply the updates from their users in the same order. Other applications may have a weaker ordering requirement.* Guaranteeing ordered message delivery in the presence of process crashes (i.e., implementing the *reliable* version of ordered multicast) is much more challenging than the basic versions. In this chapter, we will primarily focus on the *basic* versions of ordered multicasts.

The following are the three main types or orderings that have been studied in the context of ordered multicasts:

1. Local order multicast (also called single-source FIFO)

2. Causal order multicast

3. Total order multicast

Local order multicast: In local order multicast, if a process multicasts two messages in the order ($m1$, $m2$), then every correct process in the group must deliver $m1$ before $m2$. There are many applications of local order multicast: One is in the implementation of a *DSM* where the primary copy of each variable is maintained by an exclusive process, and all other processes use cached copies of it. Whenever the primary copy is updated, the owner of the primary copy multicasts the updates to the holders of the cached copies, and these copies are updated in the same order. Other applications include video distribution and software distribution.

Causal order multicast: Let $m1$ and $m2$ be a pair of messages in a group, such that $sent(m1) \prec sent(m2)$. Then causal order multicast requires that *every process* in the system

---

* Chapter 16 will discuss some of these applications.

must deliver $m1$ before $m2$. Local order multicast trivially satisfies this. Causal order multicast modifies it by imposing delivery orders among causally ordered messages from distinct senders too. Here is an example: A group of students scattered across a large campus are preparing for an upcoming quiz through a shared bulletin board. Someone comes up with a question and throws it to the entire group, and whoever knows the answer multicasts it to the entire group. The delivery of a question to each student must happen before the delivery of the corresponding answer, since these are causally related. It will be awkward (and a violation of the rules of causal ordered multicast) if some student receives the answer first and then the corresponding question!

Total order multicast: In total order atomic multicast, *every member* of the group is required to deliver *all messages* sent within the group in identical order. It implies that if every process i maintains a queue $Q \cdot i$ (initially empty) to which a message is appended as soon as it is delivered, then eventually, for any two distinct processes i and j, $Q.i = Q.j$. Note that the order in which the messages are delivered has no connection with the real time at which these messages were sent out.

The underlying abstract concept behind total order multicast is that of a replicated *state machine*. Assume that each client sends its request to a group of identical servers, the objective being that if one server crashes then another server will take over. The servers have states, and each request from a client modifies the state of a server. For consistency, it is essential that all nonfaulty servers remain in the same state after each update. This cannot be met without total order multicast. Compared with causal order multicast, total order multicast is more restrictive. However, the order in which messages will be delivered in a total order multicast need not agree with the causal order.

In real-time environment, one more class of atomic multicast is relevant—it is called *timed multicast*. A Δ-*timed multicast* is one in which every message sent at time t is delivered to the each member at or before the time $(t + \Delta)$. If a message is not delivered to at least one process within that time window, then it should not be delivered to any process within that time window. The time may be maintained either by an external observer or by the clocks local to the processes. Depending on the choice, different variations of the timed multicasts can be defined.

In the presence of omission failures, local order reliable multicast can be implemented using Stenning's protocol or an appropriate window protocol (Chapter 12). So, we will only discuss the implementation of the other two types of ordered multicasts. We focus only on the basic versions.

15.5.1 Implementing Total Order Multicast

Several implementations of total order multicast exist. Here, we present two different implementations.

15.5.1.1 Implementation Using a Sequencer

A designated member S of the group acts as a *sequencer process* [CM84]. It assigns a unique sequence number *seq* to every message m that it receives and multicasts it to every other member of that group. Each member first sends its message to the sequencer process. The variable *seq* in the sequencer defines the order in which every member will accept the messages. After receiving the multicast from the sequencer *S*, each member accepts the messages in the ascending order of *seq*. Messages received out of order are buffered (assuming that there is available space for it), until the message with the expected number arrives. Essentially, the sequencer performs a local order multicast.

```
{The sequencer S}
define seq: integer {initially seq = 0}
do receive m →    multicast (m, seq) to all members;
                  seq := seq + 1;
                  deliver m
od
```

One criticism of this simple implementation is that the sequencer process becomes a bottleneck. The next implementation illustrates a distributed version that does not use a central process as a sequencer.

15.5.1.2 Distributed Implementation

For a distributed implementation of total order multicast, it is tempting to use the real time of the individual multicasts for determining the total order. Thus, if three messages *m*1, *m*2, *m*3 are multicast at times *t*1, *t*2, *t*3, respectively, and *t*1 < *t*2 < *t*3, then every member of the group will deliver the messages in the order *m*1, *m*2, *m*3. The solution is feasible in systems where message propagation delays are bounded. In asynchronous system, message propagation delays can be arbitrarily large, so messages can reach their destinations in any order. As a result, a process that first received *m*2 has no way of guessing if another message *m*1 has been sent at an earlier time and is still in transit. This leads to the following two-phase protocol for total order multicast, which is similar to the 2PC protocol of Section 14.5.

Phase 1: To multicast a message *m*, a sender sends (*m*, *ts*) every member of the group, where *ts* is the *send time stamp* of the message. Upon receiving it, a member saves it in a holdback queue, assigns a *receive time stamp ts'* to it, and sends an ack (*a*, *ts'*) back to the sender.

Phase 2: After the sender receives (*a*, *ts'*) from *every* member in the group, it picks up the largest value of *ts'*. Let *ts''* be this value. Then it multicasts a *commit* message with the time stamp *ts''* to every member of the group. The recipients subsequently use the value of *ts''* to determine the delivery order of the messages in its holdback queue.

Figure 15.4 shows an example. Here, *p* receives three acks with time stamps 3, 4, and 10 from the members—so *p* multicasts a *commit* message with a time stamp 10 that is the *maximum* of these values. Similarly, the commit messages from processes *q* and *r* have time

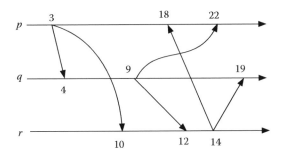

FIGURE 15.4   Every process will deliver the messages from the senders in the order (p, r, q).

stamps 22 and 19, respectively. Therefore, every process will sequentially accept messages from the senders in the order (p, r, q).

Why does it work? The argument is straightforward: Every process will eventually receive all *commit* messages, and every process will eventually learn about the unique delivery order from the values of *ts*. To guarantee liveness, let $m \cdot 1, m \cdot 2, m \cdot 3, \ldots$ be the sequence of messages currently in the holdback queue of a member p, sorted in the ascending order of the commit times. To decide if a message $m \cdot i$ is ready for delivery, member p must ascertain that (1) all messages up to $m \cdot (i-1)$ have been delivered and (2) no *commit* message with a time stamp *ts* smaller than that of $m \cdot i$ will arrive in the future. Since the logical clocks of every member continue to increase and every member learns about the values the logical clocks of its peers via messages and acks, the second condition is guaranteed at a time when the minimum of all the logical clocks exceeds the commit time of the message $m \cdot i$, assuming each channel to be FIFO.

In the previous implementation, for each message multicast by a member, two additional messages (*ack* and *commit*) are required. Therefore, to implement total order multicast in a group of size n where each member sends one message, a total number of $3n^2$ messages will be required.

15.5.2 Implementing Causal Order Multicast

The implementation of causal order multicast uses vector time stamps (introduced in Chapter 6), with a minor modification.* In a group of n members 0, 1, 2,…, $(n-1)$, a vector clock *VC* is a nonnegative integer vector of length n. The vector clock is event driven. Let $VC[i]$ denote the current value of the local vector clock of member i and $T[j]$ denote the vector time stamp associated with the incoming message sent by member j. Also, let $VC_k[i]$ represent the kth element of the vector clock $VC[i]$. To multicast a message, a member i will increment $VC_i[i]$ and append the updated $VC[i]$ to the message. If the recipient j decides to accept the message, there it updates $VC[j]$ as follows: $\forall k: VC_k[j] := \max(T_k[i], VC_k[j])$. To decide whether to accept this message, the recipient has to decide the causal order of this message with respect to other messages. In Figure 15.5, *send(m1)*<*send(m2)* so P2 should deliver these messages in that order. But *m2* from P1 arrived at P2 earlier than *m1*, so P2 must delay the delivery of *m2* until *m1* from P0 arrives (and P2 delivers it). But the important question is how will P2 decide that it cannot deliver *m2*

---

* The vector time stamp scheme is often fine-tuned to satisfy specific requirements.

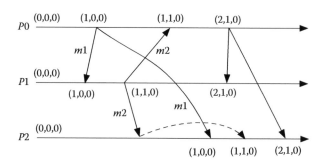

FIGURE 15.5 An example illustrating the delivery of messages per causal order: P2 postpones the delivery of $m2$ until it receives $m1$.

as soon as it receives $m2$—it has no knowledge that $m1$ will arrive in the future? The following two observations form the basis of member i's decision to deliver a message from process j:

> *Observation 1*: Member i must have received and delivered all previous messages sent by member j. So $T_j[j] = VC_j[i] + 1$.

> *Observation 2*: For every other member $k(k \neq j)$, member i must also have received all messages sent by k, which were received by member j *before* it sent the current message. This means $\forall k \neq j{:}T_k[j] \leq VC_k[i]$.

We now revisit Figure 15.5 and focus on the message $m2$ from process $P1$ to process $P2$. Here, $T[1]=1,1,0$ and $VC[2]=0,0,0$. So the first requirement is satisfied, but the second requirement is not, since $T_0[1] > VC_0[2]$. The missing link is as follows: Process 2 has not received the first message $m1$ from process 0, but process 1 has already received it prior to sending the current message. Process 2 therefore saves this message in a holdback queue, accepts the next incoming message from process 0 and delivers it (which updates its VC to 1, 0, 0), and then delivers the pending message $m2$. The algorithm for causal order multicast can thus be summarized as follows:

> *Send*: Sender j increments $VC_j[j]$ and appends $T = VC[j]$ to message m.

> *Receive*: To deliver m, receiver i waits until the conditions (1) $T_j[j] = VC_j[i] + 1$ and (2) $\forall k \neq j{:}T_k[j] \leq VC_k[i]$ hold. Thereafter, the receiver delivers m and updates $VC[i]$ as follows: $\forall k{:}0 \leq k \leq n-1{::}VC_k[i]=\max(T_k[j], VC_k[i])$.

15.6 RELIABLE MULTICAST

Compared to basic multicasts, the additional requirement of reliable multicasts is that only *correct processes* will receive the messages from all correct processes in the group. Multicasts by faulty processes* will be received either by every correct process or by

---

* These processes might have failed after the multicast is initiated.

none at all. Let n processes {0, 1, 2,..., $n-1$} form a closed group and assume that multicasts are implemented as a sequence of *unicasts*. An implementation that survives the crash failure of members is described later. Note that we consider the communication links to be reliable.

Sender's program
```
i := 0;
do i ≠ n→
    send message m to i;
    i:=i+1
od
```

Receiver's program
```
if m is a new message →
        accept it;
        multicast m to every member;
[] m is not a new message → discard m
fi
```

If the sender crashes before sending a single message, then no one receives it, and atomicity trivially holds. Otherwise, if the sender crashes at the middle of the multicast, then the receivers carry out the unfinished work of the sender by forwarding each newly received message to every other member of the group. This satisfies the requirement of atomic multicast, although at the expense of a large number of messages, the complexity of which is $O(n^2)$. The message complexity can be reduced if the maximum number t of faulty processes is known. For example, if $t = 1$, then after the initiator i sends out one round of messages, it is adequate for one more process j to send a second round of messages. Success is guaranteed since both i and j cannot crash—so every nonfaulty process must receive a copy of the message. The maximum number of messages is limited to $2n$.

15.6.1 Scalable Reliable Multicast

Practical multicast systems deal not only with process crashes but also with omission failures and network partitions. IP multicast or application layer multicast provides no guarantee for reliability. Several issues must be addressed in such systems: (1) detection of message loss and types of feedback (*ack*, *nack*, etc.) to be used, (2) sending the feedback to the sender, (3) retransmission of lost or corrupted messages, and (4) congestion control and scalability. To add to the complications, the composition of the group may change at any time. An implementation of reliable multicast for large-scale systems is addressed here.

Let $m[0]$ through $m[k-1]$ be a sequence of k messages that a sender wants to multicast to its group of size n. We first consider a minimal program for reliable multicast that works on bounded delay channels and rules out the failure of the sender or change of group membership. After each basic multicast, the sender expects $(n - 1)$ *acks* from the group members within a time-out period. If this does not arrive, then the initiator retransmits the message using another instance of basic multicast. When the sender receives $(n - 1)$ *acks*, it begins the multicast of the next message. A closer look will reveal that this is essentially an implementation of reliable local order multicast in the presence of omission failures.

```
Program Reliable Multicast
{Sender's program}
define: seq, count : integer
initially seq = 0; count = 0; {counts the number of ack's};
do (seq ≠ k − 1)∧(count = 0)→ (basic) multicast (m[seq],seq)
[] ack received → count:=count+1
[] (seq ≠ k − 1)∧(count ≠ n − 1)∧timeout→ (basic) multicast (m[seq],seq)
[] (seq ≠ k − 1)∧(count = n − 1)→  seq: = seq + 1;count:=0;
od

{Receiver's program}
define r: integer; {initially r = 0}
do seq = r → deliver m[seq];send ack; r:=r + 1
[] seq ≠ r → send ack for the last received message
od
```

When n is large, the number of acks is a clear bottleneck in the scalability of the previous scheme. If omission failures are rare, then a more scalable scheme is as follows: Receivers will only report the *nonreceipt* of messages using *nack*, instead of using positive acks for reporting the receipt of messages. This triggers selective point-to-point retransmission—and either the sender or another member that received the message will forward the missing message. Members will eventually delete the received messages from their local buffers after a time-out period. The reduction of acks is the underlying principle of *scalable reliable multicasts* (SRMs).

Floyd et al.'s [FJM+97] idea takes the principle one step further by allowing negative feedback to be combined or suppressed. If several members of a group fail to receive a message, then each such member will apparently multicast its *nack* to all the other members. However, the algorithm requires each such member to wait for a random period of time before sending its *nack*. If in the mean time one such member notices a *nack* sent by another member, then it suppresses the sending of its own *nack*. It is sufficient (and in fact, ideal) for the sender to receive only one *nack* for a given missing message, following which it multicasts the missing message to all recipients. In practice, Floyd's scheme reduces the number of *nacks* sent back to the sender for each missing message, although it is rarely reduced to a single *nack*. However, reductions in the scale of redundant retransmissions save bandwidth. Finally, upon receiving a *nack* for a missing message, another member may supply that message from its local cache, thus saving the overhead of a system-wide retransmission. The proper orchestration of these techniques helps in drastically reducing the number of retransmitted messages, which saves bandwidth and improves scalability.

15.6.2 Reliable Ordered Multicast

Some reliable versions of ordered multicasts can be implemented by replacing each basic multicast component of the implementation by its reliable version. Interestingly, total order reliable multicasts cannot be implemented on an asynchronous system in the presence of crash failures. The following theorem explains why it is so.

Theorem 15.1

In an asynchronous distributed system, total order multicasts cannot be implemented when even a single process crashes.

Proof: If we could implement total order multicasts in the presence of crash failures, then we could as well solve the asynchronous consensus problem in the following manner: Let each correct process enter the messages received by it in a local queue. If every correct process picks the head of the queue as its final decision, then consensus is reached. But we already know from Theorem 13.1 that the consensus problem cannot be solved in an asynchronous system even if a single process crashes. So it is impossible to implement total order multicasts in an asynchronous system in the presence of a crash failure. ■

15.7 OPEN GROUPS

Open groups allow members to spontaneously join and leave. Changing group sizes adds a new twist to the problems of group communication and needs a precise specification of the requirements. Consider a group g that initially consists of four members $\{0, 1, 2, 3\}$. Assume that each member knows the current membership of this group. We call this a *view* of the group and represent it as $V(g)=\{0, 1, 2\}$.

When members do not have identical *views*, problems can arise. Such situations can occur due to the local nature of the join or leave operations and the latency of information propagation. As a simple example, suppose the members of the group $\{0, 1, 2, 3\}$ have been given the responsibility of sending out emails to 144 persons. Initially, every member had the view $\{0, 1, 2, 3\}$, so they decided that each would send out $144/4 = 36$ emails to equally share the load. Due to some reason, member 3 left the group and only 2 knew about this, while others were not updated about 3's departure. Therefore, the views of the remaining three members will be as follows:

$$V(g) = \{0, 1, 2, 3\} \quad \text{for member } 0$$

$$V(g) = \{0, 1, 2, 3\} \quad \text{for member } 1$$

$$V(g) = \{0, 1, 2\} \quad \text{for member } 2$$

As a result, members 0 and 1 will send out emails 0–35 and 36–71 (the first and the second quarter of the task), member 2 will send out 97–144 (the last one-third of the task), and the emails 72–96 will never be sent out!

Membership service

A basic group membership service deals with the following tasks:

1. Handles the join and leave operations of members

2. Propagates the latest view of the group to the members

3. Detects failures

The second component requires inputs from the first and the third components. Thus, in addition to the voluntary join and leave operations, group compositions can change due to the involuntary or unannounced departure (i.e., crash) of members or network partitions. The view of a group is extremely relevant before a member executes an action. The membership service propagates view changes in identical order to all the members. As an example, let $V_0(g)=\{0, 1, 2, 3\}$ be the initial view. Assume that members 1 and 2 leave the group, and member 4 joins the group concurrently. The membership service, upon detection of these events, serializes the interim views and installs them at the various members—this enables all processes observe the view changes in the same order. There are several feasible serializations (which may depend on how and when joins and leaves were detected) in this case, such as the following:

$\{0, 1, 2, 3\}, \{0, 1, 3\}, \{0, 3, 4\}$

$\{0, 1, 2, 3\}, \{0, 2, 3\}, \{0, 3\}, \{0, 3, 4\}$

$\{0, 1, 2, 3\}, \{0, 3\}, \{0, 3, 4\}$

Views are propagated just like messages, so processes send and receive messages in specific views. This simplifies the specification of the semantics of multicast in open groups and the development of application programs. Examples of message sequences for two different members are as follows:

```
{Process 0}:   V₀(g);
                   send m1, ... ;
               V₁(g);
                   send m2, send m3, receive m6;
               V₂(g);

{Process 1}:   V₀(g);
                   send m4, send m5, receive m1;
               V₁(g);
                   receive m2, send m6;
               V₂(g);
```

where, $V_0(g) = \{0, 1, 2, 3\}, V_1(g) = \{0, 1, 3\}, V_2(g) = \{0, 3, 4\}$

Communication actions that take place between the views $V_i(g)$ and $V_{i+1}(g)$ are said to have taken place in view $V_i(g)$. The switching of views does not happen instantaneously across the entire network; rather, views are *eventually* delivered to all nonfaulty members. The following two properties hold for view delivery in open groups:

Property 1: Let $i \in V(g)$. If a member j joins group g and thereafter continues its membership in g, then eventually j appears in all views delivered by process i.

Property 2: Let $i, j \in V(g)$. If member j permanently leaves group g, then eventually j is excluded from all views delivered by process i.

15.7.1 View-Synchronous Group Communication

View-synchronous group communication (also known as *virtual synchrony*) presents to each processor a consistent total order of views and defines how messages will be delivered to the members with respect to these views. Since crashes change group composition and process views, the specifications of view-synchronous group communication takes crashes into account. Let a message m be multicast by a group member, and before it is delivered to the current members, a new member joins the group. Will the message m be delivered to the new member? Or consider the distribution of a secret key to the members of a group. While the distribution was in progress, a member left the group. Should that member receive the key? There perhaps can be many policies. View synchrony specifies one such policy. The guiding principle of view-synchronous group communication is that *with respect to each message, all correct processes must have the same view*, that is, if a message is sent in a view $V(g)$, then it should be received in the same view (Figure 15.6). Thus, there should be basic agreement about the next view as well as the set of messages delivered in the current view. This translates to a sequentially consistent (see Chapter 16) interface for the members of the group.

In the presence of crash failures, there are three key requirements of view-synchronous group communication:

Agreement: If a member k delivers a message m in view $V_i(g)$ before delivering the next view $V_{i+1}(g)$, then every nonfaulty member $j \in V_i(g) \cap V_{i+1}(g)$ must deliver m before delivering $V_{i+1}(g)$.

Integrity: If a member j delivers a view $V_i(g)$, then $V_i(g)$ must include j.

Validity: If a member k delivers a message m in view $V_i(g)$ and another member $j \in V_i(g)$ does not deliver that message m, then the next view $V_{i+1}(g)$ delivered by k must exclude j.

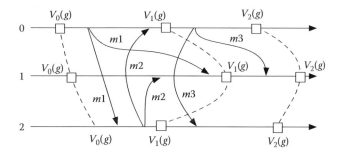

FIGURE 15.6 Visualizing view-synchronous group communication.

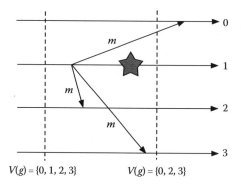

FIGURE 15.7   An unacceptable schedule of message delivery in a group of changing size.

Consider the example in Figure 15.7. Here, process 1 sends out a message m and then crashes, so the group view eventually changes from $\{0, 1, 2, 3\}$ to $\{0, 2, 3\}$. How will the surviving processes handle this message? Here are three possibilities:

Possibility 1: No one delivers m, but each delivers the new view $\{0, 2, 3\}$.

Possibility 2: Processes 0, 2, 3 deliver m and then deliver the new view $\{0, 2, 3\}$.

Possibility 3: Processes 2, 3 deliver m and then deliver the new view $\{0, 2, 3\}$, but process 0 first delivers the view $\{0, 2, 3\}$ and then delivers m.

Of these three possibilities, 1 and 2 are acceptable, but 3 is not—since it violates the agreement criteria of view-synchronous communication.

15.8 OVERVIEW OF TRANSIS

Danny Dolev of Hebrew University of Jerusalem directed the project Transis. Its goal was to support various forms of group communication. The multicast communication layer of Transis facilitates the development of fault-tolerant applications in a network of machines. Transis is derived from the earlier work in the ISIS project undertaken at Cornell University. Transis supports most of the features supported by ISIS. Among several enhancements over ISIS, Transis takes advantage of the multicast facility available in the routers, thus maintaining a high communication bandwidth. A message addressed to a group is sent only once. Atomicity of multicasts is guaranteed and the desired message delivery order is maintained, whereas message losses and transient network failures are transparent. Reliable communication is based on the *Trans protocol* devised originally by Melliar-Smith et al. Trans piggybacks acks (both positive and negative) with multicast messages. All messages and acks contain a progressively increasing sequence number that starts from 0. Message losses may be caused due to hardware faults, or buffer overflows, or (sometimes) due to the inability to retrieve messages at a high speed from the network. Each ack is sent only once, and the acks from other processes form a causal chain, from which the loss of messages can be deduced. Consider a group consisting of processes

$P,Q,R,S,\ldots$. Let process P multicast a sequence of messages $P_0,P_1,P_2,\ldots$ to the group members, and let p_k denote the ack of the message P_k. If p_iQ_j denotes the ack p_i piggybacked on the message Q_j, then the sequence of messages $P_0,P_1,P_2,Q_0,p_2Q_1,Q_2,q_3R_0\ldots$ received by a member of the group will reveal to that member that it did not receive message Q_3 so far. It will send a *nack* for Q_3 requesting its retransmission. In response to this, another group member that has Q_3 in its buffer can retransmit Q_3.

Transis handles network partitioning caused by communication failures, but guarantees that virtual synchrony is maintained within each partition. After the connectivity is restored, virtual synchrony is eventually restored in the entire system. Group communication systems developed prior to Transis (often called *first-generation* systems) made no guarantees when a partition occurred—in most cases, only the primary component was operational. Transis provides four major modes of multicast communication. These are the following:

Single-source FIFO: If a process multicasts two messages in the order ($m1$, $m2$), then every correct process in the group must deliver $m1$ before $m2$.

Causal mode: Messages that are causally related are delivered to the members according to the causal order.

Agreed mode: Messages are delivered in the same total order at all processes. It is a special version of total order communication mode with the added requirement that message delivery order has to be consistent with the causal order.

Safe mode: Messages are delivered according to the agreed mode, with the additional constraint that all other group members' machines must confirm the receipt of the message. This is possible only after the lower levels of the system have received acks from all the destination machines. Safe mode guarantees that (1) if a safe message m reaches a process P in a given configuration, then P *will* deliver that message unless it crashes and (2) every other message is ordered relative to a safe message. When a network partition occurs, some members deliver a message and some do not. The first guarantee allows a process to know for certain who has delivered a safe message (or crashed). The second guarantee means that if a safe message m was multicast and delivered in a certain configuration, then any message will be delivered by all processes either before m or after m. Of the three modes of communication, the causal mode of communication is the fastest, and the safe mode is the slowest, since it has to explicitly wait for all acks.

The concept of virtual synchrony, when extended to a partitioned environment is known as *extended virtual synchrony*. When a process moves to a different partition, it is treated as a failed process by its fellow processes in the previous configuration. One requirement is *failure atomicity*—if two processes proceed together from one configuration to the next, then they deliver an identical set of messages. An example of a safe message delivery in a partitioned environment is illustrated in Figure 15.8. Assume that A was sending a *safe message*

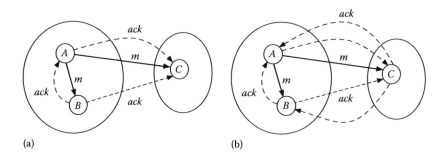

FIGURE 15.8 Two different scenarios in the delivery of a safe message. (a) All but C sent *ack* to A, (b) C sent *ack* to A, and also received a copy of the *acks* sent by A and B before the partition occurred.

m to the group of three members $\{A, B, C\}$, and the configuration partitioned from $\{A, B, C\}$ → $\{A, B\}, \{C\}$. In case (a), all but C sent *ack* to A. In order to deliver m, the members A and B must receive the new view $\{A, B\}$ first—this is a confirmation that only A and B are currently in the partition and the members should not wait for the *ack* from C [VES99]. In case (b), C sent the *ack* and also received a copy of the *ack* sent out by A and B just before the partition occurred. Now, C can deliver m before receiving the new view $\{C\}$. Otherwise, if it receives the new view $\{C\}$, then C can ignore m as a spurious message.

Process groups are dynamic. A member may voluntarily leave a group or lose its membership following a crash. In case of a partition, each component continues to operate separately, but each process identifies the processes in the other partition to have failed. The application is notified about this change in the group membership. To keep track of membership changes, ISIS used an approximate failure detector that relied on time-out: A process that fails to respond within the time-out period is presumed to be faulty and messages directed toward it are discarded. In contrast, Transis presumes that a failed machine can potentially rejoin the group later and need not give up.

15.9 CONCLUDING REMARKS

Since group-oriented activities have significantly increased in commerce and academia, the scalability and reliability of group communication services are major issues. Causal order multicasts using vector time stamps suffer from poor scalability. The situation is equally bad for total order multicasts too, unless a separate sequencer process is used. However, in that case, the sequencer itself may be a bottleneck. Practical systems like Transis achieved scalability using hardware facilities in the router, piggybacking acks, and using a combination of positive and negative acknowledgments to recover missing messages.

15.10 BIBLIOGRAPHIC NOTES

The V system by Cheriton and Zwaenepoel [CZ85] was the first to support for process groups. Chang and Maxemchuck [CM84] presented the algorithm for reliable atomic multicast. Some early implementations of group communication primitives can be found in the reports of the ISIS project [B93]. All ISIS publications are listed in http://www.cs.cornell. edu/Info/Projects/ISIS/ISISpapers.html. IP multicast was developed by Deering, a student of Cheriton as a part of the V system [DC90]. Deering and Cheriton organized a large-scale demonstration during an audiocast at the 1992 IETF meeting. Mbone was a follow-up of IP multicast. Andreev et al. [AMM+04] proposed the overlay multicast scheme. Nygren et al. [NSS10] described the Akamai approach to content distribution on the Internet. The total order multicast algorithm described here is the same as the ABCAST algorithm of ISIS. A few other implementations of ordered multicasts have been proposed by Garcia-Molina and Spauster [GS91] and by Melliar-Smith et al. [MMA90]. The causal order multicast algorithm follows the work by Birman et al. [BSP91]. Birman and Joseph [BJ87] introduced view synchrony (originally known as *virtual synchrony*) in the context of the ISIS project. Floyd et al. [FJM+97] developed the SRM protocol. The 1996 issue of *Communications of the ACM* describes several group communication systems like Transis, Totem, and Horus. Melliar-Smith and Moser [MM93] developed the Trans protocol that was later used in Transis. The reports on the Transis project are available from http://www.cs.huji.ac.il/labs/transis/publications.html. The example highlighting the delivery of safe message while the network partitions is described in detail from the Transis website.

EXERCISES

15.1 In a group, there are eight members, of which at most one can crash at any time. You have no idea about which process can crash. If the topology is a completely connected network, then in the worst case, what is the smallest number of messages needed to guarantee reliable atomic multicast from a given process to the entire group? What is the rationale behind your decision? How will you modify your answer of up to four members can crash at any time?

(Hint: The worst case corresponds to the situation when no process fails, but no one knows about it.)

15.2 Two processes 0 and 1 form a group and want to create a shared bulletin board, so they decide to implement *causal order atomic multicast* using vector clocks. At a certain point, process 1's local vector clock is (0, 2) and it received a message m (from process 0) stamped with vector clock value of (2, 1). Which one of the following steps will process 1 take?

   a. Accept m now.

   b. Accept m only after receiving a message with $VC = (1, 1)$.

   c. Accept m only after receiving two messages with $VC = (1, 0)$ $(1, 1)$.

   d. Accept m only after receiving three messages with $VC = (1, 0)$ $(1, 1)$ $(2, 0)$.

   e. None of the above is true.

15.3 Three processes 0, 1, 2 of a group communicate with one another, and the requirement is *causal order multicast*. A message from process 0 has a vector time stamp (1, 2, 0), and it reaches node 2 when its local vector clock is (0, 1, 2).

   a. Draw a diagram reconstructing the exchange of all the messages in the group.

   b. Will the message be accepted by process 2? Explain.

15.4 Consider an election in a state. The citizens cast their votes at the individual polling centers. At the end of the day when the poll closes, counting begins. Each count recorded at a center is multicast to all the other centers, so that all of them exactly know the latest count at any time when the counting is in progress. The interconnection topology of the network connecting the polling centers is a completely connected graph.

a. Assume that at any time failures can bring down some of the communication lines without creating a partition. What kind of multicast will guarantee that all centers are able to record every vote?

b. Assume that communication failure partitioned the network for an hour. Apparently, the counting must stop. What would you recommend so that the progress of counting is not affected even if the network temporarily partitions?

15.5 Consider a multiparty game of quiz involving five teams. Each team poses a question and a member of another team has to answer it. The team that answers the question first scores a point. If no team can answer a question within 30 s, then no one scores any point. The clocks are synchronized, so who answered first is decided by the time when the reply was posted. The teams can pose questions at any time and in no particular order. What kind of multicast is appropriate here?

15.6 Different applications of group communication demand different types of ordered multicast. Consider the problem of maintaining a shared calendar of appointments by four different clerks, each having a separate copy of the calendar. A clerk will make an entry into his or her own copy of the shared calendar and send it to the other three clerks. The requirement is that after any finite sequence of updates, all four copies of the calendars must be identical and meaningful (thus, if you first make an appointment at 10 a.m. on January 3, 2013, with one clerk and then change it to 9 a.m. on January 4, 2013, with another clerk, then the last entry will prevail). In case of conflicts (like two persons trying to make appointments for the same time slot, the second one will be rejected). What kind of ordered multicast will you recommend? Explain your answer.

15.7 The members of a group use *view-synchronous communication* to communicate with one another. Initially, there are four processes 0, 1, 2, 3. Process 0 sent a message m in view (0, 1, 2, 3). Processes 0, 1, and 2 delivered the message m, but process 3 did not. Is this an acceptable behavior? Justify your answer.

15.8 Bob is the president of the Milky Way club of sky watchers and also the president of the Himalayan Hikers club of nature lovers. From time to time, Bob will send out messages to the members of these groups, and these messages will be delivered in the FIFO order among the group members.

Now assume that some members of the Milky Way club also join the Himalayan Hikers club; as a result, the two groups overlapped. Argue why the FIFO-ordered multicast algorithm may not work for the members who belong to both clubs. Also, suggest modifications that will preserve the FIFO order of message delivery among members of both clubs, including those in the intersection.

15.9 Consider two groups A and B each containing a set of members. Assume that within each group, total order multicast and causal order multicast have already been implemented. Assume that some members of group *A* became members of group *B* too. Using the existing implementations

a. Will causal order multicast correctly work if the groups overlap?

b. Will total order multicast correctly work if the groups overlap?

Briefly justify your answer.

15.10 (Programming exercise) *Implementing totally ordered multicast*

Implement a distributed chat program using a programming language of your choice. When a user types in a message, it should show up at all users. Further, all users should see all messages in *the same order*. Your program should use totally ordered multicast algorithm based on a sequencer process. To multicast a message, a user **p** sends the message to the sequencer. The sequencer assigns a number to the message, returns this number to **p,** and sends the message to all machines in the group together with the sequence number.

Use process 0 as the sequencer. You may or may not allow the sequencer to be a user in the chat. You will need to implement a holdback queue at each user.

15.11 (Programming exercise) *Implementing casually ordered multicast*

Implement causal order broadcast in a group of 16 processes. You will write a program for a client class. This class will store the name of all the clients including its own. This client should also provide the interface including following two functions:

public boolean SendMessage(String strMessage)
This method should send the message to all clients.

public String ReceiveMessage()
This method should return the next message to the client from a pool of messages satisfying the causal order requirement. Remember that the underlying platform is non-FIFO.

public void AddClient(String strClientName) -
This method should add another client to this client's list.

For interclient communication, use the SimulateSocket.class that simulates non-FIFO channels on a reliable network and will garble the message order arbitrarily. The SimulateSocket.class has the following interface:

public SimulateSocket(String strClientName)
This function opens up a new communication channel for the calling client.

public int send(Object objSend, String ReceiverClientName)
This function sends required object to the recipient client.

public Object receive()
This function fetches the next message available for calling client. Remember, this message need not be in FIFO order.

Specify a distributed chat simulation scenario (e.g., label a message from *A* to *B* as (*A*, *B*) and its response as Re[1]: (*A*, *B*), the response to Re[1]: (*A*, *B*) as Re[2]: (*A*, *B*)).

All messages are sent to the entire group—there are no one-to-one messages. Thus, Re[3]: (A, B) is causally ordered before Re[5]: (A, B), but Re[12]: (A, B) is not causally ordered before Re[2]: (B, C). Your experiment should let the clients run chats with titles as earlier. At the end of say 32 messages sent by each process, stop the chat and examine the order of message delivery at each process, and observe that the causal order is not violated. Provide a copy of the client class that you have written along with the test program that validates your client class. Also, prepare a small document that will describe why and how your program works.

Replicated Data Management

16.1 INTRODUCTION

Data replication is an age-old technique for tolerating faults, increasing data availability, and reducing latency in data access. With online activities taking over our lives, our critical data, both personal and professional, are being increasingly stored in data centers somewhere in the globe. Such data are replicated across multiple data centers either proactively or reactively, so that even when some of the copies are lost due to a crash or become inaccessible due to a network partition, our data still remain intact and accessible. Cached copies of downloaded data on our personal devices enable us to use them even if the network connectivity is absent. Apart from data backup, replication is widely used in the implementation of distributed shared memory (DSM), distributed file systems, and bulletin boards.

In addition to fault tolerance, replication reduces access latency. DNS servers at the upper levels are highly replicated—without this, IP address lookup will be unacceptably slow. Accordingly, in large systems (like P2P systems and grids), how many replicas will be required to bring down the latency of access to an acceptable level and where to place them are interesting questions. Another major problem in replication management is that of replica update. The problem does not exist if the data are read-only, which is true for program codes or immutable data. When a replica is updated, every other copy of that data has to be eventually updated to maintain coherence. However, due to the finite computation speed of processors and the network latencies involved in updating geographically dispersed copies, it is possible that even after one copy has been updated, users of the other copies still access the old version. What inconsistencies are permissible and how coherence can be restored are important issues in replicated data management.

16.1.1 Reliability versus Availability

Two primary motivations behind data replication are *reliability* and *availability*. Data replicated in the cloud not only serve as a backup and safeguard against failures but also facilitate anytime anywhere availability through a simple browser. In the *WWW*, proxy servers provide service when the main server becomes overloaded. All these illustrate how replication can improve data or service availability. There may be cases in which the service

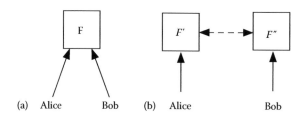

FIGURE 16.1   Two different ways of sharing a file F: (a) Users sharing a single copy and (b) each user has a local replica of the file.

provided by a proxy server is not as efficient as the service provided by the main server— but certainly, it is better than having no service at all. This is an example of graceful degradation. When all servers are up and running and client accesses are uniform, the quality of service goes up since the server loads are balanced. In mobile terminals and handheld devices, disconnected modes of operation are important, and replication of data or service offers partition tolerance and minimizes the disruption of service. Redundant array of inexpensive disks (RAID) is another example of providing reliability and improving availability through replication.

Reliability and availability address two orthogonal issues. A server that is reliable but rarely available serves no purpose. Similarly, a server that is available but frequently malfunctions or supplies incorrect or stale data causes headache for all.

Consider two users, Alice and Bob, updating a shared file F. There may be a single copy of F, or Alice and Bob each may want to maintain a private replica of F (Figure 16.1). Each user's life is as follows:

```
{Alice and Bob sharing a file F}
do true→
      read F;
      modify F;
      write F
od
```

Depending on how the file is maintained, the write operation (1) will update a central copy of F that is shared by both (2) or will update the local copies of F separately maintained by Alice and Bob. In either case, ideally, Alice's updates must reach Bob before he initiates the read operation, and vice versa. However, with separate local replicas, this may not be possible due to channel latencies, so Bob may either read a stale copy or postpone reading until the update arrives. What is an acceptable semantics of sharing F? This will depend on data consistency models.

16.2 ARCHITECTURE OF REPLICATED DATA MANAGEMENT

Ideally, replicated data management must be *transparent*. Transparency implies that users have the illusion of using a single copy of the data or the object—even if multiple copies of a shared data exist or replicated servers provide a specific service, the clients should not have any knowledge of which replica is supplying the data or which server is providing the service. It is conceptually reassuring for the clients to believe in the existence of a single server that is highly available and trustworthy, although in real life, different replicas may take turns to create the illusion of a single reliable copy of data or server. Although ideal replication transparency can be rarely achieved, approximating replication transparency is one of the architectural goals of replicated data management.

16.2.1 Passive versus Active Replication

For maintaining replication transparency, two different architectural models are widely used: *passive* replication and *active* replication. In passive replication, every client communicates with a single replica called the *primary*. In addition to the primary, one or more replicas are used as *backup* copies. Figure 16.2 illustrates this. If the primary is up and running, it provides the desired service. If the request from a client does not modify the state of the server, no further action is necessary. If, however, client actions modify the server state, to keep the states of the backup servers consistent, the primary performs an atomic multicast of the updates to the backup servers, before sending the response to the client. If the primary crashes, one of the backup servers is elected as the new primary.

The *primary-backup* replication architecture satisfies the following specifications:

1. At most, one replica can be the primary server at any time.

2. Each client maintains a variable L (leader) that specifies the replica to which it will send requests. Requests are queued at the primary server.

3. Backup servers ignore client requests.

There may be periods of time when there is no primary server—this happens during a changeover after the primary server has crashed, and a backup server is yet to be designated as the new primary. This period is called the *failover time*. When repairs are ignored, the primary-backup approach implements a service that can tolerate a bounded number of faults over the lifetime of the service. Here, unless specified otherwise, a failure implies a server crash. Since the primary server returns a response to the client after completing an atomic multicast, when a client receives a response, it is assured that each nonfaulty replica

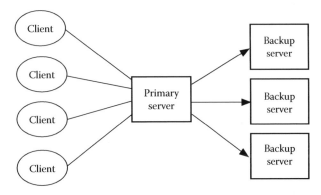

FIGURE 16.2   Passive replication of servers.

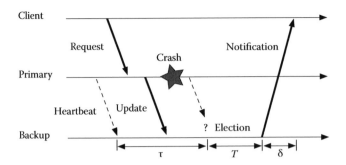

FIGURE 16.3   An illustration of the primary-backup protocol. The broken lines represent the heartbeat messages.

has received the update. The primary server periodically broadcasts heartbeat messages to the backups. If a backup server fails to receive this message within a specific window of time, it concludes that the primary has crashed and initiates an election. The new leader takes over as the new primary and notifies the clients. Figure 16.3 illustrates the basic steps of the primary-backup protocol.

To maintain *replication transparency*, the earlier switchover must be instantaneous. But in real life, this may not be true. Consider the following description of the life of a client:

```
do    request for service → receive response from the server
[]    timeout → retransmit the request to the server
od
```

If due to the crash of the primary the client does not receive a response within a time-out period, the request for service is retransmitted. In fact, multiple retransmissions may be necessary, until a backup server becomes the new primary server. This happens during the failover time. An important question is: If at most m servers can crash, then what are the smallest possible values of the *degree of replication* and the *failover time*?

At least $(m + 1)$ replicas are sufficient to tolerate the crash of m servers, since to provide service, it is sufficient to have only one server up and running. From Figure 16.3, the smallest failover time is $(\tau + 2\delta + T)$ where τ is the interval between the two consecutive heartbeat messages, T is the election time, and δ is the maximum message propagation delay between a client and a backup server. This corresponds to the case when the primary crashes immediately after sending a heartbeat message and an update message to the backup servers.

An alternative to passive replication is *active replication*. Here, each of the n clients communicates with a group of $k(1 < k \leq n)$ servers (also called *replica managers*). Unlike the primary-backup model, these servers do not have any master–slave relationship among them. Figure 16.4 shows a *bulletin board* shared by a group of members. Each member i uses her local copy Bi of the bulletin board. Whenever a member posts an update, her local copy is updated, and the update is propagated to each of the servers using *total order* multicast (so that eventually all copies of the bulletin board become identical). By combining the output of the servers in this ensemble, one can obtain the output of the fault-tolerant state machine.

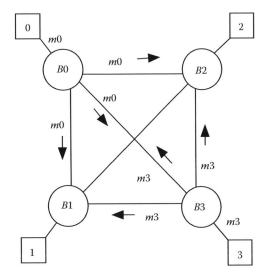

FIGURE 16.4 An example of active replication with $n = 4$ clients and $k = 4$ servers. Each client's message is multicast to every other server.

16.2.2 Fault-Tolerant State Machines

Client–server-based architecture is widely used in the implementation of distributed systems. A server can be viewed as a deterministic *state machine* that executes client actions in some sequence. The state machine has a current state; it executes a step by executing an input command from the client, which produces an output and a new state. As an example, consider a distributed banking system with the tellers as clients. The state of the state machine consists of the account balances of all users. A deposit will change the state of the state machine producing as output the old and the new balances. When implemented as a single server, the failure of that server can completely halt the banking system. To implement a fault-tolerant state machine, we therefore use a collection of servers, each one independently implementing the state machine. When started from the same initial state, in spite of the failure of a fraction of the servers, all nonfaulty servers will produce the same sequences of states and outputs. A client issuing a command can then use the output generated by any nonfaulty server. The replica coordination problem can be reduced to the consensus problem, since all state machines have to agree to the choice of the *next request* that will be used to update its state.

The degree of replication k will depend on the nature and the extent of the failure. In [Sch90], Schneider presented a theory of fault-tolerant state machines, and it captures the essence of maintaining a fault-tolerant service via active replication. Central to this theory are reliable atomic multicasts (Chapter 15). The following two requirements must be satisfied by any implementation of a fault-tolerant state machine:

Agreement: Every correct replica receives all the requests.

Order: Every correct replica receives the requests in the same order.

The *agreement* part is solved using the *reliable atomic multicast* protocol of Section 15.2. The implementation of the *order* part demands that this be further strengthened to *total order reliable multicast*. However, no deterministic solution is feasible on asynchronous distributed systems (since it will disprove the FLP impossibility result; see Theorem 15.1), so a synchronous model must be used.

With at most *m* faulty replicas, at least (*m* + 1) replicas are adequate since the faulty processes produce correct outputs until they fail. To read a value, a client's machine queries the servers one after another, and the first response that it receives is the desired value.

If instead the failures were byzantine, then to compensate for the erroneous behavior of faulty processes, *at least* (2*m* + 1) replicas would be required, so that the faulty processes become a minority, and the bad values can be voted out using a simple majority. However, this assumes that all correct replicas have identical values after every multicast, and with only (2*m* + 1) replicas, this guarantee cannot be provided. To guarantee that all correct replicas are updated to identical values after each multicast in the presence of *m* byzantine failures, one of the protocols from Section 13.3 should be used. If the oral message algorithm is used, then at least (3*m* + 1) replicas will be required. The *order* part can be satisfied by modifying the total order multicast protocols of Section 15.5.1 so that it deals with faulty replicas.

In most implementations, time stamps determine the desired order in which updates will be delivered to the replicas. An update is *stable*, if no update with a lower time stamp is expected to arrive after it. Only stable updates will be delivered to the state machines. Consider a replica *i* receiving time-stamped updates from the other replicas 1, 2, 3,... via channels (1, *i*), (2, *i*), (3, *i*),.... If nonfaulty replicas communicate infinitely often and the channels are FIFO, then each state machine replica will receive through every channel the updates in the ascending order of time stamps. The dilemma here is that some nonfaulty replicas may not have any update to send for a long time. This affects the stability test. While applying an update with time stamp *t*, how will a replica decide if an update with a time stamp smaller than *t* will not arrive in the future via some other channel? One approach to overcome this and make progress is to ask nonfaulty servers to periodically send out *null* requests when it has no update to send. It is possible to combine the null messages with the periodic heartbeat messages that are exchanged among the replica servers. An alternative is to use real time as a basis of ordering the pending updates—from the known upper bound of the message propagation delay, a receiving process can deduce that the sender did not have anything to send up to a certain time. Since crash of a replica is detectable, a replica will ignore the channel from a faulty replica. This will handle stability in the presence of crash failures.

If the failure is byzantine, then a faulty client (or a faulty replica connected to a client) may refuse to send the null message. However, since the upper bound of the message propagation delay is known, the absence of a message can be detected.

16.3 DATA-CENTRIC CONSISTENCY MODELS

Replica consistency requires all copies of data to be *eventually* identical. However, due to the inherent latency in message propagation, it is sometimes possible for the clients to receive anomalous responses. Here is an example: In a particular flight, all the seats were sold out, and two persons A and B in two different cities have been trying to make reservations. At 9:37:00, there was a cancellation by passenger C. A tried to reserve a seat at 9:37:25 but failed to reserve the seat. Surprisingly, B tried to make the reservation at 9:38:00 and could grab it! While this appears awkward at a first glance, its acceptability hinges on the type of *consistency model* that is being used. Consistency determines what responses are acceptable following an update of a replica. It is a contract between the clients and the replica management system.

A *DSM* creates the illusion of a shared memory on top of a message-passing system (Figure 16.5). It can potentially support many data consistency models. Such consistency models are relevant in various applications like distributed file systems, distributed databases, and web caching. The choice of a consistency model also influences the efficiency of concurrent programming. From the user's perspective, stronger models impose severe restrictions on the system behavior, whereas weaker models tend to relax them. In the following, we present a few well-known consistency models:

1. Strict consistency

2. Linearizability

3. Sequential consistency

4. Causal consistency

5. FIFO consistency

16.3.1 Strict Consistency

The *trace* (also called *history*) of a computation is a sequence of read (R) and write (W) operations on a shared variable x. One or more processes can execute these operations. We assume that each read and write operation is atomic.

Strict consistency corresponds to true replication transparency. If one of the processes executes $x := 5$ at real time t and this is the latest write operation, then at a real time $t' > t$, every process trying to read x will receive the value 5. Strict consistency criterion requires

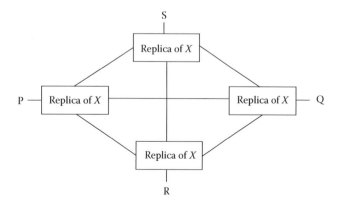

FIGURE 16.5   A DSM with four processes P, Q, R, S sharing a read–write object X.

that regardless of the number of replicas of x, every process receives a response that is consistent with the *real time*. Uniprocessor systems with a single copy of each variable trivially satisfy strict consistency.

The scenario in the previous example of two passengers A and B trying to reserve airline seats does not satisfy the strict consistency criteria, since even if a seat was released at 9:37:00 via a cancellation, the availability of the seat was not visible to A at 9:37:25. Due to the nonzero latency of the messages, strict consistency is not implementable on distributed hardware. Note that read and write operations are considered to be nonblocking (otherwise, one could allow the writer to acquire exclusive right to update all the copies before the update begins, and readers would remain blocked until the update ends).

16.3.2 Linearizability

Strict consistency is too strong and requires all replicas to be updated instantaneously. A slightly weaker version of consistency is *linearizability*.

Define a composite trace as an interleaving of the individual reads and writes into a *single total order* that respects the internal ordering of the actions of every process, as well as the external ordering of actions between processes as defined by their time stamp values. If such a composite trace is *consistent*, which means that every read returns the *latest* value written into the shared variable preceding that read operation, then the shared object is linearizable.

Figure 16.6 shows two example traces with two shared variables x and y. Initially, $x = y = 0$. For (b), the read and write operations by the processes A and B lead to the composite trace $\{init\}$ $W_A(x := 1) R_A(y = 1) W_B(y := 1) R_B(x = 1)$, which is not consistent, so linearizability is not satisfied.

16.3.3 Sequential Consistency

A slightly weaker (and more widely used) form of consistency is *sequential consistency*. To understand the notion of sequential consistency, form a composite trace as an interleaving of the individual reads and writes that respect the internal ordering of the actions of every process, but it is not necessary to respect the external ordering of actions between processes. Clearly, many such composite traces can be generated. Of all such traces, if there is at least one consistent trace, then sequential consistency is satisfied. The key concept in sequential consistency is that all processes should see all the write operations in the same order as in a consistent composite trace. Thus, if a write $x := u$ precedes another write $x := v$ in a trace, then no process reading x will read $x = v$ before $x = u$.

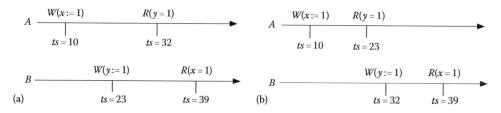

FIGURE 16.6   Two traces: (a) is linearizable but (b) is not linearizable. Here, ts denotes the time stamp of each action. Initially, $x = y = 0$.

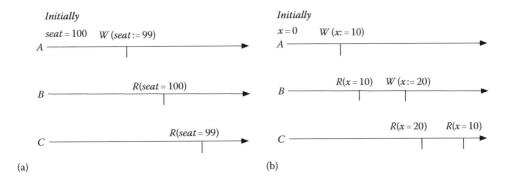

FIGURE 16.7 (a) Sequential consistency is satisfied. (b) Sequential consistency is violated.

Linearizability is stronger than sequential consistency, that is, every linearizable object is also sequentially consistent. Therefore, the trace in Figure 16.6a satisfies sequential consistency. However, consider the trace in Figure 16.6b—it is not linearizable, but one can still generate a consistent composite trace $W_A(x := 1)W_B(y := 1)R_A(y = 1)$ $R_B(x = 1)$ out of it ($W_B(y := 1)$ can be placed before $R_A(y = 1)$ since there is no obligation to respect the time stamp values when ordering action between two different processes). So sequential consistency holds.

Consider again the example of two customers B and C interacting with the airlines reservation system (Figure 16.7a), and assume that the total number of available seats in the flight is 100. This behavior is sequentially consistent, since there exists a consistent composite trace:

$$\{init\}R_B(seat = 100)W_A(seat := 99)R_C(seat = 99)$$

If client B is unhappy with this anomalous outcome of the reservation system, then she is perhaps asking for linearizability.

The scenario in Figure 16.7b does not satisfy sequential consistency. The actions of the processes A and B lead to the only feasible composite subtrace $W_A(x := 10)R_B(x = 10)$ $W_B(x := 20)$. However, process C reads x as 20 first and then reads x as 10, so the two consecutive reads violate the program order of the two writes, and it is not possible to build a consistent composite trace that satisfies the local or internal orders.

16.3.4 Causal Consistency

In the causal consistency model, every process must see all writes that are *causally related*, in the same order. The order of values returned by various read operations must be consistent with this causal order, forming a consistent composite trace. The writes may be executed by the same process or by different processes. Writes that are not causally related to one another can however be seen in any order by the different processes. Consequently, these do not impose any constraint on the order of values read by a process. Figure 16.8 illustrates a causally consistent trace, but it is not sequentially consistent.

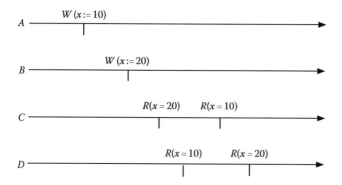

FIGURE 16.8 The trace is causally consistent, but not sequentially consistent.

In this trace, there is no causal order between $W_A(x:=10)$ and $W_B(x:=20)$. Therefore, processes C and D are free to read these values in any order, building their own perception of the history of the write operations. The trace does not satisfy sequential consistency, because that will require processes C and D to see the two writes in the same order, and that order should be reflected in the values that they read out.

Why are there so many different consistency models? True replication transparency is the ultimate target, and stronger consistency models are semantically closer to this target. However, the implementation of stronger consistency models has a higher time complexity and is thus inefficient. Weaker consistency models have fewer restrictions, are cheaper to implement, and lead to faster operations. This is particularly relevant in large-scale systems.

16.3.5 FIFO Consistency

FIFO consistency further weakens the specifications of causal consistency. It only requires that every process see all the write actions by *a single process* in the order in which they were issued. Processes are free to see the write actions by different processes in any order, even if they are causally related. Figure 16.9 shows an example. Processes C and D see the

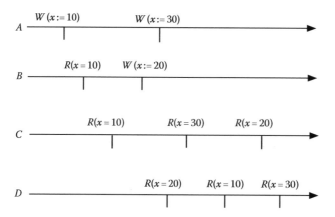

FIGURE 16.9 The trace satisfies FIFO consistency, but not causal consistency.

two writes by process A in the same order. However, although there is a causal relation between $W_A(x:=10)$ and $W_B(x:=20)$, processes C and D see them in different orders.

Consistency issues have been investigated in many different contexts. In addition to DSM on multicomputers, consistency models have also been extensively studied for problems related to cache coherence, web caching, distributed file systems, distributed databases, and various highly available services.

16.4 CLIENT-CENTRIC CONSISTENCY PROTOCOLS

Data-centric consistency models have one thing in common: When multiple clients simultaneously update a shared data, a write–write conflict results and has to be appropriately resolved. In another variation of the problem, there are no write–write conflicts to resolve. A single authority or server will update the data in the storage system, and client side consistency has to do with how and when an observer sees updates made to a data object. Consider the example of a web page in the WWW. Only the owner of the page will update it. Usually, the updates are infrequent, and the number of reads far outnumbers the number of updates. Since browsers and web proxies keep a copy of the fetched page in their local cache and return it to the viewer following the next request, quite often, stale versions of the web page may be returned to the viewer. To maintain freshness, the cached copies are periodically discarded and updated pages are pulled from the server. As a result, updates propagate to the clients in a lazy manner. There will always be a window of inconsistency. Protocols that deal with consistency issues in such cases are known as *client-centric* protocols. In the following, we present the *eventual consistency* model, which is a popular model for client-centric consistency in large distributed systems.

16.4.1 Eventual Consistency

In eventual consistency, the storage system guarantees that if there is no new update on the data, then *eventually* all clients receive the correct view of the latest data. Such a model is acceptable in many large databases—the most popular the DNS (Domain Name System), where updates on a particular domain name are performed by designated naming authorities and propagated in a lazy manner. Such updates are very infrequent and the implementation overhead is small.

Eventual consistency is acceptable as long as a client accesses the same replica of the shared data. Mobile clients can potentially access different versions of the replica and eventually consistency may be unsatisfactory. Consider a busy executive trying to update a database. She performs a part of the update, saves it in the cloud, then leaves for a meeting in a different city, and in the evening decides to finish the remaining updates. Since she may not be working on the same replica that she was working on in the morning (which of course depends on the data replication policy in the data centers), she may notice that some of the updates that she made in the morning are invisible. This means that consistency models for mobile clients need separate attention.

16.4.2 Consistency Models for Mobile Clients

When replicas are geographically distributed, a mobile user will most likely use a replica closest to her. The following four consistency models are based on the sequence of operations that can be carried out by a mobile user:

1. Read-after-read

2. Write-after-write

3. Read-after-write

4. Write-after-read

16.4.2.1 Read-After-Read Consistency

Read-after-read or *monotonic read* consistency implies that when a read of a data set from one server S is followed by a read of its replica from another server S', each read from S' should return a value that is at least as recent as the value previously read from S. Relate this to the story of the busy executive and assume that she was reading emails in Iowa City and then flew to Boston and started reading her emails from the same mailbox. She should see all the emails that she read in Iowa City in the new location too.

16.4.2.2 Write-After-Write Consistency

Write-after-write or *monotonic write* consistency requires that when a write on a replica in server S is followed by a write on another replica in a different server S', the earlier updates must be available to the replica in S' before the next write is performed. An update on a large data set or a structured data can sometimes be partial. Consider the following scenario: Alice is the president of a company. She starts updating the salary database of 100 employees. She finishes giving raises from her office in Dallas and then travels to San Francisco, and there, she starts adding a 10% year-end bonus on the new salary for her employees. If only the first half of the updates (i.e., employees 1–50) is visible in the new location, then the bonus will not be correctly added to the salaries of employees 51–100.

As another example, consider that Bob is updating a tuple from $(x1, y1)$ to $(x2, y2)$. After updating only the first element, that is, modifying $(x1, y1)$ to $(x2, y1)$, he moves to another location and updates the second element of the tuple. If the first update is not visible in the second location at the time of the second update, then after the second update, the value of the tuple will be incorrectly written as $(x1, y2)$ instead of the correct value $(x2, y2)$. When the old update arrives later, it modifies the tuple to $(x2, y1)$. Ideally, all updates must be applied on the previously updated values. Write-after-write consistency guarantees this.

This means that all replicas should be updated in the same order. In the data-centric consistency model, this is equivalent to sequential consistency that expects some total order among all writes. If writes are propagated in the incorrect order, then a recent update may be overwritten by an older update. The order of updates can be relaxed when they are commutative: for example, when the write operations update disjoint variables.

16.4.2.3 Read-After-Write Consistency

Each client must be able to see the updates in a server S following every write operation by itself on another server S'. This is also called *read your writes*.

Consider a large distributed store containing a massive collection of music. Clients set up password-protected accounts for purchasing and downloading musical items of their choice. If a client changes her password in one location, calls her spouse in a distant city, and asks him to access the collection by logging into the account using her new password, then he must be able to do so. Read-after-write consistency will guarantee that the new password propagates to the new location before log in.

16.4.2.4 Write-After-Read Consistency

Each write operation following a read should take effect on the previously read copy, or a more recent version of it. Consider that Alice checked her account balance in the Sunrise Bank Iowa, and found that her paycheck of $1500 has been credited to her account, raising her balance to $1900. She took a flight to Denver during the afternoon, went to a store there, and tried to buy an item for $700 using her bankcard from the Sunrise Bank. But to her embarrassment, the payment did not go through due to insufficient balance in the account. Alice clearly remembered that she did not spend any money from her account in the interim period. Her attempted write (deduction from her balance in the bank) did not take place on the latest balance that she read before her departure from Iowa. This is an example of a violation of the write-after read consistency.

16.5 IMPLEMENTATION OF DATA-CENTRIC CONSISTENCY MODELS

Ordered multicasts (Chapter 15) play an important role in the implementation of consistency models. In the succeeding text, we outline some of these implementations.

Linearizability: Linearizability can be implemented using *total order multicast* that forces every replica to process *all reads and writes* in the same order. Here are the steps:

1. A process that wants to read (or write) a shared variable x calls a procedure $read(x)$ (or $write(x)$) from its local server. The local server sends out the read and writes requests to all other servers using *total order multicast*.

2. Upon the final delivery of these requests, all replica servers update their copies of x in response to a write (which signals the completion of the $write(x)$ operation). For $read(x)$, the replica servers only return acks to the initiating server signaling the appropriate time (consistent with the total order) to read the copy of x from its local server.

The correctness follows from the fact that every replica sees and applies all writes and reads in the same order, which reflects a consistent composite trace.

Sequential consistency: To implement *sequential consistency*, each read operation immediately returns the local copy. Only write operations trigger a total order multicast. Following the multicast, other servers update their local copies of the shared variables. As a result, every replica sees all writes in the same order, and each read returns a value consistent with some write in this total order.

Causal consistency: To implement *causal consistency*, one approach is to use vector time stamps. As usual, every write request is tagged with the current value of the vector clock and multicast to all replicas using the causal order multicast protocol (Chapter 15). Upon receiving these writes, each replica updates its local copy. This satisfies the causality requirement. Read operations immediately return the local copy, which is consistent with some write in this causal order.

Eventual consistency: The basic method involves multicasting the updates to all the servers. Possible methods include *flooding* the update (expensive in terms of the number of messages), or using a spanning tree (works for fixed topologies, but failures and frequent disconnections limit its applicability), or using *gossip* protocols. The order in which a replica will see these updates can vary from one replica to another and can potentially lead to inconsistency that needs to be addressed. One such implementation using the gossip protocol interlaced with antientropy sessions is discussed in the description of the Bayou distributed data management service (Section 16.8.2).

Read-after-read and write-after-write: Starting from the initiating server, at each server where a user has read a data, build an RS with all the reads performed by that user. Along with data items, the set RS should also be propagated to all the other servers. When the user logs in at a different server and issues a read request, the server checks if it has received the most recent set RS and brought the read values up to date. If this is not done, then the read operation is forwarded to a server that has received the set RS, the reads are performed, and the updated RS is pulled into that server. Efficiency considerations need some refinements of these basic steps. For write-after-write, the steps are similar except that the server maintains (and propagates) a *write set* (WS) with all the writes performed by a user.

16.6 QUORUM-BASED PROTOCOLS

A classic method of managing replicated data uses the idea of a *quorum*. A quorum system consists of a family of subsets of replicas with the property that any two of these subsets overlap. To maintain consistency, read and write operations engage these subsets of replicas, leading to several benefits: First, the load is distributed and the load on each replica is minimized. Second, the fault tolerance improves by minimizing the impact of failures, since the probability that every quorum has a faulty replica is quite low. This improves the availability too. However, classic quorum systems do not support partition tolerance, although in real-life network, outages can partition the replicas into two or more subgroups. In general, there are two approaches for dealing with such partitions. One approach is to maintain internal consistency within each subgroup, although no consistency can be enforced across subgroups due to the lack of communication. Mutual consistency is eventually restored only after the connectivity is restored. The other approach is to allow updates in only one group that contains the majority of replicas and postpone the updates in the remaining replicas—so these contain out-of-date copies until the partition is repaired.

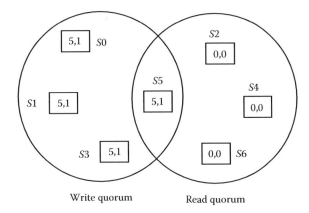

FIGURE 16.10   A quorum system with $N = 7$, $R = 4$, $W = 4$. The old (value, version) of the data is (0, 0) and the updated (value, version) is (5, 1).

Most quorum systems use the second approach. The size of the subset of replicas needed to perform a read or a write operation is called the *read quorum* or the *write quorum*. Here is a description of how it works: Let there be N servers. To complete a write, a client must successfully place the updated data item on $W > N/2$ servers. The updated data item will be assigned a new *version number* that is obtained by incrementing its current version number. To read the data, the client must read out the copies from any subset of $R > N/2$ servers. From these, it will find out the copy with the *highest version number* and accept that copy. Figure 16.10 shows an example. For performing a read or a write operation, the reader or the writer has to acquire exclusive control of the required number of replicas. For this, as in concurrency control of transactions, 2PL can be used—a reader will use read locks, and a writer will use write locks. For any given replica, writers will be granted exclusive rights, but readers can share the access to a server with other readers. The locks will be released after the operation is completed or aborted.

It is easy to make the following observations in such a quorum system:

Observation 1: The system is resilient to the crash of $f \le (N/2)-1$ servers.

Observation 2: Since a read lock does not block readers, multiple readers can concurrently read: For example, in Figure 16.10, one reader can read from the quorum {S0, S1, S2, S4}, while a second reader can read from the quorum {S1, S3, S5, S6}.

Observation 3: Two different write operations cannot proceed at the same time, so all write operations are serialized. Furthermore, the intersection of the *read quorum* and the *write quorum* is nonempty, so reads do not overlap with writes. As a result, every read operation returns the latest version that was written, and single-copy serializability* is maintained.

---

* The effect of transactions performed by clients on replicated data is the same as if they had been performed sequentially on a single set of data items.

Observation 4: During a network partition, the component containing the majority of the replicas will be active. The minority component is brought up to date after the connectivity is restored. However, in partitioned configurations, the system becomes more vulnerable. As an example, in Figure 16.10, if {$S0$, $S2$, $S4$} move to a different partition, then in the majority component, an adversary can crash any single server, and the replication system stops working.

The original idea of quorum systems is due to Thomas [T79]. The scheme presented earlier can be further generalized to allow relaxed choices of the read and write quorums. For example, it is possible to design a quorum-based protocol if the read and write quorums satisfy the following two conditions:

1. $W + R > N$

2. $W > \dfrac{N}{2}$

Thus, in a system with 10 replica servers, an example of a possible choice is $W = 9$, $R = 2$. The extreme case of $W = N$, $R = 1$ (known as *read-one-write-all*) is useful when the writes are very infrequent compared with reads. It allows faster read and better parallelism. The downside is that write will be impossible if a single server crashes. This generalization is due to Gifford [G79].

16.7 REPLICA PLACEMENT

There are many different policies for the placement of replicas of shared data. Here is a summary of some of the well-known strategies:

Mirror sites: On the WWW, a mirror site contains a replica of the original website on a different server. There may be several mirror sites for a website that expect a large number of hits. These replica servers are geographically dispersed and improve both availability and reliability. A client can choose any one of them to cut down the response time or improve availability. Sometimes a client may also be automatically connected to the nearest site. Such mirror sites are *permanent replicas*.

Server-generated replicas: Here, replication is used for load balancing. The primary server copies a fraction of its files to a replica site only when its own load increases. Such replica sites are hosted by *web-hosting* services that temporarily loan their spaces to a third party on demand. The primary server monitors the access counts for each of its files. When the count exceeds a predetermined threshold $h1$, the file is copied into a host server that is geographically closer to the callers, and all calls are rerouted. The joint access counts are periodically monitored. When the access count falls below a second threshold $h2(h2 > h1)$, the file is removed from the remote server, and rerouting is discontinued. The relation $h1 > h2$ helps overcome a possible oscillation in and out of the replication mode.

Client caches: The client maintains a replica in its own machine or another machine in the same LAN. The administration of the cache is entirely the client's responsibility, and the server has no contractual obligation to keep it consistent. In some cases, the client may request the server for a notification in case another client modified the server's copy. Upon receiving such a notification, the client invalidates its local copy and pulls a fresh copy from the server during a subsequent read. This restores consistency.

16.8 BREWER'S CAP THEOREM

Large-scale web services like high-frequency trading or auction heavily rely on replication across a vast pool of servers that are geographically dispersed. In an invited talk in the ACM PODC 2000 conference, Eric Brewer presented a conjecture—that it is impossible for a web service to provide all three of the following guarantees: *consistency, availability*, and *partition tolerance* (CAP). Individually, each of these guarantees is highly desirable; however, a web service can meet at most two of the three guarantees. The conjecture was later analyzed by Gilbert and Lynch [GL02] and is popularly called the CAP theorem. In this section, we present arguments in support of Brewer's CAP theorem.

Consistency in this case refers to atomicity as in the ACID properties of transactions. Consider an auction system in which the current price of an item is x across all servers as shown in Figure 16.11a. If the price changes from x to x', then the revised price should be reflected in all the servers in a bounded time so that linearizability holds. *Availability* implies that every request received by a nonfaulty server in the system leads to a response. Ideally, a web service should be available as long as the network on which they run is up. Finally, *partition tolerance* implies that the service should be available even if the system partitions into two or more disjoint components. This implies that the service will allow the network to lose arbitrarily many messages sent from one partition to another.

Now, consider Figure 16.11b that shows that as the price of the item is being updated from x to x', a communication failure has partitioned the system to two fragments $\pi 1$ and $\pi 2$. For consistency (read linearizability), the new value must be propagated from $\pi 1$ to $\pi 2$; otherwise, a read operation by a client in $\pi 2$ at a later time will not return the new value. But clearly, this is not possible. It does not matter if the new value x' is pushed from partition $\pi 1$ or pulled from partition $\pi 2$. This means that to support consistency and availability, partition tolerance has to be sacrificed. To tolerate network partition, one can

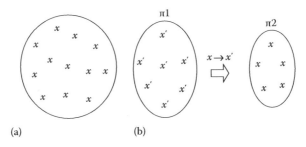

(a) (b)

FIGURE 16.11 A value x is being updated to x' across all the replicas in the system when the system partitioned: (a) before partition and (b) after partition.

however totally disregard consistency or only assure eventual consistency (since the servers in partition $\pi 2$ can be brought up to date after the connectivity is restored), which is much weaker than linearizability. Another possibility is to sacrifice availability—so when a client logs in at a server in partition $\pi 2$, the server will send a request to some server in partition $\pi 1$ asking for an update. Until that update arrives, the client will be kept waiting. When the response comes, the value in the server will be brought up to date and consistency will be restored, but there is no guarantee how long it will take. So, the service will remain unavailable, although consistency and partition tolerance are guaranteed.

16.9 CASE STUDIES

In this section, we discuss three replication-based systems and study how they manage the replicas and maintain consistency. The first is the distributed file system Coda, the second is the highly available service Bayou that is based on the gossip architecture, and the third is Amazon's Dynamo.

16.9.1 Replication Management in Coda

Coda (acronym for *constant data availability*) is a distributed file system designed by Satyanarayanan and his group in Carnegie Mellon University. It evolved from their earlier design of the Andrew file system (AFS). In addition to supporting most features of AFS, Coda also supports data availability in the *disconnected mode*.

Coda allows clients to continue even when the network is partitioned. Mobile users frequently get disconnected from the network, but still want to continue their work without much disruption. Also, network problems and communication failures can cause network partitions, making some replicas unreachable. This affects the implementation of consistency protocols, since partitions prevent update propagation. As a result, clients in disconnected components receive inconsistent values. To sustain operation and maintain availability, consistency properties will at best hold within each connected component and with nonfaulty replicas only. If applications consider this preferable to nonavailability of the service, then contracts have to be rewritten, leading to weaker consistency models. An additional requirement in disconnected operation is that after connectivity is restored, all replicas must be brought up to date and reintegrated into the system. There are various schemes to track the recentness of the updates and to identify conflicts. Coda uses vector time stamps for this purpose.

To maintain high availability, the clients' files are replicated and stored in volume storage groups (*VSGs*). Depending on the current state of these replicas and the connectivity between the client and the servers, a client can access only a subset of these so-called available VSGs (*AVSGs*). To open a file, the client downloads a copy of it from a *preferred server* in its *AVSG* and caches it in his local machine. The preferred server is chosen depending on its physical proximity or its available bandwidth, but the client also makes sure that the preferred server indeed contains the latest copy of the file (using the time stamps of the updated copies)—otherwise, a server that contains the most recent updates is chosen as the preferred server. While closing the file, the client sends its updates to *all* the servers in its *AVSG*. This is called the *read-one-write-all* strategy. When the client becomes disconnected from the network, its *AVSG* becomes empty, and the client relies on the locally cached copies to continue operation.

Files are replicated in the servers, as well as in the client cache. Coda considers the server copies to be more trustworthy (first-class objects) than the client copies (second-class objects), since clients have limited means and resources to ensure the quality of the object. There are two approaches to file sharing: *pessimistic* and *optimistic*. In a pessimistic sharing, file modifications are not permitted until the client has exclusive access to all the copies. This maintains a stricter consistency but has poor availability. The optimistic approach allows a client to make progress regardless of whatever copies are accessible and is the preferred design choice in Coda. Reintegration to restore consistency is postponed to a later stage.

The reintegration occurs as follows: Each server replica has a vector (called *Coda version vector [CVV]*) attached to it—this reflects the update history of the replica. The kth component of the *CVV* (call it $CVV[k]$) of the replica of a file F represents the number of updates made on the replica of that file at server k. As an example, consider four replicas of a file F stored in the servers $V0$, $V1$, $V2$, $V3$ shared by two clients 0 and 1. Of these four replicas, the copies $V0$, $V1$ are accessible to client 0 only, and copies $V2$, $V3$ are accessible to client 1 only. During a time period A, client 0 updates its local copy and multicasts them to $V0$ and $V1$, so the *CVV* of these versions becomes 1, 1, 0, 0. If client 1 also updates its local copy during a time period B that overlaps with A and multicasts its updates to the servers, the *CVV* of client 1's version in $V2$, $V3$ becomes 0, 0, 1, 1. When the connections are restored and the *AVSG* of both clients includes all four servers, each server compares the two *CVVs* and finds them incomparable.* This leads to a conflict that cannot always be resolved by the system and may have to be resolved manually. If however $CVV(F) = 3, 2, 1, 1$ for client 0, and $CVV(F) = 0, 0, 1, 1$ for client 1, then the first *CVV* is considered greater than the second one. In such a case, during reintegration, the *CVVs* of all four replicas are modified to 3, 2, 1, 1 and replicas $V2$ and $V3$ are updated using the copies in $V0$ or $V1$. Coda observed that write–write conflicts are rare (which is true for typical UNIX environments). The traces of the replicas reflect the serialization of all the updates in some arbitrary order and preserve sequential consistency.

Coda guarantees that a change in the size of the AVSG is reported to the clients within a specific period of time. When a file is fetched from a server, the client receives a promise that the server will call it back when another client has made any change into the file. This is known as *callback promise*. When a shared file F is modified, a preferred server sends out callbacks to fulfill its promise. This invalidates the local copy of F in the client's cache—so, to open file F, the client has to load the updated copy from an available server in the VSG. If no callback is received, then the copy in the local cache is used. However, in Coda, since the disconnected mode of operation is supported, there is a possibility that the callback may be lost because the preferred server is down or out of the client's reach. In such a case, the callback promise is not fulfilled, that is, the designated server is freed from the obligation of sending the callback when another client modifies the file.

16.9.2 Replication Management in Bayou

Bayou is a distributed data management service that emphasizes high availability. It was designed to operate under diverse network conditions, and it supports an environment for

---

* See Chapter 6 to find out how two vector time stamps can be compared.

computer-supported cooperative work, such as shared calendars, bibliographic databases, document editing for disconnected workgroups, as well as applications that might be used by individuals at different hosts at different times.

The system satisfies *eventual consistency*, which only guarantees that all replicas eventually receive all updates. Update propagation only relies on occasional pairwise communications between servers. The system does not provide replication transparency—the application explicitly participates in conflict detection and resolution. The updates received by two different replica managers are checked for conflicts and resolved during occasional *antientropy* sessions that incrementally steer the system toward a consistent configuration. Each replica applies (or discards) the updates in such a manner that eventually the replicas become identical to one another.

Supporting application-specific conflict detection and resolution is a major feature in the design of Bayou. A given replica enqueues the pending updates in the increasing order of time stamps. Initially, all updates are *tentative*. In case of a conflict, a replica may undo a tentative update or reapply the updates in a different order. However, at some point, an update must be *committed* or *stabilized*, so that the result takes permanent effect. An update becomes stable when (1) all previous updates have become stable and (2) no update operation with a smaller time stamp will ever arrive at that replica. Bayou designates one of the replicas as the primary—the commit time stamps assigned by the primary determine the total order of the updates. During the antientropy sessions, each server examines the time stamps of the replica on other servers. The number of trial updates depends on the order of arrival of the updates during the antientropy sessions. A single server that remains disconnected for some time may prevent updates from stabilizing and cause rollbacks upon its reconnection.

Bayou allows every possible input from clients with no need for blocking or locking. For example, Alice is a busy executive who makes occasional entries into her calendar, but her secretary Bob usually schedules most of the appointments. It is possible that Alice and Bob make entries into the shared calendar for two different meetings at the same time. Initially, these are tentative. Before the updates are applied, the dependency check will reveal the conflict. Such conflicts are resolved by allowing only one of the conflicting updates or by invoking application-specific merge procedures that accommodate alternative actions.

The *antientropy** sessions* minimize the degree of chaos in the state of the replicas. The convergence rate depends on the connectivity of the network, the frequency of the antientropy sessions, and the policy of selecting the partners. The protocols used to propagate the updates are known as *epidemic protocols*, reminiscent of the propagation of infectious diseases. Efficient epidemic protocols should be able to propagate the updates to a large number of replicas using the fewest number of messages. In one such approach, a server $S1$ randomly picks another server $S2$. If $S2$ lags behind $S1$, then either $S1$ can *push* the update to $S2$ or $S2$ can *pull* the update from $S1$. Servers containing updates to be propagated are called *infective*, and those waiting to receive updates are called *susceptible*. If a large fraction of the servers are infective, then only pushing the update to other replicas may not be the most efficient

---

* *Entropy* is a measure of the degree of disorganization of the universe.

approach, since in a given time period, the likelihood of selecting a susceptible server for update propagation is small. From this perspective, a combination of push and pull strategies works better. The goal is to eventually propagate the updates to all servers.

The gossip protocol is very similar to the mechanism of spreading rumors in real life. A server S1 that has recently updated a replica randomly picks up another server S2 and passes on the update. There are two possibilities: (1) S2 has already received that update via another server, or (2) S1 has not received that update so far. In the first case, S1 loses interest in sending further updates to S2 with a probability $1/p$—the choice of p is left to the designer. In the second case, both S1 and S2 pick other servers and propagate the updates and the game continues. Demers et al. [DGH+87] showed that the fraction of servers that remains susceptible (i.e., does not receive the update) satisfies the equation $s = e^{-(1+p)(1-s)}$. Thus, s is a nonzero number—as a result, eventual consistency may not be satisfied. However, as p increases, the fraction of susceptible servers decreases. A change of p from 1 to 2 reduces the number of susceptible servers from 20% to 6%.

Terry et al. [TTP+95] showed that deleting an item during the antientropy sessions of Bayou can be problematic: Suppose a server S deletes a data item x from its store and propagates it as an update to other servers. This does not prevent S from receiving older copies of x from other servers and treat them as new updates. The recommendation is to keep a record of the deletion of x, issue a *death certificate*, and spread it across the system. Stale death certificates can be flushed out of the system using time-outs.

16.9.3 Amazon Dynamo

Amazon's Dynamo is a highly scalable and highly available *key–value storage* designed to support the implementation of its various e-commerce services. Dynamo serves tens of millions of customers at peak times using thousands of servers located across numerous data centers around the world. At this scale, component failures and communication outages are expected events. In spite of these, Dynamo helps create a reliable system to provide its clients an *always-on* experience. The services using Dynamo guarantee a service-level agreement (SLA) that reflects the high availability and reliability of the system. The implementation of the key–value store should satisfy the ACID properties of transactions. However, for efficiency considerations, such a system sacrifices strong consistency properties. Conflicts frequently occur, but that does not impact the write operation—conflicts are resolved during a subsequent read operation.

Each key is stored in a selected set of servers. Dynamo uses distributed hash tables (DHTs) to map its servers in a circular key space using *consistent hashing* that is more commonly used in many P2P networks. The key K to be stored is also hashed using the same hashing function, mapping it to a point in the circular key space. Let S_F and S_G be the two neighbors of key K on the ring. Then all keys satisfying the condition $hash(S_F) < hash(K) \le hash(S_G)$ are stored in the server S_G that is a clockwise neighbor of K (Figure 16.12). Each server thus becomes responsible for the region in the ring between itself and its anticlockwise neighbor on the ring. However, the random position assignment of each node on the ring may lead to a skewed data and load distribution, as it is oblivious to the heterogeneous node capacities. To address this, Dynamo decided to use *virtual nodes* that map each physical

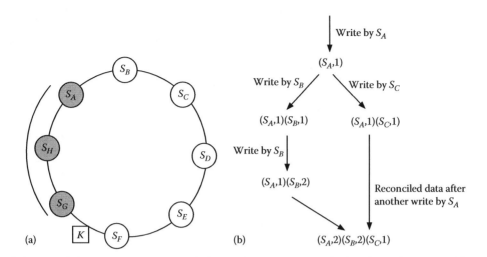

FIGURE 16.12   (a) The key K is stored in the coordinating server S_G and is also replicated in the next few servers like S_H and S_A. (b) The evolution of multiversion data is reflected by the values of the vector clocks.

server to multiple points, instead of a single point on the ring. This allows an unavailable node to shed its load to the available ones with higher capacity.

High availability is achieved by replicating each data on selected servers. The server on which a key is mapped takes up the role of a coordinator, and it replicates the key across the next T servers that are its clockwise neighbors—the value of T depends on the reliability and the accessibility of the servers. Dynamo uses a quorum system: Let W and R be the write and the read quorums. Thus, for a write to succeed, W out of the T servers must be updated. A successful read is defined in a similar way.

Vector clocks: Dynamo uses vector clocks to keep track of the history of its multiple versions of data and resolve conflicts. The multiple versions are generated when the replicas are updated by different servers at different times. A vector clock is a list $(S_i, n \cdot i)$ where S_i is the server id and $n \cdot i$ is a nonnegative integer associated with server i. The vector clock changes after every write operation on a given object (Figure 16.12b), which helps determine whether there exists a causal ordering between two versions of an object. Measurements showed that multiple versions of data are however rare. In a 24 h profile of the *shopping cart* service, 99.94% of requests saw exactly one version, 0.00057% of requests saw 2 versions, 0.00047% of requests saw 3 versions, and 0.00009% of requests saw 4 versions.

Implementing write and read: In response to a *write* request, the coordinator generates the vector clock for the new version and sends the new version to the top T reachable nodes. If at least W nodes respond, then the write is considered successful. During a read operation, the coordinator sends a request for all existing version to the top T reachable servers. If it receives R responses, then the read is considered successful—the various versions are returned to the reader for reconciliation, and the reconciled version is written back. Dynamo uses *sloppy quorum* that is a variation of the traditional quorum system—T, R, and W are limited to the first set of reachable nonfaulty servers in the consistent hashing ring—this speeds up the read and the write operations by allowing them to ignore the responses from the slow servers. In some of the common configuration, Dynamo used the values (3, 2, 2) for (T, R, W).

Hinted hand off: To maintain the spirit of *always write*, Dynamo uses hinted hand off. When a designated server S is inaccessible or down, the write is directed to a different server S' with a *hint* that this update is meant for S. S' later delivers the update to S when it recovers.

The service level agreement (SLA) is quite stringent—a typical SLA requires that 99.9% of the read and write requests execute within 300 ms. Dynamo demonstrated that decentralized techniques can be used to build a scalable and highly available system, and an eventual consistent storage system can be a building block for highly available applications.

16.10 CONCLUDING REMARKS

The notion of consistency is central to replicated data management. Weaker consistency models satisfy fewer requirements—but their implementation costs are also lower, so they are preferred in the design of large-scale distributed systems. This chapter discusses a few important consistency models that have been widely used in various applications. It is the responsibility of the application to judge whether a given consistency model is appropriate.

Client mobility adds a new dimension to the consistency requirements. Also, network partitions complicate the implementation of consistency models—Coda, Bayou, and Dynamo present three different examples of dealing with fragile network connectivity. Since the CAP theorem states that the three properties (strong) consistency, availability, and partition tolerance cannot be implemented in a single system, practical implementations have to decide which one to forego. Availability is the driving force behind successful e-commerce systems like Dynamo.

16.11 BIBLIOGRAPHIC NOTES

The primary-backup protocol follows the work by Budhiraja et al. [BMS+92]. In [L78], Lamport coined the term *state machine*, although he did not consider failures in that paper. Schneider [Sch90] extended the work by including failures and wrote a tutorial on the topic. Lamport also introduced *sequential consistency* [L79]. Herlihy and Wing [HW90] introduced linearizability in order to devise formal proofs of concurrent algorithms. Gharachorloo et al. [GLL+90] presented causal consistency and several other relaxed versions of consistency in the context of shared-memory multiprocessing (DSM). The general architecture of Coda is presented in [SKK+90]. In Kistler and Satyanarayanan [KS92] discussed the disconnected mode of operation in Coda.* Client-centric consistency models originate from Bayou and are described by Terry et al. [TPS+98]. Demers et al. [DGH+87] proposed the epidemic protocol for update propagation. The CAP theorem was first presented as a conjecture by Eric Brewer in his ACM PODC 2000 invited talk. Later, Lynch and Gilbert [LG02] analyzed it in more detail. In the Symposium on Operating Systems (SOSP) 2007 conference, DeCandia et al. [DHJ+07] presented the architecture of Amazon's Dynamo.

---

* Coda papers are listed in http://www.cs.cmu.edu/~odyssey/docs-coda.html, accessed April 9, 2014.

EXERCISES

16.1 A replication management system uses the primary-backup protocol in the presence of omission failures. Argue that in the presence of *m* faulty replicas, omission failures can be tolerated when the total number of replicas $n > 2m$.

(*Hint*: Divide the system into two groups each containing *m* replicas. With the occurrence of omission failure, each group may lose communication with the other group and end up electing a separate leader.)

16.2 Consider two shared read and write objects *X, Y* and a history of read and write operations by two processes *P, Q* as shown in Figure 16.13.

What is the strongest consistency criterion satisfied by the earlier traces? Briefly justify.

16.3 In traces satisfying *causal consistency*, different servers can see the concurrent updates of a shared object in different orders. As a result, the object replicas at various servers may not converge to the same final state after all updates stop. For example, in the traces in Figure 16.8, the final values of *x* will be 10 in server C and 20 in server D. The *causal coherence* model attempts to address this problem by requiring that updates of a given object be ordered the same way at all processes that share the object. Accesses to distinct objects can still satisfy the requirements imposed by causal consistency.

Present an example to distinguish causal consistency from causal coherence. Suggest a distributed implementation of this model and compare it with the implementation of causal consistency.

16.4 Consider two implementations of quorum systems with $N = 10$ replicas. In the first implementation, the read quorum $R = 1$ and the write quorum $W = 10$. The second implementation uses $R = 4$ and $W = 8$. Are these two equivalent? For what reason will you favor one over the other? Explain. What is the problem with $R = 7$ and $W = 5$?

16.5 Consider a quorum-based replica management system with *N* servers, a read quorum *R*, and a write quorum *W*:

a. Will sequential consistency be satisfied if $W = N/2$, $R = N/2$? Briefly justify.

b. Will linearizability be satisfied if $W = (N/2) + 1$, $R = (N/2) + 1$? Briefly justify.

16.6 To prevent an adversary from tampering the quorum system by crashing a few servers, *probabilistic quorum systems* specify a probability distribution in choosing the quorum from a given set. The rule is as follows: If N_1 and N_2 are two independently chosen quorums, then $prob \; [N_1 \cap N_2 = \emptyset] < \varepsilon$.

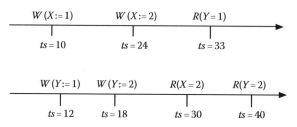

FIGURE 16.13   A history of read and writes by two processes *P, Q*.

Now, consider randomly choosing two different quorums of size $c\sqrt{n}$ from a set of n servers. Show that it will satisfy the requirements of a probabilistic quorum system with $c = \ln(1/\varepsilon)$.

16.7 Most replicated data management protocols assume that the copies of data reside at static nodes (i.e., servers). Technological growth now permits users to carry their personal servers where they keep copies of all objects that are of interest to them. Some of these objects may be of interest to multiple users and thus may be replicated at the mobile personal server nodes. Accordingly, update dissemination must find out which nodes will receive a given update.

Suggest the design of a protocol that will integrate node discovery with update dissemination.

16.8 In this exercise, we address if the composition of multiple traces preserves certain forms of replica consistency. Consider a shared queue and two processes P and Q enqueuing and dequeuing x, y as shown in Figure 16.14. Parts (a) and (b) illustrate two traces. Trace (c) is the composite trace that combines (a) and (b).

Are these three traces linearizable? What conclusions can you draw from this?

16.9 Alice changed her password for a bank account while she was in Colorado, but she could not access the bank account using that password when she reached Minneapolis. Can this be attributed to the violation of some kind of consistency criterion? Explain.

16.10 Consider the following program executed by two concurrent processes P, Q in a shared-memory multiprocessor. Here, x, y are two shared variables.

```
Process P                    Process Q
{initially x = 0}            {initially y = 0}
x: = 1;                      y: = 1;
if y = 0 → x: = 2 fi;        if x = 0 → y: = 2 fi;
display x                    display y
```

If sequential consistency is maintained, then what values of (x, y) will be displayed?

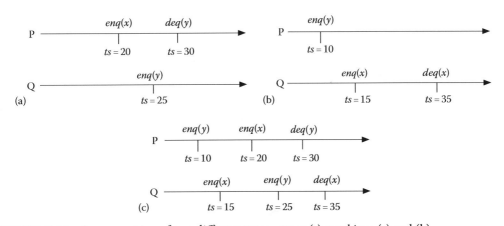

FIGURE 16.14   A composition of two different traces: trace (c) combines (a) and (b).

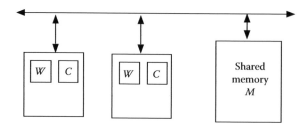

FIGURE 16.15   A shared-memory multiprocessor with write buffers.

16.11 Some shared-memory multiprocessors use *write buffers* to speed up the operation of the *write-through* cache memory (Figure 16.15). It works as follows: When a variable x is updated, the processor writes its value into the local cache C and at the same time puts the updated value into a *write buffer* W. A separate controller then transfers this value into the main memory. The advantage is that the processor does not have to wait for the completion of the *write memory* operation, which is slow. This speeds up instruction execution. For a read, the data are retrieved from the local cache.

   If the program in Exercise 16.10 runs on the multiprocessor with the write buffers, then will sequential consistency be satisfied? Explain your answer.

16.12 Consider the following three schemes for large-scale replica management. For each of these, find out which two of the three CAP properties hold:

   a. Web servers prefer weak consistency. The content of a cached page expires after a lease of time. Users still use the old content, until they are refreshed by the source, or the latest data pulled by the user.

   b. Quorum-based protocols.

   c. Replica update is treated as a transaction and replicas are updated using the 2PC protocol.

16.13 (Programming exercise) In a building, there are eight meeting rooms, which are shared by 10 departments on an *as-needed* basis. Each department has a secretary who is responsible for reserving meeting rooms. Each reservation slot is of 1 h duration, and rooms can only be reserved between 8 a.m. and 5 p.m. Your job is to implement a shared room reservation schedule that can be used by the secretaries.

   Use the following conflict resolution rule: If multiple appointments overlap, the earliest entry for that slot prevails and others are cancelled. If there is a tie for the earliest entry, the secretaries' first names will be used to break the tie. Each secretary will work on a separate replica of the schedule and enter the reservation request stamped by the clock time. The clocks are synchronized. Dependency checks will be done during the antientropy sessions as in Bayou. The goal is *eventual consistency*.

   Arrange for a demonstration of your program.

CHAPTER 17

Self-Stabilizing Systems

17.1 INTRODUCTION

In large-scale distributed systems, failures and perturbations are expected events and not catastrophic exceptions. External intervention to restore normal operation or to perform a system configuration is difficult, and it will only get worse in the future. Therefore, means of recovery have to be built in.

Fault-tolerance techniques can be divided into two broad classes: *masking* and *non-masking*. Certain types of applications call for masking type of tolerance, where the effect of the failure is completely invisible to the application; these include safety–critical systems, some real-time systems, and certain sensitive database applications in the financial world. For others, nonmasking tolerance is considered adequate. In the area of control systems, feedback control is a type of nonmasking tolerance used for more than a century. Once the system deviates from its desired state, a detector detects the deviation and sends a correcting signal that restores the system to its desired state. Rollback recovery (Chapter 14) is a type of nonmasking tolerance (known as backward error recovery) that aims at making the history of the computation correct and relies on saving intermediate states or checkpoints on a stable storage. Stabilization (also called *self-stabilization*), on the other hand, does not rely on the integrity of any kind of data storage and makes no attempt to recover lost computation but guarantees that eventually a good configuration is restored. This is why it is called *forward error recovery*.

Stabilizing systems are meant to tolerate transient failures that can corrupt the *data memory* in an unpredictable way. However, it rules out the failure or corruption of the program codes. The program codes act as recovery engines and help restore the normal behavior from any interim configuration that may be reached by a transient failure. Stabilization provides a solution when failures are infrequent and temporary malfunctions are acceptable and the MTBF is much larger than the MTTR.

The set of all possible configurations or behaviors of a distributed system can be divided into two classes: legitimate* and illegitimate. The legitimate configuration of a nonreactive

---

* Legitimate configurations are also called legal or consistent configurations.

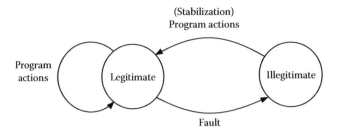

FIGURE 17.1   The states and state transitions in a stabilizing system.

system is usually represented by an invariant over the global state of the system. For example, in network routing, a legal state of the network corresponds to one in which there is no cycle in the routes between a pair of nodes. In a replicated database, a legitimate configuration is one in which all the replicas are identical. In reactive systems, legitimacy is determined not only by a state predicate but also by behaviors. For example, a token ring network is in a legitimate configuration when (1) there is exactly one token in the network, and (2) in an infinite behavior of the system, each process receives the token infinitely often. If a process grabs the token but does not release it, then the first criterion holds, but the second criterion is not satisfied, so the configuration becomes illegitimate. Figure 17.1 outlines the state transitions in a stabilizing system.

Well-behaved systems always remain in a legal configuration. This is made possible by (1) proper initialization that makes the initial configuration legal, (2) the *closure property* of normal actions that transform one legal configuration to another, and (3) the absence of failures or perturbations. However, in real life, such a system may switch to an illegal configuration due to the following reasons:

Transient failures: A transient failure may corrupt the system state (i.e., the data memory) in an unpredictable way. Some examples are the disappearance of the only circulating token from a token ring or data corruption due to radio interference or power supply variations.

Topology changes: The topology of the network changes at run time when a node crashes or a new node is added to the system. These are characteristics of dynamic networks. Frequent topology changes are common in mobile computing.

Environmental changes: The *environment* of a program may change without notice. The environment of a program or a system consists of external variables that it can only read but not modify. A network of processes controlling the traffic lights in a city may run different programs depending on the volume and the distribution of traffic. In this case, the traffic pattern acts as the environment. If the system runs the early morning programs in the afternoon rush hours, the application will perform poorly, and, therefore, so the configuration is considered illegal.

Once a system configuration becomes illegal, the closure property alone is not adequate to restore the system to a legal configuration. What is required is a *convergence* mechanism to guarantee eventual recovery to a legal configuration. A system is called *stabilizing* (or *self-stabilizing*) when the following two conditions hold:

Convergence: Regardless of the initial state, and regardless of eligible actions chosen for execution at each step, the system eventually returns to a legal configuration.

Closure: Once in a legal configuration, the system continues to be in the legal configuration unless a failure or a perturbation corrupts the data memory.

17.2 THEORETICAL FOUNDATIONS

Let *true* be the program of a stabilizing system and Q be the predicate defining a legal configuration. Using the notations of predicate transformers introduced in Chapter 5, the condition $wp(SSS, Q) = true$ holds for a stabilizing system. The weakest precondition *true* denotes that starting from *all* possible initial states, stabilizing systems recover to a correct or legal configuration. This implies that regardless of how the state is modified by a failure or a perturbation, the recovery is guaranteed.

A closely related version is an *adaptive system*, where the system spontaneously adjusts itself by reacting to environmental changes. An example of environment is a variable that represents the time of the day. Data centers expect different volumes and patterns of traffic during different times of the day and arrange to provision hardware and software resources accordingly. A network may compute the shortest path between pairs of nodes regardless of node crashes—here, the environment *crashed(i)* (reflecting the crash failure of node *i*) is a Boolean variable that the network can only read but not modify. The shortest path computation adapts to the value of *crashed(i)*. Our previous formulation of a stabilizing system did not explicitly recognize the role of the environment, but it can be accommodated with a minor modification. Let S be a system and e be a Boolean variable representing two possible states (*true* and *false*) of the environment. To be *adaptive*, S must adjust its behavior as follows: when e is *true*, S must stabilize to a legal configuration that satisfies predicate Q_1, and when e is *false*, S must stabilize to a legal configuration that satisfies predicate Q_0. These imply that starting from any initial configuration, S must stabilize to the predicate $E = (e \wedge Q_1) \vee (\neg e \wedge Q_0)$, which is formally represented as $wp(S, E) = true$. This completes the transformation.

Let SSS be a stabilizing system and the predicate Q define its legal configuration. Consider a finite behavior, in which the sequence of global states satisfies the sequence of predicates: $(\neg Q, \neg Q, \neg Q, ..., Q, Q, Q, Q)$. This behavior satisfies both *convergence* and *closure*. However, a behavior in which the global state satisfies the sequence of predicates $(\neg Q, \neg Q, ..., \neg Q, Q, \neg Q, Q, Q)$ is inadmissible, since the transition from Q to $\neg Q$ via normal program actions violates the closure property.

Finally, we address the issue of termination. Termination reflects a configuration when the system reaches a legitimate configuration and all guards or actions are disabled. Stabilizing systems sometimes relax the termination requirement by allowing infinite behaviors that lead the system to a fixed point, where all enabled actions maintain the configuration in which Q holds. For our purpose, the behavior $(\neg Q, \neg Q, ..., \neg Q, Q, Q, Q, ...)$ with an infinite suffix of states satisfying Q will be considered an acceptable demonstration of both convergence and closure.

17.3 STABILIZING MUTUAL EXCLUSION

Dijkstra's work [D74] initiated the field of self-stabilization in distributed systems. He first demonstrated its feasibility by proposing stabilizing solutions to the problem of mutual exclusion on three different types of networks. In this section, we present the solutions to two of these versions.

17.3.1 Mutual Exclusion on a Unidirectional Ring

Consider a *unidirectional ring* of n processes $0, 1, 2, \ldots, n - 1$ (Figure 17.2). Each process can remain in one of the k possible states $0, 1, 2, \ldots, k - 1$. We consider the shared-memory model of computation: A process i, in addition to reading its own state $s[i]$, can read the state $s[i - 1 \bmod n]$ of its *predecessor* process $i - 1 \bmod n$. Depending on whether a predefined guard (which is a Boolean function of these two states) is true, process i may choose to modify its own state.

We will call a process with an enabled guard a *privileged process* or a process *holding a token*. This is because a privileged process is one that can take an action, just as in a token ring network, a process holding the token is eligible to transmit or receive data. A legal configuration of the ring is characterized by the following two conditions:

Safety: The number of processes with an enabled guard is exactly one.

Liveness: During an infinite behavior, the guard of each process is enabled infinitely often.

A privileged process executes its critical section. A process that has an enabled guard but does not want to execute its critical section simply executes an action to pass the privilege to its neighbor. Transient failures may transform the system to an illegal configuration. The problem is to design a protocol, so that starting from an arbitrary initial state, the system eventually converges to a legal configuration and remains in that configuration thereafter.

Dijkstra's solution assumed process 0 to be a *distinguished process* that behaves differently from the remaining processes in the ring. All the other processes run identical programs. There is a central scheduler for the entire system. In addition, the condition $k > n$ holds. The programs are as follows:

```
Program ring;
{program for process 0}
do s[0]=s[n-1]→s[0]:=s[0]+1 mod k      od
{program for process i≠0}
do s[i]≠s[i-1]→s[i]:=s[i-1]  od
```

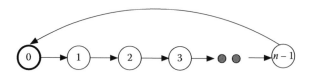

FIGURE 17.2   A unidirectional ring of n processes.

Before studying the proof of correctness of the aforementioned algorithm, we *strongly* urge the reader to study a few sample runs of the aforementioned protocol and observe the convergence and closure properties. A configuration in which $\forall i, j : s[i] = s[j]$ is an example of a legal configuration.

Proof of correctness

Lemma 17.1

[No deadlock] In any configuration, at least one process must have an enabled guard.

Proof: If every process 1 through $n - 1$ has a disabled guard, then $\forall i > 0, s[i] = s[i - 1]$. But this implies that $s[0] = s[n - 1]$, so process 0 must have an enabled guard. ▪

Lemma 17.2

[Closure] The legal configuration satisfies the closure property.

Proof: If only process 0 has an enabled guard, then $\forall i, j : 0 \leq i, j \leq n - 1 : s[i] = s[j]$. A move by process 0 will disable its own guard and enable the guard for process 1. If only process i $(0 < i < n - 1)$ has an enabled guard, then

- $\forall j < i : s[j] = s[i - 1]$
- $\forall k > i : s[k] = s[i]$
- $s[i] \neq s[i - 1]$

Accordingly, a move by process i will disable its own guard and enable the guard for process $(i + 1) \bmod n$. Similar arguments hold when only process $(n - 1)$ has an enabled guard. ▪

As a consequence of Lemmas 17.1 and 17.2, in an infinite computation, the guard of each process will be true infinitely often.

Lemma 17.3

[Convergence] Starting from any illegal configuration, the ring eventually converges to a legal configuration.

Proof: Observe that every action by a process disables its own guard and enables *at most* one new guard in a different process—so the number of enabled guards never increases. Now, assume that the claim is false, and the number of enabled guards remains constant during an infinite suffix of a behavior. This is possible if every action that disables an existing guard enables exactly one new guard.

There are n processes with $k(k > n)$ states per process. By the pigeonhole principle, in any initial configuration, at least one element $j \in \{0, 1, 2, ..., k - 1\}$ *must not* be the initial state of any process. Each action by process ($i > 0$) essentially copies the state of its predecessor, so if j is not the state of any process in the initial configuration, no process can be in state j until $s[0]$ becomes equal to j. However, it is guaranteed that at some point, $s[0]$ will be equal to j, since process 0 executes actions infinitely often, and every action increments $s[0]$ (mod k). Once $s[0] = j$, eventually every process attains the state j, and the system reaches a legal configuration. ■

The property of stabilization follows from Lemmas 17.2 and 17.3.

17.3.2 Mutual Exclusion on a Bidirectional Array

The second protocol operates on an array of processes 0 through $n - 1$ (Figure 17.3). We present here a modified version of Dijkstra's protocol, taken from [G93]

In this system, $\forall i : s[i] \in \{0, 1, 2, 3\}$ and is independent of the size of the array. The two processes 0 and $n - 1$ behave differently from the rest—they have two states each. By definition, $s[0] \in \{1, 3\}$ and $s[n - 1] \in \{0, 2\}$. Let $N(i)$ designate the set of neighbors of process i. The program is as follows:

```
program four-state;
{program for process i, i = 0 or n - 1}
do ∃j ∈ N(i):s[j]=s[i]+1mod4 → s[i]:=s[i]+2mod4 od
{program for process i, 0 < i < n - 1}
do ∃j ∈ N(i):s[j]=s[i]+1mod4 → s[i]:=s[i]+1mod4 od
```

Proof of correctness

The *absence of deadlock* can be trivially demonstrated using arguments similar to those used in Lemma 17.1. We focus on convergence only.

For a process i, call the processes $i + 1$ and $i - 1$ to be the *right* and the *left* neighbors, respectively. Define two predicates $L \cdot i$ and $R \cdot i$ as follows:

$$L \cdot i \equiv s[i-1] := s[i] + 1 \bmod 4$$

$$R \cdot i \equiv s[i+1] := s[i] + 1 \bmod 4$$

We will represent $L \cdot i$ by drawing a $\rightarrow$ from process $i - 1$ to process i and $R \cdot i$ by drawing a $\leftarrow$ from process $i + 1$ to process i. Thus, any process with an enabled guard has at least one arrow pointing toward it.

For a process i ($0 < i < n - 1$), the possible moves fall exactly into one of the seven cases in part (a) of Table 17.1. Each entry in the columns *precondition* and *postcondition* represents

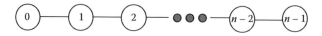

FIGURE 17.3   An array of n processes.

TABLE 17.1   Possible Changes Caused due to an Action by Process i

Case	Precondition	Postcondition
(a) $0 < i < n - 1$		
a	$x + 1 \rightarrow x, x$	$x + 1, x + 1 \rightarrow x$
b	$x + 1 \rightarrow x \leftarrow x + 1$	$x + 1, x + 1, x + 1$
c	$x + 1 \rightarrow x, x + 2$	$x + 1, x + 1 \leftarrow x + 2$
d	$x + 1 \rightarrow x \rightarrow x + 3$	$x + 1, x + 1, x + 3$
e	$x, x \leftarrow x + 1$	$x \leftarrow x + 1, x + 1$
f	$x + 2, x \leftarrow x + 1$	$x + 2 \rightarrow x + 1, x + 1$
g	$x + 3 \leftarrow x \leftarrow x + 1$	$x + 3, x + 1, x + 1$

Case	Precondition	Postcondition
(b) $i = 0$ and $i = n - 1$		
h	$x \leftarrow x + 1$	$x + 2 \rightarrow x + 1$
k	$x + 1 \rightarrow x$	$x + 1 \leftarrow x + 2$

the states of the three processes ($i - 1, i, i + 1$) before and after the action by process i. Note that $x \in \{0, 1, 2, 3\}$, and all + operations are mod 4 operations.

Case (a) represents the move when a $\rightarrow$ is transferred to the right neighbor. Case (e) shows the move when a $\leftarrow$ is transferred to the left neighbor. Cases (c) and (f) show how a $\rightarrow$ can be converted to a $\leftarrow$, and vice versa. Finally, cases (b), (d), and (g) correspond to moves by which the number of enabled guards is reduced. Note that in all seven cases, the total number of enabled guards is always nonincreasing.

Part (b) of Table 17.1 shows a similar list for the two processes 0 and $n - 1$. Case (h) lists the states of processes 0 and 1, and case (k) lists the states of processes $n - 2$ and $n - 1$. Process 0 transforms a $\leftarrow$ to a $\rightarrow$, and process $n - 1$ transforms a $\rightarrow$ to a $\leftarrow$. These two processes thus act as *reflectors*. Once again, the number of enabled guards does not increase.

Lemma 17.4

If the number of enabled guards in the processes $0..i$ ($i < n - 1$) is positive, and process $i + 1$ does not execute an action, then after at most *three moves* by process i, the total number of enabled guards in the processes $0..i$ is reduced.

Proof: By assumption, at least one of the processes $0..i$ has an enabled guard, and process $i + 1$ does not make an action. We will designate the three possible moves by process i as move 1, move 2, and move 3. The following cases exhaust all possibilities:

Case 1: If move 1 is of type (a), (b), (d) or (g), the result follows immediately.

Case 2: If move 1 is of type (e) or (f), after move 1, the condition $s[i] = s[i + 1]$ holds. In that case, move 2 must be of type (a), and the result follows immediately.

Case 3: If move 1 is of type (c), after the first move, the condition $s[i + 1] = s[i] + 1$ mod 4 holds. In this case, move 2 must be of types (b), (e), (f) or (g). If cases (b) and (g) apply, the result follows from case 1. If cases (e) and (f) apply, then we need a move 3 of type (a) as in Case 2, and the number of enabled guards in processes 0..*i* will be reduced. ■

Lemma 17.4 implies that if there is an enabled guard to the left of a process i, then after a finite number of moves by process i, a $\rightarrow$ appears at its right neighbor $i + 1$. Using similar arguments, one can demonstrate that if there is an enabled guard to the right of process $i(i > 0)$, then after a finite number of moves by process i, a $\leftarrow$ appears at its left neighbor $i - 1$.

Lemma 17.5

In an infinite behavior, every process makes infinitely many moves.

Proof: Assume that this is not true, and there is a process j that does not make any move in an infinite behavior. By Lemma 17.4 and its follow-up arguments, in a finite number of moves, the number of enabled guards for every process $i \neq j$ will be reduced to 0. However, deadlock is impossible. So process j must make infinitely many moves.

Lemma 17.6

[Closure] If there is a single arrow, then all subsequent configurations contain a single arrow.

Proof: This follows from the cases (a), (e), (h), (k) in Table 17.1.
In an arbitrary initial state, there may be more than one $\rightarrow$ and/or $\leftarrow$ in the system. To prove convergence, we need to demonstrate that in a bounded number of moves, the number of arrows is reduced to 1. ■

Lemma 17.7

[Convergence] Program *four-state* guarantees convergence to a legal configuration.

Proof: We argue that every arrow is eventually eliminated unless it is the only one in the system. We start with a $\rightarrow$. A $\rightarrow$ is eliminated when it meets a $\leftarrow$ (Table 17.1, case b) or another $\rightarrow$ (Table 17.1, case d). From Lemma 17.4, it follows that every $\rightarrow$ *eventually* moves to the right until it meets a $\leftarrow$ or reaches process $n - 1$. In the first case, two arrows are eliminated. In the second case, the $\rightarrow$ is transformed into a $\leftarrow$ after which the $\leftarrow$ eventually moves to the left until it meets a $\rightarrow$ or a $\leftarrow$. In the first case (Table 17.1, case d), both arrows are eliminated, whereas in the second case (Table 17.1, case g), one arrow disappears and the $\leftarrow$ is transformed into a $\rightarrow$. Thus, the number of arrows progressively goes down. When the number of arrows is reduced to one, the system reaches a legal configuration. ■

Figure 17.4 illustrates a typical convergence scenario. Since closure follows from Lemma 17.6 and convergence follows from Lemma 17.7, the program four-state guarantees stabilization. This concludes the proof.

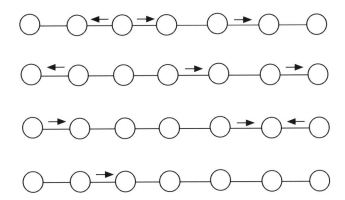

FIGURE 17.4   An illustration of convergence of the four-state algorithm.

17.4 STABILIZING GRAPH COLORING

Graph coloring is a classical problem in graph theory. Given an undirected graph $G = (V, E)$ and a set of colors C, node coloring defines a mapping from $V \to C$ such that no two adjacent nodes have the same color. In this section, we focus on designing a stabilizing algorithm for coloring the nodes of a *planar graph* using at the most *six colors*.

Section 10.4 illustrates a distributed algorithm for coloring the nodes of any *planar graph* with at the most *six* colors, but the algorithm is not stabilizing. In this section, we present the stabilizing version of it. Readers should review this algorithm before studying the stabilizing version.

The algorithm in Section 10.4.2 has two components. The first component transforms the given planar graph into a directed acyclic graph (dag) for which $\forall i \in V : outdegree(i) \leq 5$. The second component performs the actual coloring on this dag. Of the two components, the second one is stabilizing, since no initialization is necessary to produce a valid node coloring. However, the first one is not stabilizing, since it requires specific initialization (all edges were initialized to the state undirected). As a result, the composition of the two components is also not stabilizing. Our revised plan here has two goals:

1. Design a *stabilizing* algorithm A that transforms any planar graph into a dag for which the condition $P \equiv \forall i \in V : outdegree(i) \leq 5$ holds.

2. Use the dag-coloring algorithm B from Section 10.4 such that the desired postcondition Q reflecting a valid six-coloring holds.

If the actions of B do not negate any enabled guard of A, we can use the idea of *convergence stairs* [GM91] and run the two components concurrently to produce the desired coloring. We revisit algorithm B for coloring the dag. Recall that

- The color palette $C = \{0,1,2,3,4,5\}$

- $c(i)$ denotes the color of node i

- $succ(i)$ denotes the *successors* of a node i

- $sc(i) = \{c(j) : j \in succ(i)\}$

The following coloring algorithm from Section 10.4 works on a dag G' for which $\forall i \in V : outdegree(i) \leq 5$, and produces a valid six-coloring of the dag.

```
program colorme;
{program for node i}
do    ∃j ∈ succ(i):c(i)=c(j)→c(i):=b:b∈{C\sc(i)}  od
```

To recall how it works, note that the leaves of the dag have stable colors. So after at most one round, the predecessors of the leaf nodes will have a stable color. After at most two rounds, the predecessors of the predecessors will attain a stable color. In at most $|V|$ rounds, all nodes will be appropriately colored.

However, the crucial issue here is the generation of a dag G' that satisfies the condition $P \equiv \forall i \in V : outdegree(i) \leq 5$. To find a stabilizing solution, assume that initially the edges of G may be oriented in an arbitrary way (instead of being undirected), and the original algorithm of Section 10.4.2 is not designed to handle it! So we present a *stabilizing* algorithm for dag generation.

Dag-generation algorithm: To represent the edge directions, let us introduce a nonnegative integer variable $x(i)$ for every node i. Let i and j be the ids of a pair of neighboring nodes. The relationship between x and the edge directions is as follows:

- $i \rightarrow j$ iff, $x(i) < x(j)$ or $x(i) = x(j)$ and $i < j$

- $j \rightarrow i$ otherwise

Let $sx(i) = \{x(j) : j \in succ(i)\}$. Then regardless of the initial values of x, the following algorithm A will generate a dag that satisfies the condition $\forall\, i \in V : outdegree(i) \leq 5$:

```
program dag;
{program for process i}
do  |succ(i)|>5 → x(i):=max(sx(i))+1  od
```

As in the coloring algorithm, we assume large grain atomicity, so that the maximum element of the set $sx(i)$ is computed atomically. The proof of correctness relies on Euler's polyhedron formula (see [H69]), which was introduced in Chapter 10. Corollary 10.1 established earlier that every planar graph has at least one node with degree ≤ 5. It remains to show that given an arbitrary planar graph and no initial edge direction specified, the dag-generation algorithm guarantees convergence to a global configuration in which the condition $\forall i \in V : outdegree(i) \leq 5$ holds.

Lemma 17.8

The dag-generation algorithm stabilizes to a configuration in which the condition $\forall i \in V : outdegree(i) \leq 5$ holds.

Proof (by contradiction): The partial correctness is trivial, that is, if the algorithm terminates, then the condition $\forall i \in V : outdegree(i) \leq 5$ will hold. So we focus on termination only. Assume that the algorithm does not terminate. Then there is at least one node j that makes infinitely many moves. Every move by j directs *all* its edges toward j. Therefore, between two consecutive moves by node j, at least six nodes in $succ(j)$ must make moves. Furthermore, if j makes infinitely many moves, then at least six nodes in $succ(j)$ must also make infinitely many moves. Since the number of nodes is finite, it follows that (1) there exists a subgraph in which every node has a degree >5 and (2) the nodes of this subgraph make infinitely many moves. However, since every subgraph of a planar graph is also a planar graph, and there is no planar graph in which the degree of every node is >5 (Corollary 10.1), condition leads to a contradiction. ■

The combination of the two programs is straightforward. In program *colorme*, it may be impossible for node i to choose a value for $c(i)$ when $|succ(i)| > 5$, since the set $C\backslash sc(i)$ may become empty. To avoid this situation, strengthen the guard to $\big(|succ(i)| > 5\big) \wedge \big(\exists j \in succ(i) : c(i) = c(j)\big)$, without changing the corresponding action. The final version is shown as follows:

```
program six-coloring;
{program for node i}
do    {Component A: dag generation action}
| succ(i)|>5 → x(i):=max(sx(i))+1
{Component B: coloring action}
[] (|succ(i)|≤5) ∧ (∃j ∈ succ(i): c(i) = c(j)) → c(i):=b:b ∈ {C \sc(i)}
od
```

Since the predicate of component A is closed under the actions of component B, the concurrent execution of these actions will disable all guards, leading to a configuration that satisfies

$$P \equiv \forall i \in V : |\,succ(i)\,| \leq 5$$

and

$$Q \equiv \forall i, j \in V : j \in succ(i), \quad c(i) \neq c(j)$$

A drawback of the aforementioned solution is the unbounded growth of x.

17.5 STABILIZING SPANNING TREE PROTOCOL

Spanning trees have important applications in routing and multicasts. When the network topology is static, a spanning tree can be constructed using the probe-echo algorithm from Section 10.3.1 or some other algorithm. However, in real life, network topology constantly changes as nodes and links fail and come up, or their states are corrupted, affecting the spanning tree. Following such events, the spanning tree has to be reconstructed to maintain service.

In this section, we present a stabilizing algorithm for constructing a spanning tree, originally proposed by Chen et al. [CYH91]. The algorithm works on a connected undirected graph $G = (V, E)$ and assumes that failures do not partition the network. Let $|V| = n$. A distinguished process r is chosen as the root of the spanning tree. Each process picks a neighbor as its *parent P*, and we denote it by drawing a directed edge from i to $P(i)$. By definition, the root does not have a parent. The corruption of one or more P-values can create a cycle. To detect a cycle and restore the graph to a legal configuration, each process uses an integer variable $L (0 \leq L \leq n)$. Call L the *level* of a process—it defines its distance from the root via the tree edges. By definition, $L(r) = 0$. Denote the set of neighbors of process i by $N(i)$. In a legitimate configuration, (1) $\forall i \in V : i \neq r :: L(i) < n$ and (2) $L(i) = L(P(i)) + 1$.

Once the configuration becomes illegal, all processes (except the root) execute actions to restore the spanning tree. One action preserves the invariance of $L(i) = L(P(i)) + 1$ regardless of the integrity of the relation P. However, if $L(P(i)) \geq n - 1$ (which reflects something is wrong), then process i sets $L(i)$ to n, looks for a neighbor j such that $L(j) < n - 1$, and chooses j as its new parent. Note that the condition $L(i) = n$ is possible either due to a corruption of the value of $L(i)$ or due to the directed edges forming a cycle. Figure 17.5a shows a spanning tree, and Figure 17.5b shows the effect of a corrupted value of $P(2)$—a cycle is formed consisting of the nodes 2, 3, 4, 5 and the directed edges connecting them with their parents. Because of this cycle, when each node i updates $L(i)$ to $L(P(i)) + 1$, the values of $L(2), L(3), L(4), L(5)$ eventually reach the maximum value $n = 6$. For recovery, a node i with $L(i) = 6$ will look for a neighbor with level <5 to designate it as its new parent. In our example, let node 2 choose node 1 as $P(2)$ (since $L(1) = 1$). Following this, nodes 3, 4, 5 update their levels, and the stabilization is complete.

```
program stabilizing spanning tree;
{program for each node i ≠ r}
do  (L(i)≠n)∧(L(i)≠L(P(i))+1)∧(L(P(i))≠n) →L(i):=L(P(i))+1        (1)
[]  (L(i)≠n)∧(L(P(i)=n)) →L(i):=n                                  (2)
[]  (L(i)=n)∧(∃k∈N(i):L(k)<n−1) →P(i):=k;L(i):=L(k)+1              (3)
od
```

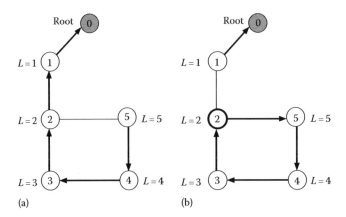

(a)   (b)

FIGURE 17.5   (a) A spanning tree with correct values of L and P for each node. (b) $P(2)$ is corrupted and a cycle is created. The edge (2, 5) is not well formed.

Correctness proof

We follow the arguments from [CYH91]. Define an edge from i to $P(i)$ to be *well formed*, when $L(i) \neq n$, $L(P(i)) \neq n$, and $L(i) = L(P(i)) + 1$. In any configuration, the nodes and the well-formed edges form a spanning forest. Delete all edges that are not well formed and designate each tree T in the forest by $T(k)$, where k is the smallest value of the level of a node in the tree. A single node with no well-formed edge incident on it represents a degenerate tree. In Figure 17.5b, there are two trees: $T(0) = \{0, 1\}$ and $T(2) = \{2, 3, 4, 5\}$.

We now examine how these multiple trees combine into a single spanning tree via the actions of the algorithm. Define a tuple $F = (F(0), F(1), F(2), ..., F(n))$, where $F(k)$ counts the number of $T(k)$s in the spanning forest. In Figure 17.5b, $F = (1, 0, 1, 0, 0, 0)$. Define a lexicographic order (>) between a pair of tuples F and F' as follows:

$$F > F' \equiv F(0) > F'(0)$$

or

$$\exists j > 0 : F(j) > F'(j) \wedge \forall i < j : F(i) = F'(i)$$

With n nodes, the maximum value of F is $(1, n - 1, 0, 0, ..., 0)$ and the minimum value is $(1, 0, 0, ..., 0)$ that represents a single spanning tree rooted at node 0.

The important thing to observe here is that with each action of the algorithm, F decreases lexicographically. With action 1, node i combines with an existing tree, so $F(i)$ decreases, but no other component $F(j), j < i$ decreases in value. With action 2, node i becomes a new tree with a single node, $F(n)$ increases, but $F(L(i))$ decreases, and $F(L(j)), L(j) > L(i)$ may increase. Therefore, F decreases. Finally, with action 3, $F(n)$ decreases, but all other components of F remain unchanged. So F decreases.

With the repeated application of the three actions, F decreases monotonically until it reaches the minimum value $(1, 0, 0, ..., 0)$, which represents a single spanning tree rooted at node 0. ■

17.6 STABILIZING MAXIMAL MATCHING

Matching is a well-studied problem in graph theory. Given an undirected connected graph $G = (V, E)$, a matching M is a subset of the set of edges E, such that no two edges in M share a common node. A matching M is called *maximal*, if no other edge can be added to the set M. If $(i, j) \in M$, then call the nodes i and j *partners* of each other. To compute maximal matching, each node i maintains a variable *partner* whose value has to be chosen from its neighbor set $N(i)$, or it can be $\bot$; the latter indicates that node i does not have a partner. $(i, j) \in M$ implies $partner(i) = j$ (denoted by $i \rightarrow j$) and $partner(j) = i$ (denoted by $j \rightarrow i$). In an arbitrary initial configuration, the values of the partner variables may be arbitrary. A stabilizing maximal matching algorithm should lead the system from an arbitrary initial configuration to a configuration in which the set M is correctly generated and becomes maximal. We present here an algorithm due to Hsu and Huang [HH92].

The state $S(i)$ of a node i can belong to one of the following five categories:

1. $S(i) = waiting \Rightarrow (i \rightarrow j) \land (j \rightarrow \perp)$.

2. $S(i) = matched \Rightarrow (i \rightarrow j) \land (j \rightarrow i)$.

3. $S(i) = chaining \Rightarrow (i \rightarrow j) \land (j \rightarrow k) \land (k \neq i)$.

4. $S(i) = dead \Rightarrow i \rightarrow \perp$, but every neighbor of node i is in the matched state.

5. $S(i) = free \Rightarrow i \rightarrow \perp$, but at least one neighbor of node i is not matched.

In a legal configuration, the state of each node is either *matched* or *dead*. The stabilizing maximal matching algorithm is based on three rules.

```
program stabilizing maximal matching
{program for process i}
do
(i → ⊥) ∧ (∃k∈N(i): k → i) → i chooses k                              (1)
(i → ⊥) ∧ (∀j∈N(i): j did not choose i) ∧ (∃k∈N(i): k → ⊥) → i chooses k  (2)
(i → j) ∧ (j → k) ∧ (k ≠ i) → i rejects j and chooses ⊥               (3)
od
```

Hsu and Huang [HH92] uses a *variant function* to prove convergence to a legal configuration. First of all, observe that if the system is not in a legal configuration, then at least one of the guards in the algorithm must be true. Define c, d, f, m, w to be the number of nodes in the states *chaining, dead, free, matched,* and *waiting*, respectively. Choose the tuple $F = (m + d, w, f, c)$ as the variant function. The claim is that F will always increase lexicographically after each of the actions 1–3 of the algorithm, until a legal configuration is reached:

- Action 1 matches i with j—this increases the value of m by 2. It is possible that due to this action, some neighbors of i or j change their state from *free* to *dead*. As a result, d never decreases, although f can decrease, so F increases.

- Action 2 changes the state of i from *free* to *waiting*, so w increases by 1, while f decreases by 1. Therefore, F increases.

- Action 3 has two possible consequences: (a) If initially $k \rightarrow l$ ($l \neq j$ or i), then i changes its state from *chaining* to *dead* or *free* (so c decreases by 1, but d or f increases by 1). Therefore, F increases. (b) If initially $k \rightarrow i$, then the state of k changes from *chaining* to *waiting*, but the state of i changes from *chaining* to *free* (so c decreases by 2, but w or f increases by 1). Therefore, F increases.

The maximum value that F can attain is $(n, 0, 0, 0)$—at this state, the system is in a legal configuration. Since the closure property is trivial, the proposed algorithm for maximal matching is stabilizing.

17.7 DISTRIBUTED RESET

Reset is a general technique for restoring a distributed system to a predefined legal configuration from any illegal configuration caused by data corruption or coordination loss. While distributed snapshot correctly *reads* the global state of a system, distributed reset correctly *updates* the global state to a designated value without freezing the application. Reset can be used as a general tool for designing stabilizing systems in the following way: Any process can take a snapshot of the system from time to time, and if the global state turns out to be illegitimate, the process can *reset* the system to a legal configuration.

To realize the nontriviality of distributed reset, consider that a system of n processes, each with an integer variable x. Let $x[i]$ be the value of the x variable of process i. Consider resetting the system to a global state $\forall i : x[i] = 0$. Since the operation is not atomic, first, let one process reset x to 0. By the time another process sets its x to 0, the value of x of the previous process may have changed based on messages received by its from other processes. By the time the last process has reset its x to 0, the states of the earlier processes may have changed in arbitrary ways. Whether the resulting global state is reachable from the reset state via a legal computation is debatable.

In [AG93], Arora and Gouda proposed an algorithm for distributed reset on a system of n processes $0, 1, 2, \ldots, n-1$. They assume that in addition to data memory corruption, processes can fail by stopping and fail-stop failures can be detected, but process failures do not partition the system. The system topology is defined in terms of nonfaulty nodes only. The components of a reset system consist of two layers of actions, which are distinct from the main application (called the application layer). The two layers are defined as follows:

Spanning tree layer: Processes first elect a leader, which is a nonfaulty process with the *largest id*. With this leader as the root, a spanning tree is constructed. All communications in the reset system take place via the tree edges only.

Wave layer: Processes detect inconsistencies and send out requests for reset. These requests propagate to the root along the tree edges. The root subsequently sends a reset wave down the tree edges to reset the entire system.

The wave layer does not influence the progress of the spanning tree construction, so the *convergence stairs* rule guarantees that the concurrent execution of the action of these two layers guides the reset actions to eventually reach the nodes at the correct time. The actions of the individual layers are described in the following:

Spanning tree layer protocol: This protocol is very similar to the stabilizing spanning tree construction protocol described in the previous section, with the added factor that it also guarantees that the process with the largest id is selected as the root. Each process i has three variables:

1. $root(i)$ represents the root of the tree, and in a legal configuration, $root(i) \geq 1$.

2. $P(i)$ represents the parent of node i. Its range is $\{i\} \cup N(i)$, where $N(i)$ is the set of neighbors of i. If i is the root node, then $P(i) = i$.

3. $d(i)$ represents process i's shortest distance from the root. When i is the root node, $d(i) = 0$, and in a legal configuration, $d(i) = d(P(i)) + 1$. In a well-formed tree, $\forall i : d(i) \leq n - 1$, so $d(i) > n - 1$ will indicate a faulty value of $d(i)$.

The program for the spanning tree layer is as follows:

```
{Spanning tree layer: program for process i}
do  (root(i) < i)∨
    P(i) = i ∧ (root(i) ≠ i∨d(i) ≠ 0)∨
    P(i) ∉ {i} ∪ N(i) ∨ d(i) ≥ n  →root(i) := i;  P(i) := i;  d(i) = 0

[]  (d(i) < n) ∧ (P(i) = j) ∧ (j ∈ N(i))∧
    (root(i) ≠ root(j) ∨d(i) ≠ d(j)+1)  →root(i) := root(j);  d(i) := d(j) +1

[]  ∃j ∈ N(i):(root(i) < root(j) ∧ d(j) < n)∨
    (root(i) = root(j) ∧ d(i) > d(j) + 1)  →root(i) := root(j);  P(i) := j;
      d(i) := d(j) + 1
od
```

The first rule resolves local inconsistencies. If there is a cycle in the initial graph, then the repeated application of $d(i): = d(P(i)) + 1$ bumps $d(i)$ up to n that enables the first guard. This can also happen due to a bad initial value of $d(i)$. The corresponding action prompts the process to designate *itself* as the root, and the cycle is broken if there is one. The second rule enforces the condition $d(i): = d(P(i)) + 1$ and corrects the value of $root(i)$. Finally, the third action sets the value of $d(i)$ to the shortest distance of node i from the root.

Wave layer protocol: The wave layer protocol uses the spanning tree to perform the reset operation. Once a node locally detects an illegal configuration, it initiates a request for reset that propagates up the spanning tree edges to the root node. The root, upon receiving such a request, initiates a reset command that propagates down the spanning tree to the leaf nodes. Each process *I* has a state variable $s(i)$ that can have three possible values: *normal, initiate, and reset*:

- $s(i) = normal$ represents the failure-free mode of operation for process i.

- $s(i) = initiate$ indicates that process i has requested a reset operation. The request eventually propagates to the root, which then starts the reset wave.

- $s(i) = reset$ implies that process i is now resetting the application program. The root, upon receiving the *initiate* command, initiates a reset command that propagates via the children toward the leaves of the spanning tree. Following the completion of the reset operation, the leaf nodes return to the *normal* state, and their parents follow suit, so that all the processes return to their normal states.

The wave layer uses a nonnegative integer variable *seq* that tags a specific round in the reset operation. The steady state of the wave layer satisfies the predicate *steady(i)* for every i (Figure 17.6), which corresponds to the following two conditions:

- $s(P(i)) \neq reset \Rightarrow s(i) \neq reset \wedge seq(P(i)) = seq(i)$

- $s(P(i)) = reset \Rightarrow s(i) \neq reset \wedge seq(P(i)) = seq(i) + 1 \vee seq(P(i)) = seq(i)$

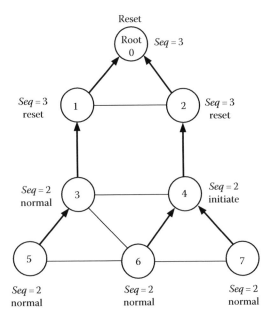

FIGURE 17.6 A steady state of the wave layer: The reset wave with *seq* = 3 has made partial progress. Node 4 now initiates a fresh request for reset.

The program for process *i* in the wave layer is as follows:

```
(Wave layer: program for process i}
do  s(i)= normal ∧∃j ∈N(i):P(j)= i ∧s(j) = initiate → s(i):= initiate
[]  s(i) := initiate ∧ P(i) = i → s(i):= reset; seq(i):=seq(i)+1;
                           Reset the application state at node i
[]  s(i)≠reset ∧ s(P(i)) = reset ∧ seq(P(i)) = seq(i)+1 →
                                s(i):= reset; seq(i):= seqP(i);
                           Reset the application state at node i
[]  s(i)=reset ∨∀j ∈N(i): P(j) = i: s(j) ≠ reset ∧ seq(i) = seq(j) →
                                s(i):= normal
[]  ¬steady(i) →          s(i):= s(P(i)); seq(i):= seq(P(i))
od
```

While the first four actions implement a diffusing computation, the last action restores the wave layer to a steady state. In the example configuration of Figure 17.6, the reset will be over when the condition $\forall i : seq(i) = 3 \wedge s(i) = normal$ holds. The proof of correctness of the composition of the two layers relies on the notion of convergence stairs and is formally described in [AG93].

Application layer protocol: The application layer of process *i* is responsible for requesting the distributed reset operation[*] by setting *s(i)* to *reset*. Any node *i* for which *s(i)* = *reset* resets the application at that node. In addition, the protocol needs to ensure that when the reset operation is in progress, no communication takes place between the nodes that have been reset

---

[*] We assume that the application periodically runs a detection algorithm to decide this.

in the current round and the nodes that are yet to be reset. To enforce this, a process j will accept a message from process i, only when $seq(i) = seq(j)$. The rationale is that the resetting of the local states of the two processes i and j must be concurrent events—however, if a message is allowed between i and j after i is reset but j is not, a causal chain (i is reset π i sends a message to j π j receives it $\prec$ j is reset) can result between the two local resets.

The role of seq is to distinguish between nodes that have been reset and nodes that are yet to be reset during the current round of the wave layer. While the proposed algorithm reflects an unbounded growth of seq, it is sufficient to limit the range of seq to $\{0, 1, 2\}$ and replace the operation $seq(i) := seq(j) + 1$ in the wave layer by $seq(i) := seq(j) + 1 \bmod 3$.

17.8 STABILIZING CLOCK PHASE SYNCHRONIZATION

We revisit the problem of synchronizing the phases of a system of clocks ticking in unison. The goal is to develop a protocol that will guarantee that regardless of the starting configuration, the phases of all the clocks eventually become identical and remain identical thereafter. Recall that Chapter 6 describes such a protocol—it is a stabilizing protocol for synchronizing the phases of an *array of clocks* ticking in unison (Section 6.5). The protocol uses three-valued clocks, and each clock ticks as 0, 1, 2, 0, 1, 2, ... ($O(1)$ space complexity means that the solution is scalable). The solution has subsequently been generalized to work on any *acyclic network* of clocks, as described in [HG95]. Such a solution has two drawbacks: (1) It uses *coarse-grained atomicity*, which implies that in one atomic action, a process can read the states of all its neighbors regardless of the size of the neighborhood (clearly this is unrealistic), and (2) the solution fails to work on cyclic topologies. In this section, we present a different protocol for synchronizing the phases of a network of clocks. This protocol, originally described by Arora et al. in [ADG91], stabilizes the system of clocks under the weaker system model of *fine-grained atomicity*: In each step, every clock in the network checks the state of only one of its neighbors and updates its own state if necessary. Note that as a process reads the state of one neighbor, the state of another neighbor may change. Apart from working under fine-grained atomicity, another advantage of the protocol is that it works on cyclic topologies too, although the space complexity per process is no more constant.

Consider a network of clocks whose topology is an undirected graph. Each clock c is an m-valued variable in the domain $0 .. m - 1$. Let $N(i)$ denote the set of neighbors of clock i and $\delta = |N(i)|$ represent the degree of node i. Define a function $next$ on the neighborhood of a process i: in each step, $next N(i)$ will return a new neighbor id, and δ successive applications of $next$ return the starting neighbor. Each clock i executes the following protocol:

```
program clock synchronization
{program for clock i, j is a neighbor of i}
do   true → c(i):= min{c(i), c(j)} + 1 mod m;
          j:= next{N(i)}
od
```

Each clock i will scan the states of its neighbors in a cyclic order and *fall back* with the value of clock j if $c(i) > c(j)$; otherwise it will keep ticking as usual.

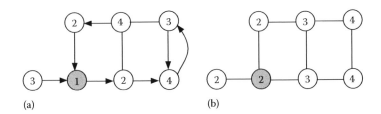

FIGURE 17.7 A sample step of the [ADG91] clock synchronization protocol: (a) initial configuration—an arrow from j to k indicates that clock j is now comparing its value with clock k. (b) The updated clock values after one step.

To understand the convergence mechanism of the protocol, consider a clock j whose value $c(j)$ is the *smallest* of all clock values in the current configuration. We will call j an *anchor* node. Note that the anchor node may not be unique. In each step, $c(j)$ is incremented, and in at most δ steps, each neighbor of j sets its clock to $c(j)$. Figure 17.7 shows a sample step of the protocol.

It *appears* that eventually every clock k will set $c(k)$ to $c(j)$ and the system will stabilize. The argument is sound if the clocks are unbounded. However, with bounded clocks, there is one catch—if the value of $c(i)$ for some clock i reaches the maximum value $(m - 1)$, then at the next step, it may roll back from $(m - 1)$ to 0. As a result, the neighbors of clock i will not follow previous anchor node j but start following clock i, throwing the issue of stabilization wide open.

One can argue that from now on, clock i will act as the *new anchor* until all clocks set their values to that of clock i. But in order to prevent the occurrence of the previous scenario before the system of clocks stabilizes, the condition $m > 2\delta_{max}D$ must hold, where δ_{max} is the maximum degree of a node and D is the diameter of the graph. This guarantees that the clock *farthest* from the anchor clock will be able to set its value to that of the anchor *before* any clock rolls over to 0. The stabilization time for this protocol is $3\delta_{max}D$. A formal proof of the protocol and the analysis of its time complexity are available in [ADG91].

17.9 CONCLUDING REMARKS

In modern large-scale distributed systems, the principle of spontaneous recovery without external help has received significant attention. Such systems are supposed to take care of themselves as far as possible. A number of related ideas conforming to the basic principle of self-management are becoming popular: These include *self-organization*, *self-optimization*, and *self-healing*. The distinction among these different concepts is somewhat blurry. Stabilization guarantees recovery from arbitrary initial states and handles the corruption of *all state variables*—in contrast, self-organizing systems appear to restore the system functionality when processes join and leave a distributed system. A distributed system may be self-organizing but may not be self-stabilizing: An example is the P2P network Chord [SML + 02]. A collection of such self-management properties will enable computer systems to spontaneously regulate themselves much in the same way our autonomic nervous system regulates and protects our bodies. This is the essence of *autonomic computing*. Mimicking the behavior of autonomic nervous systems calls for a large assortment of techniques. For example, systems that are prone to crash frequently can be given a longer lease of life, via proactive microreboot [CKF + 04] of the critical components prior to the anticipated crash.

Katz and Perry [KP89] described a generic method of transforming any distributed system into a stabilizing one using the following approach: (1) Start with a traditional design, (2) allow a node to collect the global state and check whether the configuration is legal or not, and (3) if the configuration is illegal, reset the system to a legal configuration. This approach takes into account the possibility of additional failures during the steps (2) and (3) by taking repeated snapshots. As long as such failures are not perpetual, eventually the system is restored to a legal configuration.

An interesting observation in many stabilizing systems is that even if a single process undergoes a transient failure, a large number of processes may be corrupted before the system recovers to a legal configuration. As an example, in the spanning tree protocol, the corruption of the *parent* variable of a *single* process affects the routing performance of a large number of processes, and the recovery can take $O(n^2)$ time. Ideally, recovery from a single failure should be local and should take $O(1)$ time. This guarantee is known as *fault containment* and is a desirable property of all stabilizing systems. Another closely related refinement of the stabilization property is *superstabilization*—a superstabilizing algorithm guarantees fast recovery from a single change in the network topology, like the addition or the deletion of an edge or node in the network. The implementation of fault containment or superstabilization needs extra work.

Finally, although stabilizing systems guarantee recovery, it is impossible to detect whether the system has reached a legal configuration, since any detection mechanism itself is prone to failures leading to a false conclusion. Nevertheless, the ability of spontaneous recovery from state variable corruption, changing network topology, or changing environments makes stabilizing and adaptive distributed systems an important area of study in fault-tolerant distributed systems.

17.10 BIBLIOGRAPHIC NOTES

The paper [D74] by Dijkstra initiated the field of stabilization. For this work, Dijkstra received the most influential paper award from the Principles of Distributed Computing (PODC) in 2002, and in the same year, the award was named after him. Lamport in his PODC invited address of 1982 [L82] described this work as "Dijkstra's most brilliant work." The paper has three algorithms for stabilizing mutual exclusion, of which the first one appears in this chapter. The modification of his second algorithm is due to Ghosh [G93]. Dijkstra [D74] contains a third algorithm that works on a bidirectional ring and uses three states per process. Dijkstra furnished a proof of it in [D86]. The planar graph-coloring algorithm follows the work of Ghosh and Karaata [GK93]. Chen et al. [CYH91] presented the spanning tree construction algorithm. The maximal matching algorithm is due to Hsu and Huang [HH92]. Arora and Gouda [AG94] proposed the algorithm for distributed reset. Arora et al. [ADG91] presented the stabilizing clock synchronization algorithm. Herman's PhD thesis [H91] provides a formal foundation of adaptive distributed systems—a summary appears in [GH91]. Ghosh et al. studied fault containment [GGHP96, GGHP07]—a more comprehensive treatment of the topic can be found in Gupta's PhD thesis [G96a]. Dolev and Herman [DH97] introduced superstabilization. Dolev's book on self-stabilization [D00] is a useful resource covering a wide range of stabilization algorithms.

EXERCISES

17.1 Revisit Dijkstra's mutual exclusion protocol on a unidirectional ring of size n and compute the maximum number of moves required to reach a legal configuration.

17.2 Consider Dijkstra's stabilizing mutual exclusion protocol on a unidirectional ring of size n with k states per process. Show that if $k = 2$ and $n = 4$, then the system may never converge to a legal configuration. To demonstrate this, you need to specify an initial configuration that is illegal and a sequence of moves that brings the system back to the starting configuration.

17.3 Consider a bidirectional ring of n processes numbered 0 through $n - 1$. Each process j has a right neighbor $(j + 1) \bmod n$ and a left neighbor $(j - 1) \bmod n$. Each process in this ring has three states 0, 1, 2. Let $s[j]$ denote the state of process j. These processes execute the following program:

```
{process 0}      do s[0] +1 mod 3 = s[1] → s[0]:= s[0] -1 mod 3 od
{process n - 1}  do s[0] = s[n - 2] ∧ s[n - 2] ≠ s[n - 1] - 1 mod 3 →
                        s[n - 1]:= s[n - 2] + 1 mod 3 od
{process j:0 < j < n - 1}
                 do   s[j] +1 mod 3 = s[j + 1] → s[j]:= s[j + 1]
                 []   s[j] +1 mod 3 = s[j - 1] → s[j]:= s[j - 1]
                 od
```

a. Show that the aforementioned protocol satisfies all the requirements of stabilizing mutual exclusion.

   (See [D86] for a proof of correctness but try to do your own analysis first.)

b. Show that the worst-case stabilization time of the aforementioned protocol is $O(n^2)$.

17.4 Figure 17.8 shows a network of four processes. The state $s[i]$ of each process i is either 0 or 1.

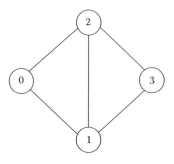

FIGURE 17.8   A network of four processes: Each process has two states 0, 1.

These processes execute the following programs:

```
{process 0}: do s[0],s[1] = (1,0) → s[0]:= 0 od
{process 1}: do s[1],s[2],s[3] = 1,1,0 → s[1]:= 0 od
{process 2}: do s[0],s[1],s[2] = 1,1,0 → s[2]:= 1 od
{process 3}: do s[2],s[3] = 1,1 → s[3]:= 0 od
```

Verify if the system of processes implements a stabilizing mutual exclusion protocol. What is the behavior of the system in a legal configuration? [See G91]

17.5 Design a stabilizing algorithm for electing a leader on a unidirectional ring of processes. Processes have distinct ids. Each process k maintains a leader variable $L(k)$. In a legal configuration, $L(k)$ = max, where max is the largest id of a nonfaulty process existing in the ring. Note that the values of $L(k)$ in different processes may be badly initialized—this includes the ids of nonexistent processes.

Provide a correctness proof to justify why your solution will work.

17.6 In an undirected graph $G = (V, E)$, the *eccentricity* of a node $i \in V$ is the maximum shortest path distance of i from any other node of the graph. A node with minimum eccentricity is called a *center* of G.

Design a stabilizing algorithm for finding the *center of a tree*. Whenever the topology of the tree changes, the center will spontaneously shift itself to the appropriate node. Your algorithm should start from an arbitrary initial state and reach a fixed point that satisfies the unique predicate for the center. Remember that sometimes a tree can have two centers. (See [BGK+99].)

17.8 Given a strongly connected directed graph and a designated node k as the root, design a self-stabilizing algorithm to generate a BFS tree. In a legal configuration, each nonroot node i must identify a *parent* node $p(i)$ from its neighborhood $N(i)$ such that the node i can send a message to the root using the smallest number of hops. In an arbitrary initial configuration, these parent variables may be badly initialized.

17.9 Consider a linear array of processes numbered 0 through $n - 1$. Let $s[j]$ denote the state of process j and $N(j)$ denote the neighbors of node j. By definition, each process has only two states: $s[j] \in \{0, 2\}$ if j is *even*, and $s[j] \in \{1, 3\}$ if j is odd. Starting from an arbitrary initial state, each process j executes the following program:

do $\quad \forall k: k \in N(j): s[k] = s[j] + 1 \bmod 4 \rightarrow s[j] := s[j] + 2 \bmod 4$ **od**

If processes execute their actions in lock-step synchrony (i.e., at each step, every process with an enabled guard executes its action), then describe the steady-state behavior of the system.

(Hint: Observe that in the steady state, (1) no two neighboring processes execute their actions at the same time and (2) maximum parallelism is reached, that is, at least $\lfloor n/2 \rfloor$ processes will change their states simultaneously. Argue why this will be so.)

17.10 Let $G = (V, E)$ represent the topology of a network of wireless terminals: Here, V is a set of nodes representing terminals, and E is the set of edges between them. An edge between a pair of terminals indicates that one is within the communication range of the other and such edges are bidirectional. The local clocks of these terminals are perfectly synchronized.

One problem in such a system is the collision in the MAC layer, which is traditionally avoided using the RTS-CTS signals. Now consider a different solution of collision avoidance using TDMA: Let processes agree to a cyclic order of time slots of equal size, numbered 0, 1, 2, ..., $\delta - 1$, where δ is the *maximum* degree of a node. The goal now is for each process to pick up a time slot that is distinct from the slots chosen by its neighbors, so that across any edge, only one terminal can transmit at any time.

Design a stabilizing TDMA algorithm for assigning time slots to the different nodes, so that no two neighboring nodes have the same time slot assigned to it. Explain how your algorithm works and provide a proof of correctness.

17.11 Revisit the clock phase synchronization protocol due to Arora et al. [ADG91]. Illustrate with an example that single faults are not contained, that is, a single faulty clock can force a large number of nonfaulty clocks to alter their states to incorrect values before stability is restored.

17.12 Some people believe that the property of stabilization is too demanding, since it always requires the system to recover from *arbitrary* configurations and no assumption can be made about initial state. Consider softening this requirement where we may allow a fraction of the variables to be initialized, while the others may assume arbitrary values. Call it *assisted* stabilization.

Show that Dijkstra's stabilizing mutual exclusion protocol on a unidirectional ring of size n will eventually return the ring to a legal configuration when (1) there are $(n - k + 1)$ states (numbered 0 through $n - k$) per process, (2) the states of the processes 0 through k are initialized to 0, and (3) the remaining processes start from arbitrary initial states.

V

Real-World Issues

CHAPTER 18

Distributed Discrete-Event Simulation

18.1 INTRODUCTION

Simulation is a widely used tool for monitoring and forecasting the performance of real systems. Before building the system, experiments are performed on the simulated model to locate possible sources of errors, which saves money and effort, prevents possible malfunction, and boosts confidence. In the area of networking, network simulators are very popular and widely used. VLSI designers use circuit simulation packages like SPICE to study the performance before actually building the system. Sensitive safety-critical applications invariably use simulation, not only to verify the design but also to find ways to deal with catastrophic or unforeseen situations. Airlines use flight simulators to train pilots. Most real-life problems are large and complex, and they take an enormous amount of time when run sequentially on a single processor. Distributing the total simulation among several processors has the potential to reduce the run time. Although it is conceptually easy, distributed simulation poses interesting synchronization challenges. In this chapter, we will study some of these challenges.

18.1.1 Event-Driven Simulation

We distinguish between the real system (often called the *physical system*) and the *simulated version* that is expected to mimic the real system. The correctness of simulation requires that the simulated version preserve the temporal order of events in the physical system. Considering the physical system to be message based, a typical event is the sending of a message m at a time t. In the simulated environment, we will represent it by a pair (t, m). The fundamental correctness criteria in simulation are as follows:

Criterion 1: If the physical system produces a message m at time t, then the simulated version must schedule an event (t, m).

Criterion 2: If the simulated version schedules an event (t, m), then the physical system must generate the message m at time t.

Central to an event-driven simulation are three data components: (1) a *clock* that tracks the progress of simulation, (2) the *state variables* that represent the state of the physical system, and (3) an *event list* that records pending events along with their start and finish times. Initially, *clock* = 0, and the event list contains those events that are ready to be scheduled at the beginning. Events are added to the event list when the completion of an event triggers new events. Similarly, events are deleted from the event list, when the completion of the current event disables another anticipated event originally scheduled for a later point of time. The following loop is a schematic representation of event-driven simulation:

```
{Event-driven simulation: basic steps of a process}
do  <termination condition> = false →
    Simulate event (t, m)with the smallest time t from the event list;
    Update the event list (by adding or deleting events, as appropriate);
    clock: = t;
od
```

As an example of discrete-event simulation, consider Figure 18.1. The customers enter the bank through door 1. A receptionist (*A*) of a bank receives customers *C*1, *C*2, *C*3, ... and directs them to one of two tellers *W*1 and *W*2 who is not busy. If both tellers are busy, then the customer waits in a queue.

Each customer takes a fixed time of 5 min to complete the transactions. Eventually, they leave via door 2. Assuming three customers *C*1, *C*2, and *C*3 arrive at *A* at times 1, 3, and 5, the events till time 6 are listed in the Table 18.1. Here, the notation *Cn* @ *v* will represent the fact customer *Cn* is currently at node *v*.

As another example, consider the following events in a battlefield. An enemy convoy is scheduled to cross a bridge at 9:30 a.m., and an aircraft is scheduled to bomb the bridge and

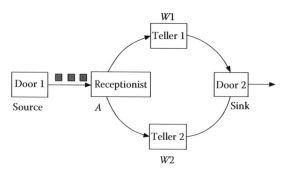

FIGURE 18.1   Two bank tellers in a bank.

TABLE 18.1   Partial List of Events in the Bank

No.	Time	Event	Comment
1	1	*C*1 @ *A*	
2	1	*C*1 @ *W*1	*W*1 will be busy till time 6.
3	3	*C*2 @ *A*	
4	3	*C*2 @ *W*2	*W*2 will be busy till time 8.
5	5	*C*3 @ *A*	Both tellers are busy.
6	6	*C*3 @ *W*1	Now, *C*3 is sent to *W*1.

destroy it at 9:25 a.m. on the same day. Unless we simulate the events in temporal order, we may observe the convoy crossed the bridge—something that will not happen in practice. This leads to the third requirement in event-driven simulation: The progress of simulation must preserve the temporal order of events. An acceptable way of enforcing the temporal order is to respect causality. Criterion 3 elaborates this:

Criterion 3 (Causality requirement) If $E1$ and $E2$ are two events in the physical system, and $E1 \prec E2$, then the event $E1$ must be simulated before the event $E2$.

Two events that are not causally ordered in the physical system can be simulated in any order. In sequential simulation, criterion 3 is trivially satisfied if the pending event with the lowest time is always picked for simulation. In distributed environments, timestamps can be used to prevent the violation of causal order. If $T1$ and $T2$ are the timestamps of the events $E1$ and $E2$, respectively, and $T1 < T2$, then $E2$ cannot be causally ordered before $E1$—so simulating events in timestamp order is a sufficient condition for preserving criterion 3.

18.2 DISTRIBUTED SIMULATION

18.2.1 Challenges

The task of simulating a large system using a single process is slow due to the large number of events that needs to be taken care of. For speedup, it makes sense to divide the simulation job among a number of processes. We will distinguish between *physical processes* (PPs) and LPs. The real system to be simulated consists of a number of PPs where events occur in *real time*. The simulated system consists of a number of *LP*—each LP simulates one or more PPs. In an extreme case, a single LP can simulate all the PPs—this would have been the case of centralized or sequential simulation. When the simulation is divided among several LPs, each LP simulates a partial list of events. An LP *LPi* simulates the events of its constituent PPs at its own speed, which depends on its own resources and scheduling policies. It may not have any relationship with the speed at which a different LP *LPj* simulates a different list of events. In this sense, distributed simulation is a computation on an *asynchronous* network of LPs.

The system of tellers in the bank can be modeled as a network of four processes $LP1$, $LP2$, $LP3$, $LP4$. Of these, $LP1$ simulates PP A, $LP2$ simulates PP $W1$, $LP3$ simulates PP $W2$, and $LP4$ simulates PP *sink* (Figure 18.2). Each *LP* maintains a local virtual clock that oversees the progress of the simulation of events assigned to it. The LP corresponding to the source is not shown. Figure 18.3 shows the event lists for the four LPs.

All correct simulations satisfy the following two criteria:

Realizability: The output of any LP at time t is a function of its current state and all messages received by it up to time t. No PP can guess the messages it will receive at a future time.

Predictability: For every LP, there is a real number $L(L > 0)$ such that the output messages (or the lack of it) can be predicted up to time $(t + L)$ from the set of input messages that it receives up to and including time t.

In the bank example, the predictability criterion is satisfied as follows: Given the input to teller 1 at time t, its output can be predicted up to time $(t + 5)$. Predictability is important

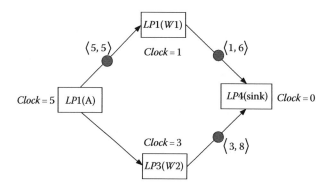

FIGURE 18.2   A network of four LPs simulating the events in the bank: The message output of each *LP* indicates the <start, end> times corresponding to the most recent event simulated by that *LP*.

Event list of receptionist (A)

No.	Start Time	Event	End Time
1	1	C1 @ A	1
2	3	C2 @ A	3
3	5	C3 @ A	5

Event list of teller 1 (W1)

No.	Start Time	Event	End Time
1	1	C1 @ W1	6

Event list of teller 2 (W2)

No.	Start Time	Event	End Time
1	3	C2 @ W2	8

Event list of sink

No.	Start Time	Event	End Time
1	6	C1 @ W1	6

FIGURE 18.3   List of events in the four LPs.

in those systems where there is a circular dependency among the PPs. As an example, consider the life of a process that alternates between the CPU and the I/O devices: A typical process spends some time in the CPU, then some time doing I/O, then again resumes the CPU operation, and so on. The corresponding network of LPs is shown in Figure 18.4. For simplicity, it assumes that the CPU, the I/O, the input switch *S1*, and output switch *S2* need *zero time* to complete their local tasks.

To make progress, each LP will expect input events with times *greater than* those corresponding to the previous event. Due to the realizability criterion, no LP can now make any progress, since the smallest input clock value of any *LP* does not exceed the local clock of that unit. The predictability criterion helps overcome this limitation. For example, if the *LP* simulating the switch *S1* can *predict* its output up to some time $(9 + \varepsilon)$, then it provides the *push* needed for the progress of the simulation.

Now, revisit the causality requirement in criterion 3. The *virtual clock T* of each *LP* is initialized to 0. The causality constraints within an LP are trivially satisfied as long as each

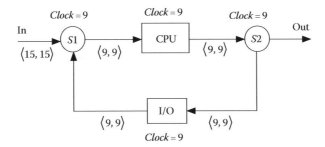

FIGURE 18.4   A network of *LPs* showing the life of a process alternating between the CPU and I/O.

LP schedules events in ascending order of its local virtual clock. To satisfy the causality requirement between distinct *LPs*, a logical clock value is assigned to each channel in the network of LPs. Consider a channel from *LPj* to *LPi*. Initially, for every channel, *clock* = 0. By sending a message ($t0$, $m0$), *LPj* makes *LPi* aware of all messages sent by *PPj* to *PPi* up to time $t0$.* As *LPj* makes progress and sends the next message ($t1$, $m1$), *LPi* interprets that *clock*(j, i) has been incremented from $t0$ to $t1$. The causality criterion for a single channel is satisfied by the following chronology requirement:

Chronology requirement: If a sequence of messages ($t0$, $m0$), ($t1$, $m1$), ($t2$, $m2$), … is sent by one LP to another across a channel, then $t0 < t1 < t2 < …$.

When an LP receives a message ($t1$, $m1$), it assumes that the corresponding PP has received all messages prior to $t1$. Only after an LP receives a message (t, m) through *every* incoming channel, it simulates the events of the physical system up to the *smallest channel clock time* among all (t, m) pairs through the incoming channels and sends out messages to other LPs. Thereafter, the LP updates its own virtual clock to the smallest clock value on its input channels. Only a source node can send a message without receiving any input message. Note that some of these incoming or outgoing messages may be blank, since there may not be any significant update from a sender or an update to a receiver. We will address this issue shortly.

The life of an LP *LPi* can thus be summarized as follows:

```
    {Program outline for a logical process LP_i}
    T_i:=0{T_i is the local virtual clock}
do  <termination condition> = false →
    {Simulate PPi up to time Ti}
        Compute the messages that PP_i would send to each neighbor;
        Send them to the corresponding LPs in ascending time order;
        Update the local event list
    {Increment the local virtual clock Ti}
        Receive messages along all incoming channels;
        Update the local event list;
        T_i: = minimum of all incoming channel clock values
od
```

---

* Ideally, we would specify a time window ⟨*x,y*⟩ with each message output. It will signal to the receiving process that the sender will not send any message with time less than *y*. For simplicity, we only mention **y** here.

The termination condition is left unspecified here. If the physical system terminates, then the simulated system of LPs will also terminate. Otherwise, the simulation may be terminated based on some other predefined criteria, for example, after a given interval of time beyond which the simulation results are of no interest to anyone.

18.2.2 Correctness Issues

The correctness of distributed simulation is specified by the following two properties:

Safety: At least one LP must be able to make a move until the simulation terminates.

Progress: Every virtual clock (of the LPs and the channels) must eventually advance until the simulation terminates.

The safety property corresponds to the absence of deadlock, and the progress (liveness) property guarantees termination. Depending on how the safety and progress properties are satisfied, the techniques of distributed simulation have been classified into two types: *conservative* and *optimistic*. Conservative methods rule out the possibility of causality errors. Optimistic methods, on the other hand, allow limited amount of causality errors but allow the detection of such errors, followed by a rollback recovery.

18.3 CONSERVATIVE SIMULATION

The life of an LP described in the previous section reflects the principle of conservative simulation. Causality errors are ruled out, since every LP schedules each event (of the corresponding PP) at time T *after* it has received the notification of the simulation of all relevant prior events up to time T. While this approach is technically sound, it fails to satisfy the safety requirement, since deadlock is possible. To realize why, revisit the example of the bank. Assume that for some reason, the receptionist (A) decides to send all the customers to teller 1. As a result, the clock on the link from the LP simulating $W2$ to the LP simulating the sink will never increase, which will disable further actions by the *sink LP*.

What is the way out? If an LP does not receive a message through one or more of its input channels, should it go ahead with the messages from the remaining input channels and maintain progress? No process can forecast if it will ever receive a message with a lower timestamp through an incoming channel in the future. One solution is to allow deadlock to occur and subsequently use some method for deadlock detection and resolution (see Chapter 9). Another method is to use *null* messages, proposed by Chandy and Misra.

A null message (t, *null*) sent out by LPi to LPj is an announcement of the *absence* of a message from PPi to PPj up to time t. It guarantees that the next regular (i.e., non-null) message to be sent by LPi, if any, will have a time component *larger* than t. Such messages do not have any counterpart in the physical system. In the bank example, if at time t the receptionist decides not to send any customer to $W2$ for the next 30 min, then the LP simulating A will send a message ($t + 30$, *null*) to the LP simulating $W2$. Such a guarantee accomplishes two things: (1) It helps advance the channel clock on the link from the LP simulating A to the LP

simulating $W2$ to be updated to $t + 30$. The local virtual clock of $W2$ is updated according to the simulation algorithm, and progress is maintained. (2) It helps avoid deadlock since every LP is eventually guaranteed to receive a message through each incoming channel process.

Conservative simulation relies on good predictability for achieving good performance: a guarantee by an LP that no event will be generated during a time window $(t, t + L)$ enables one or more neighboring LPs to safely proceed with the simulation of events scheduled between t and $t + L$. A drawback of conservative simulation is its inability to exploit potential parallelisms that may be present in the physical system. Conservative simulation is able to speed up simulation tasks by the parallel scheduling of events that are not causally related. However, causal dependencies are sometimes revealed at run time. Even if an event e seldom affects another event e' or is not likely to affect e', conservative simulation *will not risk* violating the causal order between e and e', although most of the time concurrent execution would have been feasible.

18.4 OPTIMISTIC SIMULATION AND TIME WARP

Optimistic simulation method, on the other hand, takes reasonable risks to expedite the simulation. It does not necessarily avoid causality errors—but it detects them when such an error occurs and arranges for recovery. The advantage is better parallelism in those cases in which causality errors *are rare*. An optimistic simulation protocol based on the virtual time paradigm is *Time Warp,* originally proposed by Jefferson [J85]. A causality error occurs whenever the timestamp of the message received by an *LP* is smaller than the value of that of its virtual clock. The event responsible for the causality error is known as a *straggler.* Recovery from causality errors is performed by a rollback that will *undo* the effects of all events that have been prematurely processed by that *LP*. Such a rollback can trigger rollbacks in other *LP*s too, and the overhead of implementing the rollback is nontrivial. Performance improves only when the number of causality errors is small.

To estimate the cost of a rollback, note that an event executed by an *LP* may result in (1) changing the state of that *LP* and (2) sending messages to the other *LP*s. To undo the effect of the first operation, old states need to be saved by the *LP*s, until there is a guarantee that no further rollbacks are necessary. To undo the effects of the second operation, an *LP* will send an *antimessage* or a *negative message* that will annihilate the effect of the original message when it reaches its destination *LP*.

When a process receives an antimessage, two things can happen: (1) If it has already simulated the previous message, it rolls back. (2) If it receives the antimessage before processing the original message (which is possible but less frequently), it simply ignores the original message when it arrives.

Certain operations are irrevocable, and it is not possible to undo them once they are committed. Examples include dispensing cash from an ATM or performing an I/O operation that influences the outside world. Such operations should not be committed until there is a guarantee that it is final. Uncommitted operations are tentative and can always be canceled. This leads to the interesting question of how to define a moment after which the result of an operation can be committed. The answer lies in the notion of *global virtual time* (GVT).

18.4.1 Global Virtual Time

In conservative simulation, the virtual clocks associated with the LPs and the channels monotonically increase until the simulation ends. However, in optimistic simulation, occasional rollbacks require the virtual clocks local to the LPs (LVTs) to turn back from time to time (Figure 18.5). This triggers a rollback in some channel clocks too.

In addition to the time overhead of rollback (which potentially reduces the speedup obtained via increased parallelism), the implementation of rollback has a space overhead— the states of the LPs have to be saved so that they can be retrieved after the rollback. Since each processor has finite resources, how much memory or disk space should be allocated to preserve the old states and when these can be deallocated are important issues.

At any real time, the GVT is the smallest among the local virtual times of all LPs, and the timestamps of all messages (positive and negative) in transit in the simulated system. By definition, no straggler will have a timestamp smaller than the GVT, so the storage used by events having timestamps smaller than the GVT can be safely deallocated. Furthermore, all events older than the GVT can be committed, since their rollbacks are ruled out. The task of reclaiming storage space by trashing states related to events older than the GVT is called *fossil collection*.

Theorem 18.1

The GVT eventually increases if the scheduler is fair.

Proof: Arrange all pending events and undelivered messages in the ascending order of time-stamps. Let $t0$ be the *smallest timestamp*. By definition, $GVT \leq t0$. A fair scheduler will eventually pick the event (or message) corresponding to the smallest timestamp $t0$ for processing, so there is no risk of rollback for this event. Any event or message generated by the execution of this event will have a timestamp greater than $t0$, so the new GVT will be greater than its previous value. Recursive application of this argument shows that GVT will eventually increase. ∎

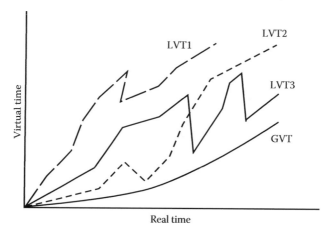

FIGURE 18.5 The progress of local and global virtual clocks in optimistic simulation.

Theorem 18.1 demonstrates that optimistic simulation satisfies the progress property. Two interesting related questions are as follows: (1) How will the *LP*s compute the GVT and (2) how often will they compute it? One straightforward way to calculate the GVT is to take a distributed snapshot (Chapter 8) of the network of *LP*s, which takes $O(n)$ time where n is the number of *LP*s. Following this, the *LP*s can free up some space. If the GVT is infrequently computed, then the fossils will tie up the space for a longer period of time. On the other hand, if the GVT is computed frequently, then the time overhead of GVT computation will slow down the progress of simulation. An acceptable performance requires a balancing act between these two conflicting aspects.

18.5 CONCLUDING REMARKS

Real-life simulation problems are increasing in both scale and complexity. As a result, the bulk of the recent work on simulation focuses on methods of enhancing the performance of simulation. While conservative simulation methods are technically sound and easier to implement, optimistic methods open up possibilities for various kinds of optimizations that enhance performance. Application-specific knowledge plays a useful role in maximizing the efficiency.

Performance optimization opportunities in conservative simulation are limited. Good predictability is a big plus: Whenever an *LP* sends out a message with a prediction window *L*, it enables the receiving *LP*s to process events *L* steps ahead. Accordingly, the larger is the value of *L*, the greater the speedup.

Lazy cancellation is a technique for speeding up optimistic simulation. If a causality error *immediately* triggers the sending of antimessages, then the cancellation is *aggressive*, and it is the essence of the original time-warp protocol. As an alternative, the process may first deal with the straggler and reexecute the computation (by taking into account the straggler, as if it was received earlier) to check if it indeed generated the *same* message. If so, then there is no need to send the antimessage, and many needless rollbacks at other *LP*s may be avoided. Otherwise, the antimessages are sent out. The downside of lazy cancellation is that if indeed the antimessages are needed, then the delay in sending them and the valuable time wasted in reexecution will cause the causality error to spread and delay recovery. This can potentially have a negative impact on the performance.

18.6 BIBLIOGRAPHIC NOTES

Chandy and Misra [CM81] and independently Bryant [B77] did much of the early work on distributed simulation. Misra's survey paper [M86] formulates the distributed simulation problem and introduces conservative simulation. In addition to the safety and liveness properties, this chapter also explains the role of *null messages* in deadlock avoidance. Jefferson [J85] proposed optimistic simulation techniques. The bulk of the work following Jefferson's paper deals with the performance improvement of Time Warp. For example, Gafni proposed *lazy cancellation* [Ga88]. Fujimoto [Fu90] wrote a survey on distributed discrete-event simulation. Further details can be found in his book on the same topic.

EXERCISES

18.1 Consider the implementation of the XOR function using NAND gates in Figure 18.6.

    a. Simulate the operation of the previous circuit using a single process. For each gate, the delay is shown in nanoseconds. Assume that the signal propagation delay along the wires is zero.

    b. Set up a conservative simulation of the preceding circuit using a separate *LP* for each NAND gate. Apply the input X,Y = (0,0) (0,1) (1,1) (0,0) and simulate the events at intervals of 1 ns.

18.2 Consider the M/M/1 queuing network as shown Figure 18.7:

Assume that the arrival of customers into the bank queue is a Poisson process with an arrival rate of 3 per minute, and the service rate by the bank teller is 2 per minute. Perform a conservative simulation of the previous network using two *LPs*: One for the *queue of customers* and the other for the *bank teller*. Run the simulation for 150 min of real time and find out the maximum length of the queue during this period.

[**Note**: The probability of n arrivals in t units of time is defined as $P_0(t) = e^{-\lambda t}(\lambda t)^n/n!$. The interarrival times in a Poisson process have an exponential distribution with a mean of λ. Review your background of queuing theory before starting the simulation.]

18.3 Consider the cancellation of messages by antimessages in optimistic simulation.

    a. What happens if the antimessage reaches an *LP before* the original message?

    b. Unless the antimessages travel faster than the original messages, causality errors will infect a large fraction of the system. Is there a way to contain the propagation of causality errors?

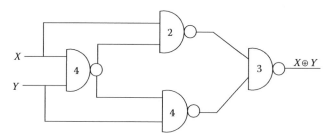

FIGURE 18.6 An implementation of XOR using NAND gates: For each gate, the delay is shown in nanoseconds.

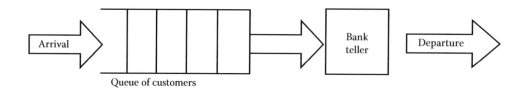

FIGURE 18.7 A M/M/1 queuing network.

18.4 In an optimistic simulation, suppose the available memory space in each *LP* is substantially reduced. What impact will it have on the speed of the simulation? Justify your answer.

18.5 Many training missions require human beings to interact with the simulated environment. Outline a method to include a human in the loop of distributed simulation. Explain your proposal with respect to a typical conservative simulation setup for a problem of your choice.

CHAPTER 19

Security in Distributed Systems

19.1 INTRODUCTION

With the rapid growth of networking and Internet-based activities in our daily lives, security has become an important issue. The security concerns are manifold: If the computer is viewed as a trustworthy box containing legitimate software only, security concerns relate to data in transit. The main concern here is that almost all these communications take place over public networks, and these networks are accessible to anyone. So how do we prevent an eavesdropper from stealing sensitive data, like our credit card numbers or social security numbers that may be in transit over a public network? How can we preserve the secrecy of a sensitive conversation between two agencies over a public network? These concerns relate to *data security*. Another aspect questions the trustworthiness of the machines that we work with. Data thieves constantly attack our computing equipment by sending virus and worms, which intrude our systems and compromise the software or the operating system—as a result, the integrity of our machines becomes a suspect. *Spywares* steal our sensitive data. *Trojan horses*, in the disguise of carrying out some useful task, indulge in illegitimate activities via the backdoor. These concerns relate to *system security*. This chapter primarily addresses data security—only a brief discussion of system security appears toward the end. There are six major requirements in security. These are as follows:

1. *Confidentiality*: Secure data must not be accessible to unauthorized persons.

2. *Integrity*: Data consistency should never be compromised. All modifications must be done via authorized means only.

3. *Authentication*: The identity of the person performing a secure transaction must be established beyond doubt.

4. *Authorization*: The user's actions must be consistent with what he or she is authorized to do. Unauthorized actions should not be allowed.

5. *Nonrepudiation*: The originator of a communication must be made accountable.

6. *Availability*: Legitimate users must have access to the data when they need them.

To understand these requirements, consider the example of Internet banking. Many of us do electronic bank transactions through secure channels from the computer at our homes. *Confidentiality* requires that your transactions should not be visible to outsiders even if the communications take place through public networks. *Integrity* requires that no one should be able to tamper with your account balance by any means whatsoever. *Authentication* requires the system to verify you are what you claim to be and only allow you and nobody else to access your account. *Authorization* requires the system to allow you to carry out those actions for which you have permissions. For example, you can transact money from your savings account, but cannot modify the interest rate. *Nonrepudiation* is a form of accountability that guarantees that if you indeed performed some transactions on your account (e.g., withdrew a large sum of money on a certain date), you should not be able to say later: "I did not do it." This is important for settling disputes. Finally, *availability* guarantees that when you need to access the account (say for paying a bill by a certain deadline), the system should be available. A secure system is useless, if it is not available.

19.2 SECURITY MECHANISMS

Three basic mechanisms are used to meet security requirements. These are as follows:

1. *Encryption*: Encryption implements a *secure data channel*, so that information is not leaked out to or stolen by outsiders.

2. *Signature*: Digital signatures provide authentication and nonrepudiation and protect integrity.

3. *Hashing*: Checksums or hash functions maintain data integrity and support authentication.

There are many algorithms to implement each of these mechanisms. These mechanisms collectively implement a *security service* offered to clients. Examples of secure services are secure socket layer (SSL) for confidential transactions and secure shell (SSH) for remote login to your computer.

Secure system design is unrealistic, unless the nature of the threat is known. In the following, we discuss some common types of security attacks or threats.

19.3 COMMON SECURITY ATTACKS

19.3.1 Eavesdropping

Unauthorized persons can intercept private communication and access confidential information. Data propagating through wires can be stolen by wiretapping. Wireless communication can be intercepted using a receiver with an appropriate antenna. Eavesdropping violates the *confidentiality* requirement.

19.3.2 Denial of Service

Denial-of-service (DoS) attack uses malicious means to make a service *unavailable* to legitimate clients. Ordinarily, such attacks are carried out by flooding the server with phony requests. For example, DHCP clients are denied service if all IP addresses have been drained out from the DHCP server using artificial requests for connection. Although it does not pose a real threat, it causes a major inconvenience and may cause revenue loss.

19.3.3 Data Tampering

This refers to unauthorized modification of data. For example, the attacker working as a programmer in a company somehow doubles his monthly salary in the salary database of the institution where he works. It violates the *integrity* requirement.

19.3.4 Masquerading

The attacker disguises himself or herself to be an authorized user and gains access to sensitive data. Sometimes authorized personnel may masquerade to acquire extra privileges greater than what they are authorized for. A common attack comes through stolen login id's and passwords. Sometimes, the attacker intercepts and *replays an old message* on behalf of a client to a server with the hope of extracting confidential information. A replay of a credit card payment can cause your credit card to be charged twice for the same transaction.

A widespread version of this attack is the *phishing* (deliberately misspelled) attack. *Phishing* attacks use spoofed emails and fraudulent websites to fool recipients into divulging personal financial data such as credit card numbers, account usernames and passwords, and social security numbers. Average citizens are targeted by imposters masquerading for financial institutions, social networking sites, or Internet auction sites. Statistics [APWG] show that the fraudulent emails are able to convince up to 5% of the users who respond and divulge their personal data to these scam artists.

19.3.5 Man in the Middle

Assume that Amy wants to send confidential messages to Bob. To do this, both Amy and Bob ask for each other's public key (an important tool for secure communication in open networks) first. The attacker Mallory intercepts messages during the public key exchange, acquires copies of these public keys, and substitutes his own public key in place of the requested one to both Amy and Bob. Now, both communicate using Mallory's public key, so only Mallory can receive the communication from both parties. To Amy, Mallory impersonates as Bob, and to Bob, Mallory impersonates as Amy. This is also known as *bucket brigade* attack.

19.3.6 Malicious Software

Malicious software (commonly called *malware*) gets remotely installed into your computer system without your informed consent. Such malware facilitates the launching of some of the attacks highlighted earlier. There are different kinds of malicious software:

19.3.6.1 Virus

A virus is a piece of self-replicating software that commonly gets attached to an executable code, with the intent of carrying out destructive activities. Such activities include erasing files, damaging applications or parts of the operating system, or sending emails on behalf of an external agent. The file to which the virus gets attached is called the *host*, and the system containing the file is called *infected*. When the user executes the code in a host file, the virus code is first executed. Common means of spreading virus are emails and file sharing. Although viruses can be benign, most have *payloads*, defined as actions that the virus will take once it spreads into an uninfected machine. One such payload is a backdoor that allows remote access to a third party.

19.3.6.2 Worms

Like a virus, a worm is also a self-replicating program with malicious intent. However, unlike a virus, a worm does not need a host. A hacker who gains control of the machine through the backdoor can control the infected machine and perform various kinds of malicious tasks. The computers that are taken over are called *bots* or *zombies*. The infected computers form a network called a botnet. Criminals use botnets to send out spam emails, spread viruses, launch distributed denial-of-service (DDoS) attacks and commit other kinds of crime and fraud. Spammers using zombies save the bandwidth of their own machines and avoid being detected. As of August 2010, an estimated number of 200 billion spam emails were sent per day.

Two well-known worms in the past decade are *Mydoom* and *Sobig*. The Sobig worm was first spotted in August 2003. It appeared in emails containing an attachment that is an executable file—as the user clicked on it, the worm got installed as a Trojan horse. Mydoom was identified on January 26, 2004. The infected computers sent junk emails. Some believe that the eventual goal was to launch a DoS attack against the SCO group who was opposed to the idea of open-source software, but it was never confirmed.

19.3.6.3 Spyware

Spyware is a malware designed to collect personal or confidential data, to monitor your browsing patterns for marketing purposes, or to deliver unsolicited advertisements. Spyware gets downloaded into a computer (mostly by nonsavvy users) when they surf the web and click on a link used as bait. Examples of baits include pop-ups asking the user to claim a prize, or offer a substantially discounted airfare to a top destination, or invite the user to visit an adult site. Users may even be lured away to download free spyware-protection software. The baits are getting refined every day. Unlike virus, spyware does not replicate itself or try to infect other machines. Spyware does not delete files, and consumes a fraction of the bandwidth causing it to run slower.

19.4 ENCRYPTION

The Old Testament (600 BC) mentions the use of reversed Hebrew alphabets to maintain secrecy of message communication. Modern cryptography dates back to the days of Julius Caesar who used simple encryption schemes to send love letters. The essential components of a secure communication using encryption are shown in Figure 19.1. Here,

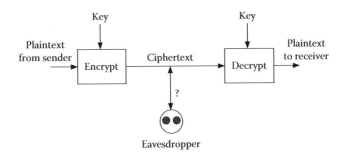

FIGURE 19.1   A scheme for secure communication.

Plaintext = raw text or data to be communicated.

Ciphertext = encrypted version of the plaintext.

Key = the secret code used to encrypt the plaintext. No one can decipher or decrypt the ciphertext without having the appropriate key for it.

Caesar's encryption method was as follows: Substitute each letter of the English alphabet by a new letter k placed ahead of it. Thus, when $k = 3$ (which is what Caesar used) "**A**" in plaintext will become "**D**" in the ciphertext, and "**Y**" in plaintext will become "**B**" in the ciphertext (the search rolls back to the beginning), and so on. Formally, we can represent the encryption and decryption mechanisms (with key k) as two functions:

$$E_k : L \to (L + k) \bmod 26$$

$$D_k : L \to (L - k) \bmod 26$$

where L denotes the numerical index of a letter and $0 \leq L \leq 25$. Such ciphers are called *substitution ciphers*.

Unfortunately, such an encryption scheme provides very little secrecy, as any motivated puzzle solver can decrypt the ciphertext in a few minutes. Nontrivial cryptographic technique should therefore have a much more sophisticated encryption scheme, so that no intruder or eavesdropper can find any clue about the encryption method, nor can he use a machine to systematically try out a series of possible decryption methods within a relatively short period of time. Note that given enough time, any ciphertext can be deciphered by an intruder using a systematic method of attack. The larger this time is, the more secure is the encryption.

A *cryptosystem* is defined by its encryption and decryption mechanisms. If P denotes the plaintext and C denotes the corresponding ciphertext, then

$$C = E_{k1}(P)$$

$$P = D_{k2}(C) = D_{k2}(C_{k1}(P))$$

where
   E_{k1} is the encryption function with key $k1$
   D_{k2} is the decryption function with key $k2$

The decryption function D_{k2} is the inverse of the encryption function E_{k1} (denoted as E_{k1}^{-1}) and vice versa. A cryptosystem is called *symmetric*, if both parties share the same key for encryption and decryption. Otherwise, the cryptosystem becomes *asymmetric*.

19.5 SECRET KEY CRYPTOSYSTEM

Secret key cryptosystem (also known as *private key cryptosystem*) is a symmetric cryptosystem. It consists of an encryption function E, a decryption function D (the inverse of E), and a *secret key k* that is shared between the sender and the receiver. The functions E and D need not be secret. All that is required is for the two parties to agree upon a *secret key* before the communication begins. There are numerous examples of secret key cryptosystems. We classify these into two different types: *block ciphers* and *stream ciphers*. In block ciphers, the data are first divided into fixed size blocks before encryption. This is applicable, when the data to be sent are available in its entirety before communication begins. In contrast, stream ciphers are used for transmitting *real-time data* that are spontaneously generated, for example, voice data.

19.5.1 Confusion and Diffusion

According to Shannon [S49], *confusion* and *diffusion* are the two cornerstones of secret-key cryptography. To break a code, one of the important tools for a cryptanalyst is to make use of the frequency of letters or phrases in human-readable messages. This is called a statistical attack. Two kinds of statistics are useful in launching attacks: In the first, the cryptanalyst has access to several ciphertexts of messages using the same key, but does not have access to the plaintext. In the second, the cryptanalyst has access to both the plaintext and the ciphertext of messages using the same key. Diffusion spreads the influence of each bit of the plaintext over several bits of the ciphertext. Thus, changing a small part of the plaintext affects a large number of bits of the ciphertext. This dissipates the statistical structure of the plaintext, making it difficult for the statisticians to utilize the statistics by analyzing the ciphertext. As an example, let $Y = y_0 y_1 y_2 ... y_{m-1}$ be the ciphertext for the plaintext $X = x_0 x_1 x_2 ... x_{m-1}$, each x and y being the indices of letters from the English alphabet. An encryption mechanism may use the function $y_j = \sum_{k=3}^{k=13} x_{j+k} \bmod 26$ to implement diffusion. An intruder will need many more ciphertexts (compared that in a simple Caesar-like substitution cipher) to launch a meaningful statistical attack on the cipher and discover the key.

Confusion, on the other hand, makes the relationship between the key and the ciphertext as complex as possible. Caesar cipher that replaced each letter by k letters ahead of it is a simple example of confusion—perhaps too simple for a modern adversary. Confusion is a complex form of substitution where every bit of the key influences a large number of bits of the ciphertext. In an effective confusion, each bit of the ciphertext depends on several parts of the key, so much so that the dependence appears to be random to an intruder. Even if the statistics about the plaintext is known, the complex use of the key makes it very hard for a cryptanalyst to deduce it.

TABLE 19.1   Lookup Table of a (6 × 4) S-Box

	0000	*0001*	***0010***	...	*1111*
00	0010	1100	0100	...	1001
01	1110	1011	0010	...	1110
10	0100	1000	**0001**	...	0011
11	1011	0010	1100	...	0110

Note: The rows represent the outer bits of the input, the columns represent the middle bits of the input, and the entries in the matrix represent the output bits.

Traditional cryptography implements confusion using *substitution boxes* (S-boxes). A $(p \times q)$ S-box takes p bits from the input and coverts it into q bits of output, often using a lookup table. Table 19.1 illustrates a lookup table for a (6×4) S-box. The row headings are the pairs consisting of the *first* and the *last* bits, and the column headings represent the middle 4 bits of the input.

Using this table, the 6-bit input **100100** is transformed as follows: The outer bit pair is 10, and the middle bits are 0010, so the output is 0001.

In *block ciphers*, diffusion propagates changes in one block to the other blocks. Substitution itself diffuses the data within one block. To spread the changes into the other blocks, *permutation* is used. In an ideal block cipher, a change of even a single plaintext bit will change every ciphertext bit with probability 0.5, which means that about half of the output bits should change for any possible change of a bit in an input block. Thus, the ciphertext will appear to have changed at random even between related message blocks. This will hide message relationships that can potentially be used by a cryptanalyst. This kind of diffusion is a necessary, but not a sufficient requirement good block cipher.

Block ciphers implement confusion by mapping each block of plaintext to the same block of ciphertext. Without a diffusion mechanism, an attacker can recognize such repetitions, do a frequency analysis, and infer the plaintext. *Cipher block chaining* (CBC) is a diffusion mechanism that overcomes this problem. Let $P_0, P_1, P_2, \ldots$ be the blocks of the plaintext and E_k be the encryption function. Then CBC encrypts the plaintext as follows:

$$C_0 = E_k(P_0)$$

$$C_1 = E_k(XOR(P_1, C_0))$$

...

$$C_m = E_k(XOR(P_m, C_{m-1}))$$

19.5.2 DES

Data encryption standard (DES) is a secret-key encryption scheme developed by IBM (under the name Lucifer) in 1977. In DES, the plaintext is divided into 64-bit blocks. Each such block is converted into a 64-bit ciphertext using a 56-bit key. The conversion mechanism is outlined in Figure 19.2. The plaintext, after an initial permutation, is fed into a cascade of 16 stages. A 56-bit secret master key is used to generate 16 keys, one for each

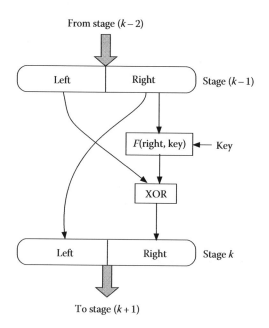

FIGURE 19.2   One stage of transformation in DES.

stage of the cascade. The right half of stage $(k − 1)$ $(k > 0)$ becomes the left half of stage k (diffusion), but the left half undergoes a transformation (confusion) before it is used as the right half of the next stage. This transformation uses a special function F (which is not a secret function) and a 48-bit key derived from the 56-bit secret master key. The output of the final stage (i.e., stage 15) is once more permuted to generate the ciphertext.

A blind attack by a code breaker will require on an average 2^{55} trials (half of 2^{56}) to find out the secret key. However, with the rapid improvement in the speed of computers and with the collective effort by a set of machines running in parallel, DES does not provide enough security against a desperate code breaker—blind attacks can crack the code in a reasonable short period of time. For example, the winner of the *RSA Challenge II* contest in 1998 cracked DES in *56 h* using a supercomputer. In 1999, distributed.net partnering with Electronic Frontier Foundation won *Challenge III* and cracked DES in *22 h*. With today's technology, for an investment of less than a few thousand dollars, dedicated hardware can be built to crack DES in less than an hour. Government agencies ruled DES as out of date and unsafe for financial applications. Longer keys are needed.

A major challenge in secret key encryption is the distribution of the secret key among legitimate users. The distribution of the secret key itself requires a secure channel (e.g., by registered post or through a trusted courier). The recipient should know ahead of time when the key is being sent. The problem of communicating a large message in secret is reduced to communicating a small key in secret. In fact, key distribution is one of the major impediments behind the use of secret-key encryption. Another concern is the large number of keys that need to be used in a network with a large number of users. Note that for confidentiality, every pair of communicating users needs to have a unique secret key. This means that in a network with n users, $n(n − 1)/2$ keys will be needed, and for a large network, the cost of distributing them could be prohibitive. In Section 19.10, we will discuss how an *authentication server* (AS) can distribute the keys to its clients.

19.5.3 3DES

3DES or Triple DES is a refinement of DES designed to reduce its vulnerability against brute-force attacks. It applies the DES transformation three times in succession (encrypt–decrypt–encrypt) using three separate 64-bit secret keys $k1$, $k2$, $k3$ to generate the ciphertext C from a plaintext P:

$$C = E_{k3}(D_{k2}(E_{k1}(P)))$$

3DES reuses DES implementations for the sake of efficiency. It has been shown that against brute-force attacks, 3DES provides resilience equivalent to a 112-bit key version of DES. However, the conversion in DES is slow, and that in 3DES is painstakingly slow. These days, 3DES is considered a legacy encryption algorithm.

Several other symmetric cryptosystems use 128-bit keys. These include tiny encryption algorithm (TEA) by Wheeler and Needham [WN94] and international data encryption algorithm (IDEA) due to Lai and Massey [LM90]. Blowfish is an encryption mechanism designed by Schneier—it uses a variable size key of 32–448 bits. Another symmetric encryption mechanism designed by Schneier et al. [SKW+98] is Twofish with a 128-bit key. The advanced encryption standard (AES) eventually superseded all these, and it has been accepted as the US government's symmetric encryption standard.

19.5.4 AES

NIST ran a public process to choose a successor of DES and 3DES. Of the many submissions (Twofish was one of them), they chose *Rijndael* developed by two Belgian cryptographers Joan Daemen and Vincent Rijmen. A restricted version of this is now known as the AES. The AES is a block cipher with a fixed block size of 128 bits and a key size of 128, 192, or 256 bits, whereas Rijndael can be specified with key and block sizes in any multiple of 32 bits, with a minimum of 128 bits and a maximum of 256 bits.

The US government approved the use of 128-bit AES to protect classified information up to the *secret* level. *Top secret* information will require use of either the 192 or 256 key lengths. However, the National Security Agency (NSA) must certify these implementations before they become official.

19.5.5 One-Time Pad

A one-time pad is a secret cryptosystem in which each secret key expires after a single use. Also called the *Vernam cipher*, one-time pad uses a string of bits generated completely at random. The length of the keystream is the same as that of the plaintext message. The random string is combined with the plaintext using bitwise XOR operation to produce the ciphertext. Since the entire keystream is randomly generated, an eavesdropper with unlimited computational resources can only guess the plaintext if he sees the ciphertext. Such a cipher is provably secure, and the analysis of one-time pad provides some insight into modern cryptography.

Despite being provably secure, one-time pad introduces serious key-management problems. Users who started out at the same physical location, but later separated, have used one-time pads. One-time pads have been popular among spies carrying out covert operations.

19.5.6 Stream Ciphers

Stream ciphers are primarily used to encrypt real-time data that are spontaneously generated. The encryption takes place on smaller size data units. Stream ciphers were developed as an approximation to the mechanism of the one-time pad. While contemporary stream ciphers are unable to provide the satisfying theoretical security of the one-time pad, they are at least practical. Stream data are encrypted by first generating a *keystream* of a very large size using a *pseudorandom number generator* and then XORing it with data bits from the plaintext. These pseudorandom numbers have very large periods and often serve as a good approximation for the perfect random number required in a one-time pad. If the recipient knows the keystream, then she can decipher it by XORing the key stream with the ciphertext (for any data bit b and keystream bit k, the ciphertext is $b \oplus k$, and $b \oplus k \oplus k = b$ returns the plaintext.)

The pseudorandom number generator is seeded with a key, and the keystream is the output of the pseudorandom number generator. To generate identical random keystreams, both senders and receivers use the same random number generator and identical seeds. This is a lightweight security mechanism targeted for mobile devices. The keystream can be computed using known plaintext attacks: The XOR of a known plaintext b and its corresponding ciphertext $b \oplus k$ reveals the keystream.

RC4: Designed by Ron Rivest of RSA Security, RC4 is a stream cipher optimized for fast software implementation. The encryption mechanism uses a 256-byte array R and runs in three phases:

1. Choose a 40–256-bit key. The *length* of a key is the number of bytes in the key.

2. Initialize register R using the chosen key.

3. Generate the keystream, a pseudorandom sequence of bits.

This keystream will be used to encrypt the plaintext. The scheme is illustrated in Figure 19.3. The first step is to initialize a byte array R as follows:

```
program Initialization of R
define R: 256-byte array, key: array of bits, j,k: integer
initially ∀i:0 ≤ i ≤ 255::R[i]:=i;
j:=0; k:=0;
do   k ≤ 255 →
     j:=(j+R[k]+key[k mod length(key)])mod 256
     swap(R[j],R[k])
     k:=k+1
od
```

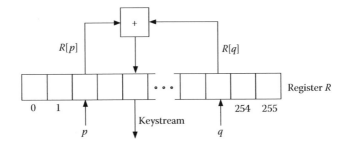

FIGURE 19.3   The encryption mechanism in RC4 stream cipher.

To generate the keystream, the byte array R is associated with two pointers p and q. After R is initialized, the following program generates the keystream:

```
program generation of the keystream in RC4
define R: 256-byte array, k: byte
       p,q : pointer variables initialized to 0
do     true →
       p := p + 1 mod 256
       q := q + R[p] mod 256;
       swap (R[p],R[q]);
       k := R[R[p]+R[q]mod256]  {k is the output}
       {k is XORed with the plaintext byte to generate the ciphertext byte}
od
```

RC4 is used in the implementations of SSL/(Transport Layer Security) TLS, WEP (for wireless networks), and several other applications. In 2013, however, a new analysis showed how RC4 can be broken [ABP+13], and the impact of this weakness is being investigated.

19.5.7 Steganography

In Greek, *steganography* means *covered writing*. It is the technique of hiding the actual communication in the body of an inconspicuous sentence or a paragraph. Unlike encryption where the transmitted ciphertext is incomprehensible and meaningless to an outsider, the transmitted text in steganography has a valid, but different meaning.

In [K67], Kahn quotes an example of a message [K96] that was actually sent by the German embassy to Berlin during WWI:

> Apparently neutral's protest is thoroughly discounted and ignored.
> Isman hard hit. Blockade issue affects pretext for embargo on byproducts, ejecting suets and vegetable oils.

Now, take the *second letter* in each word, the following message emerges:

> Pershing sails from NY June 1.

The transmitted message is called a *stego-text*. In modern times, some use spams as stego-texts to embed hidden message. Note that embedding the plaintext inside *a text message* is not the only possibility—there are other techniques too. For example, embedding the secret message inside a picture is another popular approach. Consider a GIF image that is a string of pixels—one can hide the plaintext in the LSB of the pixels of the image. Such an embedding will cause no appreciable alteration in the appearance of the picture. Retrieving the plaintext from the stego-text is quite straightforward.

While cryptography is popular in implementing secure channels, an important application of steganography is in *digital watermarking*, used for copyright protection and plagiarism detection. When the copyright (or the author's identity) is embedded in the body of the document using steganography, a plagiarized copy of the original document will easily reveal the author's true identity and expose plagiarism.

Steganography is not intended to replace cryptography, but supplement it. Concealing a message using steganographic methods reduces the chance of that message being detected. However, if the message is also encrypted, then it provides another layer of security.

19.6 PUBLIC KEY CRYPTOSYSTEMS

The need for a secure channel to distribute the secret conversation key between any two communicating parties is a major criticism against the secret-key encryption scheme. Public-key cryptosystems overcome this problem. In public-key cryptography, each user has two keys: *e* and *d*. The encryption key *e* is posted in the *public domain*, so we will call it a *public key*. Only the decryption key *d* is kept secret. No secret key is shared between its owner and any other party. The encryption function (*E*) and the decryption function (*D*) are not secret either. The main features of public-key encryption are as follows:

1. The plaintext $P = D_d(E_e(P))$.

2. The functions *E* and *D* are easily computable.

3. It is impossible to derive *d* from *e*. So public knowledge of *e* is not a security threat.

Diffie and Hellman's work [DH76] laid the foundation of public-key cryptography. To understand this scheme, we first explain what a *one-way function* (also called a *trapdoor function*) is. A function $y = f(x)$ is called a one-way function, if it is easy to compute *y* from *x*, but it is computationally intractable to determine *x* from *y* (even if *f* is known), unless a secret code is known. For example, it is easy to multiply two large prime numbers (>200 digits), but it is computationally intractable to find the prime factors of a large number containing more than 200 digits. Trapdoor functions are the cornerstones of public-key cryptography.

The next section explains the Rivest–Shamir–Adleman (RSA) cryptography that is the most popular version of public-key cryptography so far.

19.6.1 Rivest–Shamir–Adleman Cryptosystem

RSA encryption starts with the choice of an appropriate integer $N = s \times t$, *s* and *t* being two *large* prime numbers. The plaintext is divided into equal-sized blocks of *k* bits

(k is between 512 and 1024), and the numerical value P of each block is less than N. The integer N is publicly known. The pair of encryption and the decryption keys (e and d) are chosen using the following steps.

1. Choose a decryption key d, such that d and $(s - 1) \times (t - 1)$ are relative primes.

2. Choose e such that $(e \times d) \bmod (s - 1) \times (t - 1) = 1$.

Once the keys are chosen, the encryption and the decryption functions for each block are as follows (C = ciphertext, P = plaintext):

$$C = P^e \bmod N \tag{19.1}$$

$$P = C^d \bmod N \tag{19.2}$$

Here is a sample calculation. Choose $N = 143$, which is the product of two primes $s = 11$ and $t = 13$. Then, $(s - 1) \times (t - 1) = 123$. We choose $d = 7$ since 7 and 120 are relative primes. The smallest value of e that can satisfy the equation $(e \times 7) \bmod 120 = 1$ is 103. So, a possible pair of keys in this case is (103, 7).

Proof of RSA encryption
We first show that if $C = P^e \bmod N$, then the plaintext P can indeed be retrieved from the ciphertext C using the formula $P = C^d \bmod N$. Let $\Phi(N)$ be the number of positive integers that are (1) less than N and (2) relatively prime to N. Thus, if $N = 12$, then $\Phi(N) = 4$ (since there are four integers 1, 5, 7, 11 that are relatively prime to 12). Clearly, for any prime number $p > 2$, $\Phi(p) = p - 1$. The function Φ is called *Euler's totient function*. The following two are important theorems from number theory:

Theorem 19.1

For any integer p that is relatively prime to N, $p^{\Phi(N)} = 1 \bmod N$.

Theorem 19.2

If $N = p \times q$, then $\Phi(N) = \Phi(p) \times \Phi(q)$.
   Since $N = s \times t$ and s and t are prime numbers, using Theorem 19.2,

$$\Phi(N) = \Phi(s) \times \Phi(t)$$
$$= (s-1) \times (t-1)$$
$$= N - (s+t) + 1$$

Since s is a prime, using Theorem 19.1,

$$P^{(s-1)} = 1 \bmod s$$

So, $\quad P^{r(s-1)(t-1)} = 1 \bmod s \quad \{\text{where } r \text{ is an integer}\}$

Similarly, $\quad P^{r(s-1)(t-1)} = 1 \bmod t$

Therefore, $\quad P^{r(s-1)(t-1)} = 1 \bmod (s \times t) = 1 \bmod N \qquad (19.3)$

According to the encryption algorithm, $C = P^e \bmod N$. Therefore,

$$C^d = P^{(e \times d)} \bmod N$$

$$= P^{r(s-1)(t-1)+1} \bmod N \quad \{\text{since } (e \times d) = 1 \bmod (s-1)(t-1)\}$$

$$= P^{r(s-1)(t-1)} \times P \bmod N \quad \{r \text{ is an integer}\}$$

$$= P \bmod N \quad \{\text{follows from } (19.3)\} \qquad ■$$

The *encryption key* is posted in the public domain. Only the *decryption key* is kept secret. Therefore, only the authorized recipient (who has the decryption key) can decipher the encrypted message, which guarantees confidentiality. In fact, due to the symmetry, the two keys are interchangeable.

The confidentiality is based on the fact that given e and N, it is not possible to derive d, since it requires the knowledge of the prime factors of N. The size of N makes it computationally intractable for the intruder to determine s and t from N and thus to break the codes. This is a classic example of a one-way function that makes RSA a secure cipher. If somebody could design an efficient algorithm to find the factors of a huge integer, then RSA cipher could be broken. By combining the power of the state-of-the-art supercomputers (like those in NSA or FBI), 512-bit numbers can be factored in a reasonable time. Thus, to assure the security of the RSA cipher, the key length should be chosen sufficiently larger than that. Currently, a value of N in excess of 10^{200} (i.e., more than 800 bits) is recommended to guarantee security.

The speed of encryption and decryption is an important issue in real applications. Knuth presented an efficient algorithm for computing $C = P^e \bmod N$. Let $e = e_k e_{k-1} e_{k-2} \cdots e_2 e_1 e_0$ be the $(k+1)$-bit binary representation of the encryption key. Then, Knuth's exponential algorithm for efficiently computing C is as follows:

```
program Knuth's algorithm for RSA encryption
define P: plaintext, C: ciphertext, e:=(k+1)-bit encryption key
C:=1; i:=k;
do k ≠ 0 →
     C:= C² mod N;
     if eᵢ = 1 → C:= C × P mod N[] eᵢ = 0 → skip fi;
     k:= k - 1
od
```

Note that this algorithm requires at most $k = \log_2 e$ multiplications, instead of e multiplications that a straightforward computation of C would require. Due to the symmetry, a similar algorithm can be used for decryption.

19.6.2 ElGamal Cryptosystem

In 1984, ElGamal presented a new public-key encryption scheme. Unlike RSA encryption whose security relies on the difficulty of factoring large primes, the security of ElGamal encryption relies on the difficulty of solving the *discrete logarithm* problem. We first present a brief overview of the discrete logarithm problem.

Consider modulo arithmetic, specifically numbers mod q where q is a prime number. Take another positive integer $g(g < q)$, and compute $g^x \bmod q$ for various positive values of x. If these values are uniformly distributed in the set $\{0,1,2,3,\ldots,q-1\}$, then g is called a generator. For example, consider $q = 17$ and $g = 3$, and verify that the values of $3^x \bmod 17$ are uniformly distributed in the set $\{0,1,2,3,\ldots,16\}$. Here, 3 is a generator of the *multiplicative cyclic* group $\{0,1,2,3,\ldots,16\}$. The *discrete logarithm* problem asks the question: If $g^x \bmod q = y$ and the values of q, g, y are all given, then compute x. For example, $3^{19} \bmod 17 = 10$, but determining what value x will satisfy the equation $3^x \bmod 17 = 10$ is not an easy problem.* Particularly, when the prime q is 200 digits long, or even longer, the problem becomes almost intractable. This intractability is the cornerstone of ElGamal cryptosystem.

Let G be a *cyclic group* $\{0,1,2,3,\ldots,q-1\}$ and g be the *generator* of that group. To receive a message from Bob, Amy will randomly pick a number d from G as her *secret key*. She will then compute $h = g^d \bmod q$ and publish her *public key* e as (g, q, h). Bob will learn the public key of Alice.

To secretly send a plaintext message block $P(0 \leq P \leq q-1)$ to Alice, Bob will randomly pick a number y from G. He will then compute $C_1 = g^y \bmod q$, and $C_2 = P \cdot h^y \bmod q$, and send the ciphertext $C = \langle C_1, C_2 \rangle$ to Alice. Note that the length of C is double the length of the plaintext.

Amy will decrypt C and retrieve the plaintext by computing $C_2 \cdot C_1^{-d}$. To verify that it indeed retrieves the original plaintext P, observe that

$$C_2 \cdot C_1^{-d} = \frac{P \cdot h^y}{g^{y \cdot d}} \bmod q = \frac{P \cdot g^{y \cdot d}}{g^{y \cdot d}} \bmod q = P$$

The encryption scheme could be broken if one could derive Alice's secret key d from her public key $e = (g,q,h)$, given that $h = g^d \bmod q$ and g,q,h publicly known. However, for very large q, this is the intractable part in the computation of discrete logarithm! Note that ElGamal encryption is probabilistic, in as much as a single plaintext can be encrypted to many possible ciphertexts depending on the choice of h.

---

* There may be multiple solutions.

19.7 DIGITAL SIGNATURES

Digital signatures preserve the integrity of a document and the identity of the author. Signed documents are believed to be (1) authentic, (2) unforgeable, and (3) nonrepudiable. Consider the following secret communication from Amy to Bob:

> I will meet you under the Maple tree in the Sleepy Hollow Park between 11:00 PM and 11:15 PM tonight.

To authenticate the message, the receiver Bob needs to be absolutely sure that it was Amy who sent the message. If the key is indeed secret between Amy and Bob, then confidentiality leads to authentication, as no third party can generate or forge such a message.

However, this is not all. Consider what will happen if following an unfortunate sequence of events, Amy and Bob end up in a court of law, where Amy denies having ever sent such a message? Can we prove that Amy is telling a lie? This requires a mechanism by which messages can be *signed*, so that a *third party* can verify the signature. Problems like these are quite likely with contracts in the business world or with transactions over the Internet.

The primary goal of digital signatures is to find a way to bind the identity of the signatory with the message of the text. We discuss how a message can be signed in both secret and public cryptosystems.

19.7.1 Signatures in Secret-Key Cryptosystems

To generate a signed message, both Amy and Bob should have a trusted third party, Charlie. For the sake of nonrepudiation, Amy will append to her original message M an *encrypted form* of a *message digest m* (known as a *message authentication code*, or MAC). The message digest m is a fixed-length entity that is computed from M using a hash function H, which is expected to generate a *unique footprint* of M. (The uniqueness holds with very high probability against accidental or malicious modifications only.) In case of a dispute, anyone can ask the trusted third party, Charlie, to compute $m' = H(M)$ and compare the result with the signature m. The signature is verified only if $m = m'$. If the channel is not secret and confidentiality is also an issue, then Amy has to further encrypt (M, m) using her secret key k. The major weakness here is the need for a trusted third party who will know the secret key of Amy and Bob.

19.7.2 Signatures in Public-Key Cryptosystems

To sign a document in public-key cryptosystem, Amy will compute a digest $m = H(M)$ of her message M using a hash function H, encrypt the digest m with her private key d_A, and send out $(M, d_A(m))$. Now Bob (in fact, anyone) can decrypt $d_A(m)$ using the public key e_A of Amy and compare it with $m' = H(M)$. When $m = m'$, it is implied that only Amy could have sent the message M, since no one else would know Amy's secret key. Digital signatures using public-key cryptosystems is much more popular and practical, since it does not require the sharing or distribution of the private key.

19.8 HASHING ALGORITHMS

Hash functions are used to reduce a variable length message to a fixed-length fingerprint. Also known as *cryptographic hash function*, such a function H should have the following properties:

1. Computing $m = H(M)$ from a given M should be easy, but computing the inverse function that will uniquely identify M from a given m should be impossible.

2. $M \neq M' \Rightarrow H(M) \neq H(M')$. Thus, any modification of the original message M will cause its footprint to change.

The second condition is difficult to fulfill for any hash function, so ordinarily, the interpretation of M' is restricted to *malicious modifications* of M. The number of (M,M') pairs that lead to the same fingerprint is a measure of the robustness of the hash function. A well-known hash function is SHA-2 that was adopted by NIST after weaknesses were discovered in the two earlier hash functions MD5 and SHA-1. SHA-2 is a family of that includes two hash functions SHA-256 and SHA-512 with different block sizes. Many of the US government agencies now use SHA-2 for various security-related applications. SHA-2 hash function is also implemented in some widely used security applications and protocols like SSL/TLS, PGP, and SSH. In 2012, NIST selected a new hash function SHA-3 that uses a new algorithm and not derived from SHA-2. As of now, the decision about adopting SHA-3 is pending.

19.8.1 Birthday Attack

The *birthday attack* is a cryptographic attack (on hashing algorithms) that exploits the mathematics behind the birthday paradox: if a function $y = f(x)$ yields any of n different output values of y with *equal probability* and n is sufficiently large, then after evaluating the function f for about $\sqrt{n}$ different arguments, we expect to find a pair of arguments $x1$ and $x2$ such that $f(x1) = f(x2)$ with a probability $p > 0.5$—this is known as a *collision*.

Now, apply this to the birthdays of a set of people who assembled in a room. There are 365 possible different birthdays (month and day). So if there are more than $\sqrt{365}$ people in the room, then we will expect at least two persons having the same birthday. In fact, it can be shown that with more than 23 people, the probability that two of them have the same birthday is > 0.5. If the outputs of the function are distributed unevenly, then a collision can occur even faster.

Digital signatures in secret-key cryptosystems are susceptible to birthday attack. A message M is signed by first computing $m = H(M)$, where H is a cryptographic hash function, and then encrypting m with their secret key k. Suppose Bob wants to trick Alice into signing a fraudulent contract. Bob prepares two contracts: a fair contract M and a fraudulent one M'. He then finds a number of positions where M can be changed without changing the meaning, such as inserting commas, empty lines, and spaces. By combining these changes, he can create a huge number of variations on M that are all fair contracts. In a similar manner, he also creates a huge number of variations on the fraudulent contract M'. He then applies the hash function to all these variations until he finds a version of the fair contract and a version of the fraudulent contract having the same hash value. He presents the fair version to Alice for signing. After Alice has signed, Bob takes the signature and attaches it to the fraudulent contract. This signature apparently *proves* that Alice signed the fraudulent contract. For a good hash function, it should be extremely difficult to find a pair of messages M and M' with the same digest.

19.9 ELLIPTIC CURVE CRYPTOGRAPHY

Contrary to some beliefs, an elliptic curve is *not* the same as an ellipse.

An elliptic curve is defined by equation $y^2 = x^3 + ax + b$. There are more general versions of elliptic curves, but for our purpose, this will suffice. The coefficients of the equation must satisfy the condition $4a^3 + 27b^2 \neq 0$, which is a necessary and sufficient condition that the polynomial has three distinct roots, a condition required by the cryptosystem. The elliptic curve of Figure 19.4 is generated from the equation $y^2 = x^3 - 6x + 6$.

Let q, r be a pair of points on an elliptic curve. The curve is symmetric around a horizontal axis (in this case the x-axis). For any point p, designate its mirror image (reflection around the x-axis) by the notation $-p$. The *addition* operation (+) is defined as follows:

1. When $q \neq -r$, $q + r$ is computed by drawing a straight line through q and r. Let this line intersect the elliptic curve at point $-p$. Take the reflection of $-p$ around the x-axis (which is p). Then $q + r = p$.

2. When $q = -r$, the line joining them is vertical and parallel to the y-axis. This does not intersect the elliptic curve at a third point. This is addressed by including a special third point ∞ on the curve. Call it the *identity element I*. Then $q + (-q) = I$. This implies $q + I = q$.

3. When $q = r$, the sum $q + q$ (also called $2q$) is defined by drawing a tangent to the curve at point q. Let it intersect the elliptic curve at point $-r$. Then $q + q = 2q = r$ (the reflection of the point $-r$ around the x-axis).

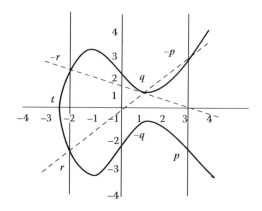

FIGURE 19.4   An elliptic curve.

The last step defines the *multiplication* (by two) operation that can be generalized to multiplication by any nonnegative scalar integer by treating this as repeated addition. Thus, $3q = 2q + q = r + q = p$. For a point like t where the tangent is parallel to the y-axis,

$$2t = t + t = I$$

$$3t = 2t + t = I + t = t$$

$$4t = 3t + t = t + t = I$$

and so on.

In cryptography, the variables and the constants are not real numbers—instead, they are always chosen from a *finite field*—as a result x, y, a, b have integer values assigned from a finite set of integers. The various values of (x, y) on the elliptic curve along with the addition operation and the identity element form an Abelian group. All computations are done modulo p, where p is a prime number.

Prior to a secret communication, both Amy and Bob must agree upon using a specific elliptic curve (which is not a secret) and a specific point F on that curve. Amy then picks a *secret random number* d_A that is her *secret key*, computes $e_A = d_A \cdot F$, and publishes it as her *public key*. In the same manner, Bob will also pick his *secret key* d_B and publish a *public key* $e_B = d_B \cdot F$.

To send the secret message M, Amy will simply compute $d_A \cdot e_B$ and use the result as the secret key to encrypt M using a conventional symmetric block cipher (like 3DES or AES). To decrypt this ciphertext, Bob has to know the secret key of the block cipher. Bob will be able to compute this by calculating $d_B \cdot e_A$, since

$$d_B \cdot e_A = d_B \cdot (d_A \cdot F)$$

$$= (d_B \cdot d_A) \cdot F$$

$$= (d_A \cdot d_B) \cdot F$$

$$= d_A \cdot (d_B \cdot F)$$

$$= d_A \cdot e_B$$

The security of the previous scheme is based on the property of elliptic curves that given F and $k \cdot F$, it is intractable to compute k when the keys are large (computationally, this is equivalent to the discrete logarithm problem discussed earlier). In the case of communication between Amy and Bob, $e_A = d_A \cdot F$, both e_A and F are known, but it is extremely difficult to compute the secret key d_A. The same applies to the computation of d_B, Bob's secret key.

19.10 AUTHENTICATION SERVER

An AS (also called a *key distribution center*) is a trusted central agent whose responsibility is to distribute conversation keys among clients, prior to initiating an authenticated conversation. It is a tricky job that has to be done right to preserve the integrity of the keys. The schemes described here are due to Needham and Schroeder [NS78].

19.10.1 Authentication Service for Secret-Key Cryptosystems

Let four users A, B, C, D be connected to an AS S. For each user i, the server maintains a unique key k_i (like a password) that is only known to user i and the server S. This is different from the secret key k_{ij} to be used in the conversation between users i and j. We will use the notation $k_i(M)$ to represent a message M encrypted by the key k_i. Obviously, i is the only user who can decrypt it. Here is a summary of the protocol that user A will use to obtain a conversation key k_{AB} between A and B:

$A \rightarrow S$ Give me a conversation key to communicate with B
$S \rightarrow A$ Retrieve k_{AB} from the message $k_A(B, k_{AB}, k_B(k_{AB}, A))$
$A \rightarrow B$ Retrieve k_{AB} from $k_B(k_{AB}, A)$ that I obtained from S
$B \rightarrow A$ Yes, can you decode $k_{AB}(n_B)$ and decrement the argument?
$A \rightarrow B$ Here is $k_{AB}(n_B-1)$

In the previous exchange, n_B (called a *nonce*) is meant for single use. When B finds A's answer to be correct, a secure communication channel is established between A and B.

To sign a document, A will first convert the original plaintext M into a digest m of a smaller size and then use the following protocol:

$A \rightarrow S$ Give me a signature block for the message m.
$S \rightarrow A$ Here is $k_S(m)$—please append it to your message.

Now, A sends $k_{AB}(M, k_S(m))$ to B. To verify the signature, B computes the digest m' from the message decrypted by it and sends it to S for signature verification. In response, S now sends back $k_S(m')$ to B. If $k_S(m) = k_S(m')$, then B verifies the signature of A and hence the authenticity of the message from A.

Safeguard from replay attacks: This problem is related to the freshness of the message that is being transmitted. It is possible for an intruder to copy and replay the ciphertext, and the recipient has no way of knowing if this is a replay of an old valid message. Imagine what will happen if an intruder unknowingly replays your encrypted message to your banker: "Transfer \$100 from my account to Mr. X's account." One way to get around this problem is to include a special identifier with each communication session. Ideally, such an identifier will be used *only once* and never be repeated—this is a safeguard against possible replay attack. For example, A, while asking for a conversation key from S, will include such an identifier n_A with the request, and S will include that in the body of its reply to A. Such an integer is called a *nonce*. Typically, it is a counter or a time stamp. Prior to each message transmission, the two parties have to agree to a nonce. The reuse of the same nonce signals a replay attack.

19.10.2 Authentication Server for Public-Key Systems

It is assumed that everyone knows the public key e_S of the server S. To communicate with B, A will first obtain the public key of B from the AS.

$A \rightarrow S$ Give me the public key e_B of B
$S \rightarrow A$ Here is $d_S(e_B)$

Now, *A* can decrypt it using the public key of the server *S*. This prevents *A* from receiving a bogus key e_B from an imposter. *B* can obtain the public key from *A* in the same way. In the next step, *A* and *B* perform a handshake:

$A \rightarrow B \quad e_B(n_A)$ Only *B* can understand it

$B \rightarrow A \quad e_A(n_B, n_A)$ Only *A* can understand it.

                                 A finds that *B* successfully received its nonce

$A \rightarrow B \quad e_B(n_B)$ *B* finds that *A* successfully decrypted its nonce

Now *A* and *B* are ready to communicate with each other.

19.11 DIGITAL CERTIFICATES

A certificate is a document issued to a user by a trusted party. The certificate identifies the user and is like a passport or a driver's license. When Amy wants to withdraw $5000 from her account in Sunrise Bank, Iowa, Sunrise Bank needs to be absolutely sure that it is Amy who is trying to access her account. So Amy will produce a certificate issued by Sunrise Bank and signed by the bank's private key. The components of a certificate are as follows:

Name	**Amy Weber**
Issued by	Sunrise Bank, Iowa
Certificate type	Checking account number
Account number	1234567
Public key	1A2B3C4D5E6F
Signature of the issuer	Signed using the private key of Sunrise Bank

In electronic banking, the web browser uses digital certificates to enhance the security of access. They are electronic counterparts to ATM cards and must be presented by the browser to the bank before the bank will permit the user access to her account. The bank ATM card stores an encrypted version of a digital certificate along with other information that may be needed by a financial institution. Used in conjunction with encryption, user id, and password, digital certificates provide an acceptable security solution.

In a slightly different scenario, when Amy wants to electronically transfer $5000 to a car dealership for buying her car, she presents the certificate to the dealership. The dealership will want to verify the signature of the bank using their public key. Once the signature is verified, the dealership trusts Amy's public key in the certificate and accepts it for future transactions.

A *public key infrastructure* (PKI) is a mechanism for the certification of user identities by a third party. It binds public keys of each user to her identity. For verifying the authenticity of the public key of Sunrise Bank, the car dealership may check Sunrise Bank's certificate issued by a higher authority. The chain soon closes in on important certification authorities. These certification authorities sign their own certificates, which are distributed in a trusted manner. An example of a certification authority is Verisign (www.verisign.com), where individuals and businesses can acquire their public-key certificates by submitting acceptable proofs of their identity. Public keys of important certification authorities are also posted on the WWW. The correctness of these keys is the basis of the trust.

A widely used certification format is X.509, which is one component of the CCITT's* standard for constructing global directories of names. It binds a public key to a distinguished name, or an email address, or a DNS entry. In addition to the public key certificate, it also specifies a certification path validation algorithm. Note that the names used in a certificate may not be unique. To establish the credentials of the owner and her certificate (or the certificate issuer's signature), references to other individuals or organizations may be necessary. In PKI, the administrative issues of *who trusts whom* can get quite complex.

19.12 CASE STUDIES

In this section, we present three security protocols for the real world. The first one uses secret-key encryption, and the two others are hybrid, since they use both public-key and secret-key encryption.

19.12.1 Kerberos

Kerberos is an authentication service developed at MIT as a part of Project Athena. It uses secret keys and is based on the Needham–Schroeder authentication protocol. The functional components of Kerberos are shown in Figure 19.5. Clients and servers are required to have their keys registered with an AS. Servers include the file server, mail server, secure login, and print server. The users' keys are derived from their passwords, and the servers' keys are randomly chosen. For enhanced security, an unencrypted password should neither travel over the network nor be stored on the client machine or even the AS database. Once a password is used, it must be immediately discarded. Client A planning to communicate with a server first contacts the AS and acquires a *session key* $k_{A,TGS}$. This allows

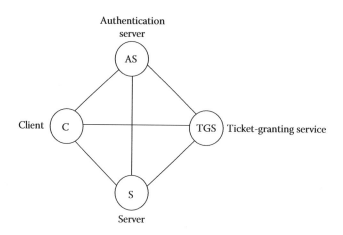

FIGURE 19.5   The components of Kerberos.

---

* Acronym for Comité Consultatif International Téléphonique et Télégraphique, an organization that sets international communications standards.

A to communicate with a ticket-granting service (TGS). A ticket is an entity that a client presents to an application server to demonstrate the authenticity of its identity. When the session key $k_{A,TGS}$ is presented to the TGS, it issues a *session key* that enables the client to communicate with a specific server for a predefined window of time.

A sample authentication and ticket-granting operation is outlined here:

$A \rightarrow AS$ Give me a key to communicate with TGS

$AS \rightarrow A$ Retrieve $k_{A,TGS}$ from $k_{A,AS}(k_{A,TGS}, k_{AS,TGS}(k_{A,TGS}, A))$
           (Here $k_{A,TGS}$ is A's *key* for TGS, and $k_{AS,TGS}(k_{A,TGS}, A)$ is the *ticket*)

$A \rightarrow TGS$ Here is my ticket $k_{AS,TGS}(k_{A,TGS}, A)$ issued by AS.
           Now grant me a session key to contact server B.

$TGS \rightarrow A$ TGS retrieves $k_{A,TGS}$ and sends $k_{A,TGS}((k_{A,B}, B, T), k_{B,TGS}(k_{A,B}, A, T))$
           (A will retrieve the session key $k_{A,B}$ from it. T is the expiration time).

$A \rightarrow B$ $k_{B,TGS}(k_{A,B}, A, T)$ (B will retrieve $k_{A,B}$ from this)

If a client asked only the AS to generate $k_{A,B}$, then the reply would have been encrypted with $k_{A,AS}$, and *A* would have to use $k_{A,AS}$ to decrypt it. This means entering the password for each session (it is a bad idea to cache passwords). In the two-step process, to obtain a session key for any other service, it is sufficient to show the ticket to TGS. The expiration time (usually 10 h) prevents the reuse of a stolen ticket at a later moment. The two-step process thus relieves the user from repeatedly entering the password, and this improves the security.

Although AS and TGS are functionally different, they can physically reside in the same machine. For small-scale systems, a single AS–TGS is adequate. However, as the scale of the system grows, the AS–TGS unit becomes a bottleneck. So Kerberos designers divide the network into *realms* across organizational boundaries. Each realm has its own AS and TGS.

After a successful login, clients use tickets to acquire separate session key for file servers, printers, remote login, or email. Initially, Kerberos 4 used DES for encryption, but it has been retired in Kerberos 5, which allows a range of more secure encryption methods—these include 3DES, AES-128, and AES-256.

19.12.2 Pretty Good Privacy

On April 17, 1991, *New York Times* reported on an unsettling US Senate proposal. It is part of a counterterrorism bill that would force manufacturers of secure communication equipments to insert special trap doors in their products, so that the government could read anyone's encrypted messages. The US government's concern was to prevent encrypted communications related to clandestine operation with entities outside the United States. The US government was quite concerned about the rising circulation of RSA public-key encryption at that time. This led Philip Zimmermann to develop PGP cryptosystem in a hurry before the bill was put to vote. Zimmermann distributed PGP as a freeware. In a way, this bill led to the birth of PGP encryption, although the US government's bill was later defeated.

PGP is primarily used to encrypt emails. It is a *hybrid cryptosystem* that combines the best features of both private and public-key cryptography. To encrypt a message, PGP takes the following steps:

1. It compresses the plaintext M. In addition to saving disk space and transmission time, data compression increases its resilience to cryptanalysis that relies on discovering patterns in the plaintext.

2. It generates a session key derived from the random movements of the sender's mouse and the sender's keystrokes. The session key is a one-time secret key.

3. It encrypts (a) the compressed data with the session key and (b) also encrypts the session key with the public key of the recipient. The two are then transmitted.

To decrypt the message, the recipient first retrieves the session key using her private key and then uses this key to decrypt (and subsequently decompress) the ciphertext.

While the use of public-key cryptography overcomes the key distribution problem, the encryption of the plaintext using secret keys is much faster than public-key encryption. PGP keys are 512–1024 bits long. Longer keys are cryptographically more secure. PGP stores the keys on the user's hard disk in files called *keyrings*. The *public keyring* holds the public keys of parties that the sender wants to communicate with, and the private keyring stores the sender's private keys.

19.12.3 Secure Socket Layer

The SSL protocol was developed by Netscape to establish secure communication between applications on the Internet. SSL 3.0 (available from 1996) received endorsements from the credit card giants Visa and MasterCard. SSL certifies the identity of the website to an online user and allows clients to communicate with servers while preserving confidentiality and integrity. The upper layer protocol is HTTP (for web service), or IMAP (for mail service), or FTP (for file transfer), and the lower layer protocol is TCP. These higher-level services process their requests through SSL. The upper layer protocol has two layers: a *handshake layer* and a *record layer*.

The handshake layer is the upper layer that provides three guarantees:

1. At least one of the peers is authenticated using public-key cryptography.

2. The shared secret key negotiated between the peers remains unavailable to eavesdroppers.

3. No intruder can transparently modify the communication.

The initial handshake uses public-key cryptography and helps establish a shared secret key, which is subsequently used for secure communication. The following steps illustrate the establishment of a secure shared key, when a client (Amy) communicates with a web server (Bob):

Amy: Sends a *ClientHello*: at this time, she presents Bob with (1) a session id, (2) protocol version (http here), (3) a list of block or stream ciphers (in order of preference) that is supported by her machine, (4) a set of message compression algorithms (like MD5, SHA-2) that is supported by her machine, and (5) a random data to be used for secret key generation.

Bob: Responds with a *ServerHello*: now Bob (1) picks a cipher of his choice and the message compression algorithm to be used, (2) echoes the session id, and (3) sends random data to be used for secret-key generation. Then Bob presents his certificate and (optionally) asks Amy for her certificate.

Amy: Verifies Bob's certificate and presents her own certificate. Then she creates a *premaster secret* and sends it to Bob using his public key obtained from his certificate.

Bob: Verifies Amy's certificate and decrypts the *premaster secret* using his private key.

Both Amy and Bob now move to the *Client Key Exchange* phase, where they use the *premaster secret* and the random numbers to create a new *master secret*. Both will use the *master secret* to generate the session keys—these are symmetric keys that will be used to encrypt and decrypt information exchanged during the SSL session. When all these are done, Amy and Bob exchange a *Change Cipher Spec* message followed by a *finished* message. This signals that they will use these keys for the rest of the session. The secure session can now begin.

The record layer is the lower layer that fragments the messages into manageable blocks, compresses the data, appends a message digest using the hash function from the agreed protocol suite, and encrypts, before transmitting it through the TCP connection. Received data are decrypted, decompressed, reassembled, and then delivered to the client.

SSL 3.0 has been later upgraded to TLS 1.0 [DA99], so the protocol is referred to as SSL 3.0/TLS 1.0. Despite many similarities, they are not interoperable. There is a potential for a *man-in-the-middle attack* in SSL, and TSL 1.0 initially claimed to address it. However, in 2009, the vulnerability of SSL 3.0/TLS 1.0 against a man-in-the-middle attack was made public. Despite its feasibility in laboratory scale experiments, there is no documented history of such industrial strength attacks. An absolute safeguard is to send the public key of the server to the client via a separate channel. The client can take advantage of browsers and some Internet software that are distributed via CD-ROMs for obtaining the public keys.

The various ciphers and the message digest functions are preloaded at each site. When a user communicates with a secure site on the WWW, she notices *https://* instead of *http://* in the URL. The word https means http using SSL.

The implementation of SSL uses hybrid encryption, where the mutual authentication is based on public keys, but final communication uses the secret keys. This is because public-key-based encryption and decryption mechanisms are computationally expensive—so their use should be minimized as much as possible. Due to its extreme popularity in e-commerce, SSL/TLS is supported by almost all browsers.

19.13 VIRTUAL PRIVATE NETWORKS AND FIREWALLS

Unlike cryptographic solutions to security that are primarily implemented above the TCP/IP layers, *virtual private networks* (*VPNs*) and *firewalls* provide security by tweaking some of the lower layers of the protocol stack.

19.13.1 Virtual Private Network

A VPN is a trusted communication tunnel between two or more devices across an untrusted public network (like the Internet). Businesses today are faced with supporting a broad variety of communications among a wider range of sites and escalating communication cost. Employees are looking to access the resources of their corporate intranets as they take to the road, telecommute, or dial in from customer sites. In addition, business partners share business information, either for a joint project of a few months' duration or for long-term strategic advantage. Wide-area networking between the main corporate network and branch offices, using dedicated leased lines or frame-relay circuits, does not provide the flexibility required for quickly creating new partner links or supporting project teams in the field. The rapid growth of the number of telecommuters and an increasingly mobile sales force gobbles up resources as more money is spent on modem banks and long-distance phone charges. VPN provides a solution to this problem without using leased lines or WAN.

VPNs rely on *tunneling* to create a private network that reaches across the Internet. *Tunneling* is the process of encapsulating an entire packet within another packet and sending it over a network. The network understands the protocol of the outer packet and knows its endpoints (i.e., where the packet enters and exits the network). There are two types of endpoints for tunnels: an individual computer or a LAN with a security gateway, which might be a router (or a firewall). The tunnel uses cryptographically protected secure channels at the IP or link level (using protocols like IPSec or L2TP) and relieves the application layer from overseeing the security requirements. Tunneling has interesting implications for VPNs. For example, one can place a packet that uses a protocol not supported on the Internet inside an IP packet and send it safely over the Internet. Or one can put a packet that uses a private IP address inside a packet that uses a globally unique IP address to extend a private network over the Internet.

An obvious question is: Do we need VPN if we use SSL? What is the difference between the two? Protocols used to implement VPN operate at a much lower level (network or link layer) in the network protocol stack. The advantage of having the crypto bits at a lower level is that they work for all applications/protocols. On the flip side, to support this capability, extra software is needed. SSL caters to client–server communication, and the extra software is already packaged into all web browsers.

19.13.2 Firewall

A *firewall* is a filter between your private network (a zone of high trust) and the Internet (a zone of low trust). A *personal firewall* provides controlled connectivity between a personal computer and the Internet, whereas a *network firewall* regulates the traffic between a local network and the Internet. Firewalls can be configured according to an individual or an organization's security requirements. They can restrict the number of open ports or determine what types of packets will pass through and which protocols will be allowed. Application layer firewalls can inspect the contents of the traffic and block inappropriate materials or known viruses. Some VPN products are upgraded to include firewall capabilities.

19.14 SHARING A SECRET

Consider the following problem: Nine members of a family have their family treasures guarded in a safe that can be opened by a secret code. No individual should know this secret code. The locking mechanism should be such that the lock can be opened if and only if five or more of the members cooperate with one another. To protect data, one can encrypt it, but to protect a secret key, further encryptions will not help. One also needs to safeguard against a single point of failure that could destroy the key. Making multiple copies of the key appears to be a solution to the problem of single point of failure, but it increases the danger of a security breach—by not taking the majority into confidence, anyone can open the safe himself or herself or with the help of a small number of accomplices.

What we are looking for is a mechanism of splitting a secret code. How to split a secret code and hand over the pieces to the members of the family, so that this becomes possible? In [S79], Shamir proposed a solution to the problem of sharing a secret key.

Shamir's solution is as follows: let D be the secret code that we want to safeguard, n be the number of members, and k be the quorum, that is, the smallest number of members who must cooperate with one another to open the safe. Without loss of generality, consider D to be an integer. Shamir used polynomial interpolation: Given k distinct points $(x_1,y_1),(x_2,y_2),\ldots,(x_k,y_k)$ on the 2D plane, there is one and only one polynomial $q(x)$ of degree $k - 1$, such that $\forall i : 1 \le i \le k : y_i = q(x_i)$. Now pick any polynomial $q(x)$ of degree $(k - 1)$:

$$q(x) = a_0 + a_1 \cdot x + a_2 \cdot x^2 + \cdots + a_{k-1} \cdot x^{k-1}$$

The only requirement here is that $a_0 = D$, the secret code. Now, pick a set of values $1, 2, 3, \ldots, n$ for x and use the above polynomial to evaluate $D_1 = q(1), D_2 = q(2), D_3 = q(3),\ldots,D_n = q(n)$. From any subset of k of these (i, D_i) values, one can find all the coefficients of $q(x)$ by interpolation. This includes finding the value of $a_0 = D$, the secret code. However, by using fewer than k of these values of D_i, no one can derive D. So these n values $D_1,D_2,D_3,\ldots,D_n$ can be distributed as *pieces* of the secret code, so that any k out of n holders of the pieces have to come together to decrypt the secret. Such a scheme is known as a (k, n)-threshold scheme for secret sharing. To avoid two disjoint quorums, it makes sense to make $[(n + 1)/2] \le k \le n$.

The keys can be made more uniform in size by using mod p arithmetic, where p is a prime number larger than D and n. All keys will be in the range $[0, p)$. The mechanism is robust: A loss of a single key or member poses no threat to the security. A new member can be added to the family by generating another key piece.

19.15 CONCLUDING REMARKS

Security is a never-ending game. The moment we think we have provided enough security to data or communication, crooks, hackers, and cryptanalysts start digging out loopholes. New algorithms are developed for operations whose apparent intractability provided the cornerstone of security. In addition, technological progress enables the development of faster machines with enormous computing power, which simplifies code breaking. Yet, electronic transactions in the business world have increased so much that we need some security—we cannot sleep with our doors open. Every web server installed at a site opens a window to the Internet. History shows that complex software usually has bugs—it needs a smart crook to discover some of them and use these as security loopholes. This reaffirms the importance of designing reliable software.

Digital certificates were proposed to authenticate transactions between parties unknown to each other. PKI was formulated to assist a client in making decisions about whether to proceed with an electronic transaction. Traditional PKI has come under criticism. The fact that online transactions are increasing rapidly without much help of PKI is paradoxical. Some find the traditional proof of identity to be intrusive. The one-size-fits-all electronic passport has been a topic of debate.

Since the past decade, *phishing* has been a major source of security fraud. According to the RSA Fraud report published by EMC, the total number of phishing attacks launched in 2012 was 59% higher than that in 2011, leading to an estimated global loss of $1.5 billion in 2012. The fraudulent websites typically last for a week or less.

19.16 BIBLIOGRAPHIC NOTES

Although cryptography dates back to thousands of years, Shannon's work laid the foundation of modern cryptography. Kahn [K67] provides the historical perspectives. Schneier's book [S96] is an excellent source of basic cryptographic techniques including many original codes. The official description of DES published by NIST is available from http://www.itl.nist.gov/fipspubs/fip46-2.htm. AES is described in [DR02]. Blowfish is described in Schneier's book [S96], and Twofish was created by Schneier et al. [SKW+98]. Diffie and Hellman [DH76] laid the foundation of public-key encryption. RSA public-key cryptography is discussed in the article by Rivest et al.'s book [RSA78]. Needham and Schroeder [NS78] contains the original description of Needham–Schroeder authentication protocol. ElGamal cryptosystem is presented in [E84]. The distributed computing project Athena in MIT spawned many technologies, including Kerberos. Steiner et al.'s book [SNS88] contains a detailed description of Kerberos. A complete description of PGP is available from Garfinkel's book [G94]. Miller [M85] (and independently Neal Koblitz) described elliptic curve cryptography.

An excellent tutorial by Certicom is available from http://www.certicom.com/index. php/ecc-tutorial. Netscape developed SSL. The specification of SSL 3.0 appears in <draft-ietf-tls-ssl-version3-00.txt>. RFC 2246 by Dierks and Allen [DA99] describes TLS 1.0.

EXERCISES

19.1 Learning how to attack is an important component of learning about security methods. A well-known technique for breaking common substitution ciphers is *frequency analysis*. It is based on the fact that in a stretch of the English language, certain letters and combinations occur with varying frequencies: For example, the letter E is quite common, while the letter X is very infrequent. Similarly, the combinations NG or TH are quite common, but the combinations TJ or SQ are very rare. The array of letters in decreasing order of frequency are E T A O I N S H R D L U ... J, Q, X, Z. Basic frequency analysis counts the frequency of each letter and each phrase in the ciphertext and tries to deduce the plaintext from the known frequency of occurrences.

Search the web to find data about frequency analysis [L00]. Then use frequency analysis to decrypt the following cipher:

WKH HDVLHVW PHWKRG RI HQFLSKHULQJ D WHAW PHVVDJH
LV WR UHSODFH HDFK FKDUDFWHU EB DQRWKHU XVLQJ D
ILAHG UXOH, VR IRU HADPSOH HYHUB OHWWHU D PDB
EH UHSODFHG EB G, DQG HYHUB OHWWHU E EB WKH OHWWHU
H DQG VR RQ.

(*Hint*: Do a frequency analysis of the single letter in the ciphertext. Then try to match it with the frequency analysis data and discover the mapping.)

19.2 A simple form of substitution cipher is the *affine cipher* that generates the ciphertext for an English language plaintext as follows: (1) The plaintext is represented using a string of integers in the range [0..25] ($a = 0, b = 1, c = 2 \ldots$ and so on), and (2) for each plaintext letter P, the corresponding ciphertext letter $C = a \cdot P + b \bmod 26$, where $0 < a, b < 26$ and a and b are relatively prime to 26.

Bob generated the ciphertext C by encrypting the plaintext P twice using two different secret keys $k1$ and $k2$. Thus, $C = E_{k1}(E_{k2}(P))$. Show that if *affine ciphers* are used for encryption, then the resulting encryption is possible using just a single affine cipher.

19.3 Your bank asked you to change your netbanking password. The previous password had six characters consisting of lowercase letters and integers. The new password will have at least eight characters—it must contain both upper and lower case letters, integers, and a special character chosen from @,*,−,^,&. If you choose a new password whose length is 10 characters, then how much more effort (measured by the number of steps) will be needed to break your password compared with the previous scenario?

19.4 What are the advantages and disadvantages of CBC over simple block ciphers?

19.5 Certain types of message digests are considered *good* for cryptographic checksums. Is the sum of all bits a good cryptographic checksum? Why or why not? Suggest a measure of this goodness.

19.6 Let W' denote the bit pattern obtained by flipping every bit of a binary integer W. Then show that the following result holds for DES encryption

$$C = E_k(P) \Rightarrow C' = E_{k'}(P')$$

19.7 To authenticate a message m using ElGamal cryptosystem, Alice picks a hash function H and computes a message digest $H(m)$. Then she (1) chooses a prime p and a random number $k(0 < k < p)$ relatively prime to $(p - 1)$, (2) finds numbers r and s such that $r = h^k \pmod{p}$ and $H(m) = x \cdot r + k \cdot s \bmod (p-1)$ (such a pair (r, s) is guaranteed to exist when k is relatively prime to $(p - 1)$), and (3) sends the signature as (r, s). Bob verifies the signature as authentic, only if $0 < r, s < p-1$ and $gH(m) = h^r \cdot r^s$. Prove that the previous condition authenticates the message from Alice [see E84].

19.8 Consider the following protocol for communication from Alice to Bob:

a. Alice signs a secret message M with her private key, then encrypts it with Bob's public key, and sends the result to Bob, that is, $A \rightarrow B : e_B(d_A(M))$.

b. Bob decrypts the message using his private key d_B and then verifies the signature using Alice's public key e_A, that is, $e_A(d_B(e_B(d_A(M)))) = M$.

c. Bob signs the message with his private key, encrypts it with Alice's public key, and sends the result back to Alice, $B \rightarrow A : e_A(d_B(M))$.

d. Alice decrypts the message with her private key and verifies the signature using Bob's public key. If the result is the same as the one she sent Bob, she knows that Bob correctly received the secret message M.

Part 1: Show that this protocol violates confidentiality. Let Mallory be an active hacker who intercepts the encrypted signed message $e_B(d_A(M))$ communicated by Alice to Bob in step 1. Show that Mallory can use a modified protocol (by changing only steps (a) and (d)) to learn the original secret M without Bob (or Alice) noticing.

Part 2: Consider ways of repairing the protocol. Will the use of different sets of keys for signing/verification and encryption/decryption fix the problem?

19.9 The security of RSA encryption is based on the apparent difficulty of computing the prime factors s, t of a *large integer* $N = s \times t$. Investigate what are some of the fast algorithms for computing the factors of N.

19.10 What are the advantages of elliptic curve cryptography over RSA?

19.11 A company stores its payroll in a local server, and there are five employees in the payroll department. Access to payroll data requires a special password. Devise a scheme so that payroll data cannot be accessed unless any *three out of five employees* reach an agreement about carrying out the task. Implement your scheme, and provide a demonstration.

19.12 A *birthday attack* refers to the observation that in a room containing only 23 people or more, there is a better than even chance that 2 of the people in the room have the same birthday even though the chances of a person having any specific birthday is 1 in 365. The point is, although it might be very difficult to find M from $m = H(m)$, it is considerably easier to find two different messages (M, M') with identical hash.

Using this fact, explain how birthday attacks can be used to get the signature of someone on a fraudulent document.

19.13 Experts are concerned about various kinds of loopholes in Internet voting. Study how Internet voting is carried out in practice, and think of how Internet voting can be misused.

Programming Exercises

19.1 Dr. Susan L. Gerhart and her team, with funding from NSF, designed a simple version of DES (called S-DES). It is a block cipher that uses 8-bit blocks and a 10-bit key. See http://security.rbaumann.net/modtech.php?sel=2 for a description.

Design an encryption and decryption system working on printable ASCII text using S-DES. Consider only the letters from A to Z, a to z, the digits 0 to 9, and the two punctuation symbols: space (writing it as underscore) and period (this accounts for 64 symbols). Give a few examples of how to use your system.

19.2 PGP is a popular cryptosystem that helps the exchange of confidential email messages. Download and install PGP from its official website* to your personal computer. Run PGPkeys (e.g., click on the PGPtray icon on your taskbar and choose Launch PGPkeys), and use the on-screen instructions to generate your initial private key. To choose the key size, select the default option. Upload your public key to the keyserver using the on-screen instructions.

PGP stores your private key in a file called *secring.skr*. PGPfreeware guards your private key by asking you to invent a secret hint that only you will know.

Obtain public keys: To send encrypted email to someone, you need to get that person's public key. Obtain the public key of the TA or a friend, and add this key to your public key ring. There are several ways to do this:

1. Look to see if the person has their public key on their web page. Copy that key in full to the clipboard, click on the PGP icon, and select "Add Key from Clipboard."

---

* PGP freeware version 8.0 can be downloaded, for both PC and Mac, from http://www.pgpi.org/.

2. Look into the central keyserver where most people put their public keys (you did this in step 1). To do this, run PGPkeys, select Keys/Search, type (any part of) the person's name or email address in the User id box, and click Search. If the key is found, then right-click on it and select "Import to Local Key ring."

3. Ask the person to send you their public key by email, and use the same method as in the prior step.

Encrypt your message: To send an encrypted message, choose a text file, right-click on it, and choose the PGP menu item that offers to encrypt it. Select (drag and drop) the name of the user for whom you want to encrypt; this will encrypt your message using the public key of the user. The result is a new file with the same name, but with an added .pgp extension.

Send email: To send the encrypted message, attach the .pgp file to your email. Check with the recipient to verify if she could decrypt it.

19.3 Team up with two other persons, and let these two run a communication across a communication channel that has not been secured. We suggest that you run the experiment on a dedicated network in a lab that is not being used by others. Now implement a man-in-the middle attack, and then divulge to the team members how you launched the attack.

19.4 Implement an application layer firewall that will block all emails containing a given topic of your choice.

CHAPTER 20

Sensor Networks

20.1 VISION

In the 21st century, the number of processors in daily use vastly outnumbers the laptops or desktop computers or personal digital assistants. Most gadgets that have become indispensable for us contain one or more processors. A state-of-the-art automobile has 50 or more processors in it. These sense various conditions and actuate devices for our safety or comfort. With the advancement of technology, the physical size of the processors has diminished, and a new breed of applications has emerged that relies on miniature postage-stamp-size processors sensing physical parameters and performing wireless communication with one another to achieve a collective goal. The primary job of a sensor network is tracking and monitoring. Estrin et al. [EGH+99] summarized the endless potential of sensor networks that range from ecological monitoring to industrial automation, smart homes, military arena, disaster management, security devices, sustainability, and health-related application. Some of these applications have been developed during the past few years, and new areas of applications are constantly being explored. In addition, with the rapid penetration of smartphones that are equipped with a few sensors, new applications are emerging almost on a daily basis. These are slowly but surely transforming our daily lives and habits and will continue to do so in the foreseeable future.

The technology of networked sensors dates back to the days of the Cold War. The SOund SUrveillance System (SOSUS) was a system of acoustic sensors deployed at the bottom of the ocean to sense and track Soviet submarines. Modern research on sensor networks started since the 1980s under the leadership of Defense Research Advanced Projects Agency (DARPA) of the US government. However, the technology was not quite ready until the late 1990s.

Imagine hundreds of sensor nodes being airdropped on a minefield. These sensor nodes form an ad hoc wireless network, map out the location of the buried mines, and return the location information to a low-flying aircraft. Nodes may be damaged or swept away in wind and rain. In fact, the loss or failure of such devices is an expected event—yet the sheer number of these devices is often enough to overcome the impact of failures. As another example in disaster management, randomly deployed sensor nodes in a disaster zone can help identify hot spots and guide rescuers toward it as quickly as possible. A flagship problem is to rescue people from one of the upper floors of a tall building where a fire breaks out, and two of the four stairwells are unusable due to carbon monoxide formation. These outline the kind of applications charted for sensor networks.

Some deployments of sensor networks are preplanned, as in automobiles or safety devices. Others are ad hoc in nature. Sensor networks can be wired or wireless. In this chapter, we will primarily focus on wireless sensor networks.

20.2 ARCHITECTURE OF SENSOR NODES

A *wireless sensor network* is a wireless network of miniature sensor nodes. These nodes sense various types of environmental parameters, execute simple instructions, and communicate with neighboring nodes within their radio range. The growth of sensor networks is largely due to the availability of miniature inexpensive sensors based on micro electro mechanical systems (MEMS) technology. With several vendors jumping in, users now have a choice of sensor nodes that can be used as the building blocks of sensor networks. In addition, many smartphones are equipped with camera, microphone, GPS, compass, and accelerometers, which have led to the development of various opportunistic and participatory sensing technologies.

20.2.1 MICA Mote

The UC Berkeley researchers and their collaborators pioneered the design of a class of sensor nodes called MICA® motes.* Several versions of these sensor nodes (MICA2, MICAZ) are now available for prototype design. A typical third-generation MICA mote consists of an 8-bit ATMEL ATMEGA 128L processor running at 4 MHz, with 128 kB of flash memory for program storage and a 4 kB SRAM for read–write memory. It also has a 512 kB flash memory for storing serial data from measurements. The data are received via 10-bit analog-to-digital converters from sensor cards attached to it. The serial port is used for downloading programs from (or uploading results to) a desktop or a laptop PC. A multichannel radio that can work at 868/916 or 433 or 315 MHz serves as the real-world communication conduit. It can send or receive data at 40 kbps. The radio range is programmable up to 500 ft, but the actual coverage depends on environmental conditions. Control signals configure the radio to either transmit or receive or the power-off mode. A schematic diagram of a MICA mote is shown in Figure 20.1.

Each mote is battery powered. Since the motes need to perform in unattended conditions for a long time, energy conservation is a major issue, even if the battery technology has been improving. The radio consumes less than 1 µA when it is off, 10 mA when it is receiving data, and 25 mA when transmitting—so conserving radio power by minimizing communication is a key to longer battery life. The processor consumes only 8 mA when it

---

* MICA mote is a product of Crossbow Technology. The data refer to MPR400CB. There is a range of products like this.

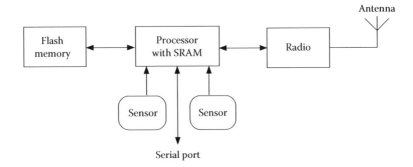

FIGURE 20.1   The architecture of a MICA mote.

is running, but only 15 µA in the sleep mode. To conserve energy, one can switch the mote to one of the several *sleep* modes. These include the following:

1. IDLE mode that completely shuts off the processor and the radio.

2. POWER DOWN mode that shuts everything off except a watchdog timer. It can help the processor set an alarm and wake up at an appropriate time.

In addition, researchers are constantly developing algorithmic solutions for energy conservation directed to specific applications. An additional technique is energy harvesting, where ambient energy in the environment is converted into electrical energy to power up the sensor nodes. Although the ambient energy is quite small (less than 100 µW/cm^3), this merits further investigation in certain forms of applications. Important sources of ambient energy are solar, mechanical energy (vibration, finger movements), thermal energy (temperature gradients), etc.

20.2.2 ZigBee-Enabled Sensor Nodes

To facilitate the growth of wireless sensor networks in *low-data-rate* and *low-power* applications, the *ZigBee alliance* proposed an open standard based on the IEEE 802.15.4 specification of the physical and the medium access control (MAC) layers. Low-data-rate applications require sensor nodes to occasionally wake up and carry out an action, but most of the time, they sleep. Numerous such applications are feasible in industrial controls, medical devices, various alarm systems, and building automation. Due to the very low power demand, such nodes are supposed to last for a year or more with a single set of alkaline batteries. This is accomplished by carefully choosing the beaconing intervals and various sleep modes. These standards have been well received by the industry—as a result, many sensor nodes manufactured today are ZigBee compliant. An example is TelosB mote developed by the University of California, Berkeley.

ZigBee nodes communicate at 2.4 GHz, 915 MHz, and 868 MHz using DSSS technology. The 2.4 GHz band (also used by Wi-Fi and Bluetooth) is available worldwide and supports a raw data rate of 250 kbps. The 915 MHz band supports applications in the United States and some parts of Asia with a raw data rate of 40 kbps. The 868 MHz band is designed for applications in Europe and supports a raw data rate of 20 kbps. ZigBee defines the lower layers of the protocol suite. Specifically, it adds network structure, routing, and security (e.g., key management and authentication) to complete the communications suite. On top of this robust wireless engine, the target applications reside. ZigBee 1.0 protocol stack was ratified in December 2004.

ZigBee and IEEE 802.15.4 support three kinds of devices: *reduced functionality devices* (RFDs), *full-functional devices* (FFDs), and *network coordinators*. Most sensor nodes in typical applications belong to the RFD class. An RFD is an end device that can only communicate with its parent node or a significant neighbor, but cannot act as a router. An FFD has the ability to act as a router. Finally, a network coordinator is the root of the network tree and has the ability to serve as a bridge connecting with other networks. For example, Stargate NetBridge is an embedded sensor network gateway device that connects the MICA motes to an existing Ethernet network.

ZigBee supports three different kinds of network topologies: (1) *star* network, (2) *cluster tree* (also known as a *connected star*), and (3) *mesh* network (Figure 20.2). A typical application of the star configuration is a home security system. The cluster tree extends the tree topology by connecting multiple star networks. Finally, the mesh network helps build a sizeable system by accommodating a large number of wireless nodes and provides multiple paths between nodes to facilitate reliable communication. ZigBee networking protocol controls the topology by computing the most reliable paths and provides networks with self-healing capabilities by spontaneously establishing alternate paths (if such a path exists) whenever one or more nodes crash or the environmental conditions change.

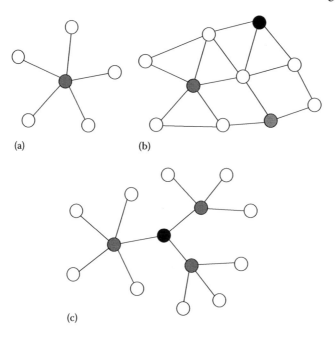

(a)

(b)

(c)

FIGURE 20.2 The three types of topologies supported by ZigBee: (a) star, (b) mesh, and (c) cluster tree. The white circles represent RFDs, the gray circles represent FFDs, and the black circles denote network coordinators (base stations).

ZigBee supports three levels of security: (1) no security, (2) access control list (ACL), and (3) 128-bit AES encryption with authentication. The choice of the appropriate security level depends on the application and the resources available in the sensor nodes.

20.2.3 TinyOS® Operating System

TinyOS is an open-source component-based operating system for wireless sensor networks and is designed to operate under severe memory constraints. It supports networking, power management, and sensor interfacing for developing application programs. TinyOS is event driven—it consists of a *scheduler* and several *components*. Each component (Figure 20.3) consists of

a. *Event handlers* to propagate hardware events to the upper levels

b. *Command handlers* to send requests to lower-level components

c. *Tasks* related to the application

A component has a set of interfaces for connecting to other components: this includes (1) interfaces that it *provides* for other components and (2) interfaces that it *uses* (there are provided by other components). Typically, *commands* are requests to start an operation and *events* signal the completion of an operation. For example, to send a packet, a component invokes the *send* command that initiates the send, and another component signals the event of *completing the send* operation. A program has two threads of execution: one executes the *tasks*, and the other handles *events*. Task scheduling policy is FIFO. A task cannot preempt another task, but an event handler can preempt a task and other event handlers too.

TinyOS is programmed in NesC, an extension of the C language that integrates reactivity to the environment, concurrency, and communication. Components are accessed via their interface points. TinyOS has two types of components: *modules* and *configurations*. A *module* provides the implementation of one or more interfaces, and a *configuration* defines how the modules are *connected together* to implement the application. As an example,

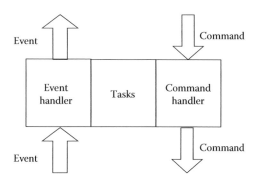

FIGURE 20.3   A component of TinyOS.

consider this problem: four different motes (1, 2, 3, 4) have to periodically send messages to a base station (0). Here is a sample program for an application:

```
// Author: Kajari Ghosh Dastidar
// This application periodically sends messages from four motes to
   the base station.
// The TOS Address of the Base Station is 0. The Address of the
   other motes are 1,2,3,4.

module SendMsgM {
provides interface StdControl;
uses {// the interfaces wired into the StdControl interface.
     interface Timer as Timer;
     interface SendMsg as SendMsg;
     interface ReceiveMsg as ReceiveMsg;
     interface StdControl as RadioControl;
     interface Leds;
   }
 }

implementation {
bool free;  // Boolean variable indicating when a message is
              received in buffer
TOS_Msg buffer; // reserves memory for a message structure
     // (see $TOSDIR/types/AM.h for details on this)
uint8_t moteval[4]; // array contains data received from each mote.
                    // moteval[i] contains the latest message
                      from mote i.
uint8_t counter = 0;  // some data to send (each mote is sending
                         a data
                  // incrementing counter to the base station)

// This will be called by main to initialize the application
     command result_t StdControl.init() {
     call RadioControl.init();  // initialize the radio
     call Leds.init();
     free = TRUE;
     return SUCCESS;
   }

// This will be called by main to start the application
     command result_t StdControl.start() {
     call Timer.start(TIMER_REPEAT,1024);  // set up the timer
     call RadioControl.start();  // start the radio
     return SUCCESS;
   }
```

```
// This will be called by main to stop the application.
// Here, the program is designed to run forever till user breaks
   out of the application.
    command result_t StdControl.stop() {
    call RadioControl.stop();
    return SUCCESS;
  }


// scheduling a message here. This task will be executed each time
   the Timer is fired.
  task void doMessage() {
      buffer.data[0] = counter;  // incremented counter in the message
      counter++;              //and increment for next time
      dbg(DBG_USR1, "*** Sent message from %d\n", TOS_LOCAL_ADDRESS);
                              // debug statement to check a program when
                              // running the simulation in TOSSIM
      call SendMsg.send(0, 1, &buffer);
    }


  // The following is required by the SendMsg interface
       event result_t SendMsg.sendDone(TOS_MsgPtr whatWasSent,
         result_t status) {
       return status;
  }
event TOS_MsgPtr ReceiveMsg.receive( TOS_MsgPtr m ) {
      uint8_t i, k;
      i = m->data[0];                  // get data from the message
      k = m->addr;

    free=FALSE;
    if (TOS_LOCAL_ADDRESS == 0)      // check the freshness of the
                                         data in the BASE.
    {
          if(moteval[k] != data[0]) moteval[k] = data[0];
          call Leds.redToggle();
          dbg(DBG_USR1, "*** Got message with counter = %d from
            mote = %d\n", i, j );
    }
    free = TRUE;
    return m;              // give back buffer so future TinyOS
                              messages have some
                           // place to be stored as they are
                              received.

  }
}
```

```
        event result_t Timer.fired() {
        post doMessage();    // task doMessage is executed.
        return SUCCESS;
    }
  }
```

```
// This NesC component is a configuration that specifies
// how the SendMsgM module should be wired with other components
// for a complete system/application to send Active Messages to
   the Base Station.
```

```
configuration SendMsgC {
}
implementation {
      components Main, SendMsgM, TimerC, LedsC, GenericComm as Comm;
      // rename GenericComm to be Comm
Main.StdControl -> SendMsgM.StdControl;
SendMsgM.RadioControl -> Comm.Control; // the local name of
StdControl
             // is just "Control" in GenericComm
SendMsgM.Timer -> TimerC.Timer[unique("Timer")];  // an unique
instance of Timer is used here
SendMsgM.SendMsg -> Comm.SendMsg[3];    // msg type 3 will be sent
SendMsgM.ReceiveMsg -> Comm.ReceiveMsg[3];  // and also received
SendMsgM.Leds -> LedsC;              // LED is used here to show
the transmission and
                             //the reception of the messages
}
```

20.3 CHALLENGES IN WIRELESS SENSOR NETWORKS

This section highlights some of the important challenges faced by applications that use wireless sensor networks.

20.3.1 Energy Conservation

Algorithms running on sensor networks should be energy-aware for maximum battery life. Sensing, computation, and (radio) communication are three cornerstones of sensor network technology. Of these, sensing and computation are power thrifty, but communication is not.

The *radio model* for a sensor node is as follows: if E_d is the minimum energy needed to communicate with a node at a distance d, then $E_d = K \cdot d^n$. Here, n is a parameter whose value ranges between 2 and 4 depending on environmental parameters, and K depends on the characteristics of the transmitter.

For a given transmission energy, the radio range forms a disk (Figure 20.4a). All nodes within the radio range are neighbors of the sending node. The *disk model* is however somewhat simplistic—variations in the environmental characteristics (like the presence of objects

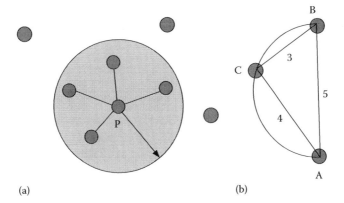

FIGURE 20.4 (a) The radio range of a sensor node P is a disk, and (b) if $E_d = K \cdot d^n$ and $n > 2$, then the path ACB is more energy efficient than the shortest path AB.

or obstructions) can distort the contour. Also, there are sensor nodes equipped with directional antennas—these are more efficient in transmitting in certain directions. However, unless otherwise mentioned, we will ignore these refinements and stick to the disk model.

As a consequence of the $E_d = K \cdot d^n$ formula, the shortest Euclidean path between a pair of nodes is not necessarily the minimum-energy path. For example, in Figure 20.4b, if $n > 2$, then the path ACB between A and B will consume lesser energy compared to the direct path AB between them. The task of identifying an appropriate topology that reduces energy consumption and satisfies certain connectivity or routing requirements is the goal of *topology control.*

Studies on existing sensor hardware reveal that communicating a single bit across 20 ft costs as much energy as required in the execution of 1000 instructions. Therefore, *minimizing communication* is a major focus in energy conservation. Conventional textbook models and algorithms are often inadequate for many sensor network applications. For example, shared memory algorithms require each process to constantly read the states of the neighbors, which steadily drains battery power. Message-passing algorithms optimize performance in terms of number of point-to-point messages, but do not use (or rarely use) *local multicasts* to the immediate neighborhood, which are clearly more energy efficient. This is because each point-to-point message is as expensive (in terms of power consumption) as a local multicast. Another technique for energy conservation is *aggregation.* An *aggregation point* collects sensor readings from a subset of nodes and forwards a single message by combining these values. For example, if multiple copies of the same data are independently forwarded to a base station, then an intermediate node can act as aggregation point by suppressing the transmission of duplicate copies and thus reduce energy consumption. Furthermore, energy-aware algorithms must utilize the various power-saving modes supported by its operating system.

20.3.2 Fault Tolerance

Disasters not only affect a geographic region but also affect the monitoring infrastructure that includes the sensor nodes and the network. Node failures and environmental hazards cause frequent topology change, communication failure, and network partition, adding to the fragility of wireless sensor networks. Such perturbations are far more frequent than those found in traditional LAN or WAN. Due to the ad hoc nature of the network, self-stabilization is a promising method for restoring consistency. Tolerating failures and perturbations and maintaining the fidelity of information in spite of the fragile nature of the environment are fundamental goals of sensor networks.

20.3.3 Routing

Routing strategies in sensor networks differ from those in ordinary networks. Conventional routing is *address-centric*, where data are directed to or retrieved from certain designated nodes. Routing in sensor networks, however, is mostly *data-centric*—the base station queries the network for a particular type of data (and may not care about *which node* generates that data), and the appropriate sensor nodes route the data back to the base station. Reliable routing requires identification of reliable data channels, energy-efficient routing requires data aggregation, and fault-tolerant routing relies on the network's ability to discover alternative routes and self-organize when an existing route fails.

20.3.4 Time Synchronization

Several applications on sensor networks require synchronized clocks with high precision. For example, the precise online tracking of fast-moving objects requires a precision of 1 μs or better. Unfortunately, low-cost sensor nodes are resource thrifty and do not have a built-in precision clock—a clock is implemented by incrementing a register at regular intervals driven by the built-in oscillator. With such simple clocks, the required precision is not achievable by traditional synchronization techniques (like NTP). Furthermore, clock synchronization algorithms designed for LAN and WAN are not energy-aware. GPS is expensive and not affordable by the low-cost nodes. Also GPS does not work in an indoor setting since a clear sky view is absent.

20.3.5 Location Management

The deployment of a sensor network establishes a physical association of the sensor nodes with the objects in the application zone. Identifying the spatial coordinates of the objects, called *localization*, has numerous applications in tracking. The challenge is to locate an object with high precision. GPS is not usable inside large buildings and lacks the precision desired in some applications.

20.3.6 Middleware Design

Applications interact with the sensor network through appropriately designed middleware. A query like *What is the carbon monoxide level in room 1739?* is easy to handle with an appropriate location management infrastructure. However, a data-centric query *Which area has temperatures between 55 and 70 degrees?* needs to be translated into low-level actions by the individual sensors, so that the response is generated fast and with minimum energy consumption. This is the job of the middleware.

20.3.7 Security

Radio links are insecure. This means that an adversary can steal data from the network, inject data into the network, or replay old packets. Adversarial acts include surreptitiously planting malicious nodes that can alter the goals of the network. Such nodes can either be new nodes that did not belong to the original network, or these can be existing nodes that were captured by the adversary, and their memory contents altered with malicious codes. Another tool for the attack is a laptop-class node with high-quality wireless communication links—it can hoodwink other nodes into false beliefs about the network topology and force them to forward data to the attacker. The threats are numerous and need to be countered using lightweight tools, since sensor nodes have limited resources to implement countermeasures.

20.4 ROUTING ALGORITHMS

In wireless sensor networks, a base station (sometimes called a *sink* node) sends commands to and receives data from the sensor nodes. A primitive method of routing (data or commands) is *flooding*, which is unattractive from the energy efficiency point of view. A better alternative is *gossiping*, where intermediate nodes forward data to their neighbors with a certain probability. Compared to flooding, gossiping uses fewer messages, and so consumes less energy. For reliable message delivery and energy efficiency, numerous routing methods have been proposed so far. In this section, we present some well-known routing algorithms.

20.4.1 Directed Diffusion

Directed diffusion was proposed by Intanagonwiwat et al. [IGE+02] for data monitoring and collection in response of data-centric queries. Assume that a sensor network has been deployed to monitor intrusion in a sensitive area. A sink node sends out queries about its *interests* down a sensor network to the appropriate nodes. The intermediate nodes cache these interests. A typical interest has a default monitoring rate and an expiration time. An example is *monitor the northwest quadrant of the field for intrusion every minute until midnight*. This dissemination of interests sets up *gradients* (a mechanism for tagging preferred paths based on their responsiveness) in the network (Figure 20.5). Data are named using *attribute–value* pairs. As sensor nodes generate data, the various paths transfer data with matching interests toward the originator of the interest. Depending on the responsiveness of these paths (or the importance of the event), a receiving node reinforces only a small fraction of the gradients. It does so by resending the interest with a higher asking rate. The reinforced gradients define the preferred links for data collection. This also prunes some neighbors since the interests in their caches will expire after some time. For example, if node *i* does not detect any intrusion but node *j* does, then the gradient toward node *j* is reinforced. Accordingly, in directed diffusion, with the progress of time, all data do not propagate uniformly in every direction. The *gradient* is represented by a tuple (*rate*, *duration*), where *rate* denoted the frequency at which data are desired and *duration* designates the expiration time of the request. A higher rate encourages data transmission, and a lower rate inhibits data transmission.

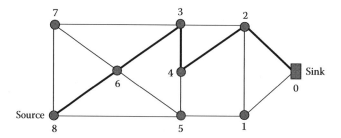

FIGURE 20.5 Directed diffusion in a sensor network. The route in bold lines has links with the highest gradient and is the preferred route for data transfer from a source to the sink.

Implementers can use various heuristics for quantifying gradient. During the initial propagation of an interest from the sink to the sources, all gradients (i.e., their rates) have the same value. In Figure 20.5, when the sink disseminates the interest, all links have a gradient with rate 1. When node 6 discovers that link (8,6) provides the desired data with a lower delay, it increases the gradient of that link to a higher value. Thereafter, node 3 discovers that the data arrive through link (6,3) with a lower delay compared to the other links incident on it. So, it increases the gradient of link (6,3). In this process, eventually links (3,4), (4,2), and (2,0) are reinforced. New paths transmitting high-quality data get spontaneously reinforced, and poor-quality paths automatically drop out from the scene.

That directed diffusion consumes less energy compared to flooding is no surprise. Intanagonwiwat et al. [IGE+02] reports that it also consumes less energy compared to omniscient multicast.

20.4.2 Cluster-Based Routing

Cluster-based routing, also known as *hierarchical routing*, uses a two-level approach for routing data from the sensor nodes to the base station. The network is partitioned into clusters: each cluster has a cluster head, and a few nodes associated with it. The cluster heads receive data from the nodes in the cluster, aggregate them, and send them to neighboring cluster heads. Eventually, the data get forwarded to the base station. If the clocks are synchronized, then intercluster communication can be scheduled at predefined time slots. *Cluster-based routing* is energy efficient, scalable, and robust.

20.4.2.1 LEACH

Low-Energy Adaptive Clustering Hierarchy (*LEACH*) is a self-organizing routing protocol that uses the idea of hierarchical routing (Figure 20.6a). It uses randomization to distribute the energy load evenly among the sensors. The protocol runs in phases. The first phase *elects the cluster heads*. The second phase *sets up the cluster*: here, each cluster head sends out advertisements inviting other nodes to join its cluster. A noncluster node makes the decision depending on the energy needed to communicate with a cluster head—the cluster head that is reachable using the minimum energy is its best choice for that node. In the third phase, the cluster heads agree to a time schedule for transmission—the schedule helps avoid conflicts caused by overlapped transmission. The nodes that are not cluster

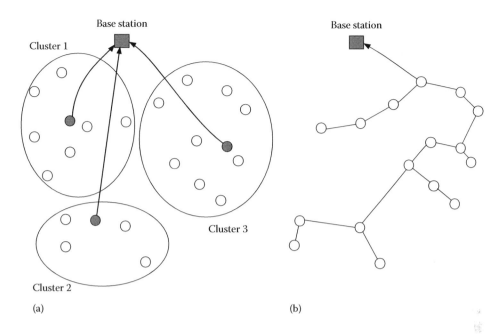

FIGURE 20.6 (a) Cluster-based routing in LEACH: the cluster heads are shown as dark circles. (b) Data transmission in PEGASIS.

heads are notified about this schedule, which enables them to transmit data at appropriate times. The cluster heads are responsible for data aggregation and data compression, so that multiple data are bundled into a single message. LEACH saves energy via a reduction in the number of transmissions, and this extends the life of the network.

The cluster heads transmit the data directly to the base station (Figure 20.6a), so this is a high-energy transmission. They spend more energy than the rest of the nodes, running the risk of draining their battery sooner than others and causing a network partition. To prevent this imbalance in energy drainage, cluster heads are rotated via periodic reelection. When the residual power in a cluster head reaches below a certain level, a new cluster head is elected. Nodes with significant residual energy are expected to volunteer for becoming new cluster heads. This helps with energy load balancing and increases the life of the application.

20.4.2.2 PEGASIS

Power-Efficient GAthering in Sensor Information Systems (*PEGASIS*) is an improvement over LEACH in the sense that it requires less energy per round of data transmission. The sensor nodes form a chain, so that each node communicates with a close neighbor by spending a small amount of energy. Gathered data move from node to node and get fused, and eventually, a designated node (called the leader) transmits the data packet to the base station (Figure 20.6b). Nodes take turns to be the leader—this reduces the average energy spent by each node per round and balances the load. The task of building a chain that will expend the minimum energy to collect data can be reduced to the traveling salesman problem, which is known to be intractable; PEGASIS therefore uses a greedy protocol to form such chains, makes use of a variety of aggregation methods, and claims to reduce the per-round energy consumption of LEACH by a factor of 2 or better.

20.4.3 Metadata-Based Routing: SPIN

Sensor protocol for information via negotiation (SPIN) defines a family of protocols that overcomes redundant data transmission (the major weak point of flooding or gossiping) by using *metadata* for negotiation before any actual data are transmitted. Two major problems with flooding or gossiping are as follows:

1. *Implosion*: A node receives multiple copies of the same data via different channels.

2. *Conflict*: Multiple senders simultaneously send data to the same destination node.

Although there is 1-1 correspondence between real data and metadata, the protocol assumes that metadata are much smaller in size—therefore, less energy is expended in transmitting or receiving metadata compared to real data. Three types of data packets are used in SPIN:

1. *ADV*: It is a metadata advertisement for new data to be shared.

2. *REQ*: A node sends a metadata REQ (request for data) when it wishes to receive an advertised data.

3. *DATA*: This is the real data with a metadata header.

SPIN uses a simple handshake protocol for data transmission. This is based on the sequence ADV–REQ–DATA. Data are sent only when a request is received based on metadata. If a node already received a copy of that data, then it does not send REQ, which suppresses data transmission. This naturally eliminates implosion. Conflict is avoided by sending REQ to only one sender at any time. In case of data loss, the protocol allows nodes a second chance to retrieve the data by sending a REQ to a duplicate ADV.

An enhanced version of SPIN allows the application to adapt to the current energy level of the node. It adapts itself based on amount of residual energy—participation in sending ADV or REQ is restricted when the residual energy level becomes low. Simulation results show that it is more energy efficient than flooding or gossiping while distributing data at the same rate or faster.

Finally, *geometric ad hoc routing* characterizes a class of routing algorithms where the nodes are location-aware, that is, each node knows its own location, the locations of its immediate neighbors, and that of the destination. There are several algorithms belonging to this class (e.g., see GOAFR+ [KWZ+03]).

20.5 TIME SYNCHRONIZATION USING REFERENCE BROADCAST

A few applications of wireless sensor networks rely on accurate synchronization among the local clocks for their success. The desired synchronization is much tighter than what we need for machines on the Internet. A protocol like NTP can synchronize clocks within an accuracy of a few milliseconds, whereas some critical sensor network applications need clocks to be synchronized within 1–2 ms. Examples of critical applications that require precise time synchronization are as follows:

- *Time of flight of sound*: How much time did it take for sound to reach from point A to B? Such measurements are important for echo depth sounding that accurately maps out the bottom of a lake or an ocean.

- *Velocity and trajectory estimate*: A sniper fired a bullet toward a particular target in a busy location. From which window of the nearly multistoried building was the bullet possibly fired? The accuracy of the computation will depend on how closely the clocks of the sensor nodes sensing the bullet are synchronized.

- *TDMA schedule*: To avoid frame collisions and the consequent loss of message and energy, it is important for a cluster of sensor nodes to agree on a common TDMA schedule. To enforce such a schedule, all nodes have to agree to a common time frame.

For external synchronization (i.e., synchronization with a precise external time source), the nodes in a sensor networks can use GPS. However, apart from the additional cost, it is not feasible to access GPS data in indoor applications, or certain urban locations, or in hostile territories where GPS signals may be jammed.

20.5.1 Reference Broadcast

Reference broadcast (RBS)–based time synchronization provides a solution to time synchronization when external synchronization is not important, and only internal synchronization is adequate. RBS uses a broadcast message to synchronize the clocks of a set of receivers with one another. This is in contrast with traditional protocols that synchronize a receiver with the sender of the message. In the simplest form, RBS has three steps:

1. A transmitter broadcasts a reference packet to a set of nodes.

2. Each receiver records its *local time* when the broadcast is received.

3. The receivers exchange these local times with one another.

RBS recognizes that a message broadcast on the physical layer arrives at a set of receiving nodes with very little variability in propagation delay (Figure 20.7a). Four components of time are taken into consideration during time synchronization:

Send time: Time to construct and transfer the message to the network interface.

Access time: Time spent in waiting for the transmission channel.

Propagation time: Actual flight time of the signal from the source to the destination node.

Receive time: Time required by the network interface to signal message arrival to the host.

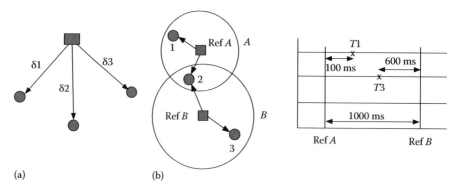

FIGURE 20.7 (a) Ref 0 broadcasts to sensors 1, 2, and 3 ($\delta 1 \approx \delta 2 \approx \delta 3$). (b) RBS-based time synchronization over two broadcast zones A and B: node 1 receives a message at time $T1 = 100$ ms after receiving the broadcast from Ref A, node 3 receives a message at time $T3 = 600$ ms before receiving the broadcast from Ref B, and node 2 receives the broadcast from Ref A 1000 ms before receiving the broadcast from Ref B.

Unlike the Internet, message *propagation time* in sensor networks is of the order of a few nanoseconds and is therefore negligible. The other three components are more or less the same for all receiving nodes. The phase offset can be accurately deduced from a series of m RBS over n distinct sensor nodes. Let $T_{i,k}$ and $T_{j,k}$ be the local times of a pair of sensor nodes (i, j) when they receive the RBS from a node k. Then the average offset between nodes i and j from the reception of the RBS from m different nodes is

$$\frac{1}{m} \cdot \sum_{k=0}^{m-1} (T_{i,k} - T_{j,k})$$

The largest of these values represents the *group dispersion*. The mean dispersion is used to reset the clocks to the correct value.

Sensor nodes do not have built-in precise clocks—once the clocks are synchronized, due to the disparity in the oscillator frequencies, the clock skew will get worse over time. To keep the skew within tolerable limits, the clocks have to be periodically resynchronized. Experiments showed that with only two sensor nodes, by using 30 RBS, RBS could achieve an accuracy of 1.6 ms, beyond which it reaches a zone of diminishing return. Details of these experiments are available in [EGE02].

The previous method works for a single broadcast domain only. What about multiple broadcast domains? Figure 20.7b shows two broadcast domains overlapping with each other. Let Refs A and B send out the RBS in their respective domains. To relate the timings in these two domains, observations made by some node 2 that is common to both domains are crucial. As an example, let node 1 receive a message (at time $T1$) 100 ms after receiving the RBS from A ($T1 = \text{Ref } A + 100$ ms), and let node 3 observe an event (at time $T3$) 600 ms before receiving the RBS from B ($T3 = \text{Ref } B - 600$ ms). Both 1 and 3 will consult node 2 that received both RBS and find out that node 2 received the broadcast from A 1000 ms before the broadcast from node B (Ref $A = \text{Ref } B - 1000$ ms). From these, it follows that $T3 - T1 = 1000 - 600 - 100 = 300$ ms.

Although designed to provide internal synchronization, RBS can also achieve external synchronization when one of the sensor nodes is equipped with a GPS receiver that can receive the UTC signal.

20.6 LOCALIZATION ALGORITHMS

Localization binds spatial coordinates with sensed data and is an important component in many applications of sensor networks. The central issue is to answer queries like "Where is this signal coming from, or where is an intruder currently located in this area?" Spatial coordinates are also useful for collaborative signal processing algorithms that combine data from multiple sensor nodes for the purpose of target tracking. The required coordinates might be absolute or relative. It is implied that GPS is not attractive due to its cost, or physical size, or accuracy, or power consumption, or the absence of a clear sky view. In this section, we outline the principles of a few location management systems.

20.6.1 RSSI-Based Ranging

The simplest method of computing the distance of one node from another uses received signal strength indicator (RSSI). While sending out a signal, the sending node appends the strength of the sending signal. Given a model of how the signal strength fades with distance, the receiving node can compute its distance from the sender. One major problem here is the poor correlation between the RSSI and the distance. The accuracy is highly affected by the variability of the wireless medium. The problem is further compounded by the manufacturing variances of the radio devices and multipath effect (caused by reflection from neighboring objects).

20.6.2 Ranging Using Time Difference of Arrival

If the local clocks are synchronized and the sender sends a time-stamped signal, then the receiver can potentially compute the time of flight and deduce the distance separating them. In practice, this does not work well, since clocks are rarely synchronized with an accuracy <1 μs, whereas the time of flight is of the order of nanoseconds. A much better accuracy is achieved using a combination of radio and acoustic waves. The *Active Bat* system is the first implementation of this concept. Each bat is tagged with a unique id, a radio receiver, and ultrasound transducers. The interrogator sends a query via radio "Bat 32, send (ultrasound) signal now." Bat 32 complies, the time of flight of the ultrasound is recorded by the interrogator, and the distance of the bat is computed from it. The technique is accurate and does not require time synchronization as long as clocks are stable over short periods of time. The indoor location system CRICKET [PCB00] at MIT used this approach to locate and track indoor objects with an accuracy of a few centimeters.

20.6.3 Anchor-Based Ranging

A straightforward scheme for localization uses a coordinate system defined by a set of powerful nodes called *beacons*. The beacons serve as anchors and are positioned at *known points* in the area of interest. These beacons periodically broadcast their current coordinates. The spatial location of a sensor node is determined by (1) how many distinct broadcasts it

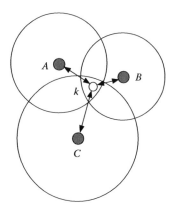

FIGURE 20.8   The sensor k receives signals from the beacons A, B, and C.

can receive and (2) the strength of the signals received for each broadcast. It is important that the sensor node receives at least three broadcasts from distinct beacons (Figure 20.8).

Using a direct ranging method like RSSI, the sensor node estimates its distances d_A, d_B, d_C from the three anchors A, B, and C, respectively. If (x, y) is the coordinate of the sensor node in a given coordinate system and (x_A, y_A), (x_B, y_B), (x_C, y_C) are the coordinates of the anchors A, B, and C, respectively, then the following equations hold:

$$(x - x_A)^2 + (y - y_A)^2 = d_A^2$$

$$(x - x_B)^2 + (y - y_B)^2 = d_B^2$$

$$(x - x_C)^2 + (y - y_C)^2 = d_C^2$$

Computation of (x, y) by solving these equations is known as *triangulation* or *lateration*. Lateration produces good results only when the distance estimates are accurate. Unfortunately, simple RSSI-based ranging method is not accurate. The accuracy of measurement can be increased if there are more than three anchors within the radio range of a sensor. This generalization is known as *multilateration*.

If the network contains multiple broadcast domains and the density of anchors is low, then the sensor node will obtain its distance from the anchors over multiple hops. To use lateration in such cases, each anchor maintains a shortest path tree with itself as the root. Distance from the anchors is estimated via the shortest paths.

20.7 SECURITY IN SENSOR NETWORKS

Most simple applications of sensor networks are vulnerable to attacks, since they have not been originally designed with security as a goal. The description of threats relates to a typical setup where a base station collects data from a bunch of sensor nodes in the physical space. We assume that the base station is always trustworthy.

One can classify adversaries into three different classes: *passive, active,* and *malicious*. A *passive adversary* quietly steals unprotected data, or tampers data in transit, or launches a replay attack. *Active adversaries* can inflict much more damage. They can (physically) capture a node, extract the codes and keys, steal protected information using the stolen keys, or launch an attack by planting malicious codes into the captured nodes. PC-class adversaries can remotely influence many nodes and launch *sinkhole* or *wormhole attacks* by faking route information. *Malicious* adversaries try to harm the network. This includes (1) draining one or more nodes of energy and thus causing network partitions, (2) tampering with the data (possibly in critical services like power plants, water supplies, hospitals) that could potentially harm the intended application, or (3) jamming the signals and disrupting communication in critical applications like anti-theft monitoring or military surveillance applications, so that an enemy can launch future attacks on the application without much difficulty. This section reviews a few security measures for sensor networks.

20.7.1 SPIN for Data Security

Due to limited resources, the implementation of conventional cryptographic protocols is impractical on sensor nodes: for example, effective public keys are long, and the communication overhead as well as the computation effort for its verification is very high. Only fast secret-key cryptography can be sparingly used. In [PSW+02], Perrig et al. introduced a cryptographic protocol (called SPIN) to secure sensor networks against passive adversaries. SPIN was designed to preserve confidentiality, integrity, authentication, and freshness (of data). The protocol works on a traditional setup where a base station communicates with the sensor nodes via source routing. Some sensor nodes may not be trustworthy. Messages may be corrupted in transit, but all messages are eventually delivered to the destination node.

SPIN consists of two components: a *sensor network encryption protocol* (SNEP) and μ*TESLA* (microversion of *Timed Efficient Stream Loss-Tolerant Authentication*). A brief description follows:

20.7.1.1 Overview of SNEP

SNEP provides confidentiality (privacy), two-party data authentication, integrity, and freshness. Each node j shares a unique master key K_j with the base station. This master key is used to derive all other keys: these include the data encryption key, the MAC key, and a key for random number generation. SNEP derives a *one-time encryption key* using the value of a monotonically increasing *message counter* and the master key. This key is XOR-ed with the message bits and sent out. The message counter value is not explicitly transmitted, but the communicating processes independently keep track of it, and the eavesdropper has no knowledge about it. The recipient generates an identical key and XORs it with the ciphertext to retrieve the clear text. This preserves *confidentiality*. SNEP has the following properties:

1. Since the same message gets transformed to different ciphertext in different transmissions, cryptanalysis via plaintext attack is ruled out. This helps achieve *semantic security*.

2. *Replay* attacks can be identified due to the use of the shared counter in the transmitted message.

3. Message delivery guarantees *weak freshness*. Since the counter monotonically increases, the recipient only knows that the current message is more recent than the previous one, but does not know who transmitted it. The use of a nonce (using a random number generator) will guarantee strong freshness.

After some evaluation, RC5 was chosen as the block cipher due to the small size of the code. The communication overhead of SNEP was 8 bytes per message.

20.7.1.2 Overview of μTESLA

This is a lightweight version of the TESLA protocol for authenticated broadcast that was designed for more heavy-duty platforms. Traditional authentication (mostly) uses asymmetric key cryptography, which is not feasible for the resource-constrained sensor nodes. μTESLA uses symmetric key and authenticates messages by introducing asymmetry through a novel method that involves delayed disclosure of the symmetric keys. Each MAC key is an element of a key chain that is generated using a public *one-way function F*. Consider a communication from the base station to the participant nodes in the network, and assume that the local clocks are approximately synchronized. The sender generates keys at regular time intervals, and there is a 1-1 correspondence between keys and time slots* (Figure 20.9). μTESLA generates the MAC key $K^{(m)}$ for interval m ($m>0$) using the formula $K^{(m)} = F(K^{(m-1)})$. Here, F is a one-way function—everybody can compute $K^{(m-1)}$ from $K^{(m)}$, but only the base station can derive $K^{(m)}$ from $K^{(m-1)}$. When the base station sends out a packet at interval 0 using the MAC key $K^{(0)}$, the receiving node cannot authenticate it since it does not have the verification key. However, it is also true that no eavesdropper knows about it, so no one else could have generated the data. The receiving node simply buffers it.

Each verification key is disclosed after a couple of time intervals. For example, in Figure 20.9, the key $K^{(0)}$ has been disclosed after one time interval, after which the receiving node(s) can authenticate the buffered message sent with MAC key $K^{(0)}$. The loss of some of the packets disclosing the keys is not a problem. For example, if both $K^{(0)}$ and $K^{(1)}$ are lost, but the packet disclosing $K^{(2)}$ is received, then the receiving node(s) can easily generate $K^{(1)}$ and $K^{(0)}$ from it and complete the authentication.

20.7.2 Attacks on Routing

An active adversary can alter routing information (or plant fake routing information) to create routing loops, attract network traffic toward compromised nodes, or divert traffic through one or more target nodes for draining their energy and partitioning the network. There are several

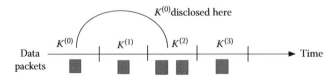

FIGURE 20.9   Broadcasting in μ*TESLA* using a chain of MAC keys.

---

* The clocks are synchronized with a reasonable degree of accuracy.

different types of attack that seem feasible. In a *selective forwarding* attack, the adversary (a compromised node) drops important data packets to cause damage to the application. *Sinkhole attacks* lure network traffic toward compromised nodes, so that they can do whatever they wish with the data. To launch a sinkhole attack, a compromised node will falsely send out (or replay) an advertisement of a high-quality path to the base station. *Wormhole attacks* create the illusion of a high-quality route by tunneling the data from one part of the network to another remote part via a low latency link. This link will use an out-of-bound channel that is only accessible to the attacker. The low latency path will attract traffic and create a wormhole. Wormhole attacks are likely to be combined with eavesdropping or selective forwarding.

20.7.2.1 Hello Flood

Many applications require nodes to periodically broadcast heartbeat (HELLO) messages. A PC-class adversary broadcasting such a message to a large number of nodes in the network can convince every such node that it is a neighbor. When this adversary advertises a high-quality route leading to the base station, other nodes will adopt this route and send their data to the adversary, which may never be forwarded to a destination. It effectively creates a sinkhole (also called a *black hole*) using a slightly different method.

While all of the previous attacks take place in the network layer, protocols in other layers are also susceptible to attack. For example, jamming attacks the physical layer, and the use of DSSS helps avoid it (unless the attacker knows the precise hopping sequence). In addition to these general attacks, specific algorithms can also be attacked. Some of these attacks are easy to defend, but for many others, effective countermeasures are necessary. Finding effective countermeasures is an important topic of research.

20.8 APPLICATIONS

In recent times, wireless sensor networks are being increasingly used, in many cases on an experimental basis, in health-care, sustainability, and surveillance-related applications. There is a growing effort to build *smart* applications that require less human intervention, yet achieve the same or a higher level of efficiency in terms of end goal or energy usage. A small fraction of these applications is summarized in the following.

20.8.1 Health-Care Applications

Wearable ambulatory medical sensors enable people to monitor important physiological parameters while engaged in the activities of everyday life. These target fitness enthusiasts as well as aging populations who are concerned about their health. Such embedded sensors are capable of communicating with smartphones or PDAs using short-range technologies like Bluetooth. Even prosthetic devices with embedded sensors have emerged. For the aging population, wireless communication enables medical data to be transmitted to caregivers, facilitating ubiquitous real-time sensing. As an example, Harvard University's CodeBlue project [GPS+08] integrated various medical sensors with mote-class devices. A publish-/subscribe-based network architecture supports data dissemination with different priorities and remote sensor control. CodeBlue also allows victims of disasters to be tracked and localized using radio-frequency-based localization techniques.

In epidemiology, the failure to comply with appropriate hand hygiene by the health-care workers in the hospitals leads to millions of infections that are preventable. At the University of Iowa, the computational epidemiology group has developed a wireless-mote-based data-collection system to capture health-care worker hand-hygiene behavior along with health-care worker interactions over time and space [HNS+12].

20.8.2 Environment Monitoring and Control

With the growth of cloud computing and cloud-based services, the number of data centers is rapidly increasing. Each data center has thousands of servers, and these servers consume a substantial amount of energy. This generates heat, and to prevent overheating, cooling becomes essential, which costs energy. In fact, energy is the single largest operating expense for most data centers. Improving the energy performance of data center systems reduces the operating costs as well as cuts down greenhouse gas emission. Sensors placed near the servers relay the data about energy consumption and climatic conditions to a base station that initiates appropriate measures for fine-grained climatic control to conserve energy. Studies sponsored by the US Department of Energy (DOE) and the US Environmental Protection Agency (EPA) have shown that energy consumption can be reduced by 25% through implementation of best practices and commercially available technologies.

20.8.3 Citizen Sensing

Citizen sensing or people-centric sensing aims at sensing and collecting various kinds of environmental data that are of interest to the citizens of a community in their daily lives. Several projects have addressed this issue. These include the urban sensing project at the Center for Embedded Network Sensing (CENS), the University of California at Los Angeles, the Hourglass project at Harvard, the CarTel project at MIT [HBC+06], and Dartmouth's MetroSense project [CEL+08]. Each one of these developed an infrastructure for general purpose sensing at Internet scale. Sustainable design, healthy living, and effective stewardship of the world's limited resources require a solid understanding of how countless individual actions generate global effects and how individuals relate to their local environments—both natural and man-made. Citizen sensing targets technologies and applications that increase our capacity to help individuals, families, and communities monitor and improve their health, monitor pollution, adopt sustainable practices in resource consumption, and participate in civic processes. Data collection and analysis uses, in addition to sensors embedded at various locations, everyday technologies like mobile phones and automobiles, and some of these projects allow individuals to decide what, where, and when to sense.

20.8.4 Pursuer–Evader Game

Pursuer–evader game is an online tracking system relevant to disaster management. Here, the rescuers pursue or track the hot spots in a disaster zone (these are the *evaders*). The goal of the *pursuer* is to *catch* the evader using the information gathered by a sensor network. If the evader is successfully tracked down, then the rescue/recovery begins. Demirbas et al. [DAG03] proposed the first set of solutions to the problem. This section outlines the problem specifications and presents one of their solutions.

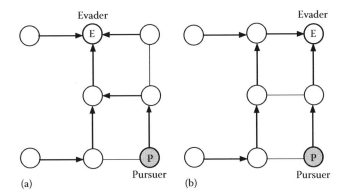

FIGURE 20.10 Two stages (a) and (b) of the pursuit as the evader moves to a new location.

Assume that the topology of the sensor network is a connected graph. The *pursuer* and the *evader* are two distinguished entities moving around in the Euclidean space that is constantly being monitored by the sensor nodes. In each step, these entities are able to move from the radio range of one sensor node to the radio range of a neighboring node* (Figure 20.10). Nodes may crash, or their states may be altered by transient failures. Exactly one node can sense the presence of the pursuer or the evader at any time. We further assume that the evader is omniscient—it knows the network topology as well as the current location of the pursuer and can pick appropriate moves to distance itself from the pursuer. However, the pursuer does not have much knowledge beyond its immediate neighborhood, and nobody has any knowledge about the strategy of the evader. The pursuer *catches* the evader when both of them reach the radio range of the same sensor node. The reaction time of the sensor nodes is much smaller than the time needed by the evader or the pursuer to move—so we assume that each sensor node executes an action or detects the evader within its range in zero time. Furthermore, the pursuer moves faster than the evader, and the local clocks of the sensor nodes are synchronized.

The *evader-centric* solution proposed in [DAG03] is as follows: the motes collectively maintain a *tracking tree* rooted at the evader. As the evader moves, the motes detect it and reconfigure the tracking tree. The pursuer moves up the tree edges to reach the evader at the root. The two activities run concurrently.

For each node k, we use the following notations:

- *Evader @k* (or *pursuer @k*) designates that currently, the evader (or the pursuer) is residing in the radio range of node k.

- $N(k)$ represents the neighbors of node k.

- $P(k)$ parent of a node k in the tracking tree.

- $T(k)$ designates the time when the evader was last seen. This information can be obtained from direct observation or through indirect observation via a neighbor.

- $d(k)$ represents that node's hop distance from the root via tree edges.

---

* In real life, this is not necessarily true—it is a simplifying assumption only.

The programs of the evader, the pursuer, and the sensor nodes are presented as follows:

```
{The evader's program}
{Evader moves from one node to another}
do evader@j → evader@k : k ∈ N(j) od
{The program of a sensor node j}
initially T(j) = 0{This is a simplifying assumption only}
do      evader@j → P(j) := j ; T(j) = clock of j;
[]      ∃k ∈ N(j) : (T(j) < T(k) ∨ T(j) = T(k) ∧ d(j) > d(k) + 1) →
              P(j) := k; T(j) := T(P(j)); d(j) := d(P(j)) + 1
od
{The pursuer's program}
{Pursuer moves to the parent node in the tree}
do pursuer@j→pursuer@P(j) od
```

Assume that the system runs under a distributed scheduler that allows maximal parallelism, so that all eligible nodes execute their actions in each step. Then the following results hold for the previous algorithm:

Lemma 20.1

After a node detects the evader, a tracking tree is formed in at most D steps, where D is the diameter of the sensor network.

Proof outline: We first argue that the edges joining a sensor node with its parent induce a spanning tree in the network. Since $\forall k : d(k) = d(P(k)) + 1$, in the steady state, there will be no cycle involving the edges between the nodes k and $P(k)$. Also, $\forall k : T(P(k)) \geq T(k)$, and no time stamp can exceed that of the root. Therefore, any node will have a directed path from itself to the root by following the parent pointers. Once a node detects the evader and becomes the root, the farthest node is guaranteed to adjust its d and P values within D steps. ■

Theorem 20.1

Let M be the initial separation between the pursuer and the evader and α be the ratio between the speed of the evader and that of the pursuer ($\alpha < 1$). Then, the pursuer catches the evader in at most $M + 2M \times \lceil \alpha/(1-\alpha) \rceil$ steps.

Proof outline: As a consequence of Lemma 20.1, the pursuer takes at most M steps to orient its parent pointer in the right direction. By that time, the evader may move at most M steps away, so the distance between the pursuer and the evader may grow to at most $2M$.

Once the separation grows to the maximum, the hop distance between the pursuer and the evader can never increase thereafter. To see why, consider a path $j, j + 1, j + 2, \ldots, k - 1, k$ between the evader j and the pursuer k. Actions taken by either of them can only reduce the length of the path—if j moves away, then eventually, k closes in, since it is faster than j. So, in reality, the hop distance will eventually decrease. If x is the number of steps until the evader is caught after the chase begins, then in the same time, the purser will cover $2M + x$ steps, whereas the evader can take at most x steps. Thus, $\alpha = x/(2M + x)$. So, $x = 2M \cdot \alpha/(1-\alpha)$. Add to it the initial number of M steps that the pursuer took before it correctly oriented its parent pointer, and the result follows.   ■

This algorithm is however not energy efficient. In each step, every node has to broadcast to each of its neighbors (the shared memory simplifies program writing, but true communication takes place via message passing). The original paper by the authors contains algorithms that are more energy efficient than this, but the pursuit is slower.

20.9 CONCLUDING REMARKS

Sensor network technology is growing at a rapid pace. Energy conservation remains a challenge in computationally intensive applications, and the quest for more powerful sensor nodes, better batteries, and better sensors continues. Energy harvesting from the environment remains an attractive option for solving the energy problem in certain classes of applications. Also, security issues are receiving increased attention as the technology is making inroads into sensitive areas like medical applications or surveillance-related activities.

A related device that plays a supporting role in some embedded systems is the *radio-frequency identification tag* (RFID). An RFID is a small tag that can be attached to a physical object—the tag contains the description of that object (like manufacturer, type, serial number). The antenna on the tag enables it to receive and respond to radio-frequency signals from an RFID reader. Passive tags have no internal power source. They are cheaper and have a smaller range (a few feet), but active tags have internal power sources and a much larger range (a 100 ft or more). RFIDs can potentially enhance the development of some sensor network applications in embedded systems.

20.10 BIBLIOGRAPHIC NOTES

Networks of sensors placed at the bottom of the ocean have been used to track submarines during the Cold War: a history of the early developments can be found in [CK03]. The genesis of modern wireless sensor networks is the Active Badge system by Want et al. [WHF+92]. The current version of MICA motes is based on the research at the University of California, Berkeley. These are now commercially available from Crossbow Technology. Hill designed the operating TinyOS [HSW+00] that runs on MICA motes. The original version was only 172 bytes in size. Hill's MS thesis contains a complete description of it. Harter and his associates [HHS+99] led the Active Bat project. The paper by Priyantha et al. [PCB00] contains the first report on MIT's indoor location system CRICKET. They developed this indoor GPS for tracking mobile robots in the laboratory using a combination of radio and acoustic waves and achieved an angular precision of 2°–3° and a linear precision of a few centimeters.

In routing, the paper by Intanagonwiwat et al. [IGE+02] describes directed diffusion. The cluster-based routing protocol LEACH is due to Heinzelman et al. [HCB00]. Lindsey and Raghavendra [LR02] proposed the energy-efficient routing protocol PEGASIS.

Elson et al. [EGE02] showed how RBS can be used to synchronize clocks using off-the-shelf wireless Ethernet components.

The security protocol SPIN was presented by Perrig et al. [PSW+02]. The article by Karlof and Wagner [KW03] provides a summary of various security concerns in routing, along with some possible countermeasures. The CodeBlue project by Gao et al. [GPS+08] illustrates a range of medical monitoring applications. Pursuer–evader games were listed as a challenge problem by DARPA. The solution presented here is due to Demirbas et al. [DAG03].

EXERCISES

20.1 Consider the placement of the sensor nodes on a 2D grid (Figure 20.11):

Let $E_d = K \cdot d^3$ where E_d is the energy needed to send data at a distance d and the energy needed to receive data is negligible. Then determine how the data from A to B should be routed so that the total energy spent by all the nodes in the path is the minimum. How will the route change if the energy equation is $E_d = K \cdot d^{1.9}$?

20.2 Consider a tree construction algorithm with the base station as the root. The base station initiates the construction by sending out beacons. Each node chooses another node from which it receives a beacon packet with the *lowest hop count* as its parent node. The plan is that sensor data will be forwarded towards the base station via the parent node via a BFS tree.

Unfortunately, links are not always bidirectional: if node A receives a signal from node B, then B may not receive the transmission from A. As a result, this tree will not ensure reliable data collection from all the nodes in the tree to the root. Modify the algorithm to construct a tree that enables reliable data collection from all the sensors nodes.

20.3 If sensor nodes do not physically move, then what can cause the topology of a sensor network to change? List all possible reasons.

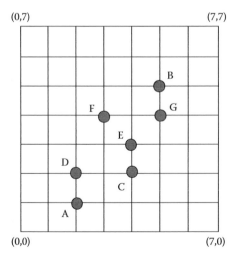

FIGURE 20.11   Seven sensor nodes placed on a 7 × 7 grid.

20.4 Uncoordinated transmissions can interfere with one another, causing conflicts in the MAC layer. When two neighbors concurrently transmit or a node receives concurrent transmissions from two other nodes, the messages are garbled. This triggers message retransmission and wastes energy.

Consider the network of Figure 20.12a. Assuming that the local clocks are synchronized, consider coordinating the transmissions to avoid such conflicts using TDMA. Your answer should specify which time slots can be used by a node to transmit data. Your goal should be to maximize the transmission rate.

a. If time is divided into five slots—0, 1, 2, 3, and 4—as shown in Figure 20.12b, then find an assignment of the slots for the nodes in the network shown in part (a).

b. Relate this exercise to the problem of graph coloring.

20.5 Four beacons, A, B, C, and D, are placed at the coordinates (20, 30), (20, 60), (50, 10), and (50, 60), respectively (Figure 20.13). Using RSSI, a sensor node finds out that its distances from A, B, C, and D are 20, 40, 35, and 45 units, respectively. Show how the sensor node will compute its *location*, if it knows the coordinates of A, B, C, and D.

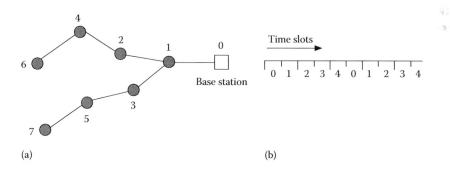

FIGURE 20.12 (a) A set of sensor nodes using TDMA to avoid MAC level interference. (b) The available time slots.

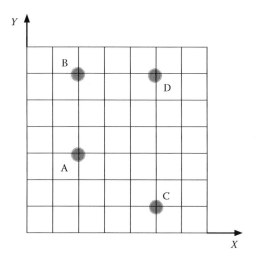

FIGURE 20.13 Four beacons—A, B, C, and D—used for the localization of a sensor node.

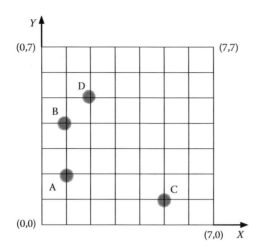

FIGURE 20.14   Find the best location of the base station here.

20.6   Four sensor nodes and a base station placed on a 2D area form an ad hoc network (Figure 20.14). These nodes will sense physical parameter and send them to a base station, which will transmit the collected data to a remote laboratory. Assuming $E_d = K \cdot d^3$, where E_d is the energy needed to send data at a distance d,

a.   Determine the best location for placing the base station so that the energy spent by all the nodes is as small as possible.

b.   Identify the data transmission paths from the sensor nods to the base station.

20.7   Given a network $G = (V, E)$, *topology control* generates a subgraph $G' = (V, E')$, such that (1) $E' \subseteq E$, and (2) less energy is needed to route packets between a pair of nodes in G' (than in G). The XTC algorithm for topology control by Wattenhofer and Zollinger [WZ04] has three steps: (1) Each node generates a ranking of its neighbors based on the strength of the signals received from them. (2) Each node exchanges the ranking with its neighbors. (3) Based on the information collected so far, nodes discard some of the links.

The strategy for including (or discarding) an edge is as follows: for a pair of nodes u and v that are neighbors of node w in the original graph G, $u \prec_w v$ implies that the signal strength from node u is *weaker than* the signal strength from node v as perceived by node w.

Per XTC, node u will want node v as a neighbor (i.e., the signal strength is good) if node v wants u as a neighbor. As an example, consider a subnetwork of four nodes u, v, w, and x and their local rankings of the neighbors are shown in Figure 20.15. Here, (v, x) is a preferred edge for both v and x since they locally rank each other at the highest order. Then prove that

a.   If $(u \rightarrow v)$ is a preferred edge for node u, then so is $(v \rightarrow u)$ for node v

b.   The topology of the resulting graph is triangle-free

c.   For disk graphs, the degree of each node is at most 6

(*Hint*: See [WZ04] for a solution.)

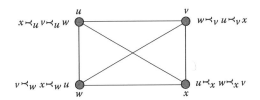

FIGURE 20.15   Four sensor nodes in a topology control exercise.

20.8   In a sensor network, the battery of some highly active nodes can run out quickly and cause a network partition. Assume that each node k has a variable $R(k)$ that records its residual battery power and each node can access the variable $R(i)$ of every neighbor i. To extend the life of the sensor network, nodes with low battery need to go to sleep for a specific period of time.

Propose an algorithm using which a node can reroute traffic toward the base station before it goes to sleep. For scalability, the time complexity of the rerouting algorithm must be low (preferably $O(1)$). Explain how your algorithm will work.

20.9   Broadcasting of data is an important activity in sensor networks. A sensor initiates the broadcast with a certain energy level E_d that is able to reach all nodes at distance $\leq d$ and $E_d = K \cdot d^2$. The recipients forward these data via additional broadcasts, and this process continues until all sensors receive the data. These operations form a broadcast tree, where the initiator is the root and for every other node, the closest sender sending the broadcast to it is the parent. The goal of the exercise is to complete the broadcast using the minimum amount of energy.

In the sensor network of Figure 20.16, identify the minimum-energy broadcast tree with the base station as the root. Assuming that the grid consists of unit squares, and $K = 1$, compute the energy spent by all the nodes in completing the broadcast.

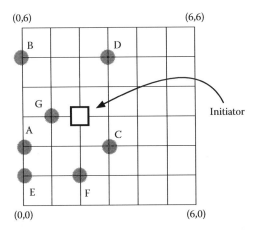

FIGURE 20.16   Identify the minimum-energy broadcast tree with the initiator as the root.

20.10 Consider a simple averaging method for time synchronization in a wireless sensor network. Assume that in each sensor node, the clock is counterdriven by a 1 MHz clock (accuracy 1 in 10^4). Every 10 s, each node executes the following steps:

a. Locally broadcasts its own clock

b. Receives the broadcasts from its neighbors

c. Computes the average, discards the outliers, and resets its own clock

Assuming that the signal propagation delay is negligibly small and ignoring message collision, estimate the expected accuracy of clock synchronization.

Programming Exercises

If you have a laboratory with working sensor nodes, then implement your solutions to the following exercises. Otherwise, obtain a simulator that provides an authentic simulation of sensor networks, and complete the exercises in the simulated environment. One such simulator is TOSSIM. Here are some references that may be relevant:

1. http://webs.cs.berkeley.edu/tos/ contains a tutorial on TinyOS.

2. http://www.tinyos.net/tinyos-1.x/doc/tython/manual.html describes Tython: a Python-based scripting extension to TinyOS's TOSSIM simulator.

You will need a few days to get ready with the required tools.

20.11 *Multicasting on a Sensor Network*

Implement multicasting on an 8 × 8 grid of sensor nodes using the gossip protocol. Here is a description of a simple version of *gossip*(*p*):

> A source sends the message to each neighbor with probability *p*. When a node first receives the message, with probability *p* it forwards the message to each neighbor. All duplicate requests are discarded.

Implement the protocol for various values of *p*. Draw a graph showing how long it took for all the nodes to receive the message as a function of *p*. Simulate the protocol on a large grid and study how many messages it took to complete the broadcast.

20.12 *Localization Using Sensor Networks*

The radio chips on the sensor nodes provide an RSSI value with each received message. Using RSSI, determine the location of a given node within a floor/laboratory. Assume that a set of fixed sensor nodes is mounted in accessible locations, and these send out beacons at regular intervals. RSSI-based measurements generally do not exhibit good accuracy—nevertheless, compare the accuracy of your measurement with the actual location of the node.

CHAPTER 21

Social and Peer-to-Peer Networks

21.1 INTRODUCTION TO SOCIAL NETWORKS

A social network depicts a social structure. It is commonly represented by a graph where the nodes are entities, and edges denote a relationship between a pair of entities. Some examples of entities are persons, organizations, various forms of living beings, web pages, cities, and airports. When nodes represent people, an edge may denote a friendship relation between them. When nodes are web pages, an edge from *page x* to *page y* may represent the existence of a link from the former to the latter page. When nodes denote businesses, an edge may denote the existence of a business relationship between two businesses. A social network like Facebook® or Twitter® is a modern Internet-based platform to create a social structure, and these have become extremely popular for communication and content sharing. But social networks existed from the dawn of civilization. Beyond human society, social network–like structures have been observed in many complex networks present in nature, pointing to the existence of a science.

The mechanism behind the evolution of social structures has intrigued people for nearly a century. Measurements conducted in various social networks have generated a large volume of data, leading to important and sometimes surprising results. The theory of social networks has evolved from such measurements, with the analysis of graphs as the primary tool.

21.1.1 Milgram's Experiment

Historically, Milgram's experiment [M67] triggered interest in the discovery of how connected social beings are. A measure of the connectedness between two randomly chosen individuals is the length of the shortest chain of acquaintances between them, often referred to as the *distance* or the *degree of separation* between them. As an example, consider Figure 21.1, where each node represents a person and an edge denotes that the nodes at the two endpoints are friends. Thus, the distance between B and H is 4, and the distance between E and C is 1.

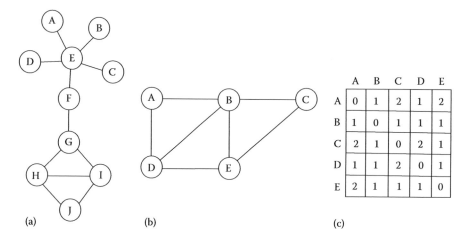

FIGURE 21.1   (a) A friendship graph with 10 persons: the distance between A and H is 4. (b) Another friendship graph with 5 nodes and (c) the distances between the different pairs of persons for the friendship group of (b).

Milgram arranged to send 160 envelopes to a group of randomly selected people from Wichita, Kansas, and Omaha, Nebraska and asked them to deliver the envelopes to a specific target person. Each envelope contained the following:

1. A document with the logo of Harvard (Milgram's home institution) on it.

2. The name, address, and occupation (stockbroker) of a friend of Milgram living in Boston, MA. This person is the intended recipient of all the envelopes.

3. Instructions to get the package to the target person following specific rules where each person could only send the package to an acquaintance. An acquaintance was defined as being on *first-name basis*.

The initiator will send the letter to an appropriate acquaintance who will forward it to the next one, until the letter reaches the target person. Milgram had a pressing concern about whether any one of these letters would reach the target. However, after a couple of modifications in the rules of package delivery, 42 of the 160 letters made it to the target person via up to a dozen intermediate persons. The median value of the number of intermediaries was 5.5, and the mean was 5.9, which can be rounded off to 6 leading to the famous term *six degrees of separation*.

Milgram's experiment had several limitations. Consider the following:

1. How did it account for the letters that did not reach the target person? Mathematically, we should consider the number of intermediaries to be infinite, but considering the imperfect nature of human beings, they were ignored.

2. The sample size was small—it was done on a small fraction of the residents of the United States only. All packages were sent to only one target person in one city, and these originated only from two cities.

It would be fair to question if these results can be generalized to a population of much larger size. Interestingly, recent experiments conducted on much larger sample sizes vindicated the myth of six degrees of separation. For example, a 2007 study done by Karl Bunyan on the Facebook platform with 5.8 million users showed that the average distance separating two participants of the application is 5.73 with the maximum distance being only 12.

These results are significant. It shows that human society is tightly knit, and there exists a way for any person to reach any other person in the planet via a short chain of acquaintances. Interestingly, small degrees of separation have been discovered in many other cases too. In the following, we present some significant ones:

World Wide Web: The web can be represented by a directed graph for which each node represents a web page, and an edge (i, j) indicates that there exists at least one pointer from page i to j. In 1998, when the number of web pages was estimated to be 800 million, Barabasi's experiment led to the conclusion that the average separation between any two web pages on the web is 19.

Film actor network: A film actor network is an undirected graph in which the nodes are actors, and an edge between a pair of nodes indicates that the two actors acted together in at least one film. Using the Internet Movie Database (IMDb) with a population of 225,226 actors, it was found that the mean distance between a pair of actors was only 3.65, and the average degree per node is 61. The most connected actors were the hubs of the graph, with Rod Steiger having the smallest distance of 2.53 from everyone else, Charlton Heston has a slightly longer distance of 2.57, and Kevin Bacon's distance is 2.79.

Electrical power grid: The electrical power grid of Western United States has been studied as a social network, with the nodes as the generating stations and the edges as high-voltage links connecting pairs of generating stations. This network has 4941 nodes, with an average distance of only 18.7 between pairs of generating stations and an average degree of 2.67 per node.

Caenorhabditis elegans: *C. elegans* is a simple worm of length 1.2 mm and lives in soil. It is one of the simplest organisms with a nervous system. Its nervous system consists of only 302 neurons, and the connections among these neurons have been completely mapped by biologists, with the nodes being the neurons and the edges being synapses connecting a pair of neurons.

21.2 METRICS OF SOCIAL NETWORKS

Many different metrics have been introduced to study the structural properties of a social network. This section describes some of these metrics.

21.2.1 Clustering Coefficient

Consider the edges of a friendship network, where each node is a person and each undirected edge between a pair of nodes i and j denotes the fact that i and j are friends with each other. For any given node i of this graph, the *clustering coefficient* measures what fraction of the various pairs of friends of node i are friends with each other. This is also the probability that two randomly chosen friends of a given node i are friends with each other. In Figure 21.1, consider B, who has four friends {A, C, D, E} from which six pairs can be constructed. Of these, (A, D) are friends with each other and so are (D, E) and (E, C). So, the clustering coefficient of B is 3/6 = 1/2. Note that the value of clustering coefficient always ranges between 0 and 1. For a tree topology, the clustering coefficient of every node is 0, whereas for a clique, the clustering coefficient of every node is 1. The clustering coefficient values of the nodes of a graph intuitively reflect how close-knit the community is.

21.2.2 Diameter

Let $d(i, j)$ denote the distance of the shortest path between a pair of nodes i and j. It is also known as the *geodesic distance*. For all such pair of nodes, the largest value of $d(i, j)$ is known as the *diameter* of a social network. Thus, it is the largest degree of separation between any pair of nodes.

21.3 MODELING SOCIAL NETWORKS

Given a number of isolated nodes representing actors, how does a social network evolve? There are several models about the formation of social networks. Erdös and Renyi proposed one of the earliest models based on random graphs. Their model is known as the *Erdös–Rényi model* or the ER model.

21.3.1 Erdös–Rényi Model

The ER model starts with a set of n isolated nodes. Between each pair nodes, an edge is added with a probability p. This results in the ER graph $G(n, p)$. It represents the formation of a social network as a random process. Note that the graph $G(n, p)$ is different from another kind of random graph known as $G(n, m)$, which is a graph that is randomly chosen from the set of all possible graphs with n nodes and m edges.

The connectivity of an ER graph $G(n, p)$ undergoes interesting changes as the value of p is increased. When p is much smaller than $1/n$, the graph consists of a large number of disjoint components—each component is a tree or a cycle of size $O(\log n)$. As p reaches or exceeds $1/n$, a *giant component* emerges—this is comparable to a phase change in systems that evolve in the nature. The size of the giant component is $O(n^{2/3})$, whereas the smaller components still have a size $O(\log n)$. Finally, when p exceeds $\log n/n$, $G(n, p)$ is almost always connected.

Some of the useful properties of the ER graph $G(n, p)$ are summarized in the following:

Property 1: The expected degree of a node in $G(n, p)$ is $(n - 1) \cdot p$. This immediately follows from the fact that a node chooses its neighbors from the pool of $(n - 1)$ nodes with probability p.

Property 2: The expected number of edges of $G(n, p)$ is $(n(n - 1)/2) \cdot p$. This follows from the fact that there are $\binom{n}{2} = \dfrac{n(n-1)}{2}$ pairs of nodes, and an edge is established with a probability p.

Property 3: The expected diameter of $G(n, p) \approx \log_k n$, where $k = (n-1) \cdot p$, is the expected degree of a node. It can be justified as follows: Let x be the diameter. Start a BFS from any node, and the farthest ones must be at a distance $\leq x$. With degree k, the total number of nodes within distance x must be $\leq 1 + k + k^2 + k^3 + \cdots + k^x$.

Therefore, $1 + k + k^2 + k^3 + \cdots + k^x \leq n$.

So, $x \leq \log_k n$.

Property 4: The expected value of the clustering coefficient of $G(n, p)$ is p.

Property 5: The number of nodes $N(k)$ with degree k in $G(n, p)$ follows a binomial distribution $\binom{n-1}{k} \cdot p^k \cdot (1-p)^{n-1-k}$. This can be understood as follows: (1) The probability that a given node connects to a given set of k nodes and does not connect to the remaining $(n - 1 - k)$ nodes is $p^k \cdot (1-p)^{n-1-k}$, and (2) there are $\binom{n-1}{k}$ different ways of choosing k nodes from the set of $(n - 1)$ nodes. A binomial distribution is represented by a *bell curve*, which is one of the signatures of a random graph.

The ER model of social networks held its ground for a long time but was later questioned. There are some networks like the US highway network that fits closely with the random network model, but most others do not. One of the characteristics of human society is living in clusters—so any model used to represent such a social structure must have a high clustering coefficient. However, the clustering coefficient of ER graphs is very small, although on the positive side, the diameter is small. The quest for a better model for such networks led Watts and Strogatz to propose the small-world model.

21.3.2 Small-World Model

In 1998, Watts and Strogatz reverse engineered Milgram's observations and proposed a model of social networks. The starting point is a regular clustered graph—specifically, they started with a *regular ring lattice* of n nodes in which every node had a degree $k (n \gg k > \ln n)^{\star}$ as shown in Figure 21.2a. This graph has a high clustering coefficient, but a high diameter too ($\approx n/2k$). To minimize the diameter, they rewired the regular graph by replacing the neighbors of each node by *randomly chosen neighbors* with a very low probability p (Figure 21.2b). The regular links represented the local contacts and maintained the clustered structure, whereas the randomly picked neighbors represented the occasional long-range contacts. They demonstrated that when $p \simeq 0.01$, the resulting graph still has a fairly large clustering coefficient, but the diameter substantially decreases, almost matching the diameter of random graphs. These satisfy the requirement of the social networks of human acquaintances. They called these *small-world graphs*.

Jon Kleinberg [K00] followed up with Watts and Strogatz's small-world model and tried to separate the issues of the *existence* of short chains among random peers and the *discovery* of such short paths during an actual search. He argued that although Watts–Strogatz's

---

* The condition $k > \ln n$ prevents the network from being partitioned.

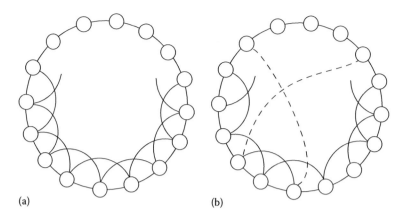

FIGURE 21.2 Watts–Strogatz construction of a sample small-world graph: (a) A regular ring lattice with n nodes, each of degree k. (b) Each node replaces an existing neighbor by a randomly chosen long-range neighbor (see the broken lines) with a very low probability $p \approx 0.1$. The resulting graph has a low diameter but a high clustering coefficient, which are the characteristics of a large class of social networks.

construction of small-world graphs only proves the existence part, no decentralized algorithm is able to discover the short paths between arbitrary pairs of nodes. He suggested a modified construction of small-world graphs, which will facilitate the discovery of short paths between pairs of nodes. Kleinberg's model suggests that the long-distance neighbors should not be chosen using uniform random probability, but with a skewed probability distribution where the probability of choosing a node as a long-range neighbor decreases with the distance of that node from the current node. He demonstrated that by carefully defining the probability of choosing the long-distance neighbors, it is possible to route a message to any other node in a very small number of hops, which explains the existence of short hops between pairs of nodes.

Kleinberg demonstrated his construction on a 2D lattice. His construction has two parts:

1. Each node maintains local contacts with every node upto a *lattice distance* of p—this reflects the local clustering of the nodes.

2. Each node randomly picks q long-range neighbors. Their addition involves a new parameter r: the probability of choosing a long-range neighbor at a *lattice distance d* is proportional to d^{-r}.

Kleinberg showed that when $r = 2$, which is the dimension of the lattice, there exists an algorithm using which each node can route a message in only $O(\log^2 n)$ hops. The algorithm requires that in each step, the current message holder send the message to a node that is as close to the target as possible.

21.3.3 Power-Law Graphs

While the small-world model is able to explain the presence of short path between pairs of nodes, many real-life social networks do not fit into this model. Unstructured P2P

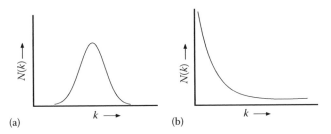

FIGURE 21.3 Degree distributions in (a) ER graph and (b) power-law graph.

networks like Gnutella, US power grids, the WWW, and the US airport network all exhibit a *power-law distribution* of node degrees. In a power-law distribution, the number of nodes $N(k)$ having degree k satisfies the condition $N(k) = C \cdot k^{-r}$, where C and r are constants. Such networks are also known as *scale-free networks*. For a rich class of systems that evolve in the nature, the condition $2 < r < 3$ holds. Figure 21.3 compares the degree distributions of a random ER graph and a power-law graph.

Barabási and Albert [BA99] described a method for generating a subclass of graphs that satisfy the power-law distribution. Their method uses the concept of *rich getting richer* that is also known as *preferential attachment*. The guiding principle is that when nodes join an existing network, they are likely to connect to existing nodes that already have a high degree, because such nodes are considered more influential and are likely to provide better connectivity to the rest of the network. More precisely, if $\delta(i)$ is the current degree of node i in the existing network, then the incoming node will connect to node i with a probability proportional to $\delta(i)$. This means that the high-degree nodes are expected to attract more neighbors than the low-degree nodes. We demonstrate here that this policy leads to the creation of a power-law network.

Consider the creation of a network as follows: At time $t = 1$, a single node appears from nowhere. Thereafter, at each time step, exactly one node is added to the existing network. Let $G = (V, E)$ denote the topology of the graph. A new node j can connect with an existing network via $m (m \geq 1)$ edges. The probability that the new node j will connect to the existing node i via edge (j, i) is $C \cdot \delta(i)$ where C is a constant (Figure 21.4). As a simple case, consider $m = 1$ (the resulting topology will be a tree). Since $\sum_{i \in V} C \cdot \delta(i) = 1$, $\sum_{i \in V} \delta(i) = 2|V|$, and at time t, $|V| = t$, it follows that $C = 1/2t$.

At time step t, let the number of nodes with degree k be $n(k, t)$. At time step $(t + 1)$, a new node will join, which will modify the number of nodes with degree k. To compute $n(k, t + 1)$, observe the following:

1. If the incoming node connects to an existing node with degree $(k - 1)$, then its degree will increase to k. The probability of this event is $(k-1)/2t$.

2. If there is an existing node with degree k but the incoming node *does not* connect with it, then its degree remains unchanged at k. The probability of this event is $1-(k/2t)$.

3. There is no other event that can influence the value of $n(k, t + 1)$.

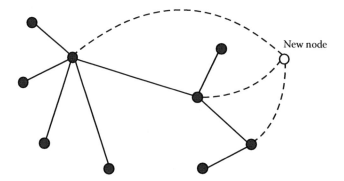

FIGURE 21.4 The evolution of power-law distribution via the rich gets richer model: the probability of the new node connecting to one of the existing nodes is proportional to the degree of that node.

This revision of the degrees is reflected in the following equation:

$$n(k,t+1) = n(k,t) \cdot \left(1 - \frac{k}{2t}\right) + n(k-1,t) \cdot \frac{k-1}{2t} \tag{21.1}$$

Let $f(k,t) = n(k,t)/|V|$ denote the fraction of the total nodes with degree k at time t. Since one node is being added at every time step, $|V| = t$ at time step t, and $|V| = t + 1$ at time step $(t + 1)$. Therefore, from (21.1),

$$(t+1) \cdot f(k,t+1) = t \cdot f(k,t) + \frac{1}{2}\left[(k-1) \cdot f(k-1,t) - k \cdot f(k,t)\right] \tag{21.2}$$

However, as $t \to \infty, f(k,t+1) \to f(k,t)$, and we designate it by $f(k)$. Accordingly, it follows from (21.2) that after a very long time,

$$(t+1) \cdot f(k) = t \cdot f(k) + \frac{1}{2}\left[(k-1) \cdot f(k-1) - k \cdot f(k)\right]$$

i.e., $\quad f(k) = \frac{1}{2}\left[(k-1) \cdot f(k-1) - k \cdot f(k)\right]$

So, $\quad 2f(k) = \left[(k-1) \cdot f(k-1) - k \cdot f(k)\right]$

i.e., $\quad f(k) = \frac{k-1}{k+2} \cdot f(k-1)$

$$= \frac{k-1}{k+2} \cdot \frac{k-2}{k+1} \cdot \frac{k-3}{k} \cdot \frac{k-4}{k-1} \cdots \frac{4}{7} \cdot \frac{3}{6} \cdot \frac{2}{5} \cdot \frac{1}{4} f(1)$$

$$= \frac{3 \cdot 2 \cdot 1}{(k+2) \cdot (k+1) \cdot k} \cdot f(1) \tag{21.3}$$

To compute $f(1)$, consider (21.1) for the case $k = 1$. Notice that there is currently no node with $k = 0$ except the one that is being added. After the addition, its degree will change to 1. So,

$$n(1,t+1) = n(1,t) + 1 - n(1,t) \cdot \frac{1}{2t}$$

that is, $\quad (t+1) \cdot f(1,t+1) = t \cdot f(1,t) + 1 - \frac{f(1,t)}{2}$. But as $t \to \infty, f(1,t+1) = f(1,t) = f(1)$

This leads to $f(1) = 2/3$, and (21.3) can be rewritten as $f(k) = 4/k(k + 1)(k + 2)$. Therefore, after a long period of addition of nodes, the network will stabilize to a topology for which the fraction of its node with degree k will be of the order of $1/k^3$, and power-law distribution will hold.

In power-law networks, short paths exist between arbitrary pairs of nodes—it becomes feasible due to the presence of the hubs. Consider the US airport network again, which is a power-law network. In traveling from one airport to another via a shortest path, in how many occasions do we need to change plane more than two to three times? Another important property of such networks is that they can tolerate the failure of random nodes very well. Only targeted attacks on major hubs have the potential to cripple the network.

While these models help understand the structures of a large class of social networks, the evolution of social structures may have many other mechanisms. One such mechanism is to selectively adopt the neighbor of a neighbor as a neighbor forming *triadic closures*. Another mechanism is to establish connections beyond local neighbors—this is possible when actors meet based on common hobbies or travel in common transports like a bus or a train or in a carpool. Both of these are mechanisms for growth in current Internet-based social networks like Facebook, Twitter, or LinkedIn.

21.4 CENTRALITY MEASURES IN SOCIAL NETWORKS

Centrality is a measure of the importance of a node (or an edge) in a social network. There are different aspects of centrality based on how you assess the importance. We present here three forms of centrality measures.

21.4.1 Degree Centrality

The more neighbors a given node has, the greater is its influence. In human society, a person with a large number of acquaintances is believed to be in a favorable position with more opportunities. This leads to the idea of *degree centrality*, which refers to the degree of a given node in the graph representing a social network. In Figure 21.1a, the degree centrality of node E is 5 and that of G, H, I, and J are 3 each.

21.4.2 Closeness Centrality

The power or influence of a node can also come from its ability to act as a reference point and by being a center of attention so that its influence is felt by a large number of nodes.

Nodes that are able to reach other nodes via shorter paths, or that are more reachable by other nodes via shorter paths, are in more favored positions. This structural advantage can be translated into power, and it leads to the notion of *closeness centrality*. Consider the network in Figure 21.1b, for which the table in Figure 21.1c lists the distances between the various pairs of nodes. Let V denote the set of nodes, and for $\forall i, j \in V, d(i, j)$ represent the distance between i and j. A yardstick of the closeness of a node i from the other nodes is $\sum_{j \in V} d(i, j)$. The smaller it is, the closer is node i to the other nodes. Closeness centrality is sometimes expressed as a normalized value (ranging between 0 and 1) with respect to the node that is the closest of all. In Figure 21.1b, it is node B and $\sum_{j \in V} d(B, j) = 4$, and its closeness centrality is 1. For node E, $\sum_{j \in V} d(E, j) = 5$, so its closeness centrality is $4/5 = 0.8$. Similarly, for node A, $\sum_{j \in V} d(A, j) = 6$, so its closeness centrality is $4/6 = 0.66$.

21.4.3 Betweenness Centrality

Communication between nonneighboring nodes is channeled through intermediaries. For each pair of nodes in a social network, consider one of the shortest paths—all nodes in this path are intermediaries. The node that falls in the shortest paths between the maximum number of such communications is a special node—it is a potential deal maker and is in a special position since most other nodes have to channel their communications through it. This leads to the notion of betweenness centrality.

To estimate the betweenness centrality of a node k, let V denote the set of nodes, $N(i, j)$ be the number of shortest paths between a pair of nodes $i, j \in V$, and $N_k(i, j)$ be the number of such shortest paths that include the node k. The fraction $N_k(i, j)/N(i, j)$ is a normalized measure of the betweenness centrality of node k with respect to the pair of nodes i, j.

Consider Figure 21.5. The shortest paths between the nodes A and G are ACEG and ABFG. Therefore, the betweenness centrality of node C with respect to (A, G) is 0.5. The overall betweenness centrality of a node is considered over all possible pairs of nodes in the network. Here, none of the shortest paths among pairs of nodes in {B, D, E, G, H} includes node C, but one of the two shortest paths between node {A, D} and {A, G} and the only shortest path between {A, E} include node C. Assume that each source node pushes 1 unit of flow to a

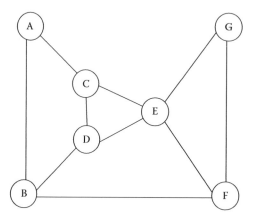

FIGURE 21.5   Example of betweenness centrality.

destination node via shortest paths. In the presence of multiple shortest paths the flow is evenly split, as node C will route (1 + 0.5 + 0.5), that is, 2 units of flow. The nodes that handle the largest volume of flow have the highest betweenness centrality in the graph. The largest volume of flow can be used to normalize the betweenness centrality of the other nodes in the network.

Betweenness centrality not only applies to nodes but also applies to edges. For a given edge, it is defined in a similar way by counting how many shortest paths between various pair of nodes include that edge and what fraction of the overall flows between pairs of nodes is routed through that edge.

The computation of betweenness centrality involves examining the shortest paths between every pair of node. There exist deterministic algorithms for community detection with time complexity of $O(m \cdot d \cdot \log n)$ where n is the number of nodes, m is the number of edges, and d is the depth of the dendrogram* [CNM04].

21.5 COMMUNITY DETECTION

Relations among entities within a social network usually have a significant amount of heterogeneity. In many such networks, there is a higher concentration of edges between a fraction of the nodes that represent a community structure and a relatively lower concentration of edges between nodes belonging to different communities. The goal of community detection is to analyze the graphs representing social networks and identify such communities. Note that communities can also be divided into subcommunities—so digging into this hierarchy will end up in finer classifications. For example, within a social network, the clients of a particular national store chain may form a community, and within such a community, the senior citizens may form a subcommunity that is different from the subcommunity formed by teenagers. In a given academic community, both biologists and computer scientists are engaged in doing research and publishing research papers. Some of these papers will be authored by only biologists, some will be authored by only computer scientists, and perhaps only a smaller number of papers will have both biologists and computer scientists as coauthors. Given an unlabeled coauthorship graph, a community detection algorithm should be able to identify the existence of two distinct communities.

21.5.1 Girvan–Newman Algorithm

In 2002, Girvan and Newman [GN02] proposed an algorithm for community detection in social networks. Given a graph G representing a social network, it identifies the edges connecting distinct communities and removes them in an iterative manner, until the communities were isolated. The edges connecting distinct communities are detected by measuring their betweenness centrality. An outline of the algorithm follows:

```
{Girvan-Newman algorithm for community detection: the first step}
do G is a single connected component →
   Detect the edge(s) of highest betweenness centrality and
   remove those edges
od
```

---

* A *dendrogram* is a tree representing a hierarchical clustered structure.

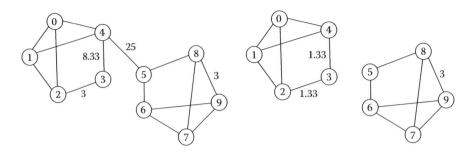

FIGURE 21.6 Illustration of a step of Girvan–Newman algorithm: the removal of the edge (4, 5) of highest betweenness splits the network into two partitions.

To detect the nested structure of the communities, the algorithm must run on each partition until all edges are removed. The tree structure showing the communities and the sub-communities in a social network is a *dendrogram* with the individual nodes at the leaves.

Since the time complexity of computing the betweenness centrality is substantial, it may be tempting to avoid recalculation of the betweenness centrality of the remaining edges after an edge is removed, but this will lead to serious error since the flows within the graph adjust to the new topology. As an example, consider the graph in Figure 21.6. The edge (4, 5) has the highest betweenness centrality of $5 \times 5 = 25$, since every shortest path from any of the nodes {0, 1, 2, 3, 4} to any other node in {5, 6, 7, 8, 9} must pass through that edge. In contrast, the betweenness centrality of the edge (2, 3) is only 4, since one-third of the shortest paths from anyone of {4, 5, 6, 7, 8, 9} to 2 will pass through the edge (2, 3), half of the shortest paths from any in {0, 1} to node 3 will pass through the edge (2, 3), and the shortest path from node 3 to 2 will pass through (2, 3).* This also applies to the edge (5, 8). Similar calculations will reveal that the betweenness centrality of the edge (3, 4) is 8.33. After the edge (4, 5) is removed, the graph is partitioned and the partitions expose the community structures. Note that the betweenness centralities of the edges (2, 3) and (3, 4) have changed after the removal of the edge (4, 5). The algorithm has a high time complexity—it runs in $O(n^3)$ time on sparse graphs of size $n = |V|$, so is clearly not scalable.

A more recent algorithm by Raghavan et al. [RAK07] solves the problem in $O(|E|)$ time, where E denotes the set of edges. It uses the concept of *label propagation*. The basic idea is as follows: Initially, each node is assigned a unique label. Thereafter, at every step, each node acquires the label that the majority of its neighbors currently have. Eventually, densely connected groups of nodes form a consensus on a unique label, and nodes with identical labels identify a community.

21.6 INTRODUCTION TO PEER-TO-PEER NETWORKS

P2P networking is a paradigm where a set of user machines at the edge of the Internet communicates with one another to share resources without the help of any central authority. Geographical boundaries become irrelevant, and the absence of any central authority promises spontaneous growth, as well as freedom from censorship. Peers include friends,

---

* We avoid double counting caused by the swapping of the source and the destination nodes of each route.

collaborators, and competitors, and the resource sharing has to be implemented through decentralized protocols. Scalability is an integral part of this concept—no P2P system is worth looking at unless it scales to millions of machines around the globe.

As an illustration, consider that you have several hundred movies in the digital format, but not enough storage space in your computer to hold all of them. Some of your friends might have surplus storage space in their computers, so they volunteer to help you with storing some of your movies at their space. When these friends acquire new movies, they also do the same thing, that is, use the storage space of their friends. In this way, a digital library of movies is formed, which is distributed over a geographic region. Now, when you want to access the movie *Life of Pi*, you would like to know where it is located. For this, you apply a lookup function on the name of the movie (or use some other tool to map names into locations), and the result gives you the location of the movie. You arrange to send a request to that location through the shortest possible path and start downloading the movie. The social networking giant Facebook uses P2P storage to build their Cassandra file system that handles their *Inbox* search problem, the physical storage being scattered around thousands of data centers around the globe.

P2P is one of the technologies that started with music sharing over the Internet and was pioneered by *Napster*. A Napster client could download any MP3 music from another client who has a copy of it. There could be multiple copies of the same song at different sites, and a client could download the desired music from a next-door neighbor or from another host halfway round the globe. After attaining significant popularity, Napster was closed by the government for copyright infringement.

Regardless of these legal ramifications or ethical issues, P2P has led users to a new form of freedom in collaborative resource sharing. For example, hundreds of small laboratories in the world generating *genomic data* about newly discovered proteins now share one another's discoveries using P2P technology. Facebook and Twitter started using BitTorrent technology for content distribution. This chapter presents the underlying principles behind the various kinds of P2P networks.

21.7 FIRST-GENERATION P2P SYSTEMS

21.7.1 Napster

All P2P networks are *overlay networks*. An overlay network is built on top of an existing network, where the set of nodes is a subset of the set of nodes of the original network, and the edges correspond to paths between distinct nodes. Each node has an IP address, and each edge can be traversed by one or more hops on the underlying IP network.

Current P2P systems are broadly classified into three different categories: *centralized*, *unstructured*, and *structured*. The centralized architecture followed by Napster does not strictly fit the profile of P2P systems since it used a central index server. However, Napster has historic significance. Another early P2P system is *Gnutella*, which belongs to the unstructured category: objects are located by flooding the queries. This section provides a brief outline of these two well-known first-generation P2P systems. We will use the terms songs and files interchangeably.

The old Napster had a *centralized directory* containing the indices of multiple files of MP3 music stored at its clients' machines. Each Napster client registers with the Napster service and exports the names of the MP3 songs that reside locally on that host. Note that Napster only stores the indices of the files but not the actual files. The directory may contain the indices of multiple files with the same name. The server keeps track of all clients currently connected to it. To locate a song, the user logs in and enters the title of the music or name of the singer, which prompts the Napster utility to query the index server. If a match is found, then a list of all matches is sent back to the client. The client can ping one or more of these sites to estimate the download speed and then *directly* download the song from one of these sites.

The new Napster is fully legal, and the company has changed its business model that includes subscription service, as well as selling individual MP3 tracks and albums.

21.7.2 Gnutella

Gnutella started after the demise of the old Napster, and it used a fully distributed alternative architecture. Unlike the old Napster, there is no central server that holds the indices of all the songs that are available on the network of Gnutella clients.* Instead, these reside in the clients' machines. A new client must know the IP address of at least one other Gnutella client. To facilitate the bootstrap operation, each client receives a list of the addresses of working nodes. When a client P connects to an existing client Q, Q sends P its list of its current neighbors. P will now try to connect not only to some of its own list of nodes but also to one or more nodes from the list of nodes that it received from Q.

Once connections are established, queries for objects are propagated down the network. The early search protocols used flooding—each client checks if the desired file is locally available—if so, then the index and the IP address are made available to the originator of the query. Otherwise, the query is forwarded to all other clients that it knows of. To guarantee termination, each query is assigned a TTL, which reflects the maximum number of levels that the query is allowed to propagate. Once this limit expires, the query is discarded. The Gnutella protocol uses the following five descriptors:

Ping: Discovers hosts on the network by asking: *Are you there?*

Pong: This is the response to a ping, and it includes its IP address of the responder and the amount of data that it wants to share.

Query: The primary mechanism for searching an object: *I am looking for XYZ* (the message is forwarded until XYZ is located or the search is abandoned).

QueryHit: It is the response to a query in case a host has that object in its local store. The descriptor with the IP address, port number, etc., required to download XYZ is propagated to the client via the return path.

Get/Push: Initiate download. If the source is firewall protected, then the source gets a request to push the object to the client.

---

* Since every client is also a server, Gnutella calls them *servent* (server + client).

One concern about the old Napster as well as the original Gnutella is their scalability. For Gnutella, the flooding of the queries hogs network bandwidth. Even though the old Gnutella kept some users happy at the scale at which it was operating, it certainly took away useful bandwidth from other useful applications. Search traffic was taking a quarter of the net bandwidth and users had limited visibility of what they could find.

As far as the old Napster was concerned, the centralized index server could have been a possible bottleneck. But apparently, there was not much complaint about it since servers were replicated. The claim was that there were 1.5 million simultaneous users at its peak, and this demand was adequately handled via replication. To feel the pinch, Napster would have had to cater to a much larger clientele, perhaps is debatable considering, but this is how efficiently Google or Yahoo now provides service. Before these issues could be examined, the government shut down the old Napster.

Another issue in P2P networks is their resistance to attacks and censorship. Did the old Napster and Gnutella live up to that promise? Clearly, the old Napster did not! Once the server sites were blocked, Napster became crippled. In contrast, Gnutella is a truly distributed architecture—so blocking a small number of sites was not much of a disruption. To disrupt Gnutella, one of the following two approaches* appears feasible:

1. Flood the network with bogus queries. This is not quite a DoS attack (as there is no central server), but it can significantly slow down the sharing process and discourage the clients.

2. Store bogus files or spams at many sites through malicious clients. The spam will frustrate and possibly discourage the users from sharing.

21.8 SECOND-GENERATION P2P SYSTEMS

The lessons from old Napster and Gnutella led to the design of new breeds of P2P systems that tried to overcome many of the limitations of the first-generation systems. There are numerous candidates among the second-generation P2P systems. Prominent among them are KaZaA, Chord, content-addressable network (CAN), Kademlia, Pastry, Tapestry, and BitTorrent. Each of these systems addresses four primary issues central to P2P file sharing:

File placement: Where to publish the file to be shared by others?

File lookup and download: Given a named item, how to find or download it? How fast can it be downloaded? Such downloads can be from the original server or from a proxy server holding a cached copy.

Scalability: How is the performance affected when the network scales to millions of nodes? No P2P network is worth looking at unless performance guarantees are provided at large scales.

Self-organization: How does the network handle the join-and-leave operation of the clients? The existing members should be made aware of the presence of the new nodes and

---

* We do not recommend any of these on an actual system!

the departure of those who left the system. Also, the new members must eventually know about the existing members.

Additionally, some of the popular systems address a few secondary issues that make them attractive to users. These are discussed in the following:

Censorship resistance: How does the network continue to offer its services in spite of potential authoritarian measures that can shut down a fraction of the nodes? Such measures are common in the face of differences in political or ideological views. One view of censorship resistance is that even if a substantial fraction of the nodes is blocked, almost all the remaining nodes should still be able to access almost all the original data items.

Anonymity: How to keep the name of the owner of an object or the location of its publication secret?

Fault tolerance: How to prevent significant performance degradation in spite of node failures?

Free-rider elimination: A fraction of peers always uses resources (by downloading stuff from others) but never contributes to the resources (uploading objects so that others can use). This is ethically unfair. How can free riding be eliminated or at least free riders be discouraged?

Many of these issues are related. For the lookup problem, a central table as used in Napster is not acceptable primarily from fault tolerance. A similar lookup is routinely done by the DNS on the Internet. DNS has a hierarchical structure. The shutdown of a reasonable fraction of nodes sufficiently high up in the hierarchy of DNS can be catastrophic.

Gia—an attempt to improve Gnutella's performance: The *Gia* network by Chawathe et al. [CRB+03] is an improvement over Gnutella in as much as it addresses the scalability problem. Four proposals form the cornerstone of the improvement:

1. Gia replaces the flooding of Gnutella by *random walk*. One or more random walkers can be engaged to locate the desired object. This lowers network congestion and expedites object location.

2. Gia keeps track of the heterogeneity of the network by identifying which nodes have higher capacity and bandwidth. Such nodes can handle a larger number of queries. The principle of *one-hop replication* enables each node to maintain an index of the content of its neighbors. Accordingly, high-degree nodes are likely to hold more clues about the object being searched in their immediate neighborhood.

3. A *dynamic topology adaptation protocol* converts the high-capacity nodes into high-degree nodes by encouraging them to adopt other nodes as neighbors. This helps guide the search in a meaningful way—a biased version of random walk, where the probability of directing a search to a node of degree δ is proportional to δ, helps the queries gravitate toward high-degree nodes and improves the efficiency of the search.

4. Gia uses flow control tokens that are predistributed according to the capacities of the nodes. Queries are not dropped but pushed to a node only when it is ready to accept and handle it.

The topology of Gnutella (or Gia) depends on when and where the peers joined the network and does not strictly conform to predefined specifications. These are examples of *unstructured networks*. In contrast, a class of P2P networks requires the peers to store or publish objects following specific guidelines and maintain their neighborhood in a predefined manner. Such networks provide uniform guarantees regarding search latency and load balancing. These are called *structured* P2P networks and are based on *DHTs*.

21.8.1 KaZaA

In 2001, the Dutch company Consumer Empowerment introduced KaZaA following the demise of Napster. KaZaA improves upon the performance of Napster by using FastTrack technology. A fraction of the client nodes that have powerful processors, fast network connections, and more storage space are used as *supernodes*. The supernodes are spontaneously designated, and they serve as temporary indexing servers for the slower clients. Since the indexing service is no more centralized, KaZaA has much improved scalability and fault resilience. A KaZaA client stores the IP address of a set of supernodes and picks one of these supernodes as its upstream and uploads to it the indices of a list of files it intends to share (with other peers). All search requests are directed to this supernode, which communicates with other supernodes to locate the desired file. After the file is located, the client directly downloads it from the peer. While the indexing servers retain the flavor of Napster, the supernode–supernode communication used for forwarding queries is reminiscent of Gnutella. KaZaA allows download from multiple sources and uses a lightweight hash algorithm to checksum large files.

21.8.2 Chord

Searching files or objects in unstructured networks like Gnutella, in a way, amounts to groping in the dark. Before initiating the search, the initiator of a query has no clue about where the file might be located or if the file is at all present. This results in poor utilization of the network bandwidth. *Structured* P2P networks use a hashing function to precisely map objects to machines. This mapping function is used to publish and locate the object. Accordingly, given the name of an object, every peer knows where it is available. As an example, consider an m-cube. Assume that objects are stored into various machines located at the $n = 2^m$ corners of the cube, and given an object, the identity of the machine storing that object can be looked up using a hashing function. This is the essence of DHT. To access any object from a remote machine, the user will forward the query along the edges of the m-cube. The propagation of both queries and replies will take up to $m = \log_2 n$ hops. To make it possible, each machine will store the address of only $m = \log_2 n$ neighbors.

The earlier principle is the main idea behind *Chord* designed by Ian Stoica and his colleagues [SML+02]. Each node in Chord maintains the IP addresses of a small number of other nodes, called its *neighbors* or *fingers*. The physical adjacency of neighbors is not relevant. A set of peers anywhere in the globe can form an overlay network, as long as their routing tables clearly indicate how to reach one node from another either directly or via other peers. Routing data from one peer to its neighbor is counted as one hop. Both node

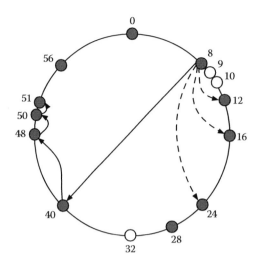

FIGURE 21.7   Node with key 8 queries for an object hosted by a node with key 51. No real machine maps to the keys 9 and 10 represented by blank circles.

identifiers and object names are mapped into the same key space $0..\ n - 1$, and the name-to-key conversion is done using a *consistent hashing* function like SHA-1. Conceptually, Chord maps the keys on the periphery of a circle, which defines the key space. Unless otherwise mentioned, we will assume that the keys are mapped in the ascending order along the clockwise direction of the circle (Figure 21.7). There may not be a physical machine for every key position. An object with a key K is mapped to the machine with key *successor*(K)—where *successor*(K) is the first node in the clockwise direction with key $\geq K$ that corresponds to an existing machine in the network. Consistent hashing [KLL+97] helps with load balancing across different machines so that (1) every node stores roughly the same number of objects with a high probability, and (2) when a new node joins or leaves the network, only $O(1/n)$ fraction of the keys are shuffled across machines.

Routing: Each node has a routing table (called a *finger* table) with $m = \log_2 n$ entries. Each entry is called a *finger* that points to a neighbor—the rth finger of the node with key K points to the node with key *successor*($K + 2^{r-1} \bmod n$). Figure 21.7 illustrates a Chord network with keys 0–63. For the machine with key 8, its first two fingers will point to the machine with key 12, since no physical machine maps to the keys 9, 10 or 11. If the first finger of node i points to node j, then node j is the *successor* of node i, and node i is the *predecessor* of node j.

Lookup: To look up an object, first generate its key K by hashing the object name. Now follow a greedy search policy by taking the first hop using a finger that will lead to a machine with a key *closest to* (but not exceeding) K. Repeat this step to route the query until you reach the machine containing the desired object. With *high probability*, each hop reduces the distance by at least half, so it takes $O(\log n)$ hops to complete the lookup.

Join and leave: Every P2P network is dynamic—from time to time, a fraction of the existing nodes *leave* the network and new nodes *join* the network. The object placement rule may be violated when an existing node leaves the network or a new node joins the network. Accordingly, the leave and join protocols must restore the invariants by appropriately modifying the fingers and moving the objects across the nodes. A consistent configuration of the Chord ring satisfies the following two invariants:

1. For each node *v*, the *successor* and *predecessor* pointers are correct and satisfy the condition *predecessor(successor(v))* = *v*. The predecessor pointer helps check the integrity of the current topology.

2. Each object with a *key K* is stored in *successor(K)*.

We first illustrate the handling of the *leave* operation first. In Figure 21.7, assume that the machine with key 50 plans to leave the network. For this, it has to offload the object(s) held by it to another machine. After these objects are relocated to the machine with key 51, all fingers pointing to the machine with key 50 should be updated and redirected to the machine with key 51. This concludes the *leave* operation.

For the *join* operation, each new node *v* will first contact an existing node *v'*. Node *v'* will help the new node find its place in the Chord ring. The following three steps are needed to restore consistency:

1. Node *v'* will help node *v* find its *successor* node and *initialize* its finger table.

2. Node *v* will be added to the finger tables of the appropriate existing nodes.

3. Any object with key *K* such that *successor(K)* = *v* will be relocated to node *v* from its current holder.

The time complexity of a join or leave operation is $O(\log^2 n)$. The join and the leave protocols of Chord are designed to handle one such event at a time. However, in practice, the churn rate can be quite high, and the routing tables can become inconsistent when multiple nodes concurrently join or leave. A *stabilization protocol* therefore periodically runs at the background and restores consistency. The essence of the stabilization protocol is for each node *v* to check if the invariant *predecessor(successor(v))* = *v* holds. If it does not hold, then stabilization protocol will modify some of the pointers and enforce this invariant. As an example, let node *v* join the system, and assume that its id falls between two nodes *v1* and *v2* (i.e., *v1* < *v* < *v2*). Clearly, node *v* sets its successor to node *v2*, while the predecessor of node *v2* is still node *v1*. When the stabilization protocol runs, node *v* finds out *predecessor(successor(v))* = *v1*. So, node *v* sets its predecessor pointer to *v1*, contacts *v1*, and asks it to set its successor pointer to *v*. Packets that are in transit while the network is not stable may temporarily reach a wrong host. However, using the stabilization protocol, each node restores its routing table. As a result, the subsequent hops will eventually lead the packet to the correct machine.

Finally, nodes can occasionally crash instead of voluntarily leaving the system. This causes objects with no backup to be permanently lost to the rest of the peers. To deal with failure of a node, each machine keeps track of the IP addresses of the next $r = 2 \log n$ nodes in the key space and replicates its objects in these nodes. If the probability of failure is ≤50%, then with high probability, at least one of the replicas will survive and become available to the users.

21.8.3 Content-Addressable Network

CAN is a structured P2P network developed by Ratnasamy et al. [RFH+01]. Unlike Chord that uses a ring of keys to implement the DHT abstraction, CAN uses a d-dimensional Cartesian coordinate space for the same purpose. For illustrating the basic architecture, we will assume that $d = 2$. Objects are hashed into points in a 2D torus $[0, 1] \times [0, 1]$ as shown in Figure 21.8. Each subspace of this 2D space is assigned to a physical machine, and this machine hosts all the objects stored in its space. Initially, there is only one machine A that will store all the objects. Later, when machine B joins the network, it contacts an existing machine (in this case, it is A), which splits its own area into two halves and allocates one of the two halves to B. All objects whose keys belong to the zone allocated to B will be transferred to B from A, and the two machines become neighbors in the P2P network. Each join is handled in the same way—an existing machine splits its own area into two halves and gives one of them to the new machine. When a node leaves, its zone is taken over by one of its neighbors. Sometimes, this leads to the creation of one larger zone. After a join or a leave operation, the routing tables are appropriately updated. Nodes can also crash from time to time, creating dead zones and causing fragmentation in the coordinate space. A node reassignment algorithm running in the background merges some of the fragments into a valid zone and assigns it to a nonfaulty node.

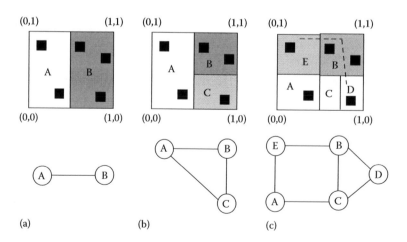

FIGURE 21.8 Three stages in CAN: The square boxes are objects. In (a), machine B joins the network and takes over three objects from machine A. In (b), machine C joins the network. In (c), two more nodes, E and D, join the network. The corresponding interconnection networks are shown in the bottom row.

The *neighborhood relationship* among machines is defined as follows: On a 2D plane, any two machines whose regions share an edge are neighbors. In the *d*-dimensional space, two machines are neighbors of each other, if their regions share a $(d-1)$-dimensional hyperplane. Figure 21.8 shows three stages of CAN. The routing table for each machine has $O(d)$ entries. Since *d* is a constant for a given implementation, the space requirement is independent of the size of the network. Routing of queries takes place via the *shortest route* from the source machine to the destination object. Figure 21.8c shows a sample route from the machine *E* to an object in machine *D*. In a 2D space, the *maximum routing distance* between a pair of nodes is $n^{1/2}$—this holds when the interconnection network is a square grid, and the source and the destination machines have coordinates (0, 0) and $(n^{1/2}/2, n^{1/2}/2)$. In general, in a *d*-dimensional torus, the routing distance between a pair of nodes is $O(d \cdot n^{1/d})$ hops.

21.8.4 Pastry

Pastry, developed by Microsoft Research in 2001, is a substrate for a variety of P2P applications. In addition to file sharing, such applications include global persistent storage utility (PAST), group communication (SCRIBE), and cooperative web caching (SQUIRREL). Pastry is a DHT-based P2P system. Each node is assigned a 128-bit id that is generated by applying a cryptographic hash function on the node's public key or its IP address. This id defines the position of a node in a circular key space of 2^{128} nodes. In an *n*-node system, any node can route messages to any other node in $O(\log_r n)$ hops, where $r = 2^b$ $(b > 0)$. A typical value of *b* is 4. Compared to Chord or CAN, the hop count in Pastry is in general much lower. In addition to routing with a low hop count, Pastry takes into account the physical proximity between nodes during neighbor selection, which minimizes the electrical distance between the source and the destination and leads to faster message delivery and query processing.

Routing: Pastry routes messages using *prefix routing*. To implement prefix routing, each Pastry node stores three types of information:

> *Leaf set L*: Each node with id *i* maintains a list of *L*/2 nodes with numerically closest larger ids and *L*/2 nodes numerically closest smaller ids. Typically, $|L| = 2^b$.

> *Routing table R*: Each node *i* has a routing table of size $(2^b - 1) \times \lceil \log_{2^b} n \rceil$. Row *j* of the table points to a node whose id shares the first *j* prefix digits with node *i*, but whose $(j + 1)$th prefix digit is different from that of node *i*.

> *Neighborhood set M*: This set contains the nodes that are nearest to *i* with respect to the *network distance* (like *round-trip delay*).

A node can directly forward (i.e., in a single hop) a message to any member of its *leaf set*. If the destination node is not in the leaf set, then the message is forwarded to a node whose id shares the *largest common prefix* with the destination id. The routing table stores information about such nodes. Thus, to route a message from X to Y, the source node X first sends the message to a node P whose id has a larger number of prefix digits in common with the id of the destination node Y. Thereafter, node P forwards the message to node Q whose id has even more prefix digits in common with the key of the destination node Y. This process is repeated until the final destination is reached (or a node in the leaf set of the final destination is reached—it takes one more hop to reach the destination node). A typical routing table for a network with $n = 2^{12}$ nodes and $b = 2$ (i.e., $r = 4$) is shown in Figure 21.9a. It has $\log_r n = 6$ rows and $(r - 1) = 3$ columns. The symbol X denotes a wildcard entry—thus, 21XXXX means some node whose id has the prefix is 21. Note that there is no guarantee that such a node can be found, and therefore, some of the entries in the routing tables are likely to remain blank. Figure 21.9b shows a possible route from node 203310 to node 130102. Define the *distance* between a pair of ids as the number of digit positions where they differ. Then in each hop, this distance is reduced. The expected routing distance between a pair of nodes is $O(\log n)$. In case a suitable intermediate node is not found in the routing table, the message is forwarded to a node that shares a prefix with the key at least as long as the local node and is numerically closer to the key than the present node's id. Only in *rare cases*, the final destination is not in the leaf set of the last node reached by prefix routing (i.e., it is still more than one hop away but there is no suitable entry in the routing table). If the distribution of the node ids is uniform, then with $|L|/2 = 2^b$, the probability of this event is less than 0.02. It becomes much lower when $|L| = 2^{b+1}$. Should this happen, the routing cost increases by one hop with high probability.

A larger value of b improves the efficacy of routing. When $b = 4$ (i.e., $r = 16$), and there are a billion nodes, then between any two Pastry nodes, a message will be routed in at most $\lceil \log_{16} 1,000,000,000 \rceil = 7$ hops. This assumes that the routing tables are fully populated.

Leaf set of node 203310

Smaller id	203302	203300
Larger id	203311	203323

Routing table of node 203310

0XXXXX	1XXXXX		3XXXXX
	21XXXX	22XXXX	23XXXX
200XXX	201XXX	202XXX	
2030XX	2031XX	2032XX	
20330X		20332X	20333X
	203311	203312	203313

Neighborhood set of 203310

100230	211201
213112	300100

(a)

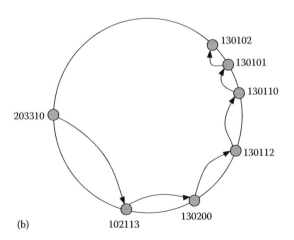

(b)

FIGURE 21.9 (a) The routing table of a hypothetical Pastry node. X denotes a wildcard entry. (b) An example of routing in Pastry.

Routing performance marginally degrades when some entries in the routing table are blank. Routing performance is also adversely affected by node failures. However, eventual message delivery is guaranteed unless $|L|/2$ nodes with consecutive ids simultaneously fail. Since the ids are randomly distributed in the key space, the possibility of this event, even for a small value of L, is very low.

Finally, Pastry routing pays attention to locality by keeping track of the physical proximity between nodes in a neighborhood set M. Physical proximity is estimated by the round-trip delay of signals. A built-in function keeps track of this proximity. When a new node joins, it updates the neighborhood sets of the nodes in its routing table. Using the proximity relationship, each step routes the message to the *nearest node* with a longer prefix match. Experiments have shown that the routing delay between a pair of nodes is only 60% higher than the routing delay in a completely connected network.

Pastry routing is inspired by the work due to Plaxton et al. [PRR99]—commonly referred to *Plaxton routing*. In both cases, the routing is based on address prefixes, which can be viewed as a generalization of hypercube routing.

21.9 KOORDE AND DE BRUIJN GRAPH

The quest for a graph topology with constant node degree and logarithmic diameter has two answers in the P2P community: the *butterfly network* (used in Viceroy) and *De Bruijn graph* (used in Koorde, the Dutch name for Chord). To explore the lower bound of the diameter, we first prove the following theorem:

Theorem 21.1

In a graph G with n nodes and a constant degree $k > 1$ per node, the diameter

$$D > \lceil \log_k n \rceil - 1$$

Proof. When $k > 1$, $1 + k + k^2 + k^3 + \cdots + k^D \geq n$.

Thus, $\dfrac{k^{D+1} - 1}{k - 1} \geq n$.

So, $D + 1 \geq \log_k(n(k - 1) + 1)$.

That is, $D + 1 > \lceil \log_k n \rceil$.

Therefore, $D > \lceil \log_k n \rceil - 1$ ■

The routing distance between the farthest pairs of nodes must at least be equal to the diameter. Directed De Bruijn graphs come closest to this bound, sometimes known as *Moore bound*. For $n = 10^6$ and $k = 20$, the diameter of De Bruijn graph is 5, when the diameter of classic butterfly network is 8, and the diameter of unidirectional Chord is 20. This led Kaashoek and Karger to propose Koorde. A De Bruijn graph with $k = 2$ can be generated as follows (Figure 21.10a): From every node i of a graph with n nodes 0 through $n - 1$, draw two outgoing edges directed to the nodes $i_0 = 2i \bmod n$

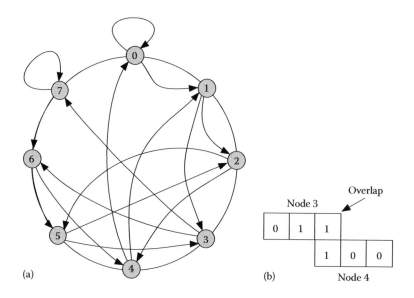

FIGURE 21.10   (a) A De Bruijn graph with $n = 8$ and $k = 2$. (b) A route from 011 to 100.

and $i_1 = (2i + 1)$ mod n. Call these the *0-link* and the *1-link of* node *i*. A message from node *i* to node *j* can be routed as follows:

- Shift the log$n$ bits of *j* so that its *leading r* bits tally with the *last r* bits of *i* as shown in Figure 21.10b.

- Forward the query along paths corresponding to the last (log$n - r$) bits of *j*: each 0 bit will define a hop along the *0-link* and each 1 bit will need a hop along the *1-link*.

For $k > 2$, the earlier construction can easily be generalized. From each node *i*, there will be k routing fingers pointing to the nodes $k \cdot i, k \cdot i + 1, k \cdot i + 2, \ldots, k \cdot i + k - 1$ (additions mod *n*).
   In spite of the promise, the use of De Bruijn graphs in P2P networking is very limited so far. This is due to the problem of dealing with the dynamic environment involving node join and leave operations. Each such operations will lead to changes in the routing tables of every node across the network, which is not very practical.

21.10 SKIP GRAPH

Bill Pugh introduced a randomized data structure called *skip list*—its goal is to accelerate the searching of objects in a sorted linked list by creating random bypass links. Figure 21.11a shows an example to illustrate the main idea. There are eight nodes in the linked list—all of them are all at level 0 (*L0*). Compare them to the stations in a subway with a single line from left to right. From these nodes, randomly pick a subset (with 50% probability) and add them to a next level. The linked list at this level is the level 1 list (*L1*). The links in level 1 are essentially *bypass links*, which are like express lines connecting selected pairs of stations. Each list is fenced off by two special nodes $+\infty$ and $-\infty$ at the right and the left ends, respectively. The construction is recursive—one can add a subset of

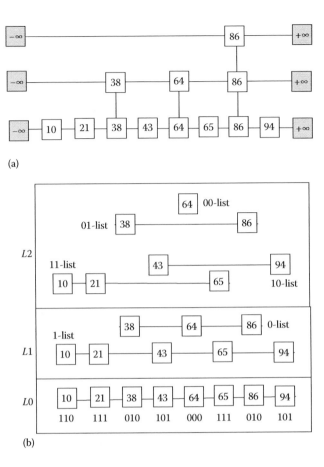

(a)

(b)

FIGURE 21.11   (a) A skip list. (b) A skip graph—only three levels are shown. Under each node in level 0, its membership vector is shown.

nodes from level 1 with 50% probability to another linked list in level 2 (*L2*). The construction ends when there is a single node at the uppermost level.

Consider the example of searching objects in a skip list. Assuming each link to cost one unit of time, a skip list can speed up search operations. At each level of the linked list, let *v.right* and *v.left* denote the elements to the right and to the left of a node *v*. To search for a node *x*, start with the leftmost node (*v* = −∞) at the highest level and follow these steps:

```
program search {search for node x in a skip list)
{initially v = ∝, level = max};
do   v.right > x → move to the lower level
[]   v.right < x → move to the right
[]   v.right = x → node x is found
od
```

The search fails when the node is not found and no lower level exists.

A query for a node 65 will succeed as $(L1)38 \rightarrow 64 \rightarrow L0(65)$, whereas a query for node 95 will fail as $(L2)86 \rightarrow (L1{:}L0)86 \rightarrow 94 \rightarrow ?$ In a standard linked list, the first search would take six steps, and the second search would take nine steps.

One can use a bidirectional skip list (which uses a doubly linked list at every level) for the efficient routing of messages between pairs of nodes. Such a route will try to make the best use of the upper-level linked lists. For example, in Figure 21.11a, a query from 21 to 94 will be routed in four hops as $(L0)21 \to 38 \to (L1)64 \to 86 \to (L0)94$.

If each list is doubly linked, then the expected number of edges in a skip list with n nodes is $2n + n + \dfrac{n}{2} + \dfrac{n}{4} + \cdots = 4n$. This means that the average degree per node is only 4, a constant. Furthermore, the expected search latency is reduced from $O(n)$ to $O(\log n)$, and the expected routing latency is also reduced from $O(n)$ to $O(\log n)$. These make skip list an attractive data structure for large-scale applications.

Now consider using a skip list for query routing in P2P networks. Although routing takes only $O(\log n)$ steps, a major problem occurs when multiple nodes start sending queries, causing too much congestion at the top-level node. To balance the load, Aspnes and Shah transformed the bidirectional skip list into a *skip graph* [AS03]. Figure 21.11b shows a skip graph. Unlike skip lists, in a skip graph, every node has a presence in all the levels, but they may join different doubly linked lists. From level 0, nodes that flip a coin get a 0 and join the *0-list* in level 1, and the remaining nodes join the *1-list*. The next higher level further refines these lists and generates four lists for 00, 01, 10, 11 (some of which may possibly be empty). The random bit string (resulting from successive coin flips) for a node defines its *membership vector*. The construction stops when at the topmost level there are lists of singleton nodes only. Essentially, a skip graph is a superposition of several skip lists that share a common linked list at the base level 0.

The routing of a message from a source node to a destination node begins at the topmost level using a greedy approach. When the next hop is likely to meet a dead end, or overshoot the destination, the message is routed through the links at the next lower level in a recursive manner. The lowest level linked list contains all the nodes—so message delivery to any destination node is guaranteed if that node exists. However, by utilizing the upper-level links as much as possible, routing latency is reduced. A summary of the performance of a skip graph is presented without proof:

1. The expected number of routing hops between any pair of nodes in $O(\log n)$.

2. The expected number of links per node $O(\log n)$.

3. A node can join and leave the skip list in an expected number of $O(\log n)$ steps using $O(\log n)$ messages.

4. The probability that a query from a source node i to a destination node j passes through a node k at a distance d from j is at most $2/(d + 1)$. This inverse relationship demonstrates good load balancing property—the presence of a hot spot affects its immediate neighborhood only.

5. Skip graphs preserve the locality of objects and thus are excellent in resolving range-based queries. One can, for example, easily locate all the publications of Daily Iowan during the period May 15–31, 2013. This is difficult in structured P2P networks like Chord or Pastry where hashing destroys the locality.

6. Skip graphs have excellent fault-tolerance properties. The expected search involves only $O(\log n)$ nodes, and most searches succeed as long as the fraction of *randomly failing* nodes is substantially less than $1/\log n$. A carefully targeted *adversarial failure* of f nodes can disconnect up to $O(f \cdot \log n)$ nodes from a skip graph. Clearly, for random failures, the resilience is much better—experiments have shown that most of the large skip graphs remain connected even if up to 70% of the nodes undergo random failure, and messages eventually reach their destination.

These properties show the promise of skip graphs for P2P applications. The details of skip graphs can be found in Shah's PhD thesis.

21.11 REPLICATION MANAGEMENT

Data replication reduces the access time and thus improves query processing in P2P networks. Ideally, if everyone maintains a local copy of every data object, then sharing will not be necessary—data will be instantly available and costly bandwidth will be saved. But this drastically increases the space requirement of active processes. Efficient replication strikes a balance between space and time complexities. Also, replication provides fault tolerance—when some machines crash or are shut down, other machines storing the replicas of the objects stored in the crashed machines maintain availability. For writeable objects, replication leads to data consistency issues—we address here read-only objects.

Replication can be *proactive* or *reactive*. The FastTrack protocol (used in KaZaA) supports proactive or explicit replication (creates additional copies regardless of the demand at a predetermined rate), and much of its success can be attributed to this policy. In contrast, replication in Gnutella is implicit and reactive—all replicas are generated from the results of previous queries. In [SC02], Shenker and Cohen investigated the issue of explicit replication and its impact on the performance of query processing in unstructured P2P networks. They formulated the problem as follows: Given an object, its *search size* is the number of machines that will be visited to locate that object. The *expected search size* (ESS) is the expected value of the search size for a set of objects and a set of queries. This depends on a number of factors, like the number of replicas of the objects available in the system, the location of the peers searching the object, number of queries made for each object, and the search strategies (like flooding or random walk). The important issue here is: given a set of objects, a search strategy, and a fixed amount total space (we assume this to be 1 unit) available for storing all the replicas across the entire system, how many replicas of each object should be generated so that the ESS is the smallest? Note that excessive replication of one object will reduce the space available to the replicas of the other objects.

Consider m objects 1, 2, 3, …, m, all of the same size, in an unstructured Gnutella-like P2P system, where objects are searched using random walk. For a single object, the ESS is inversely proportional to the scale of replication. With m distinct objects, let the normalized *query rate* of object i be q_i and the fraction of total space allocated to object i be p_i. By definition, $\sum_{i=1}^{m} p_i = 1$ and $\sum_{i=1}^{m} q_i = 1$. Without loss of generality, assume that $q_1 \geq q_2 \geq q_3 \geq \cdots \geq q_m$. To minimize the ESS, two replication policies are quite intuitive:

- *Uniform replication*: Each object is allocated the same amount of space, regardless of the query rate. Thus, $\forall i : p_i = 1/m$.

- *Proportional replication*: The space allocated to the replicas of an object is proportional to the query rate, that is, $p_i = q_i$. This makes sense because this is likely to lead to faster access to the objects that are queried more often.

For insoluble queries, the search size is either equal to the size of the system or equal to the maximum that is permitted by the TTL parameter of the system.

Cohen and Shenker [CS02] proved the surprising result that for soluble queries, the ESS is identical for both of these replication policies. The optimal strategy is somewhere in-between uniform and square root replication. It is the new policy of *square root replication*, where p_i is proportional to $\sqrt{q_i}$.

In replication management, not only the number of copies but also their placement is important. Apart from *owner replication* where the replication sites are determined by the owner of the object, two other approaches are in use: *path replication* and *random replication*. Path replication creates replicas on all nodes in the path from the provider to the requesting node. Interestingly, path replication spontaneously implements the square root replication policy. Random replication, on the other hand, places replicas on a number of randomly selected nodes in the search paths.

21.12 BITTORRENT AND FREE RIDING

BitTorrent is an efficient protocol for content distribution using the idea of *file swarming*. The protocol is particularly attractive for distributing large files like video or movie files. BitTorrent does not perform all the functions of a typical P2P system. The main idea in BitTorrent is as follows: The file to be distributed is split into a large number of pieces and a SHA-1 hash is appended for each piece. To allow sharing of the file or group of files, the initiator first creates a *.torrent* file, a small file that contains the *metadata* about the files to be shared, which includes (1) the length of the file, (2) piece size, (3) a mapping of the pieces to files, (4) the SHA-1 hashes of all pieces, and (5) information about the *tracker*, a computer that coordinates the file distribution. The tracker maintains the set of all peers participating in the sharing of that file—this set is called a *swarm*. Each downloader first accesses the *.torrent* file and then connects to the specified tracker that provides information about other peers downloading the pieces of the same file. BitTorrent peers play two different roles: *seeders* and *leechers*. A seeder is a peer that provides a complete copy of the file. Initially, there is only one seeder, which is the peer that wants to distribute the file. A leecher is a peer that downloads pieces of the file from other peers. As a leecher downloads pieces of the file, replicas of the pieces are created. More downloads mean more replica pieces are available in the swarm, and other peers that have not yet acquired these

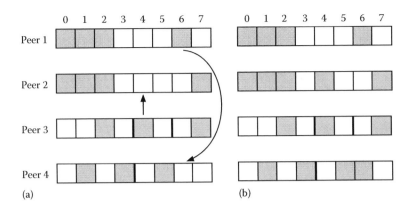

FIGURE 21.12   The states of the peers in BitTorrent: the shaded boxes denote the pieces that have been downloaded. (a) Each of the four peers has acquired a few of the eight pieces of the file. (b) The states of the peers after peer 2 downloads piece 4 from peers 3 and 4 downloads piece 6 from peer 1.

pieces can download them. As soon as a leecher acquires a complete piece, it becomes a seeder. Eventually, each leecher becomes a seeder by obtaining all the pieces and assembles the file. The checksum associated with a piece is used to check the integrity of the piece.

The order in which the pieces are downloaded is critical for good performance. If an inefficient policy is used, then peers may end up in a situation where each peer has all identical set of easily available pieces and none of the missing ones. If the original seed is prematurely taken down, then the file cannot be completely downloaded. Figure 21.12 shows the state of the peers in a swarm.

Two of the *good* policies for piece selection are (1) *rarest first*, so that a rare piece does not become a bottleneck for completing the download, and (2) *random first piece*, at the beginning of the downloading process—the eventual goal is to reduce the overlap among the sets of downloaded pieces by the various peers, which expedites the download by opening up more opportunities in piece selection and thus better progress. Near the end, missing pieces are requested from every peer containing them. This ensures that a download is not prevented from completion due to a single peer with a slow transfer rate. Some bandwidth is wasted, but in practice, this is insignificant compared to its benefits.

According to an older study [AH00] on the Gnutella system, 70% of peers only download files without uploading any, and 50% of the queries are served by only 1% of the hosts. Users who only download but never upload any files are known as *free riders*, or *freeloaders*. Freeloading is an inappropriate behavior that affects the healthy functioning of P2P systems, and many solutions have been proposed and debated. An apparent solution is to charge a fee and limit the time of downloads. Other approaches include rewarding the donors by allowing them more downloads or faster downloads. BitTorrent's anti-freeloading mechanism is based on the principle of reciprocation— downloaders barter for fragments of a file by uploading some of the pieces that they have already downloaded. This *tit-for-tat* mechanism discourages freeloaders. To cooperate, peers must upload; otherwise, they will be *choked*. The system lets users discover a set of more desirable peers who comply with this principle. It approximates *Pareto efficiency* in game theory, in which pairs of counterparts see if they can improve their lots together.

Using the BitTorrent protocol, Facebook is now able to send several hundred MB of code updates to thousands of servers under a minute. The internal Facebook swarm treats every server as a peer—this substantially accelerates the distribution of the new code. Twitter also uses the same idea.

21.13 CENSORSHIP RESISTANCE, ANONYMITY

Most nations have a body that decides what information to censor and what information to allow. What may be acceptable to one group of people may be offensive to another group. Fiat and Saia [FS02] defined a censorship resistance network as one, where even if an adversary (read *censorship authority*) deletes up to half of the nodes, $(1 - \varepsilon)$ fraction of the remaining nodes should still be able to access $(1 - \varepsilon)$ fraction of all the data items, where ε is a fixed error parameter. His solution for designing such networks involves modifying the interconnection network of DHT-based systems using redundant links.

Freenet (launched in 2001) was designed to remove the possibility of any group imposing their beliefs or values on any other group. The goal was to encourage tolerance to each other's values and freedom of speech (but not copyright infringements). Freenet preserves *anonymity* using a complex protocol. Objects stored in the system are encrypted and replicated across a large number of anonymous machines around the world, and their identities continuously change. Each file is encrypted and broken up into several pieces. Not only potential intruders but also the peers themselves have no clue about which peers are storing one of their files or a fragment of it.

21.14 CONCLUDING REMARKS

The rapid proliferation of modern social networks has renewed interest in understanding the mechanisms of growth and decay of these structures. With the membership of some of the popular social networks reaching close to a billion, efficient analytical tools for these networks are in high demand.

In P2P networks, structured or unstructured, which is better is a common debate. Both have their strong and weak points. Unstructured networks need little management overhead, although lookups are, in general, slower. On the other hand, in structured networks, lookups are faster. However, objects need to be published following a stringent mapping rule, and due to the high churn rate, the pointers as well as objects need to be constantly moved to maintain consistency.

Since P2P networks are large-scale networks involving untrusted machines, various security measures are important for serious applications. For example, a single malicious node fielding multiple ids can disrupt the normal functioning of a P2P network. This is called a Sybil attack. Another issue is resistance to spams. This will guarantee that users are not fooled into retrieving fake copies of the object despite the malicious behavior of a handful of machines.

Uniform distribution of keys is important for load balancing in DHT-based networks. The *consistent hashing* algorithm of Chord and Pastry takes care of this issue. Consistent hashing guarantees that if there are k keys in a network with an id space of size n, then the addition or deletion of a single node can lead to the reshuffling of at most k/n nodes with high probability.

Chord does not take into consideration the geographic distribution of neighbors—thus, a neighbor can be a machine in the next building or a machine halfway around the globe. Since the routing time depends on the geographic distance, lookup will be faster if the geographic location is taken into consideration while the neighborhood is defined. In contrast, Pastry routes messages via geographically close neighbors. This expedites lookup. Finally, the ability of handling high *churn rate* (the rate at which new nodes join and existing nodes drop out) and the ability to deal with *flashcrowd effect* (sudden popularity of a file) may play major roles in determining the usability of a given P2P system.

P2P networks have successful presence in the world of streaming data. Apart from the distribution of videos through social networking sites, Skype is a proprietary P2P architecture that powers our audio and video chats.

P2P networks are useful for building large amounts of storage. UC Berkeley's Oceanstore [KBC+00] and Facebook's Cassandra file system [LM10] are two prominent examples. However, in storage applications, cloud computing is a serious contender with its steeply rising popularity. One needs to carefully weigh the pros and cons of these two technologies.

21.15 BIBLIOGRAPHIC NOTES

Erdös and Rényi introduced the ER random graph model for a class of social networks. Their eight papers that established the theory of random graphs are listed by Karonski and Rucinski in [KR97]. Barabási and Albert [BA99] proposed the rich gets richer model for the generation of power-law graphs. Milgram [M67] first observed the *small-world phenomenon* in 1967. Later, Watts and Strogatz [WS98] presented a model of social networks as a possible explanation for the small-world phenomenon. Jon Kleinberg's refinement of the Watts–Strogatz model can be found in [K00].

Eighteen-year-old Shawn Fanning was the creator of Napster in 1999. The music industry sued the company, claiming losses of millions in royalties. Napster lost the case in 2000 and ordered to be shut down. In 2002, Napster filed for bankruptcy. How the old Napster worked is documented in [N02]. In 2011, Rhapsody acquired Napster.

Justin Frankel of Nullsoft, an AOL-owned company, developed Gnutella in 2001. Currently, Gnutella Developer's Forum is the sole group responsible for all Gnutella-related protocols. The official documentation of Gnutella is available in [G02].

Soon after the shutdown of Napster, Niklas Zennstrom and Janus Friis launched KaZaA [K01]. Now KaZaA survives through some of its variations. DHT-based P2P network Chord follows the work by Stoica et al. [SML+03]. Ratnasamy et al. wrote the paper on CAN [RFH+01]. Pastry was proposed by Rowstron and Druschel [RD01]. John Kubiatowicz led the Oceanstore project in Berkeley [KBC+00]—a brief summary of the main goals of the project appears in their CACM article [K03].

Pugh invented skip lists [P90]. Gauri Shah's PhD thesis contains the details of skip graphs. A brief summary of it can be found in [AS03]. Uniform, proportional, and square root replications have been discussed in Cohen and Shenker's work [CS02]. Karl Aberer's P-grid project [ACD+03] describes a structured P2P network that provides good load balancing in spite of arbitrary key distributions in the key space. Fiat and Saia [FS02] studied how the DHT-based P2P networks can be modified for censorship resistance. Freenet uses a different approach to censorship resistance and is based on the paper by Ian Clarke and his associates [CSW+00]. Ian Clarke was selected as one of the top 100 innovators of 2003 by MIT's Technology Review magazine. Bram Cohen designed the BitTorrent P2P system [C03]. Cassandra file system is described in [LM10].

EXERCISES

21.1 Consider the degree distribution (k = degree and $N(k)$ = number of nodes with degree k) of four networks. The scales are linear and approximate. Which one of these four graphs *possibly* represents a power-law distribution (Figure 21.13)?

21.2 Consider three different network topologies with n nodes in each: (1) a *ring topology* where each node k has a predecessor node $(k - 1)$ mod n and a successor node $(k + 1)$ mod n, (2) a *clique* (a completely connected topology), and (3) an *ER* graph (where with uniform probability $p \ll 1$, an edge is added between each pair of nodes). Are any of these acceptable as the topology of a P2P network? Why or why not? Briefly justify your answer for each case.

21.3 What are the values of the clustering coefficient of nodes C and D in the following two networks of Figure 21.14?

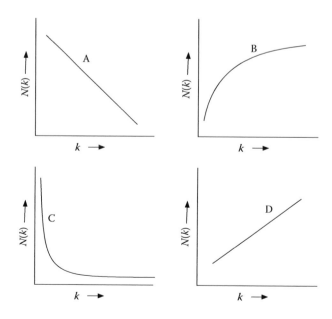

FIGURE 21.13   Four possible degree distributions of nodes.

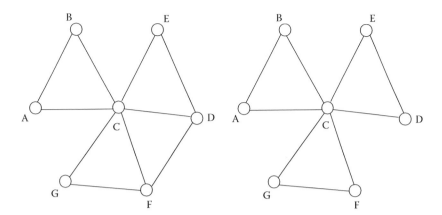

FIGURE 21.14   Two networks.

21.4   Consider search and routing in a small-scale Pastry P2P network with 256 nodes. It uses the base-4 notation for node naming and routing table construction. The following is an incomplete primary neighbor table of node 3012. Fill in the missing entries in the routing table.

L3	L2	L1	L0
	X012		XXX0
		XX12	XXX1
2012		XX22	XXX2
	X312		XXX3

(X denoted a wildcard entry)

21.5   DHT-based P2P networks aim at reducing the size of its local routing tables, as well as the maximum number of hops needed to retrieve an object. Let us focus on the architectures that use a constant size routing table per node and $O(\log n)$ hops to retrieve any object. One such system is Koorde that uses De Bruijn graph as the interconnection network. Another possibility is to use a *butterfly network*. Explore the design of a P2P system that will use the butterfly network. You have to address routing, as well as node join-and-leave protocols.

(Hint: See Viceroy [MNR02].)

21.6   Consider a unidirectional Chord network with $n = 2^{16}$ nodes numbered 0 through $n - 1$. Each hop routes every query in the forward direction, until the object is found.

a.   How many hops will it take to route a query from node 0011 0011 1111 1100 to 0100 0000 0000 0000?

b.   Now modify Chord routing to accommodate bidirectional query forwarding. Show that this reduces the number of hops, and calculate the number of hops for the example in part (a).

21.7 Consider a 3D CAN with N nodes in it. What is the smallest size of the network, beyond which the number of routing hops in CAN will be larger than that in Chord?

21.8 Caching is a well-studied mechanism for the improvement of lookup time in P2P networks. Consider *route caching* that profiles the most recent accesses and stores them in a local buffer. To the routing table of each node, add a buffer of size $\log n$ to cache the IP address of the most frequently accessed objects by that node (call them *preferred objects*), so that they are reachable in a single hop. We will call it a *selfish cache*.

The selfish cache will shorten the routing distance in many cases. When a query cannot reach its destination in a single hop, the original routing table entries will be used to forward it. Assuming that the preferred objects are randomly distributed, to what extent will the average routing distance be reduced compared to the original Chord? Assume that the hot spots are randomly distributed around the key space.

21.9 Experiments with access patterns in unstructured P2P networks have revealed *clustering effect*, where each node's communication is mostly limited to a small subset of its peers. The identity of the nodes in each cluster depends on common interest, race, nationality, or geographical proximity.

Assuming that such clusters can be easily identified through a continuous profiling mechanism, explore how the clustering effect can be used to reduce the average lookup time. (*Hint*: Use a two-level approach and speed up the common case.)

21.10 Concurrent join/leave operations or occasional transient failures can leave a Chord network in a bad configuration. For the self-stabilization of such networks, local checkability is an important requirement. Local checkability implies that for every bad configuration, at least one node must be able to detect it by checking its immediate neighborhood. Is Chord locally checkable?

21.11 Why is the geographic proximity of nodes an issue in routing queries on P2P networks?

21.12 Why can a large number of BitTorrent clients download videos faster than what is feasible in a client–server system or in the Gnutella network?

21.13 There are some strong similarities between search for domain names in DNS and object search in P2P network. How is DNS currently implemented? Are there roadblocks in implementing DNS using the P2P technology?

21.14 a. What kind of failures can partition the Chord P2P network? Are there remedies to prevent this kind of partitioning?

b. What kind of failures can make objects inaccessible even if the network is not partitioned? Are there remedies to prevent this problem?

21.15 When data are stored into unknown or untrusted machines, security of the data becomes a key concern for serious applications. Describe all the methods that you consider appropriate for safeguarding data in a P2P network.

21.16 Compare the advantages and disadvantages of storing persistent data in the cloud versus storing data in a P2P network.

21.17 (*Programming exercise*) Download NetworkX, a Python package for the creation, manipulation, and study of the structure, dynamics, and functions of complex networks. This package has a library of various kinds of graphs as well as the implementation of various useful algorithms for social network analysis.

Part 1: The goal of this homework is to generate a social network. Recall how a power-law graph is generated using the idea of *preferential attachment*. Starting with a single node, in each step, add a node to the network using *a single* edge. The probability of choosing the node to which it will attach will be proportional to the degree of that node.

a. Using this rule, generate three graphs, n = 1000, 2000, 5000 nodes.

b. Show the degree distribution of the generated graph: plot $\log N(k)$ versus $\log k$, where k is the degree of a node and $N(k)$ is the number of nodes having degree k. Verify that the power law ($N(k) = c \cdot k^{-r}$) holds, and estimate the value of r.

Part 2: Consider the *Karate Club* network described in http://haystack.csail.mit.edu/blog/2010/07/12/data-by-the-people-for-the-people/.
Your tasks are as follows:

a. Compute the edge(s) of highest betweenness in this network.

b. Remove the edge(s) of highest betweenness, and visualize the graph and see if it has been partitioned into disjoint communities. If not, then repeat the earlier two steps, until the graph partitions, and then stop.

c. List the edges that you removed.

21.18 (*Programming exercise*) Study the construction of small-world graphs by Watts and Strogatz. Then simulate a ring lattice with N = 5000 and k = 20 (the number of short-range neighbors). Compute the routing hops between fifty different pairs of peers chosen at random. Explain your observations.

The code can be written in Java/C++. The Java standard library or C++ STL will simplify life because you will have to use a lot of varying data structures to implement the system. However, you are free to use a programming language of your choice.

21.19 (*Programming project*) Use a web crawler to explore the topology of the web of your institution and check if it satisfies the power-law distribution.

References

[A00] Andrews, G., *Concurrent Programming: Principles and Practice*, 2nd edn. Benjamin Cummings, Redwood City, CA, 2000.

[AAKK+00] Araragi, T., Attie, P., Keidar, I., Kogure, K., Luchanugo, V., Lynch, N., and Mano, K., On formal modeling agent computations. In *NASA Workshop on Formal Approaches to Agent-Based Systems*, Greenbelt, MD, 2000.

[AB01] Albitz, P. and Liu, C., *DNS and BIND*, 4th edn. O'Reilly & Associates, Sebastopol, CA, 2001.

[ABP+13] AlFardan, N., Bernstein, D., Paterson, K., Poettering, B., and Schuldt, J., On the security of RC4 in TLS and WPA, March 13, 2013. http://www.isg.rhul.ac.uk/tls/ (accessed on April 30, 2014).

[ACD+03] Aberer, K., Cudré-Mauroux, P., Datta, A., Despotovic, Z., Hauswirth, M., Punceva, M., and Schmidt, R., P-Grid: A self-organizing structured P2P system. *SIGMOD Record*, 32 (3), 29–33, 2003.

[ADG91] Arora, A., Dolev, S., and Gouda, M. G., Maintaining digital clocks in step. *Parallel Processing Letters* 1, 11–18, 1991.

[AG93] Arora, A. and Gouda, M. G., Closure and convergence: A formulation of fault-tolerant computing. *IEEE Transactions on Software Engineering* 19 (11), 1015–1027, 1993.

[AG94] Arora, A. and Gouda, M. G., Distributed reset. *IEEE Transactions on Computers* 43 (9), 1026–1038, 1994.

[AH00] Adar E., Huberman, B.A., Free Riding on Gnutella. *First Monday* 5, 10, 2000.

[AK98] Arora, A., and Kulkarni, S. S., Component based design of multitolerance. *IEEE Transactions on Software Engineering* SE-24, 63–78, 1998.

[AM71] Ashcroft, E. A. and Manna, Z., Formalization of the properties of parallel programs. *Machine Intelligence* 6, 17–41, 1971.

[AM98] Alvisi, L. and Marzullo, K., Trade-offs in implementing optimal message logging protocols. In *ACM PODC*, Philadelphia, PA, 1996.

[AMM+04] Andreev, K., Maggs, B., Meyerson, A., and Sitaraman, R., Designing overlay multicast networks for streaming. In *Proceedings of the 15th Annual ACM Symposium on Parallel Algorithms and Architectures (SPAA)*, San Diego, CA, June 2003.

[AOSW+99] Arnold, K., O'Sullivan, B., Scheifler, R. W., Waldo, J., and Wollrath, A., *The Jini Specification*. Addison-Wesley, Reading, MA, 1999.

[APWG] http://www.antiphishing.org/ (accessed on April 30, 2014).

[AS03] Aspenes, J. and Shah, G., Skip graphs. In *14th Symposium on Distributed Algorithms (SODA)*, Baltimore, MD, pp. 384–393, 2003.

[AS85] Alpern, B. and Schneider, F., Defining liveness. *Information Processing Letters* 21 (4), 181–185, 1985.

[ASSC02] Akyildiz, I. F., Su, W., Sankarasubramanium, Y., and Cayrici, E., Wireless sensor networks: A survey. *Computer Networks* 38 (4), 393–422, 2002.

[Aw85] Awerbuch, B., Complexity of network synchronization. *Journal of the ACM* 32 (4), 804–823, 1985.

[B06] Burrows, M., The Chubby lock service for loosely-coupled distributed systems. In *Proceedings of the Seventh Symposium on Operating Systems Design and Implementation*, Berkeley, CA, pp. 335–350, 2006.

[B58] Bellman, R., On a routing problem. *Quarterly of Applied Mathematics* 16, 87–90, 1958.

[B77] Bryant, R. E., Simulation of packet communications architecture for computer systems. MIT-LCS-TR-188, Massachusetts Institute of Technology, Cambridge, MA, 1977.

[B82] Ben-Ari, M., *Principles of Concurrent Programming*. Prentice Hall, Englewood Cliffs, NJ, 1982.

[B83] Ben-Or, M., Another advantage of free choice: Completely asynchronous agreement protocols. In *ACM PODC*, Montreal, Quebec, Canada, pp. 27–30, 1983.

[B93] Birman, K. P., The process group approach to reliable distributed computing. *Communications of the ACM* 36 (12), 36–53, 1993.

[BA99] Barabási, A.-L. and Albert, R., Emergence of scaling in random networks. *Science* 286 (5439), 509–512, 1999.

[BGH87] Bernstein, P., Goodman, N., and Hadzilacos, V., *Concurrency Control and Recovery in Database Systems*. Addison-Wesley, Reading, MA, 1987.

[BGK+99] Bruell, S. C., Ghosh, S., Karaata, M. H., and Pemmaraju, S. V., Self-stabilizing algorithms for finding centers and medians of trees. *SIAM Journal on Computing* 29 (2), 600–614, 1999.

[BH73] Brinch Hansen, P., *Operating System Principles*. Prentice Hall, Englewood Cliffs, NJ, 1973.

[BJ87] Birman, K. P. and Joseph, T. A., Exploiting virtual synchrony in distributed systems. In *SOSP*, Austin, TX, pp. 123–138, 1987.

[BLT91] Bakker, E. M., van Leeuwen, J. and Tan, R. B., Linear interval routing schemes. Tech. Report RUU-CDS-7, Department of Computer Science, University of Utrecht, 1991.

[BMS+92] Budhiraja, N., Marzullo, K., Schneider, F., and Toueg, S., In *Distributed Systems* (Mullender, S., ed.), Chapter 8, pp. 199–216. Addison-Wesley, New York, 1993.

[BN84] Birrell, A. D. and Nelson, B. J., Implementing remote procedure calls. *ACM Transactions on Computer Systems* 2 (1), 39–59, 1984.

[BSP91] Birman, K. P., Schiper, A. and Stephenson, P., Lightweight causal and atomic group multicast. *ACM Transactions on Computer Systems* 9 (3), 272–314, 1991.

[BT93] Babaõglu, O. and Toueg, S., Non-blocking atomic commitment. In *Distributed Systems* (Mullender, S., ed.), pp. 147–168. Addison-Wesley, New York, 1993.

[C03] Cohen, B., Incentives build robustness in bittorrent. *First Workshop on Economics of Peer-to-Peer Systems*, June 2003.

[C82] Chang, E. J. H., Echo algorithms: Depth parallel operations on general graphs. *IEEE Transactions on Software Engineering* SE-8 (4), 391–401, 1982.

[C89] Cristian, F., Probabilistic clock synchronization. *Distributed Computing* 3 (3), 146–158, 1989.

[CCGZ90] Chou, C. T., Cidon, I., Gopal, I. S., and Zaks, S., Synchronizing asynchronous bounded delay networks. *IEEE Transactions on Communications* 38 (2), 144–147, 1990.

[CDD+85] Coan, B. A., Dolev, D., Dwork, C., and Stockmeyer, L. J., The distributed firing squad problem. In *ACM STOC*, Providence, RI, pp. 335–345, 1985.

[CDK11] Coulouris, G., Dollimore, J., and Kindberg, T., *Distributed Systems: Concepts and Design*, 5th edn. Addison-Wesley, Reading, MA, 2011.

[CEL+08] Campbell, A. T., Eisenman, S. B., Lane, N. D., Miluzzo, E., Peterson, R. A., Lu, H., Zheng, X., Musolesi, M., Fodor, K., and Ahn, G-S. The rise of people-centric sensing, *IEEE Internet Computing: Mesh Networking*, 12 (4), 12–21, 2008.

[CG89] Carriero, N. and Gelernter, D., Linda in context. *Communications of the ACM* 32 (4), 444–458, 1989.

[Ch82] Chang, E. J.-H., Echo algorithms: Depth-parallel operation on general graphs. *IEEE Transactions on Software Engineering* SE-8, 391–401, 1982.

[CHT96] Chandra, T. D., Hadzilacos, V., and Toueg, S., The weakest failure detector for solving consensus. *Journal of the ACM* 43 (4), 685–722, 1996.

[CK03] Chong, C.-Y. and Kumar, S. P., Sensor networks: Evolution, opportunities and challenges. *Proceedings of the IEEE* 91 (9), 1247–1256, 2003.

[CKF+04] Candea, G., Kawamoto, S., Fujiki, Y., Friedman, G., and Fox, A., Microreboot—A technique for cheap recovery. In *Sixth Symposium on Operating Systems Design and Implementation (OSDI)*, San Francisco, CA, 2004.

[CL85] Chandy, K. M. and Lamport, L., Distributed snapshots: Determining global states of distributed systems. *ACM Transactions on Computer Systems* 3 (1), 63–75, 1985.

[CLR+01] Cormen, T. H., Leiserson, C. E., Rivest, R. L., and Stein, C., *Introduction to Algorithms*, 2nd edn. MIT Press and McGraw Hill, 2001.

[CM81] Chandy, K. M. and Misra, J., Asynchronous distributed simulation via a sequence of parallel computations. *Communications of the ACM* 24 (11), 198–205, 1981.

[CM82] Chandy, K. M. and Misra, J., Distributed computation on graphs: Shortest path algorithms. *Communications of the ACM* 25 (11), 833–837, 1982.

[CM84] Chang, J.-M. and Maxemchuk, N. F., Reliable broadcast protocols. *ACM Transactions on Computer Systems* 23, 251–273, 1984.

[CM88] Chandy, K. M. and Misra, J., *Parallel Program Design*. Addison-Wesley, Reading, MA, 1988.

[CMH83] Chandy, K. M., Misra, J., and Haas, L. M., Distributed deadlock detection. *ACM Transactions on Computer Systems*, 11 (2),144–156, 1983.

[CNM04] Clauset, A., Newman, M. E., and Moore, C., Finding community structure in very large networks. *Physical Review* E70, 066111, 2004.

[CR79] Chang, E. G. and Roberts, R., An improved algorithm for decentralized extrema finding in circular configuration of processors. *Communications of the ACM* 22 (5), 281–283, 1979.

[CR83] Carvalho, O. S. F. and Roucairol, G., On mutual exclusion in computer networks. *Communications of the ACM*, 26 (2), 146–147, 1983.

[CRB+03] Chawathe, Y., Ratnasamy, S., Breslau, L., Lanham, N., and Shenker, S., Making Gnutella-like P2P systems scalable. In *ACM SIGCOMM*, Karlsruhe, Germany, pp. 407–418, 2003.

[CRZ00] Chu, Y.-H., Rao, S. G., and Zhang, H., A case for end system multicast. In *ACM SIGMETRICS*, Santa Clara, CA, 2000.

[CS02] Edith Cohen, E. and Shenker, S., Replication strategies in unstructured peer-to-peer networks. In *ACM SIGCOMM*, Pittsburgh, PA, pp. 177–190, 2002.

[CSW+00] Clarke, I., Sandberg, O., Wiley, B., and Hong, T. W., Freenet: A distributed anonymous information storage and retrieval system. In *Workshop on Design Issues in Anonymity and Unobservability*, Berkeley, CA, pp. 46–66, 2000.

[CT96] Chandra, T. D. and Toueg, S., Unreliable failure detectors for reliable distributed systems. *Journal of the ACM* 43 (2), 225–267, 1996.

[CV86] Cole, R. and Vishkin, U., Deterministic coin tossing with applications to optimal parallel list ranking. *Information and Computation* 70, 32–56, 1986.

[CV90] Chandrasekharan, S. and Venkatesan, S., A message-optimal algorithm for distributed termination detection. *Journal of Parallel and Distributed Computing*, 8 (3), 245–252, 1990.

[CYH91] Chen, N.-S., Yu, H.-P., and Huang, S.-T., A self-stabilizing algorithm for constructing spanning trees. *Information Processing Letters* 39, 147–151, 1991.

[CZ85] Cheriton, D. R. and Zwaenepoel, W., Distributed process groups in the V kernel. *ACM Transactions on Computer Systems* 3 (2), 77–107, 1985.

[D00] Dolev, S., *Self-Stabilization*. MIT Press, Cambridge, MA, 2000.

[D09] Duhigg, C., Stock traders find speed pays, in milliseconds. New York Times, July 23, 2009 (online edition). http://www.nytimes.com/2009/07/24/business/24trading.html?r=0 (accessed on April 13, 2014).

[D65] Dijkstra, E. W., Solution to a problem in concurrent programming control. *Communications of the ACM* 8 (9), 569, 1965.

[D68] Dijkstra, E. W., *Co-Operating Sequential Processes.* Academic Press, New York, 1968.

[D74] Dijkstra, E. W., Self-stabilization in spite of distributed control. *Communications of the ACM* 17 (11), 643–644, 1974.

[D75] Dijkstra, E. W., Guarded commands, nondeterminacy, and formal derivation of programs. *Communications of the ACM* 18 (8), 453–457, 1975.

[D76] Dijkstra, E. W., *A Discipline of Programming.* Prentice Hall, Englewood Cliffs, NJ, 1976.

[D82] Dolev, D., The Byzantine generals strike again. *Journal of Algorithms* 3 (1), 14–30, 1982.

[D84] Dijkstra, E. W., The distributed snapshot of K.M. Chandy and L. Lamport. Tech. Rept. EWD 864a, University of Texas at Austin, Austin, TX, 1984.

[D86] Dijkstra, E. W., A belated proof of self-stabilization. *Distributed Computing* 1 (1), 1–2, 1986.

[DA99] Dierks, T. and Allen, C., The TLS Protocol Version 1.0. RFC 2246, 1999.

[DAG03] Demirbas, M., Arora, A., and Gouda, M. G., A pursuer-evader game for sensor networks. In *Symposium on Self-Stabilizing System*, San Francisco, CA, pp. 1–16, 2003.

[DC90] Deering, S. and Cheriton, D., Multicast routing in datagram internetworks and extended LANs. *ACM Transactions on Computer Systems* 8 (2), 85–110, 1990.

[De96] Dega, J.-L., The redundancy mechanisms of the Ariane 5 operational control center. In *FTCS*, Sendai, Japan, pp. 382–386, 1996.

[DFG83] Dijkstra, E. W., Feijen, W. H. J., and Gasteren, A. J. M., Derivation of a termination detection algorithm for distributed computation. *Information Processing Letters* 16 (5), 217–219, 1983.

[DG04] Dean, J and Ghemawat, S., MapReduce: Simplified data processing on large clusters. *OSDI*, 137–150, 2004.

[DGH+87] Demers, A., Greene, D., Hauser, C., Irish, W., and Larson, J., Epidemic algorithms for replicated database maintenance. In *ACM PODC*, Vancouver, British Columbia, Canada, pp. 1–12, 1987.

[DH76] Diffie, W. and Hellman, M. E., New directions in cryptography. *IEEE Transactions on Information Theory* IT-22, 644–654, 1976.

[DH97] Dolev, S. and Herman, T., Superstabilizing protocols for dynamic distributed systems. *Chicago Journal of Theoretical Computer Science* 1997, 1997.

[DHJ+07] DeCandia, G., Hastorun, D., Jampani, M., Kakulapati, G., Lakshman, L., Pilchin, A., Sivasubramanian, S., Vosshall, P., and Vogels, W., Dynamo: Amazon's highly available key-value store. In *SOSP*, Stevenson, WA, pp. 205–220, 2007.

[DIM91] Dolev, S., Israeli, A., and Moran, S., Uniform dynamic self-stabilizing leader election (extended abstract). In *Workshop on Distributed Algorithms (WDAG)*, Delphi, Greece, pp. 167–180, 1991.

[DOSW96] Dongarra, J., Otto, S. W., Snir M., Walker, D. M., A message passing standard for MPP and workstations. *Communications of the ACM* 39 (7), 84–90, 1996.

[DR02] Daemen, J. and Rijmen, V., *The Design of Rijndael: AES—The Advanced Encryption Standard.* Springer-Verlag, Berlin, Heidelberg, New York, 2002.

[DS80] Dijkstra, E. W. and Scholten, C. S., Termination detection in diffusing computation. *Information Processing Letters* 11 (1), 1–4, 1980.

[DS82] Dolev, D. and Strong, H. R., Polynomial algorithms for multiple processor agreement. In *ACM STOC*, San Francisco, CA, pp. 401–407, 1982.

[E84] Elgamal, T., A public-key cryptosystem and a signature scheme based on discrete logarithms. *IEEE Transactions on Information Theory* IT-31 (4), 469–472, 1985 (also CRYPTO 1984:10–18, Springer-Verlag 1984).

[EGE02] Elson, J., Lewis, G., and Estrin, D., Fine-grained network time synchronization using reference broadcasts. In *OSDI*, Boston, MA, 2002.

[EGH+99] Estrin, D., Govindan, R., Heidemann, J.S., and Kumar, S., Next century challenges: Scalable coordination in sensor networks. In *MOBICOM*, Seattle, WA, pp. 263–270, 1999.

[EGT76] Eswaran, R. L. K. P., Gray, J. N., and Traiger, I., The notion of consistency and predicate lock in a data base system. *Communications of the ACM* 19 (11), 1976.

[EJZ92] Elnozahy, E. N., Johnson, D. B., and Zwaenepoel, W., The performance of consistent checkpointing. In *IEEE Symposium on Reliable Distributed Systems*, Pisa, Italy, pp. 39–47, 1991.

[ES86] Ezhilchelvan, P. D. and Srivastava, S., A characterization of faults in systems. In *Fifth Symposium on Reliability in Distributed Software and Database Systems*, Los Angeles, CA, pp. 215–222, 1986.

[F67] Floyd, R. W., Assigning meanings to programs. *Proceedings of the American Mathematical Society of Symposium on Applied Mathematics* 19, 19–31, 1967.

[F82] Franklin, W. R., On an improved algorithm for decentralized extrema finding in circular configuration of processors. *Communications of the ACM* 25 (5), 336–337, 1982.

[F86] Francez, N., *Fairness*. Springer-Verlag, Berlin, Germany, 1986.

[F88] Fidge, C. J., Timestamps in message-passing systems that preserve the partial ordering in K. Raymond (ed.). *Proceedings of the 11th Australian Computer Science Conference (ACSC'88)*, pp. 56–66, 1988.

[FJM+97] Floyd, S., Jackobson, V., McCane, S., Liu, C.-G., and Zhang, L., A reliable multicast for lightweight sessions and application level framing. *IEEE/ACM Transactions on Networking* 5 (9), 784–803, 1997.

[FLM86] Fischer, M. J., Lynch, N., and Merritt, M., Easy impossibility proofs for distributed consensus problems. *Distributed Computing* 1 (1), 26–39, 1986.

[FLP85] Fischer, M. J., Lynch, N., and Paterson, M. S., Impossibility of distributed consensus with one faulty process. *Journal of the ACM* 32 (2), 374–382, 1985.

[FS02] Amos Fiat, A. and Saia, J., Censorship resistant peer-to-peer content addressable networks. In *Symposium on Distributed Algorithms (SODA)*, San Francisco, CA, pp. 94–103, 2002.

[Fu90] Fujimoto, R., Parallel discrete event simulation. *Communications of the ACM* 33 (10), 30–41, 1990.

[G00] Ghosh, S., Agents, distributed algorithms, and stabilization. In *Computing and Combinatorics (COCOON 2000)*, Sydney, New South Wales, Australia, LNCS 1858, pp. 242–251, 2000.

[G02] RFC Gnutella 0.6. http://rfc-gnutella.sourceforge.net/src/rfc-O6-draft.html (accessed on April 13, 2014), 2002.

[G78] Gray, J., Notes on database operating systems. In *Operating Systems an Advanced Course* (Bayer, R., Graham, R., and Seegmuller, G., eds.), LNCS 60. Springer-Verlag, New York, 1978.

[G79] Gifford, D., Weighted voting for replicated data. In *ACM Symposium on Operating System Principles*, Pacific Grove, CA, pp. 150–162, 1979.

[G81] Gries, D., *The Science of Computer Programming*. Springer-Verlag, New York, 1981.

[G82] Garcia-Molina, H., Elections in a distributed computing system. *IEEE Transactions on Computers* C-31 (1), 48–59, 1982.

[G85] Gelernter, D., Generative communication in Linda. *ACM TOPLAS* 7 (1), 80–112, 1985.

[G85a] Gray, J., Why do computers stop and what can be done about it? Tech. Rept. 85.7 Tandem Computers, 1985.

[G91] Ghosh, S., Binary self-stabilization in distributed systems. *Information Processing Letters* 40, 153–159, 1991.

[G93] Ghosh, S., An alternative solution to a problem on self-stabilization. *ACM TOPLAS* 15 (7), 327–336, 1993.

[G94] Garfinkel, S., *PGP: Pretty Good Privacy*. O'Reilly, Sebastopol, CA, 1994.

[G96] Gray, R., A flexible and secure mobile agent system. In *Fourth Tcl/Tk Workshop, USENIX*, Monterey, CA, pp. 9–23, 1996.

[G96a] Gupta, A., Fault-containment in self-stabilizing distributed systems. PhD thesis, Department of Computer Science, University of Iowa, Iowa City, IA, 1996.

[G97] Gray, J. S., *Interprocess Communications in Unix: The Nooks and Crannies*, 2nd edn. Prentice Hall, Englewood Cliffs, NJ, 1998.

[G99] Gärtner, F. C., Fundamentals of fault-tolerant distributed computing in asynchronous environments. *ACM Computing Surveys* 31 (1), 1–26, 1999.

[Ga88] Gafni, A., Rollback mechanisms for optimistic distributed simulation systems. *Proceedings of SCS Multiconference on Distributed Simulation* 9 (3), 61–67, 1988.

[GBDW+] Geist, A., Beguelin, A., Dongarra, J., Jiang, W., Manchek, R., and Sunderam, V. S., *PVM: Parallel Virtual Machine: A User's Guide and Tutorial for Network Parallel Computing*, MIT Press, 1994.

[GGHP07] Ghosh, S., Gupta, A., Herman, T., Pemmaraju, S. V., Fault-containing self-stabilizing distributed protocols. *Distributed Computing* 20 (1), 53-73, 2007.

[GGHP96] Ghosh, S., Gupta, A., Herman, T., and Pemmaraju, S. V., Fault-containing self-stabilizing algorithms. In *ACM PODC*, Philadelphia, PA, pp. 45–54, 1996.

[GH91] Gouda, M. G. and Herman, T., Adaptive programming. *IEEE Transactions on Software Engineering* 17, 911–921, 1991.

[GHS83] Gallager, R. G., Humblet, P. A., Spira, P. M., A distributed algorithm for minimum-weight spanning trees, *ACM Transactions on Programming Languages and Systems* 5 (1), 66–77, 1983.

[GK93] Ghosh, S. and Karaata, M. H., A self-stabilizing algorithm for coloring planar graphs. *Distributed Computing* 7 (1), 55–59, 1993.

[GKC+98] Gray, R. S., Kotz, D., Cybenko, G., and Rus, D., D'Agents: Security in a multiple-language, mobile-agent system. In *Mobile Agents and Security* (Vigna, G., ed.), LNCS 1419, pp. 154–187. Springer-Verlag, Berlin, Germany, 1998.

[GL02] Gilbert, S. and Lynch, N., Brewer's conjecture and the feasibility of consistent, available, partition-tolerant web services. *SIGACT News* 33 (2), 51–59, 2002.

[GLL+90] Gharachorloo, K., Lenoski, D., Laudon, J., Gibbons, P. B., Gupta, A., and Hennessy, J. L., Memory consistency and event ordering in scalable shared-memory multiprocessors. *International Science Congress Association*, 15–26, 1990.

[GM91] Gouda, M. G. and Multari, N., Stabilizing communication protocols. *IEEE Transactions on Computers* C-40 (4), 448–458, 1991.

[GN02] Girvan, M. and Newman, M. E. J., Community structure in social and biological networks. *Proceedings of National Academic Science USA* 99, 7821–7826, 2002.

[GPS+08] Gao, T., Pesto, C., Selavo, L., Chen, Y., Ko, J. G., Lim, J. H., Terzis, A. et al., Wireless medical sensor networks in emergency response: Implementation and pilot results. In *2008 IEEE International Conference on Technologies for Homeland Security*, Waltham, MA, May 2008.

[GR93] Gray, J. and Reuter, A., *Transaction Processing: Concept and Techniques*. Morgan Kaufmann Publishers Inc., San Mateo, CA, 1993.

[GS91] Garcia-Molina, H. and Spauster, A., Ordered and reliable multicast communication. *ACM Transactions on Computer Systems* 9 (3), 242–271, 1991.

[GZ89] Gusella, R. and Zatti, S., The accuracy of clock synchronization achieved by TEMPO in Berkeley Unix 4.3BSD. *IEEE Transactions on Software Engineering* SE-15 (7), 847–853, 1989.

[H00] Hill, J., A software architecture supporting networked sensor. MS thesis, University of California, Berkeley, CA, 2000.

[H69] Hoare, C. A. R., An axiomatic basis of computer programming. *Communications of the ACM* 12 (10), 576–583, 1969.

[H72] Hoare, C. A. R., Towards a theory of parallel programming. In *Operating System Techniques* (Hoare, C. A. R. and Perrot, R. H., eds.), Academic Press, New York, 1972.

[H78] Hoare, C. A. R., Communicating sequential processes. *Communications of the ACM* 21 (8), 666–677, 1978.

[H91] Herman, T., PhD dissertation, Adaptivity through distributed convergence, Department of Computer Sciences, University of Texas at Austin, Austin, TX, 1992.

[Ha69] Harary, F, *Graph Theory* (1969), Addison–Wesley, Reading, MA.

[Ha72] Harary, F., *Graph Theory*. Addison-Wesley, Reading, MA, 1994.

[HBC+06] Hull, B., Bychkovsky, V., Chen, K., Goraczko, M., Miu, A., Shih, E., Zhang, Y., Balakrishnan, H., and Madden, S., CarTel: A distributed mobile sensor computing system. *Proceedings of 4th ACM Conference on Embedded Networked Sensor Systems*, Boulder, CO, 2006.

[HCB00] Heinzelman, W. R., Chandrakasan, A., and Balakrishnan, H., Energy-efficient communication protocol for wireless microsensor networks. In *HICSS*, Maui, HI, 2000.

[HG95] Herman, T. and Ghosh, S., Stabilizing phase clocks. *Information Processing Letters* 54, 259–265, 1995.

[HH92] Hsu, S.-C. and Huang, S.-T., A self-stabilizing algorithm for maximal matching. *Information Processing Letters* 43 (2), 77–81, 1992.

[HHS+99] Harter, A., Hopper, A., Steggles, P., Ward, A., and Webster, P., The anatomy of a context aware application. In *MobiCom*, Seattle, WA, pp. 59–68, 1999.

[HNS+12] Hornbeck, T., Naylor, D., Segre, A. M., Thomas, G., Herman, T., Polgreen, P. M., On hand hygiene compliance and diminishing marginal returns: An empirically-driven agent-based simulation study. *Vaccine* 2012.

[HP11] Hennessy, J. and Patterson, D., *Computer Architecture: A Quantitative Approach*, 5th edn., Morgan Kaufmann, 2011.

[HP99] Hennessy, J. and Patterson, D., *Computer Architecture: A Quantitative Approach*, 3rd edn. Morgan Kaufmann, San Francisco, CA, 1999.

[HR83] Härder, T. and Reuter, A., Principles of transaction-oriented database recovery. *ACM Computing Surveys* 15 (4), 287–317, 1983.

[HS80] Hirschberg, D. S. and Sinclair, J. B., Decentralized extrema finding in circular configuration of processors. *Communications of the ACM* 23 (11), 627–628, 1980.

[HSW+00] Hill, J., Szewczyk, R., Woo, A., Hollar, S., Culler, D. E., and Pister, K. S. J., System architecture directions for networked sensors. In *ASPLOS*, Cambridge, MA, pp. 93–104, 2000.

[HW90] Herlihy, M. and Wing, J. M., Linearizability: A correctness condition for concurrent objects. *ACM Transactions on Programming Languages and Systems* 12 (3), 463-492, 1990.

[HW91] Herlihy, M. and Wing, J. M., Specifying graceful degradation. *IEEE Transactions on Parallel and Distributed Systems* 2 (1), 93–104, 1991.

[IGE+02] Intanagonwiwat, C., Govindan, R., Estrin, D., Heidemann, J., and Silva, F., Directed diffusion for wireless sensor networking. *ACM/IEEE Transactions on Networking* 11 (1), 2–16, 2002.

[IJ90] Israeli, A. and Jalfon, M., Token management schemes and random walks yield self-stabilizing mutual exclusion. In *ACM PODC*, Quebec City, Quebec, Canada, pp. 119–130, 1990.

[J85] Jefferson, D., Virtual time. *ACM Transactions on Programming Languages and Systems* 7 (3), 404–425, 1985.

[J98] Joung, Y.-J., Asynchronous group mutual exclusion (extended abstract). *ACM Symposium on Principles of Distributed Computing*, 51–60, 1998.

[JB90] Joseph, T. and Birman, K., The ISIS project: Real experience with a fault tolerant programming system. In *ACM SIGOPS*, Bologna, Italy, pp. 1–5, 1990.

[K00] Kleinberg, J., The small-world phenomenon: An algorithm perspective. In *ACM STOC*, Portland, OR, pp. 163–170, 2000.

[K01] What is Kazaa? http://computer.howstuffworks.com/kazaa1.htm (accessed on April 13, 2014).

[K03] Kubiatowicz, J., Extracting guarantees from Chaos. *Communications of the ACM* 46 (2), 33–38, 2003.

[K66] Knuth, D. E., Additional comments on a problem in concurrent programming control. *Communications of the ACM*, 9 (5), 321–322, 1966.

[K67] Kahn, D., *The Codebreakers: The Story of Secret Writing*. Macmillan, New York, 1967.

[K87] Knapp, E., Deadlock detection in distributed databases. *ACM Computing Surveys*, 19 (4), 303–328, 1987.

[K96] Kahn, D., *The Codebreakers*. Scribner, 1996.

[KBC+00] Kubiatowicz, J., Bindel, D., Chen, Y., Czerwinski, S., Eaton, P., Geels, D., Gummadi, R. et al., OceanStore: An architecture for global-scale persistent storage. In *ASPLOS*, Cambridge, MA, 2000.

[KC03] Kephart, J. and Chess, D. M., The vision of autonomic computing. *IEEE Computer*, 4150, January 2003.

[KGNT+97] Kotz, D., Gray, R., Nog, S., Rus, D., Chawla, S., and Cybenko, G., Agent TCL: Targeting the needs of mobile computers. *IEEE Internet Computing* 1 (4), 58–67, 1997.

[KK03] Kaashoek, M. F. and Karger, D. R., Koorde: A simple degree-optimal distributed hash table. In *IPTPS*, Berkeley, CA, pp. 98–107, 2003.

[KLL+97] Karger, D., Lehman, E., Leighton, T., Panigrahy, R., Levine, M., and Lewin, D., Consistent hashing and random trees: Distributed caching protocols for relieving hot spots on the world wide web. *ACM Symposium on Theory of Computing (STOC)*, 654–663, 1997.

[KP89] Katz, S. and Perry, K. J., Self-stabilizing extensions for message-passing systems. *Distributed Computing* 7 (1), 17–26.

[KR81] Kung, H. T. and Robinson, J. T., On optimistic methods for concurrency control. *ACM Transactions on Database Systems* 6 (2), 213–226, 1981.

[KR97] Karonski, M. and Rucinski, A., *The Origin of the Theory of Random Graphs. The Mathematics of Paul Erdos*. Springer, Berlin, Germany, 1997.

[KRB99] Kulik, J., Rabiner, W., and Balakrishnan, H., Adaptive protocols for information dissemination in wireless sensor networks. In *MobiCom*, Seattle, WA, 1999.

[KS92] Kistler, J. J. and Satyanarayanan, M., Disconnected operation in the coda file system. *ACM TOCS* 10 (1), 3–25, 1992.

[KW03] Karlof, C. and Wagner, D., Secure routing in wireless sensor networks: Attacks and counter-measures. *Ad Hoc Networks* 1 (2–3), 293–315, 2003.

[KWZ+03] Kuhn, F., Wattenhofer, R., Zhang, Y., and Zollinger, A., Geometric Ad-hoc routing: Of theory and practice. In *ACM PODC*, Boston, MA, 2003.

[L00] Lewand, R., Cryptological mathematics. The Mathematical Association of America, Washington, DC, 2000.

[L01] Lamport, L., Paxos made simple. *ACM SIGACT News* 32 (4), 18–25, December 2001.

[L74] Lamport, L., A new solution of Dijkstra's concurrent programming problem. *Communications of the ACM* 17 (8), 453–455, 1974.

[L77] Lamport, L., Proving the correctness of multiprocess programs. *IEEE Transactions on Software Engineering* SE-3 (2): 125–143, 1977.

[L78] Lamport, L., Time, clocks, and the ordering of events in distributed systems. *Communications of the ACM* 21 (7), 558–565, 1978.

[L79] Lamport, L., How to make a multiprocessor computer that correctly executes multiprocess programs. *IEEE Transactions on Computers* C28 (9), 690–691, 1979.

[L82] Lamport, L. Solved problems, unsolved problems, and non-problems in concurrency. Invited Address in ACM PODC 83. In *ACM PODC*, Vancouver, British Columbia, Canada, pp. 1–11, 1984.

[L83] Lamport, L., The weak byzantine generals problem. *Journal of the ACM*, 30 (3), 668–676, 1983.

[L85] Lamport, L., A fast mutual exclusion algorithm. *ACM TOCS* 5 (1), 1–11, 1987.

[L94] Lamport, L., The temporal logic of actions. *ACM TOPLAS* 16 (3), 872–923, May 1994.

[Le77] LeLann, G., Distributed systems: Towards a formal approach. In *IFIP Congress*, Toronto, Ontario, Canada, pp. 155–160, 1977.

[LG02] Lynch, N. and Gilbert, S., Brewer's conjecture and the feasibility of consistent, available, partition-tolerant web services. *ACM SIGACT News*, 33 (2), 51–59, 2002.

[LL90] Lamport, L. and Lynch, L., Distributed computing: Models and methods. In *Handbook of Theoretical Computer Science* (van Leewuen, J., ed.). Elsevier, Amsterdam, the Netherlands, 1990.

[LLS+92] Ladin, R., Liskov, B., Shrira, L., and Ghemawat, S., Providing availability using lazy replication. *ACM TOCS* 10 (4): 360–391, 1992.

[LM10] Lakshman, A. and Malik, P., Cassandra: A decentralized structured storage system. *Operating Systems Review* 44 (2), 35–40, 2010.

[LM85] Lamport, L. and Melliar-Smith, M., Synchronizing clocks in the presence of faults. *Journal of the ACM* 32 (1), 52–78, 1985.

[LM90] Lai, X. and Massey, J.L., A proposal for a new block encryption standard. In *EUROCRYPT*, Aarhus, Denmark, pp. 389–404, 1990.

[LO98] Lange, D. B., and Oshima, M., Programming and Deploying Java™ Mobile Agents with Aglets™, Addison-Wesley, ISBN 0-201-32582-9, August 1998.

[LPS81] Lampson, B., Paul, M., and Siegert, H., *Distributed Systems Architecture and Implementation*, LNCS 105, pp. 246–265 and 357–370. Springer Verlag, 1981.

[LR02] Lindsey, S. and Raghavendra, C. S., PEGASIS: Power-efficient gathering in sensor information systems. In *IEEE Aerospace Conference*, Big Sky, MT, March 2002.

[LRS02] Lindsey, S., Raghavendra, C. S., and Sivalingam, K. M., Data gathering algorithms in sensor networks using energy metrics. In *IEEE Transactions on Parallel and Distributed Systems* 13 (9), 924–935, 2002.

[LSP82] Lamport, L., Shostak, R., and Pease, M., The Byzantine generals problem. *ACM TOPLAS* 4 (3), 382–401, 1982.

[LT87] van Leeuwen, J. and Tan, R. B., Interval routing. *Computer Journal* 30 (4), 298–307, 1987.

[Luby86] Luby, M., A simple parallel algorithm for the maximal independent set problem. *SIAM Journal on Scientific Computing* 15 (4), 1036–1053, 1986.

[Ly68] Lynch, W., Reliable full-duplex file transmission over half-duplex telephone lines. *Communications of the ACM*, 11 (6), 407–410, June 1968.

[Ly96] Lynch, N., *Distributed Algorithms*. Morgan Kaufmann, San Francisco, CA, 1996.

[LY87] Lai, T. H. and Yang, T. H., On distributed snapshots. *Information Processing Letters*, 25, 153–158, 1987.

[M09] Miller, R., Data centers move to cut water waste. Digital Realty. http://www.datacenterknowledge.com/archives/2009/04/09/data-centers-move-to-cut-water-waste/, 2009.

[M67] Milgram, S., The small world problem. *Psychology Today* 1 (1), 60–67, May 1967.

[M83] May, D., OCCAM. *SIGPLAN Notices* 18 (4), 69–79, May 1983.

[M85] Miller, V., Use of elliptic curves in cryptography. In *CRYPTO 85*, Santa Barbara, CA, 1985.

[M86] Misra, J., Distributed discrete-event simulation. *ACM Computing Surveys* 18 (1), 39–65, 1986.

[M87] Mattern, F., Algorithms for distributed termination detection. *Distributed Computing* 2 (3), 161–175, 1987.

[M88] Mattern, F., Virtual time and global states of distributed systems, *Proceedings of the Workshop on Parallel and Distributed Algorithms*, Chateau de Bonas, Elsevier, France, pp. 215–226, 1988.

[M89] Mattern, F., Message complexity of ring-based election algorithms. In *ICDCS*, Newport Beach, CA, pp. 4–100, 1989.

[M89a] Mattern, F., Global quiescence detection based on credit distribution and recovery. *Information Processing Letters* 30 (4), 195–200, 1989.

[M89b] Muellender, S. J. (ed.), *Distributed Systems*. ACM Press, New York, 1989.

[M91] Mills, D. L., Internet time synchronization: The network time protocol. *IEEE Transactions on Communications* 39 (10), 1482–1493, 1991.

[Mi83] Misra, J., Detecting termination of distributed computations using markers. *ACM Symposium on Principles of Distributed Computing*, 290-294, 1983.

[MM93] Melliar-Smith, P. M. and Moser, L. E., Trans: A reliable broadcast protocol. *IEEE Transactions on Communication, Speech and Vision* 140 (6), 481–493, 1993.

[MMA90] Melliar-Smith, P., Moser, L., and Agrawala, V., Broadcast protocols for distributed systems. *IEEE Transactions on Parallel and Distributed Systems* 1 (1), 17–25, 1990.

[MN91] Marzullo, K. and Neiger, G., Detection of global state predicates. In *Proceedings of the Fifth Workshop on Distributed Algorithms* (Toueg, S., Spirakis, P., and Kirousis, L., eds.), pp. 254–272, Springer, 1991.

[MNR02] Malkhi, D., Naor, M., and Ratajczak, D., Viceroy: A scalable and dynamic emulation of the butterfly. In *ACM PODC*, Monterey, CA, 2002.

[MP92] Manna, Z. and Pnueli, A., *The Temporal Logic of Reactive and Concurrent Specifications*, Springer-Verlag, 1992.

[MRC80] McQuillan, J. M., Richer, I., and Rosen, E. C., The new routing algorithm for the ARPANet. *IEEE Transactions on Communications*, 28 (5), 711–719, 1980.

[N02] How the old Napster worked? http://computer.howstuffworks.com /napster.htm (accessed on April 13, 2014).

[N93] Needham, R., Names. In *Distributed Systems* (Mullender, S., ed.). Addison-Wesley, Reading, MA, 1993.

[NS78] Needham, R. and Schroeder, M. D., Using encryption for authentication in large network of computers. *Communications of the ACM* 21, 993–999, 1978.

[NSS10] Nygren, E., Sitaraman, R. K., and Sun, J., The Akamai network: A platform for high-performance Internet applications (PDF). *ACM SIGOPS Operating Systems Review* 44 (3), 2–19.

[OG76] Owicki, S. S. and Gries, D., An axiomatic proof technique for concurrent programs. *Acta Informatica* 6, 319–340, 1976.

[OL82] Owicki, S. S. and Lamport, L., Proving liveness properties of concurrent programs. *ACM TOPLAS* 4 (3), 455–495, 1982.

[P77] Pneuli, A., The temporal logic of programs. In *18th FOCS*, Providence, RI, pp. 46–57, 1977.

[P79] Papadimitriou, C., The serializability of concurrent updates. *Journal of the ACM* 24 (4), 631–653, 1979.

[P81] Peterson, G. L., Myths about the mutual exclusion problem. *Information Processing Letters*, 12 (3), 115–116, 1981.

[P82] Peterson, G. L., An O(n log n) unidirectional algorithm for the circular extrema problem. *ACM TOPLAS* 4 (4), 758–762, 1982.

[P90] Pugh, W., Skip lists: A probabilistic alternative to balanced trees. *CACM* 33 (6), 668–676, 1990.

[PCB00] Priyantha, N. B., Chakraborty, A., and Balakrishnan, H., The cricket location-support system. In *Sixth ACM MOBICOM*, Boston, MA, August 2000.

[PD96] Peterson, L. and Davie, B. S., *Computer Networks: A Systems Approach*, Morgan Kaufmann, San Francisco, CA, 1996.

[Peleg00] Peleg, D., Distributed computing: A locality sensitive approach. *SIAM Monographs on Discrete Mathematics and Applications*, 2000.

[Pr91] De Prycker, M., *Asynchronous Transfer Mode: Solutions for Broadband ISDN*. Ellis Horwood, Chichester, England, 1991.

[PRR99] Plaxton, C. G., Rajaraman, R., and Richa, A. W., Accessing nearby copies of replicated objects in a distributed environment. *Theory of Computing Systems* 32 (3), 241–280, 1999.

[PSW+02] Perrig, A., Szewczyk, R., Wen, V., Culler, D., and Tygar, J. D., SPINS: Security protocols for sensor networks. *Wireless Networks Journal (WINET)* 8 (5), 521–534, September 2002.

[R75] Randell, B., System structure for software fault-tolerance. *IEEE Transactions on Software Engineering* SE-1 (2), 1975.

[R83] Rana, S. P., A distributed solution of the distributed termination detection problem. *Information Processing Letters* 17, 43–46, 1983.

[R89] Raymond, K, A tree-based algorithm for distributed mutual exclusion. *ACM Transactions on Computer Systems*, 7 (1), 61–77, 1989.

[R92] Rivest, R., The MD5 Message-Digest Algorithm, 1992. http://www.ietf.org/rfc/rfc1321.txt (accessed on April 13, 2014).

[R94] Rushby, J., Critical system properties: Survey and taxonomy. *Reliability Engineering and System Safety*, 13 (2), 189–219, 1994.

[RA81] Ricart, G. and Agrawala, A. K., An optimal algorithm for mutual exclusion in computer networks. *Communications of the ACM* 24 (1), 9–17, 1981.

[RAK07] Raghavan, U. N., Albert, R., and Kumara, S., Near linear time algorithm to detect community structures in large-scale networks. *Physical Review* E 76, 036106, 2007.

[RD01] Rowstron, A. and Druschel, P., Pastry: Scalable, distributed object location and routing for large-scale peer-to-peer systems. In *IFIP/ACM International Conference on Distributed Systems Platforms (Middleware)*, Heidelberg, Germany, pp. 329–350, November 2001.

[RFH+01] Ratnasamy, S., Francis, P., Handley, M., Karp, R. M., and Shenker, S., A scalable content-addressable network. In *SIGCOMM*, San Diego, CA, pp. 161–172, 2001.

[RSA78] Rivest, R., Shamir, A., and Adleman, L., A method of obtaining digital signatures and public key cryptosystems. *Communications of the ACM* 21 (2), 120–126, 1978.

[S49] Shannon, C. E., Communication theory of secrecy systems. *Bell System Technical Journal* 28 (4), 656–715, 1949.

[S79] Shamir, A., How to share a secret. *Communications of the ACM* 22 (11), 612–613, 1979.

[S83] Skeen, D., A formal model of crash recovery in a distributed system. *IEEE Transactions on Software Engineering* SE-9 (3), 219–228, May 1983.

[S87] Sanders, B. A., The information structure of distributed mutual exclusion algorithms. *ACM Transactions on Computer Systems*, 5 (3), 284–299, 1987.

[S96] Schneier, B., *Applied Cryptography*. John Wiley, New York, 1996.

[S98] Siegel, J., OMG overview: CORBA and the OMA in enterprise computing. *Communications of the ACM* 41 (10), 37–43, 1998.

[Sch90] Schneider, F. B., Implementing fault-tolerant services using the state machine approach: A tutorial. *ACM Transactions on Computer Systems* 22 (4), 299–319, 1990.

[Se83] Segall, A., Distributed network protocols. *IEEE Transactions on Information Theory* 29 (1), 1983.

[SET02] SETI @ home project. http://setiathome.ssl.berkeley.edu/ (accessed on April 13, 2014).

[SK85] Santoro, N. and Khatib, R., Labeling and implicit routing in networks. *Computer Journal* 28 (1), 5–8, 1985.

[SKK+90] Satyanarayanan, M., Kistler, J. J., Kumar, P., Okasaki, M. E., Siegel, E. H., and Steere, D. C., Coda: A highly available file system for a distributed workstation environment. *IEEE Transactions on Computers* 39 (4), 447–459, 1990.

[SkS83] Skeen, D. and Stonebraker, M., A formal model of crash recovery in a distributed system. *IEEE Transactions on Software Engineering* 9 (3), 219–228, 1983.

[SKW+98] Schneier, B., Kelsey, J., Doug Whiting, D., Wagner, D., and Hall, C., On the twofish key schedule. *Selected Areas in Cryptography* LNCS 1556, Springer-Verlag, Berlin, Heidelberg, 27–42, 1998.

[SMC+03] Stoica, I., Morris, R., Libert-Nowell, D., Karger, D. R., Frans Kaashoek, M. F., Dabek, F., and Balakrishnan, H., Chord: a scalable peer-to-peer lookup protocol for internet applications. *IEEE/ACM Transactions on Networking* 11 (1), 17–32, 2003.

[SML+02] Stoica, I., Morris, R., Libben-Nowell, D., Karger, D., Kasshoek, M., Dabek, F., and Balakrishnan, H., Chord: A scalable peer-to-peer lookup protocol for Internet applications. *IEEE/ACM Transactions on Networking* 11 (1), 17–32, 2003.

[SNS88] Steiner, J., Neuman, C., and Schiller, J., Kerberos: An authentication service for open network systems. In *Proceedings of Usenix Conference*, Berkeley, CA, 1988.

[SS83] Schlichting, R. D. and Schneider, F., Fail-stop processors: An approach to designing computing systems. *ACM TOCS* 1 (3), 222–238, August 1983.

[SS94] Singhal, M. and Shivaratri, N., *Advanced Concepts in Operating Systems*. McGraw Hill, New York, 1994.

[SY85] Strom, R. and Yemini, S., Optimistic recovery in distributed systems. *ACM TOCS* 3 (3), 204–226, August 1985.

[T00] Tel, G., *Introduction to Distributed Systems*, 2nd edn. Cambridge University Press, Cambridge, U.K., 2000.

[T79] Thomas, R., A majority consensus approach to concurrency control for multiple copy databases. *ACM Transactions on Database Systems* 4 (2), 180–209, 1979.

[T94] Tannenbaum, A., *Distributed Operating Systems*. Prentice Hall, Upper Saddle River, NJ, 1994.

[Ta1895] Tarry, G., Le problème des labrynthes. *Nouvelles Annales de Mathematiques* 14, 1895.

[TKZ94] Tel, G., Korach, E., and Zaks, S., Synchronizing ABD networks. *IEEE Transactions on Networking* 2 (1), 66–69, 1994.

[TPS+98] Terry, D. B., Petersen, K., Spreitzer, M., and Theimer, M., The case of non-transparent replication: Examples from Bayou. *IEEE Data Engineering* 21 (4), 12–20, 1998.

[TS07] Tanenbaum, A. and van Steen, M., *Distributed Systems: Principles and Paradigms*, 2nd edn., Pearson Prentice Hall, 2007.

[TTP+95] Terry, D. B., Theimer, M. M., Petersen, K., Demers, A., Spreitzer, M. J., and Hauser, C., Managing update conflicts in Bayou, a weakly connected replicated storage system. In *15th ACM SOSP*, Frisco, CO, pp. 172–183, 1995.

[VES99] Virtual & extended virtual synchrony, 1999. http://www.cs.huji.ac.il/labs/transis/lab-projects/guide/chap3.html (accessed on April 30, 2014).

[VN56] Von Neumann, J., Probabilistic logics and the synthesis of reliable organisms from unreliable components. In *Automata Studies, Annals of Mathematics Studies* (Shannon, C. E. and McCarthy, J., eds.), No. 34, pp. 43–98. Princeton University Press, Princeton, NJ, 1956.

[W98] Waldo, J., Remote procedure calls and java remote method invocation. *IEEE Concurrency* 6 (3), 5–6, July 1998.

[WHF+92] Want, R., Hopper, A., Falcao, V., and Gibbons, J., The active badge location system. *ACM Transactions on Information Systems* 10 (1), 91–102, 1992.

[WLG+78] Wensley, J. H., Lamport, L., Goldberg, J., Green, M. W., Levitt, K. N., Melliar-Smith, P. M., Shostak, R. E., and Weinstock, C. B., SIFT: Design and analysis of a fault-tolerant computer for aircraft control. *Proceedings of IEEE*, 66 (10), 1240–1255, 1978.

[WN94] Wheeler, D. J. and Needham, R. M., TEA, a tiny encryption algorithm. *Fast Software Encryption* (Preneel, B., ed.), LNCS 1008, pp. 363–366. Springer, Heidelberg, Germany, 1994.

[WP98] White, T. and Pagurek, B., Towards multi-swarm problem solving in networks. In *Proceedings of the Third International Conference on Multi-Agent Systems (IC-MAS'98)*, Paris, France, pp. 333–340, 1998.

[WS98] Watts, D. and Strogatz, S., Collective dynamics of small-world networks. *Nature* 393, 440–442, 1998.

[WZ04] Wattenhofer, R. and Zollinger, A., XTC: A practical topology control algorithm for ad-hoc networks. In *IPDPS*, Santa Fe, NM, 2004.

[Y2K] http://www.britannica.com/EBchecked/topic/382740/Y2K-bug, accessed on April 28, 2014.

[YG99] Yan, T. W. and Garcia-Molina, H., The SIFT information dissemination system. *ACM Transactions on Database Systems* 24 (4), 529–565, 1999.

[Z99] Zimmermann, P., *The Official PGP User's Guide*. MIT Press, Cambridge, MA, 1999.

Index